Reinforced Concrete

Design theory and examples

Third edition

Prab Bhatt, Thomas J. MacGinley and Ban Seng Choo

Taylor & Francis
Taylor & Francis Group

LONDON AND NEW YORK

First published 1978 by E&FN Spon

Second edition 1990

Third edition published 2006 by Taylor & Francis
2 Park Square, Milton Park, Abingdon, Oxon OX14 4RN

Simultaneously published in the USA and Canada
by Taylor & Francis
270 Madison Ave, New York, NY 10016, USA

Taylor & Francis is an imprint of the Taylor & Francis Group

Publisher's Note
This book has been prepared from camera-ready-copy supplied
by the authors
Printed and bound in Great Britain by
MPG Books Ltd, Bodmin, Cornwall

British Library Cataloguing in Publication Data
A catalogue record for this book is available from the British Library

Library of Congress Cataloging in Publication Data
Bhatt, P.
   Reinforced concrete : design theory and examples / P. Bhatt,
T.J. MacGinley, and B.S. Choo.—3rd ed.
     p. cm.
   Rev. ed. of : Reinforced concrete / T.J. MacGinley, B.S. Choo.
London ; New York: E & F. Spon, 1990.
   ISBN 0–415–30796–1 (pbk. : alk. paper) — ISBN 0–415–30795–3
(hardback : alk. paper)
   1. Reinforced concrete construction. I. MacGinley, T.J.
(Thomas Joseph) II. Choo, B.S. III. MacGinley, T.J.
(Thomas Joseph). Reinforced concrete. IV. Title.
TA683.2.M33 2005
624.1′834—dc22

                                                    2005021534

ISBN10: 0–415–30795–3   ISBN13: 978–0–415–30795–6 (hbk)
ISBN10: 0–415–30796–1   ISBN13: 978–0–415–30796–3 (pbk)

Reinforced Concrete

Books are to be returned on or before
the last date below.

7−DAY LOAN

The third ... d
expanded ... or
concrete el ... .

Reinfor ... es
and over (... id
Eurocode ... n
the codes ... ig
structures ... n
problems i ... le
on an asso

This bo ... in
the princi ... s,
and is als

Prab Bh ... vil
Engineeri

Thomas ... ty,
Singapor

Ban Seng ... ol
of Built E

LIBREX-

Dedicated with love and gratitude to
my mother Srimati Sharadamma
who taught us to 'never disown the poor'.

CONTENTS

15   Tall buildings   **513**

Modified version of initial contribution by J.C.D. Hoenderkamp, formerly of Nanyang Technological Institute, Singapore

16   Prestressed concrete   **529**

Preface

The third edition of the book has been written to conform to BS 8110 1997 the code for structural use of concrete and BS 8007:1987 the code for Design of structures for retaining aqueous liquids. The aim remains as stated in the first edition: to set out design theory and illustrate the practical applications of code rules by the inclusion of as many useful examples as possible. The book is written primarily for students on civil engineering degree courses to assist them to understand the principles of element design and the procedures for the design of concrete buildings. The book will also be of assistance to new graduates starting on their career in structural design.

The book has been thoroughly revised to conform to the updated code rules. Many new examples and sections have been added. In particular the chapter on Slabs has been considerably expanded with extensive coverage of Yield line analysis, Hillerborg's strip method and design for predetermined stress fields. In addition, four new chapters have been added to reflect the contents of university courses in design in structural concrete. The new chapters are concerned with design of prestressed concrete structures, design of water tanks, a short chapter comparing the important clauses of Eurocode 2 and finally a chapter on the fundamental theoretical aspects of design of statically indeterminate structures, an area that is very poorly treated in most text books.

The importance of computers in structural design is recognized by analysing all statically indeterminate structures by Matrix stiffness method. However, as design offices nowadays extensively use Spread Sheets type calculations rather than small scale computer programs, the chapter on computer programs has been deleted.

Grateful acknowledgements are extended to:

- The British Standard Institution for permission to reproduce extracts from BS 8110 & BS 8007. British Standards can be obtained from BSI Customer Services, 389, Chiswick High Road, London W4 4AL, Tel: +44(0)20 8996 9001. e-mail: cservices@bsi-global.com
- Professor Alan Ervine, Head of Civil Engineering department for all the facilities.
- Mr. Ken McCall, computer manager of Civil Engineering department for help with computing matters.
- Dr. T.J.A. Agar, former colleague for carefully reading the draft, correcting errors and making many useful alterations which have greatly improved the final version.
- Mrs. Tessa Bryden for occasional help with secretarial matters.
- Sheila Arun, Sujaatha and Ranjana for moral support.

P. Bhatt
2nd October 2005 (Mahatma Gandhi's birthday)

illustrated in Fig.1.2 which shows typical *cast-in-situ* concrete building construction.

A *cast-in-situ* framed reinforced concrete building and the rigid frames and elements into which it is idealized for analysis and design are shown in Fig.1.3. The design with regard to this building will cover

1. one-way continuous slabs
2. transverse and longitudinal rigid frames
3. foundations

Various types of floor are considered, two of which are shown in Fig.1.4. A one-way floor slab supported on primary reinforced concrete frames and secondary continuous flanged beams is shown in Fig.1.4(a). In Fig.1.4(b) only primary reinforced concrete frames are constructed and the slab spans two ways. Flat slab construction, where the slab is supported by the columns without beams, is also described. Structural design for isolated pad, strip and combined and piled foundations and retaining walls (Fig.1.5) is covered in this book.

1.3 STRUCTURAL DESIGN

The first function in design is the planning carried out by the architect to determine the arrangement and layout of the building to meet the client's requirements. The structural engineer then determines the best structural system or forms to bring the architect's concept into being. Construction in different materials and with different arrangements and systems may require investigation to determine the most economical answer. Architect and engineer should work together at this conceptual design stage.

Once the building form and structural arrangement have been finalized the design problem consists of the following:

1. idealization of the structure into load bearing frames and elements for analysis and design
2. estimation of loads
3. analysis to determine the maximum moments, thrusts and shears for design
4. design of sections and reinforcement arrangements for slabs, beams, columns and walls using the results from 3
5. production of arrangement and detail drawings and bar schedules

1.4 DESIGN STANDARDS

In the UK, design is generally to limit state theory in accordance with
BS8110: 1997: *Structural Use of Concrete Part 1: Code of Practice for Design and Construction*

The design of sections for strength is according to plastic theory based on behaviour at ultimate loads. Elastic analysis of sections is also covered because this is used in calculations for deflections and crack width in accordance with BS 8110: 1985: *Structural Use of Concrete Part 2: Code of Practice for Special Circumstances*

The loading on structures conforms to

BS 6399-1:1996 *Loading for buildings. Code of Practice for Dead and Imposed Loads*

BS 6399-2:1997 *Loading for buildings. Code of Practice for Wind Loads*

BS 6399-3:1988 *Loading for buildings. Code of Practice for Imposed Roof Loads*

The codes set out the design loads, load combinations and partial factors of safety, material strengths, design procedures and sound construction practice. A thorough knowledge of the codes is one of the essential requirements of a designer. Thus it is important that copies of these codes are obtained and read in conjunction with the book. Generally, only those parts of clauses and tables are quoted which are relevant to the particular problem, and the reader should consult the full text.

Only the main codes involved have been mentioned above. Other codes, to which reference is necessary, will be noted as required.

1.5 CALCULATIONS, DESIGN AIDS AND COMPUTING

Calculations form the major part of the design process. They are needed to determine the loading on the elements and structure and to carry out the analysis and design of the elements. Design office calculations should be presented in accordance with

Higgins, J.B and Rogers, B.R.,1999, Designed and detailed. British Cement Association.

The need for orderly and concise presentation of calculations cannot be emphasized too strongly.

Design aids in the form of charts and tables are an important part of the designer's equipment. These aids make exact design methods easier to apply, shorten design time and lessen the possibility of making errors. Part 3 of BS 8110 consists of design charts for beams and columns, and the construction of charts is set out in this book, together with representative examples. Useful books are

Reynolds, C.E. and Steedman, J.C.,1988, Reinforced concrete designers handbook, (Spon Press).

Goodchild, C.H.,1997, Economic concrete frame elements, (Reinforced Concrete Council).

The use of computers for the analysis and design of structures is standard practice. Familiarity with the use of Spread Sheets is particularly useful. A useful reference is

Goodchild, C.H. and Webster, R.M., 2000, Spreadsheets for concrete design to BS 8110 and EC2, (Reinforced concrete council).

In analysis exact and approximate manual methods are set out but computer analysis is used where appropriate. However, it is essential that students understand the design principles involved and are able to make manual design calculations before using computer programs.

1.6 TWO CARRIAGE RETURNS DETAILING

The general arrangement drawings give the overall layout and principal dimensions of the structure. The structural requirements for the individual elements are presented in the detail drawings. The output of the design calculations are sketches giving sizes of members and the sizes, arrangement, spacing and cut-off points for reinforcing bars at various sections of the structure. Detailing translates this information into a suitable pattern of reinforcement for the structure as a whole. Detailing is presented in accordance with the
Standard Method of Detailing Structural Concrete. Institution of Structural Engineers, London, 1989.

It is essential for the student to know the conventions for making reinforced concrete drawings such as scales, methods for specifying steel bars, links, fabric, cut-off points etc. The main particulars for detailing are given for most of the worked exercises in the book. The bar schedule can be prepared on completion of the detail drawings. The form of the schedule and shape code for the bars are to conform to
BS 8666: 2000: *Specification for Scheduling, Dimensioning, Bending and cutting of steel for Reinforcement for Concrete*

It is essential that the student carry out practical work in detailing and preparation of bar schedules prior to and/or during his design course in reinforced concrete. Computer detailing suites are now in general use in design offices.

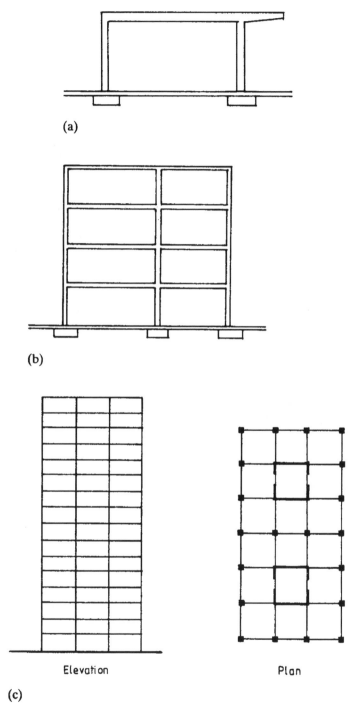

Fig 1.1 (a) Single storey portal; (b) medium-rise reinforced concrete framed building; (c) reinforced concrete frame and core structure

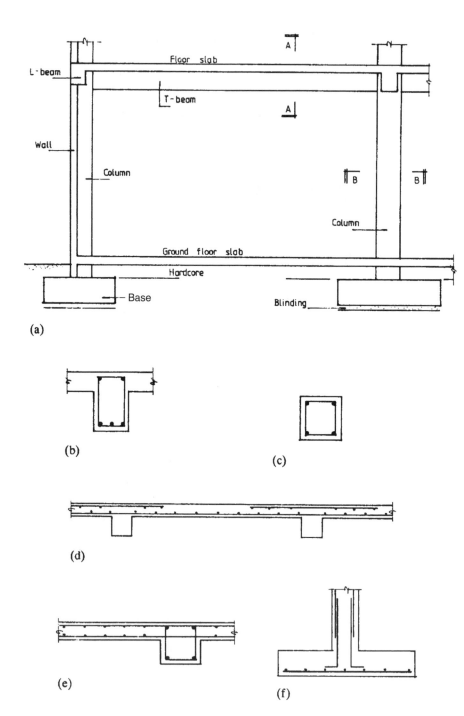

Fig 1.2 (a) Part elevation of reinforced concrete building; (b) section AA, T-beam ;
(c) section BB; (d) continuous slab; (e) wall; (f) column base

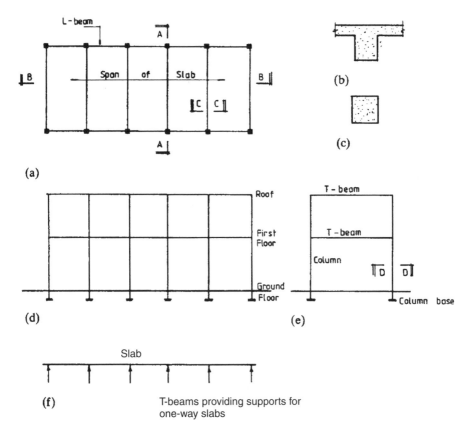

Fig 1.3 (a) Plan of roof and floor; (b) section CC, T-beam; (c) section DD, column; (d) side elevation, longitudinal frame; (e) section AA, transverse frame; (f) continuous one-way slab.

BS EN 197-1:2000: *Cement. Composition, specifications and conformity criteria for common cements*

BS EN 197-2:2000: *Cement. Conformity evaluation*

These are an initial setting time which must be a minimum of 45 min and a final set which must take place in 10 h.

Cement must be sound, i.e. it must not contain excessive quantities of certain substances such as lime, magnesia, calcium sulphate etc. that may expand on hydrating or react with other substances in the aggregate and cause the concrete to disintegrate. Tests are specified for soundness and strength of cement mortar cubes.

Many other types of cement are available some of which are:

1. Rapid hardening Portland cement: the clinker is more finely ground than for ordinary Portland cement. This is used in structures where it is necessary for the concrete to gain strength rapidly. Typical example is where the formwork needs to be removed early for reuse.

2. Low heat Portland cement: this has a low rate of heat development during hydration of the cement. This is used in situations such as thick concrete sections where it is necessary to keep the rate of heat generation due to hydration low as otherwise it could lead to serious cracking.

3. Sulphate-resisting Portland cement: this is often used for foundation concrete when the soil contains sulphates which can attack OPC concrete.

A very useful reference is

Adam M Neville, Properties of Concrete, Prentice-Hall, 4th Edition, 1996.

2.2.2 Aggregates

The bulk of concrete is aggregate in the form of sand and gravel which is bound together by cement. Aggregate is classed into the following two sizes;

1. coarse aggregate: gravel or crushed rock 5 mm or larger in size

2. fine aggregate: sand less than 5 mm in size

Natural aggregates are classified according to the rock type, e.g. basalt, granite, flint. Aggregates should be chemically inert, clean, hard and durable. Organic impurities can affect the hydration of cement and the bond between the cement and the aggregate. Some aggregates containing silica may react with alkali in the cement causing the some of the larger aggregates to expand which may lead to the concrete disintegrating. This is the alkali–silica reaction. Presence of chlorides in aggregates, e.g. salt in marine sands, will cause corrosion of the steel reinforcement. Excessive amounts of sulphate will also cause concrete to disintegrate.

To obtain a dense strong concrete with minimum use of cement, the cement paste should fill the voids in the fine aggregate while the fine aggregate and cement paste fills the voids in the coarse aggregate. Coarse and fine aggregates are graded by sieve analysis in which the percentage by weight passing a set of standard sieve sizes is determined. Grading limits for each size of coarse and fine aggregate are set out in

BS EN 12620: 2002: *Aggregates for Concrete*

The grading affects the workability; a lower water-to-cement ratio can be used if the grading of the aggregate is good and therefore strength is also increased. Good grading saves cement content. It helps prevent segregation during placing and ensures a good finish.

2.2.3 Concrete Mix Design

Concrete mix design consists in selecting and proportioning the constituents to give the required strength, workability and durability. Mixes are defined in

BS 8500-1:2002: *Concrete. Methods of Specifying and guidance for the specifier*

BS 8500-2:2002: *Specifications for constituent materials and concrete*

The five types are

1. Designated concretes: This is used where concrete is intended for use such as plain and reinforced foundations, floors, paving, and other given in Table A.6 or A.7 of the code.

2. Designed concretes: This is the most flexible type of specification. The environment to which the concrete is exposed, the intended working life of the structure, the limiting values of composition are all taken account of in selecting the requirements of the concrete mix.

3. Prescribed concretes: This is used where the specifier prescribes the exact composition and constituents of the concrete. No requirements regarding concrete strength can be prescribed. This has very limited applicability.

4. Standardised prescribed concretes: This is used where concrete is site batched or obtained from a ready mixed concrete producer with no third party accreditation.

5. Proprietary concretes: Used where concrete achieves a performance using defined test methods, outside the normal requirements for concrete.

The water-to-cement ratio is the single most important factor affecting concrete strength. For full hydration cement absorbs 0.23 of its weight of water in normal conditions. This amount of water gives a very dry mix and extra water is added to give the required workability. The actual water-to-cement ratio used generally ranges from 0.45 to 0.6. The aggregate-to-cement ratio also affects workability through its influence on the water-to-cement ratio, as noted above. The mix is designed for the 'target mean strength' which is the characteristic strength required for design plus a specified number of times the standard deviation of the mean strength.

Several methods of mix design are used. The main factors involved are discussed briefly for mix design according to

Teychenne, R.E. Franklin and Entroy, H.C., 1988, Design of Normal Concrete Mixes. (HMSO, London).

1. Curves giving compressive strength versus water-to-cement ratio for various types of cement and ages of hardening are available. The water-to-cement ratio is selected to give the required strength.

2. Minimum cement contents and maximum free water-to-cement ratios are specified in BS8110: Part 1, Table 3.3, to meet durability requirements. The maximum cement content is also limited to avoid cracking due mainly to shrinkage.

3. In *Design of Normal Concrete Mixes,* the selection of the aggregate-to-cement ratio depends on the grading curve for the aggregate.

Trial mixes based on the above considerations are made and used to determine the final proportions for designed mixes.

2.2.4 Admixtures

Advice on admixtures is given in

BS EN 934-2: 1998 *Admixtures for concrete, mortar and grout.*

The code defines admixtures as 'Materials added during the mixing process of in a quantity not more than 5% by mass of the cement content of the concrete, to modify the properties of the mix in the fresh and/or hardened state'.

Admixtures covered by British Standards are as follows:

1. set accelerators or set retarders

2. water-reducing/plasticizing admixtures which give an increase in workability with a lower water-to-cement ratio

3. air-entraining admixtures, which increase resistance to damage from freezing and thawing

4. high range water reducing agents/super plasticizers, which are more efficient than (2) above.

5. hardening accelerators which increases the early strength of concrete.

The general requirements of admixtures are given in Table 1 of the code. The effect of new admixtures should be verified by trial mixes. A useful publication on admixtures is

Hewlett, P.C (Editor). 1988, Cement Admixtures: Uses and Applications,
(Longman Scientific and Technical).

2.3 CONCRETE PROPERTIES

The main properties of concrete are discussed below.

2.3.1 Compressive Strength

The compressive strength is the most important property of concrete. The characteristic strength that is the concrete grade is measured by the 28 day cube strength. Standard cubes of 150 or 100 mm for aggregate not exceeding 25 mm in size are crushed to determine the strength. The test procedure is given in

BS EN 12390:2: 2000: *Testing Hardened Concrete: Making and curing specimens for strength tests*
BS EN 12390:3: 2000: *Testing Hardened Concrete: Compressive strength of test specimens*

2.3.2 Tensile Strength

The tensile strength of concrete is about a tenth of the compressive strength. It is determined by loading a concrete cylinder across a diameter as shown in Fig.2.1 (a). The test procedure is given in
BS EN 12390:6: 2000: *Testing Hardened Concrete: Tensile splitting strength of test specimens*

2.3.3 Modulus of Elasticity

The short-term stress–strain curve for concrete in compression is shown in Fig.2.1 (b). The slope of the initial straight portion is the initial tangent modulus. At any point P the slope of the curve is the tangent modulus and the slope of the line joining P to the origin is the secant modulus. The value of the secant modulus depends on the stress and rate of application of the load.
BS 1881-121: 1983 *Testing concrete. Methods for determination of Static modulus of elasticity in compression.*
specifies both values to standardize determination of the secant or static modulus of elasticity.

The dynamic modulus is determined by subjecting a beam specimen to longitudinal vibration. The value obtained is unaffected by creep and is approximately equal to the initial tangent modulus shown in Fig.2.1 (b). The secant modulus can be calculated from the dynamic modulus.

BS 8110: Part 1 gives the following expression for the short-term modulus of elasticity in Fig.2.1, the short-term design stress–strain curve for concrete.

$$E_c = 5.5 \sqrt{\frac{f_{cu}}{\gamma_m}} \text{ kN/mm}^2$$

where f_{cu} = cube strength and γ_m = material safety factor taken as 1.5. A further expression for the static modulus of elasticity is given in Part 2, section 7.2. (The idealized short-term stress–strain curve is shown in Fig.2.1.)

2.3.4 Creep

Creep in concrete is the gradual increase in strain with time in a member subjected to prolonged stress. The creep strain is much larger than the elastic strain on loading. If the specimen is unloaded there is an immediate elastic recovery and a slower recovery in the strain due to creep. Both amounts of recovery are much less than the original strains under load.

BS EN 12350-2: *Testing fresh concrete-Part 2: Slump Test*

(b) Compacting factor test
The degree of compaction achieved by a standard amount of work is measured. The apparatus consists of two conical hoppers placed over one another and over a cylinder. The upper hopper is filled with fresh concrete which is then dropped into the second hopper and into the cylinder which is struck off flush. The compacting factor is the ratio of the weight of concrete in the cylinder to the weight of an equal volume of fully compacted concrete. The compacting factor for concrete of medium workability is about 0.9. The following British standard covers slump test.
BS EN 12350-4: *Testing fresh concrete-Part 4: Degree of compactibility*

(c) Other tests
Other tests are specified for stiff mixes and super plasticized mixes. Reference should be made to specialist books on concrete.

2.5 TESTS ON HARDENED CONCRETE

2.5.1 Normal Tests

The main destructive tests on hardened concrete are as follows.
(a) Cube test: Refer to section 2.3.1 above.
(b) Tensile splitting test: Refer to section 2.3.2 above.
(c) Flexure test: A plain concrete specimen is tested to failure in bending. The theoretical maximum tensile stress at the bottom face at failure is calculated. This is termed the modulus of rupture. It is about 1.5 times the tensile stress determined by the splitting test. The following British standard covers testing of flexural strength.
BS EN 12390:5: 2000: *Testing Hardened Concrete: Flexural strength of test specimens*
(d) Test cores: Cylindrical cores are cut from the finished structure with a rotary cutting tool. The core is soaked, capped and tested in compression to give a measure of the concrete strength in the actual structure. The ratio of core height to diameter and the location where the core is taken affect the strength. The strength is lowest at the top surface and increases with depth through the element. A ratio of core height-to-diameter of 2 gives a standard cylinder test. The following British standard covers testing of cores.
BS EN 12504-1: *Testing concrete in structures-Part 1 Cored specimens-Taking examining and testing in compression.*

2.5.2 Non-Destructive Tests

The main non-destructive tests for strength on hardened concrete are as follows.

(a) Rebound hardness test

The Schmidt hammer is used in the rebound hardness test in which a metal hammer held against the concrete is struck by another spring-driven metal mass and rebounds. The amount of rebound is recorded on a scale and this gives an indication of the concrete strength. The larger the rebound number is the higher is the concrete strength. The following British standard covers testing by Rebound hammer.

BS EN 12504-2: *Testing concrete in structures-Part 2: Non-destructive testing-Determination of rebound number.*

(b) Ultrasonic pulse velocity test

In the ultrasonic pulse velocity test the velocity of ultrasonic pulses that pass through a concrete section from a transmitter to a receiver is measured. The pulse velocity is correlated against strength. The higher the velocity is the stronger is the concrete.

(c) Other non-destructive tests

Equipment has been developed to measure
1. crack widths and depths
2. water permeability and the surface dampness of concrete
3. depth of cover and the location of reinforcing bars
4. the electrochemical potential of reinforcing bars and hence the presence of corrosion

A useful reference on testing of concrete in structures is

Bungey, J.H. and Millard, S.G., 1996, Testing Concrete in Structures (Blackie Academic and Professional), 3rd Edition.

2.5.3 Chemical Tests

A complete range of chemical tests is available to measure
1. depth of carbonation
2. the cement content of the original mix
3. the content of salts such as chlorides and sulphates that may react and cause the concrete to disintegrate or cause corrosion of the reinforcement.

The reader should consult specialist literature

2.6 REINFORCEMENT

Reinforcing bars are produced in two grades: hot rolled mild steel bars have yield strength f_y of 250 N /mm^2; hot rolled or cold worked high yield steel bars have yield strength f_y of 460 N/mm^2. Steel fabric is made from cold drawn steel wires welded to form a mesh. It has a yield strength f_y of 460 N/mm^2.

The stress–strain curves for reinforcing bars are shown in Fig.2.2. Hot rolled bars have a definite yield point. A defined proof stress is recorded for the cold worked bars. The value of Young's modulus E is 200 kN/mm^2. The idealized design stress–strain curve for all reinforcing bars is shown in BS8110: Part 1 (see Fig.2.2). The behaviour in tension and compression is taken to be the same.

Mild steel bars are produced as smooth round bars. High yield bars are produced as deformed bars in two types defined in the code to increase bond stress:
Type 1 Square twisted cold worked bars. This type is obsolete.
Type 2 Hot rolled bars with transverse ribs

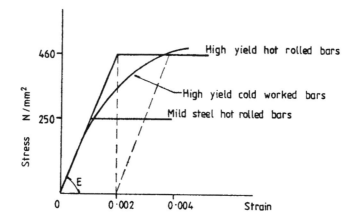

Fig.2.2 Stress–strain curves for reinforcing bars.

2.7 FAILURES IN CONCRETE STRUCTURES

2.7.1 Factors Affecting Failure

Failures in concrete structures can be due to any of the following factors:
1. incorrect selection of materials
2. errors in design calculations and detailing
3. poor construction methods and inadequate quality control and supervision
4. chemical attack
5. external physical and/or mechanical factors including alterations made to the structure

The above items are discussed in more detail below.

2.7.1.1 Incorrect Selection of Materials

The concrete mix required should be selected to meet the environmental or soil conditions where the concrete is to be placed. The minimum grade that should be used for reinforced concrete is grade 30. Higher grades should be used for some foundations and for structures near the sea or in an aggressive industrial environment. If sulphates are present in the soil or ground water, sulphate-resisting Portland cement should be used. Where freezing and thawing occurs air entrainment should be adopted. Further aspects of materials selection are discussed below.

2.7.1.2 Errors in Design Calculations and Detailing

An independent check should be made of all design calculations to ensure that the section sizes, slab thickness etc. and reinforcement sizes and spacing specified are adequate to carry the worst combination of design loads. The check should include overall stability, robustness and serviceability and foundation design.

Incorrect detailing is one of the commonest causes of failure and cracking in concrete structures. First the overall arrangement of the structure should be correct, efficient and robust. Movement joints should be provided where required to reduce or eliminate cracking. The overall detail should be such as to shed water.

Internal or element detailing must comply with the code requirements. The provisions specify the cover to reinforcement, minimum thicknesses for fire resistance, maximum and minimum steel areas, bar spacing limits and reinforcement to control cracking, lap lengths, anchorage of bars etc.

2.7.1.3 Poor Construction Methods

The main items that come under the heading of poor construction methods resulting from bad workmanship and inadequate quality control and supervision are as follows. BS 8110, clause 6.2 gives guidance on many of the aspects discussed below.

(a) Incorrect placement of steel
Incorrect placement of steel can result in insufficient cover, leading to corrosion of the reinforcement. If the bars are placed grossly out of position or in the wrong position, collapse can occur when the element is fully loaded.

(b) Inadequate cover to reinforcement
Inadequate cover to reinforcement permits ingress of moisture, gases and other substances and leads to corrosion of the reinforcement and cracking and spalling of the concrete.

(c) Incorrectly made construction joints
The main faults in construction joints are lack of preparation and poor compaction. The old concrete should be washed and a layer of rich concrete laid before pouring is continued. Poor joints allow ingress of moisture and staining of the concrete face.

(d) Grout leakage

Grout leakage occurs where formwork joints do not fit together properly. The result is a porous area of concrete that has little or no cement and fine aggregate. All formwork joints should be properly sealed.

(e) Poor compaction

If concrete is not properly compacted by ramming or vibration, the result is a portion of porous honeycomb concrete. This part must be hacked out and recast. Complete compaction is essential to give a dense, impermeable concrete.

(f) Segregation

Segregation occurs when the mix ingredients become separated. It is the result of
1. dropping the mix through too great a height in placing. Chutes or pipes should be used in such cases.
2. using a harsh mix with high coarse aggregate content
3. large aggregate sinking due to over-vibration or use of too much plasticizer
   Segregation results in uneven concrete texture, or porous concrete in some cases.

(g) Poor curing

A poor curing procedure can result in loss of water through evaporation. This can cause a reduction in strength if there is not sufficient water for complete hydration of the cement. Loss of water can cause shrinkage cracking. During curing the concrete should be kept damp and covered. See BS 8110, clause 6.2.3 on curing.

(h) Too high a water content

Excess water increases workability but decreases the strength and increases the porosity and permeability of the hardened concrete, which can lead to corrosion of the reinforcement. The correct water-to-cement ratio for the mix should be strictly enforced.

2.7.1.4 Chemical Attack

The main causes of chemical attack on concrete and reinforcement can be classified under the following headings.

(a) Chlorides

High concentrations of chloride ions cause corrosion of reinforcement and the products of corrosion can disrupt the concrete. Chlorides can be introduced into the concrete either during or after construction as follows.
(i) Before construction Chlorides can be admitted in admixtures containing calcium chloride, through using mixing water contaminated with salt water or improperly washed marine aggregates.
(ii) After construction Chlorides in salt or sea water, in airborne sea spray and from de-icing salts can attack permeable concrete causing corrosion of reinforcement.

(b) Sulphates

Sulphates are present in most cements and some aggregates. Sulphates may also be present in soils, groundwater and sea water, industrial wastes and acid rain. The products of sulphate attack on concrete occupy a larger space than the original material and this causes the concrete to disintegrate and permits corrosion of steel to begin. Sulphate-resisting Portland cement should be used where sulphates are present in the soil, water or atmosphere and come into contact with the concrete. Super sulphated cement, made from blast furnace slag, can also be used. This cement can resist the highest concentrations of sulphates.

(c) Carbonation

Carbonation is the process by which carbon dioxide from the atmosphere slowly transforms calcium hydroxide into calcium carbonate in concrete. The concrete itself is not harmed and increases in strength, but the reinforcement can be seriously affected by corrosion as a result of this process.

Normally the high pH value of the concrete prevents corrosion of the reinforcing bars by keeping them in a highly alkaline environment due to the release of calcium hydroxide by the cement during its hydration. Carbonated concrete has a pH value of 8.3 while the passivation of steel starts at a pH value of 9.5. The depth of carbonation in good dense concrete is about 3 mm at an early stage and may increase to 6–10 mm after 30–40 years. Poor concrete may have a depth of carbonation of 50 mm after say 6-8 years. The rate of carbonation depends on time, cover, concrete density, cement content, water-to-cement ratio and the presence of cracks.

(d) Alkali–silica reaction

A chemical reaction can take place between alkali in cement and certain forms of silica in aggregate. The reaction produces a gel which absorbs water and expands in volume, resulting in cracking and disintegration of the concrete. The reaction only occurs when the following are present together:

1. a high moisture level in the concrete
2. cement with a high alkali content or some other source of alkali
3. aggregate containing an alkali-reactive constituent

The following precautions should be taken if uncertainty exists:

1. Reduce the saturation of the concrete;
2. Use low alkali Portland cement and limit the alkali content of the mix to a low level;
3. Use replacement cementitious materials such as blast furnace slag or pulverized fuel ash. Most normal aggregates behave satisfactorily.

(e) Acids

Portland cement is not acid resistant and acid attack may remove part of the set cement. Acids are formed by the dissolution in water of carbon dioxide or sulphur dioxide from the atmosphere. Acids can also come from industrial wastes. Good

causes movement of the concrete which can cause cracking if restraint exists. Detail should be such as to shed water and the concrete may also be protected by impermeable membranes.

(d) Freezing and thawing
Concrete nearly always contains water which expands on freezing. The freezing-thawing cycle causes loss of strength, spalling and disintegration of the concrete. Resistance to damage is improved by using an air-entraining agent.

(e) Overloading
Extreme overloading will cause cracking and eventual collapse. Factors of safety in the original design allow for possible overloads but vigilance is always required to ensure that the structure is never grossly overloaded. A change in function of the building or room can lead to overloading, e.g. if a class room is changed to a library the imposed load can be greatly increased.

(f) Structural alterations
If major structural alterations are made to a building, the members affected and the overall integrity of the building should be rechecked. Common alterations are the removal of walls or columns to give a large clear space or provide additional doors or openings. Steel beams are inserted to carry loads from above. In such cases the bearing of the new beam on the original structure should be checked and if walls are removed the overall stability may be affected.

(g) Settlement
Differential settlement of foundations can cause cracking and failure in extreme cases. The foundation design must be adequate to carry the building loads without excessive settlement. Where a building with a large plan area is located on ground where subsidence may occur, the building should be constructed in sections on independent rafts with complete settlement joints between adjacent parts.

Many other factors can cause settlement and ground movement problems. Some problems are shrinkage of clays from ground dewatering or drying out in droughts, tree roots causing disruption, ground movement from nearby excavations, etc.

(h) Fire resistance
Concrete is a porous substance bound together by water-containing crystals. The binding material can decompose if heated to too high a temperature, with consequent loss of strength. The loss of moisture causes shrinkage and the temperature rise causes the aggregates to expand, leading to cracking and spalling of the concrete. High temperature also causes reinforcement to lose strength. At 550°C the yield stress of steel has dropped to about its normal working stress and failure occurs under service loads.

Concrete, however, is a material with very good fire resistance and protects the reinforcing steel. Fire resistance is a function of member thickness and cover. The code requirements regarding fire protection are set out below in section 2.9.2.

2.8 DURABILITY OF CONCRETE STRUCTURES

2.8.1 Code References to Durability

Frequent references are made to durability in BS8110: Part 1, section 2. The clauses referred to are as follows.

(a) Clause 2.1.3
The quality of material must be adequate for safety, serviceability and durability.

(b) Clause 2.2.1
The structure must not deteriorate unduly under the action of the environment over its design life. i.e. it must be durable.

(c) Clause 2.2.4
This states that 'integration of all aspects of design, materials and construction is required to produce a durable structure'. The main provisions in the clause are the following:
1. Environmental conditions should be defined at the design stage;
2. The design should be such as to ensure that surfaces are freely draining;
3. Cover must be adequate;
4. Concrete must be of relevant quality. Constituents that may cause durability problems should be avoided;
5. Particular types of concrete should be specified to meet special requirements;
6. Good workmanship, particularly in curing, is essential. Guidance on concrete construction such as placing and compaction, curing, etc. are set out in section 6.2 of the code.

2.9 CONCRETE COVER

2.9.1 Nominal Cover against Corrosion

The code states in section 3.3.1 that the actual cover should never be less than the nominal cover minus 5 mm. The nominal cover should protect steel against corrosion and fire. The cover to a main bar should not be less than the bar size or in the case of pairs or bundles the size of a single bar of the same cross-sectional area.

The cover depends on the exposure conditions given in Table 3.2 in the code. These are as follows.

Mild: concrete is protected against weather

Moderate:

        concrete is sheltered from severe rain

        concrete under non-aggressive water

        concrete in non-aggressive soil

Severe: concrete exposed to severe rain, alternate wetting and drying or occasional freezing or severe condensation

Very severe: concrete occasionally exposed to sea water, de-icing salts or corrosive fumes

Most severe: concrete frequently exposed to sea water, de-icing salts or corrosive fumes

Abrasive: concrete exposed to abrasive action

Limiting values for nominal cover are given in Table 3.3 of the code and Table 2.1. Note that the water-to-cement ratio and minimum cement content are specified. Good workmanship is required to ensure that the steel is properly placed and that the specified cover is obtained.

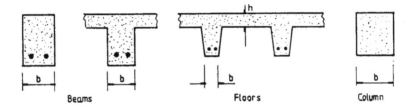

Fire resistance (hour)	Min. Beam b	Rib b	Min. floor Thickness h	Column width Fully exposed b	Min. wall Thickness 0.4%<p<1.0%
1.0	200	125	95	200	120
1.5	200	125	110	250	140
2.0	200	125	125	300	160

Fig.2.4 Minimum dimensions for fire resistance

2.9.2 Cover as Fire Protection

Nominal cover to all reinforcement to meet a given fire resistance period for various elements in a building is given in Table 2.2 and Table 3.4 in the code. Minimum dimensions of members from Fig.3.2 in the code are shown in Fig.2.4. Reference should be made to the complete tables and figures in the code.

Table 2.1 Nominal cover to all reinforcement including links to meet durability requirements

Conditions of exposure	Nominal cover (mm)		
Mild	25	20	20
Moderate		35	30
Severe			40
Very severe			50
Most severe	-	-	-
Abrasive	Nominal cover + allowance for loss of cover due to brasion.		
Maximum free water-to-cement ratio	0.65	0.60	0.55
Minimum cement content (kg/m^3)	275	300	325
Lowest grade of concrete	C30	C35	C40

Table 2.2 Nominal cover to all reinforcement including links to meet specified periods of fire resistance

Fire Resistance	Nominal Cover -mm						
	Beams		Floors		Ribs		Columns
Hour	SS	C	SS	C	SS	C	
1.0	20	20	20	20	20	20	20
1.5	20	20	25	20	35	20	20
2.0	40	30	35	25	45	35	25

SS Simply supported, C Continuous

2.10 REFERENCES

Kay, Ted. 1992, *Assessment and renovation of concrete structures*, (Longman Scientific and Technical).

Perkins, Philip H. 1986, *Repair, protection and waterproofing of concrete structures*, (Elsevier Applied Science Publishers).

Allen, R.T.L, Edwards, S.C. and Shaw, J.D.N. (Eds.), 1993, *The repair of concrete structures*, (Blackie Academic & Professional).

Campbell-Allen, Denison. and Roper, Harold. 1991, *Concrete structures: Materials, maintenance and repair*, (Longman Scientific and Technical).

Day, Ken W. 1995, *Concrete mix design, quality control and specification*, (E. & F.N. Spon).

CHAPTER 3
LIMIT STATE DESIGN AND STRUCTURAL ANALYSIS

3.1 STRUCTURAL DESIGN AND LIMIT STATES

3.1.1 Aims and Methods of Design

The code BS 8110, part 1 in clause 2.1.1 states that the aim of design is the achievement of an acceptable probability that the structure will perform satisfactorily during its life. It must carry the loads safely, not deform excessively and have adequate durability and resistance to the effects of misuse and fire. The clause recognizes that no structure can be made one hundred percent safe and that it is only possible to reduce the probability of failure to an acceptably low level.

Clause 2.1.2 states that the method recommended in the code is limit state design where account is taken of theory, experiment and experience. It adds that calculations alone are not sufficient to produce a safe, serviceable and durable structure. Correct selection of materials, quality control and supervision of construction are equally important.

3.1.2 Criteria for a Safe Design: Limit States

The criterion for a safe design is that the structure should not become unfit for use, i.e. that it should not reach a limit state during its design life. This is achieved, in particular, by designing the structure to ensure that it does not reach

1. **The ultimate limit state (ULS)**: the whole structure or its elements should not collapse, overturn or buckle when subjected to the design loads

2. **Serviceability limit states (SLS)**: the structure should not become unfit for use due to excessive deflection, cracking or vibration

The structure must also be durable, i.e. it must not deteriorate or be damaged excessively by the environment to which it is exposed or action of substances coming into contact with it. The code places particular emphasis on durability (see the discussion in Chapter 2). For reinforced concrete structures the normal practice is to design for the ultimate limit state, check for serviceability and take all necessary precautions to ensure durability.

The code BS 8110 states that nominal earth loads E_n are to be obtained in accordance with normal practice. Reference should be made to

BS 8004: 1986: *Code of Practice for Foundations*

and textbooks on Geotechnics. A useful work is

Bowles, Joseph E.,1995, Foundation analysis and design, (McGraw-Hill), 5th Edition.

The structure must also be able to resist the notional horizontal loads defined in clause 3.1.4.2 of the code. The definition for these loads was given in section 3.1.3(c) above.

design load = characteristic load × partial safety factor for load

$$= F_k \, \gamma_f$$

The partial safety factor γ_f takes account of

1. possible increases in load
2. inaccurate assessment of the effects of loads
3. unforeseen stress distributions in members
4. the importance of the limit state being considered

The code states that the values given for γ_f ensure that serviceability requirements can generally be met by simple rules. The values of γ_f to give design loads and the load combinations for the ultimate limit state are given in BS 8110: Part 1, Table 2.1. These factors are given in Table 3.1. The code states that the adverse partial safety factor is applied to a load producing more critical design conditions. The beneficial factor is applied to a load producing a less, critical design condition.

Table 3.1 Load combinations

Load combination	*Load type*					
	Dead load		*Imposed load*		*Earth and Water pressure*	*Wind*
	Adverse	*Beneficial*	*Adverse*	*Beneficial*		
1. Dead and imposed (and earth and water pressure)	1.4	1.0	1.6	0	1.4	-
2. Dead and wind (and earth and water pressure)	1.4	1.0	-	-	1.4	1.4
3. Dead, wind and imposed (and earth and water pressure)	1.2	1.2	1.2	1.2	1.2	1.2

For example in the case of a beam with an overhang as shown in Fig.3.1, maximum upward reaction at the left hand support and maximum bending moment in the main span occur when there is minimum load on the overhang and maximum load in the main span. On the other hand, possibility of uplift and maximum bending moment in the main span causing tension at top occurs when there is minimum load on the main span and maximum load on the overhang.

In considering the effects of exceptional loads caused by misuse or accident γ_f can be taken as 1.05. The loads to be applied in this case are the dead load, one-third of the wind load and one-third of the imposed load except for storage and industrial buildings when the full imposed load is to be used.

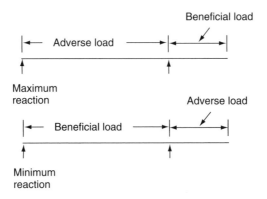

Fig.3.1 Beam with an overhang.

3.3 MATERIALS: PROPERTIES AND DESIGN STRENGTHS

The characteristic strengths or grades of materials are as follows:

Concrete: f_{cu} is the 28 day cube strength in Newtons per square millimetre. The minimum grades for reinforced concrete are given in Table 3.3 in the code. These are grades C30, C35, C40, C45 and C50 in Newtons per square millimetre.

Reinforcement: f_y is the yield or proof stress in Newtons per square millimetre. The specified characteristic strengths of reinforcement given in Table 3.1 in the code are

Hot rolled mild steel $f_y = 250$ N/mm^2
High yield steel, hot rolled or cold worked $f_y = 460$ N/mm^2

Clause 3.1.7.4 of the code states that a lower value may be used to reduce deflection or control cracking. The reason for this is that a lower stress in steel at SLS reduces the number and widths of cracks.

The resistance of sections to applied stresses is based on the design strength which is defined as

$$\frac{characteristic\ strength}{partial\ factor\ of\ safety\ of\ materials} = \frac{f_k}{\gamma_m}$$

3. the transformed section (the compression area of concrete and the transformed area of reinforcement in tension and compression based on the modular ratio are used)

The code states that a modular ratio of 15 may be assumed. It adds that a consistent approach should be used for all elements of the structure.

3.4.2 Methods of Frame Analysis

The complete structure may be analysed elastically using a matrix computer program adopting the basis set out above. It is normal practice to model beam elements using only the rectangular section of T-beam elements in the frame analysis (Fig.1.3). The T-beam section is taken into account in the element design.

Approximate methods of analysis are set out in the code as an alternative to a rigorous analysis of the whole frame. These methods are discussed below.

3.4.3 Monolithic braced frame
Shear walls, lifts and staircases provide stability and resistance to horizontal loads. A braced frame is shown in Fig.3.3. The approximate methods of analysis and the critical load arrangements are as follows.

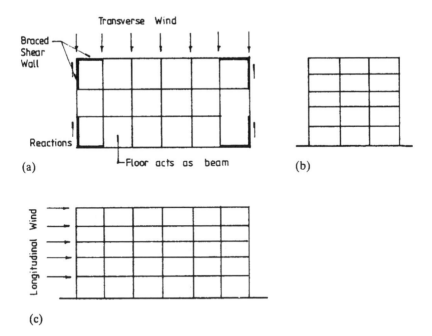

Fig.3.3 Braced multi-storey building: (a) plan; (b) rigid transverse frame; (c) side elevation.

(a) Division into subframes
The structural frame is divided into sub–frames consisting of the beams at one level and the columns above and below that level with ends taken as fixed. The moments and shears are derived from an elastic analysis (Figs 3.4(a) and 3.4(b)).

(b) Critical load arrangement
The critical arrangements of vertical load are
1. all spans loaded with the maximum design ultimate load of $1.4G_k + 1.6Q_k$
2. alternate spans loaded with the maximum design ultimate load of $1.4G_k + 1.6Q_k$ and all other spans loaded with the minimum design ultimate load of 1.0Gk
where G_k is the total dead load on the span and Q_k is the imposed load on the span. The load arrangements are shown in Fig.3.4(b).

(c) Simplification for individual beams and columns
The simplified sub–frame consists of the beam to be designed, the columns at the ends of the beam and the beams on either side if any. The column and beam ends remote from the beam considered are taken as fixed and the stiffness of the beams on either side should be taken as *one–half* of their actual value (Fig.3.4(c)).

The moments for design of an individual column may be found from the same sub-frame analysis provided that its central beam is the longer of the two beams framing into the column.

(d) Continuous beam simplification
The beam at the floor considered may be taken as a continuous beam over supports providing no restraint to rotation. This gives a more conservative design than the procedures set out above. Pattern loading as set out in (b) is applied to determine the critical beam moments and shear for design (Fig.3.4(d)).

(e) Asymmetrically loaded column
The asymmetrically loaded column method is to be used where the beam has been analysed on the basis of the continuous beam simplification set out in (d) above. The column moments can be calculated on the assumption that the column and beam ends remote from the junction under consideration are fixed and that the beams have one–half their actual stiffnesses. The imposed load is to be arranged to cause maximum moment in the column (Fig.3.4(e)). Examples of the application of these methods are given Chapter 14.

3.4.4 Rigid Frames Providing Lateral Stability

Where rigid frames provides lateral stability, they must be analysed for horizontal and vertical loads. Clause 3.1.4.2 of the code states that all buildings must be capable of resisting a notional horizontal load equal to 1.5% of the characteristic dead weight of the structure applied at roof level and at each floor.

The complete structure may be analysed for vertical and horizontal loads using computer analysis program. As an alternative the code gives the following

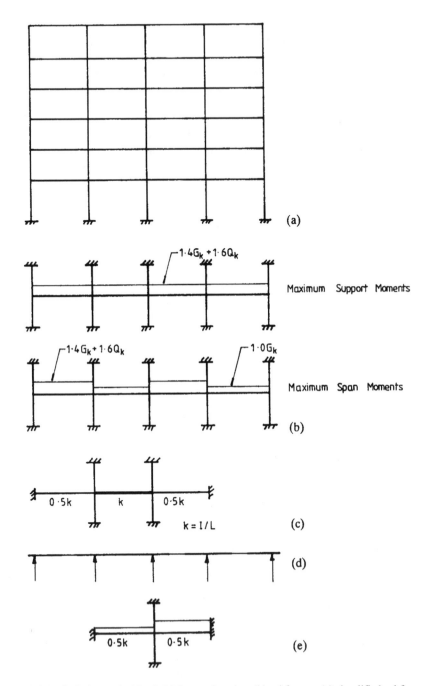

Fig 3.4 Analysis for vertical load: (a) frame elevation; (b) subframes; (c) simplified subframe; (d) continuous beam simplification; (c) column moments analysis for (d)

method for sway frames of three or more approximately equal bays (the design is to be based on the more severe of the conditions):
1. elastic analysis for vertical loads only with maximum design load $1.4G_k + 1.6Q_k$ (refer to sections 3.4.3(a) and 3.4.3(b) above)
2. or the sum of the moments obtained from
      (a) elastic analysis of subframes as defined in section 3.4.3(a) with all beams loaded with $1.2G_k + 1.2Q_k$ (horizontal loads are ignored)
      (b) elastic analysis of the complete frame assuming points of contra–flexure at the centres of all beams and columns for wind load $1.2W_k$ only

The column bases may be considered as pinned if this assumption gives more realistic analyses. A sway frame subjected to horizontal load is shown in Fig.3.5. Method of analysis for horizontal load, the portal method, is discussed in Chapter 13. Examples in the use of these methods are also given.

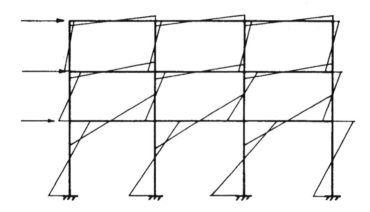

Fig.3.5 Horizontal loads.

3.4.5 Redistribution of Moments

Plastic method of analysis for steel structures based on the stress–strain curve shown in Fig.3.6(a), which gives the moment–rotation curve in Fig.3.6(b), can be used for the analysis of reinforced concrete structures provided due attention is paid to the fact reinforced concrete sections have limited ductility. In order to prevent serious cracking occurring at serviceability limit state, the code adopts a method that gives the designer control over the amount of redistribution and hence of rotation that is permitted to take place. In clause 3.2.2 the code allows a reduction of up to 30% of the peak elastic moment to be made whilst keeping internal and external forces in equilibrium. The conditions under which this can carried out are set out later in Chapter 13.

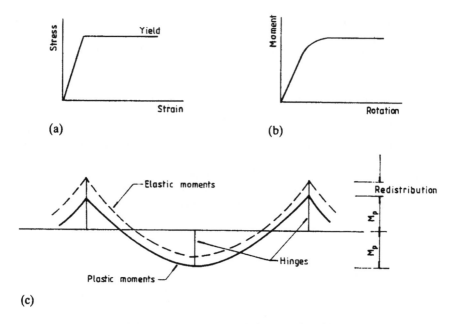

Fig.3.6 (a) Stress–strain curve; (b) moment-rotation curve; (c) elastic and plastic moment distributions.

CHAPTER 4

SECTION DESIGN FOR MOMENT

4.1 TYPES OF BEAM SECTION

The three common types of reinforced concrete beam section are
a. rectangular sections with tension steel only (this generally occurs when designing a given width of slab as a beam.)
b. rectangular sections with tension and compression steel
c. flanged sections of either T or L shape with tension steel and with or without compression steel
Beam sections are shown in Fig.4.1. It will be established later that all beams of structural importance must have steel top and bottom to carry links to resist shear.

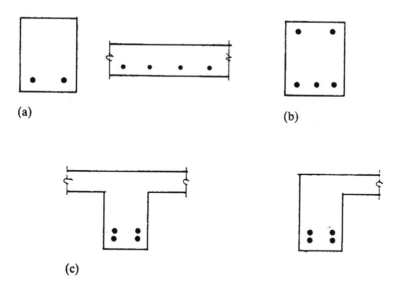

Fig.4.1 (a) Rectangular beam and slab, tension steel only; (b) rectangular beam, tension and compression steel; (c) flanged beams.

4.2 REINFORCEMENT AND BAR SPACING

Before beginning section design, reinforcement data and code requirements with regard to minimum and maximum areas of bars in beams and bar spacing are set out. This is to enable sections to be designed with practical amounts and layout of steel. Requirements for cover were discussed in section 2.9.

4.2.1 Reinforcement Data

In accordance with BS8110: Part 1, clause 3.12.4.1, bars may be placed singly or in pairs or in bundles of three or four bars in contact. For design purposes the pair or bundle is treated as a single bar of equivalent area. Bars are available with diameters of 6, 8, 10, 12, 16, 20, 25, 32 and 40 mm and in two grades with characteristic strengths f_y:

$$\text{Hot rolled mild steel } f_y = 250 \text{ N /mm}^2$$
$$\text{High yield steel } f_y = 460 \text{ N/mm}^2$$

For convenience in design calculations, areas of groups of bars are given in Table 4.1. Table 4.2 gives equivalent diameter of bundles of bars of same diameter.

Table 4.1 Areas of groups of bars

Diameter of bar in mm	Numbers of bars in group							
	1	2	3	4	5	6	7	8
6	28	57	85	113	141	170	198	226
8	50	101	151	201	251	302	352	402
10	79	157	236	314	393	471	550	628
12	113	226	339	452	566	679	792	905
16	201	402	603	804	1005	1206	1407	1609
20	314	628	943	1257	1571	1885	2109	2513
25	491	982	1473	1964	2454	2945	3436	3927
32	804	1609	2413	3217	4021	4826	5630	6434

Table 4.2 Equivalent diameters of bars in groups

Diameter in mm of bars in group	Number of bars in group			
	1	2	3	4
6	6	8.5	10.4	12
8	8	11.3	13.9	16
10	10	14.1	17.3	20
12	12	17.0	20.8	24
16	16	22.6	27.7	32
20	20	28.3	34.6	40
25	25	35.4	43.3	50
32	32	45.3	55.4	64

Detailed drawings should be prepared according to

Standard Method of Detailing Structural Concrete. Institution of Structural Engineers, London, 1989.
Bar types are specified by letters:

R mild steel bars
T high yield bars

Bars are designated on drawings as, for example, 4T25, i.e. four 25 mm diameter bars of grade 460. This system will be used to specify bars in figures.

4.2.2 Minimum and Maximum Areas of Reinforcement in Beams

The minimum areas of reinforcement in a beam section to control cracking as well as to resist tension or compression due to bending in different types of beam section are given in BS 8110: Part 1, clause 3.12.5.3 and Table 3.25. Some commonly used values are shown in Fig.4.2 and Table 4.3. Other values will be discussed in appropriate parts of the book.

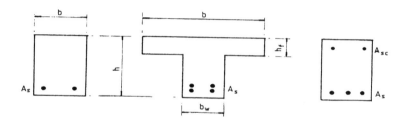

Fig.4.2 Minimum tension and compression steels

Table 4.3: Minimum steel areas

	Percentage	$f_y = 250$ N/mm^2	$f_y = 460$ N/mm^2
Tension reinforcement			
Rectangular beam	$100A_s/A_c$	0.24	0.13
Flanged beam –Web in tension: $b_w/b < 0.4$	$100A_s/b_w\,h$	0.32	0.18
Flanged beam –Web in tension: $b_w/b \geq 0.4$	$100A_s/b_w\,h$	0.24	0.13
Compression reinforcement			
Rectangular beam	$100A_{sc}/A_c$	0.2	0.2
Flanged beam –flange in compression:	$100A_{sc}/b_wh_f$	0.2	0.2

A_c = total area of concrete, A_s = minimum area of reinforcement, A_{sc} = area of steel in compression, b, b_w, h_f = beam dimensions.

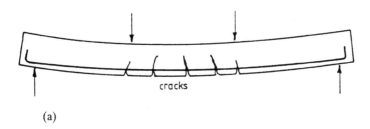

(a)

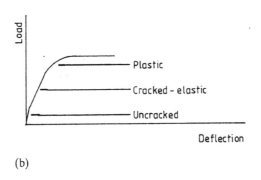

(b)

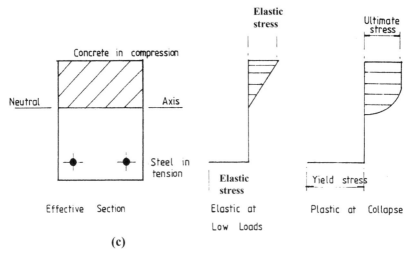

(c)

Fig.4.4 (a) Flexural cracks at collapse; (b) load-deflection curve; (c) effective section and stress distribution.

2. The stresses in the concrete in compression are derived using either:
  (a) the design stress–strain curve given in Fig.4.5(a) with $\gamma_m = 1.5$ or
  (b) the simplified stress block shown in Fig.4.6(d) where the depth of the stress block is 0.9 of the depth to the neutral axis denoted by x.
  Note that in both cases the maximum strain in the concrete at failure is 0.0035;

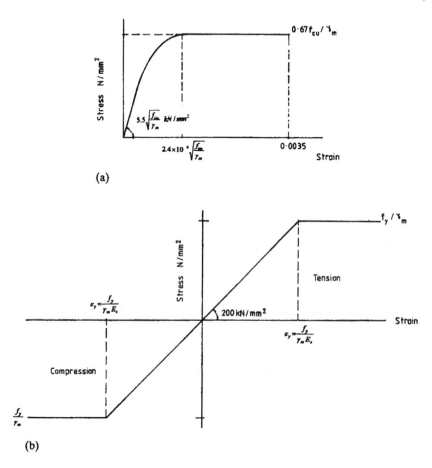

Fig.4.5 Stress–strain diagrams (a) Concrete; (b) Steel.

3. The tensile strength of the concrete is ignored;
4. The stresses in the reinforcement are derived from the stress–strain curve shown in Fig.4.5(b) where $\gamma_m = 1.05$;
5. Where the section is designed to resist flexure only, the lever arm should not be assumed to be greater than 0.95 of the effective depth, d. This is because of the fact that at the top face during compaction water tends to move to the top and causes a higher water cement ratio than the rest of the beam. In addition weathering also affects the strength. Because of that a layer of concrete at the top is likely to be weak

and by limiting the value of the lever arm z, one avoids the possibility of expecting a weak layer of concrete to resist the compressive stress due to bending.

On the basis of these assumptions the strain and stress diagrams for the two alternative stress distributions for the concrete in compression are as shown in Fig.4.6, where the following symbols are used:

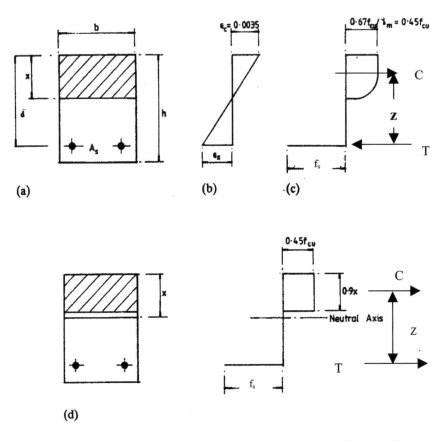

Fig.4.6 (a) Section; (b) strain; (c) rectangular parabolic stress diagram; (d) simplified stress diagram.

h   overall depth of the section
d   effective depth, i.e. depth from the compression face to the *centroid* of tension steel
b   breadth of the section
x   depth to the neutral axis
f_s stress in steel
A_s area of tension reinforcement
ε_c maximum strain in the concrete (0.0035)
ε_s strain in steel

The alternative stress distributions for the compressive stress in the concrete, the rectangular parabolic stress diagram and the simplified stress block, are shown in Figs 4.6(c) and 4.6(d) respectively.

The maximum strain in the concrete is 0.0035 and the strain ε_s in the steel depends on the depth of the neutral axis. Stress–strain curves for concrete and for steel are shown in Figs 4.5(a) and 4.5(b) respectively.

4.4.2 Moment of Resistance: Rectangular Stress Block

Fig.4.6 (d) shows the assumed stress distribution. The concrete stress is
$$0.67\, f_{cu}/\gamma_m = 0.67\, f_{cu}/(\gamma_m = 1.5) = 0.447 f_{cu}$$
which is generally rounded off to $0.45 f_{cu}$.
The total compressive force C is given by
$$C = 0.447\, f_{cu} \times b \times 0.9x = 0.402\, b \times x \times f_{cu}$$
The lever arm z is
$$z = d - 0.9x/2 = d - 0.45x$$
If M is the applied moment, then
$$M = C \times z = 0.402\, b \times x \times f_{cu} \times (d - 0.45x)$$
Setting
$$k = M/(b\, d^2\, f_{cu})$$
$$k = 0.402(x/d)\,(1 - 0.45\,(x/d))$$
Rearranging,
$$0.1809(x/d)^2 - 0.402\,(x/d) + k = 0$$
Solving for x/d
$$x/d = \{0.402 - \sqrt{(0.1616 - 0.7236\, k}\}/\, 0.3618$$
$$= 1.11 - \sqrt{(1.2345 - 5.5279k)}$$
$$z/d = 1 - 0.45(x/d)$$
$$z.d = 0.5 + \sqrt{(0.25 - k/0.9)}$$
Total tensile force T in steel is
$$T = A_s \times f_s$$
For internal equilibrium, total tension T must be equal to total compression C. The forces T and C form a couple ata lever arm of z.

$$M = T\, z = A_s\, f_s\, z$$
$$A_s = M/(f_s\, z)$$

The stress f_s in steel depends on the strain ε_s in steel. As remarked in section 4.3, it is highly desirable that final failure is due to yielding of steel rather than due to crushing of concrete. It is useful therefore to calculate the maximum neutral axis depth in order to achieve this. Assuming that plane sections remain plane before and after bending, an assumption validated by experimental observations, if as shown in Fig.4.6(b), the maximum permitted strain in concrete at the compression face is 0.0035, then the strain ε_s in steel is calculated from the strain diagram by

$$\varepsilon_s = \frac{(d-x)}{x} 0.0035$$

Strain ε_s in steel at a stress of f_y/γ_m is given by

$$E_s \, \varepsilon_s = \frac{f_y}{\gamma_m}$$

where f_y = yield stress, γ_m = material safety factor and E_s is Young's modules for steel.

Taking f_y = 460 N/mm², γ_m = 1.05, f_y/γ_m = 438 N/mm², E_s = 200 kN/mm², ε_s = 0.0022

For ε_s = 0.0022, the depth of neutral axis x is given by

$$\varepsilon_s = 0.0022 = \frac{(d-x)}{x} 0.0035$$

$$x/d = 0.6140$$

However in order to ensure that failure is preceded by steel yielding well before the strain in concrete reaches 0.0035 resulting in the desirable ductile form of failure, in clause 3.4.4.4, the code limits the ratio x/d to a maximum of 0.5. If x = 0.5 d, then

$$C = 0.447 \, f_{cu} \times b \times 0.9x = 0.402 \times b \times 0.5d \times f_{cu}$$
$$C = 0.201 \times f_{cu} \times b \times d$$
$$z = d - 0.45x = d - 0.45 \times 0.5d = 0.775 \, d$$
$$M = C \times z = 0.156 \times b \times d^2 \times f_{cu}$$
$$k = M/(b \, d^2 \, f_{cu}) = 0.156$$

This is the maximum value of the applied moment that the section can resist because it utilises fully the compression capacity of the cross section. This formula can be used to calculate the *minimum effective depth* required in a singly reinforced rectangular concrete section.

$$d_{min} = \sqrt{\frac{M}{0.156 \, b \, f_{cu}}}$$

In practice the effective depth d is made larger than the required minimum consistent with the required headroom.

$$d \geq \sqrt{\frac{M}{0.156 \, b \, f_{cu}}}$$

The reason for this is that with a larger depth, the neutral axis depth is smaller and hence the lever arm is larger leading for a given moment M, to a smaller amount of reinforcement. It has the additional advantage that in the event of unexpected overload, the beams will show large ductility before failure.

If $x/d \leq 0.5$, steel will always yield,

$$f_s = f_y/1.05 = 0.95\ f_y$$
$$M = T\ z = A_s\ 0.95\ f_y\ z,$$
$$A_s = M/(0.95\ f_y\ z)$$

4.4.3 Procedure for the Design of Singly Reinforced Rectangular Beam

The steps to be followed in the design of singly reinforced rectangular beams can be summarised as follows.

- From the minimum requirements of span/depth ratio to control deflection (see Chapter 6), estimate a suitable effective depth d.
- Assuming the bar diameter for the main steel and links and the required cover as determined by exposure conditions, estimate an overall depth h.
  $h = d +$ bar diameter + Link diameter + Cover
- Assume breadth as about half the overall depth.
- Calculate the self-weight.
- Calculate the design live load and dead load moment using appropriate load factors. The load factors are normally 1.4 for dead loads and 1.6 for live loads.
- In the case of singly reinforced sections, calculate the minimum effective depth using the formula

$$d_{min} = \sqrt{\frac{M}{0.156\ b\ f_{cu}}}$$

- Adopt an effective depth greater than the minimum depth in order to reduce the total tension reinforcement.
- Check that the new depth due to increased self-weight does not drastically affect the calculated design moment. If it does, calculate the revised ultimate moment required.
- Calculate $k = M/(b\ d^2\ f_{cu})$
- Calculate the lever arm z

$$z = d\{0.5 + \sqrt{(0.25 - k/0.9)}\} \leq 0.95d$$
Note that $z/d \leq 0.95$ if $k \geq 0.0428$

- Calculate the required steel A_s

$$A_s = M/\{0.95\ f_y\ z\}$$

- Check that the steel provided satisfies the minimum and maximum steel percentages specified in the code.

4.4.4 Examples of Design of Singly Reinforced Rectangular Sections

Example 1: A simply supported reinforced rectangular beam of 8 m span carries uniformly distributed characteristic dead load, which includes an allowance for self-weight of 7 kN/m and characteristic imposed load of 5 kN/m.

The breadth b = 250 mm. Design the beam at mid-span section. Use grade 30 concrete and high yield steel reinforcement, f_y = 460 N/mm^2.

$$\text{Design load} = (1.4 \times 7) + (1.6 \times 5) = 17.8 \text{ kN/m}$$

Design ultimate moment M at mid-span:

$$M = 17.8 \times 8^2/8 = 142.4 \text{ kNm}$$

Minimum effective depth to avoid any compression steel is given by

$$d_{min} = \sqrt{\frac{M}{0.156 \times b \times f_{cu}}} = \sqrt{\frac{142.4 \times 10^6}{0.156 \times 250 \times 30}} = 348.9 \text{ mm}$$

Using this value of d,

$$x = 0.5d$$
$$z = d - 0.45x = 0.775 \, d.$$

The area of steel required is

$$A_s = \frac{M}{0.775 \, d \times 0.95 f_y} = \frac{142.4 \times 10^6}{0.775 \times 348.9 \times 0.95 \times 460} = 1206 \text{ mm}^2$$

However, if a value of d equal to say 400 mm, which is larger than the minimum value is used, then one can reduce the area of steel required.

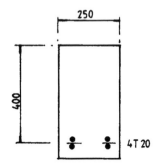

250

400

4T 20

Fig.4.7: Mid-span section of the beam.

Assuming d = 400 mm

$$k = \frac{M}{bd^2 f_{cu}} = \frac{142.4 \times 10^6}{250 \times 400^2 \times 30} = 0.119 < 0.156$$

$$\frac{z}{d} = 0.5 + \sqrt{\left(0.25 - \frac{k}{0.9}\right)} = 0.843 < 0.95$$

$$A_s = \frac{M}{0.843\,d \times 0.95 f_y} = \frac{142.4 \times 10^6}{0.843 \times 400 \times 0.95 \times 460} = 967 \text{ mm}^2$$

Provide four 20 mm diameter bars in two layers as shown in Fig.4.7. From Table 4.1, A = 1257 mm². Assuming cover of 30 mm and link diameter of 8 mm, the overall depth h of the beam is

$$h = 400 + 30 + 8 + 20 = 458, \text{ say } 460 \text{ mm.}$$

Check that the percentage steel provided is greater than the minimum of 0.13.

$$100 \, A_s/(bh) = 100 \times 1257/(250 \times 450) = 1.12 > 0.13.$$

Note that this is only one of several possible satisfactory solutions.

Example 2: Determination of tension steel cut-off.

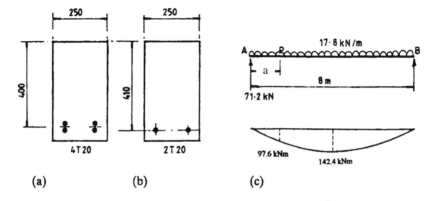

Fig 4.8 (a) Section at mid-span; (b) section at support; (c) loading and bending moment diagram.

In simply supported beams bending moment decreases towards the supports. Therefore the amount of steel required towards the support region is much less than at mid-span. For the beam in Example 4.1, determine the position along the beam where theoretically two of the four 20 mm diameter bars may be cut off.

The section at cut-off has two 20 mm diameter bars continuing: $A_s = 628$ mm². The effective depth here is 410 mm (Fig.4.8(b)). The neutral axis depth can be determined by equating total compression in concrete to total tension in the beam.

$$T = 0.95 \, f_y \, A_s = 0.95 \times 460 \times 628 \times 10^{-3} = 274.44 \text{ kN}$$
$$C = (0.445 \, f_{cu} \, b \, 0.9x)10^{-3}$$
$$C = (0.445 \times 30 \times 250 \times 0.9x) \times 10^{-3} = 3x \text{ kN}$$

Equating C = T,

$$x = 91 \text{ mm}$$
$$z = d - 0.45x = 369 \text{ mm}$$
$$z/d = 369/410 = 0.90 < 0.95$$

Moment of resistance M_R

$$M_R = T\,z = 264.44 \times 369 \times 10^{-3} = 97.6 \text{ kNm}$$

Determine the position of p along the beam such that M = 97.6 kN m (Fig.4.8c).

4.4.5.1 Examples of use of design chart

Example 1: Use the design chart to calculate the area of steel for the beam in Example 1, section 4.4.4.

$$\frac{M}{bd^2 f_{cu}} = \frac{142.4 \times 10^6}{250 \times 400^2 \times 30} = 0.119$$

From Fig.4.9,

$$100 \frac{A_s}{bd} \frac{f_y}{f_{cu}} = 100 \frac{A_s}{250 \times 400} \frac{460}{30} = 15.0$$

$A_s = 978$ mm^2 compared with 967 mm^2 previously calculated.

Example 2: Calculate the moment of resistance for the beam in Example 2, section 4.4.4, for the section where steel is curtailed to 2T20.

$$100 \frac{A_s}{bd} \frac{f_y}{f_{cu}} = 100 \frac{628}{250 \times 410} \frac{460}{30} = 9.4$$

From Fig.4.10,

$$\frac{M}{bd^2 f_{cu}} = \frac{M \times 10^6}{250 \times 410^2 \times 30} = 0.08$$

M = 100.9 kNm (Exact answer M= 97.6 kNm)

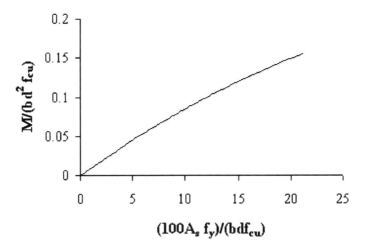

Fig.4.10 Design chart for singly reinforced rectangular concrete beams.

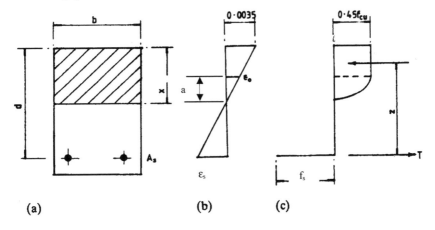

Fig.4.11 (a) section; (b) Strain diagram; (c) stress diagram.

4.4.6 Moment of Resistance Using Rectangular Parabolic Stress Block

In the previous sections, the simplified rectangular stress block was used to derive design equations. In this section, the stress–strain curve for concrete shown in Fig.4.5a will be used to derive the corresponding design equations. As shown in Fig.4.11(b), the maximum strain at the top is 0.0035. The strain ε_0 is where the parabolic part of the stress strain–strain curve ends. If the neutral axis depth is x, the distance 'a' from the neutral axis to where the strain is ε_0 is given by

$$a = x \, (\varepsilon_0 / 0.0035)$$

where ε_0 from Fig.4.5(a) is given by

$$\varepsilon_0 = 2.4 \times 10^{-4} \sqrt{\frac{f_{cu}}{\gamma_m}}$$

The compressive force C_1 in the rectangular portion of depth $(x - a)$ of the stress block is given by

$$C_1 = 0.67 \frac{f_{cu}}{\gamma_m} b (x - a)$$

The lever arm z_1 for C_1 from the centroid of steel area is

$$z_1 = d - 0.5(x - a)$$

Using the 'well-known' result that the area of a parabola is equal to two-thirds the area of the enclosing rectangle, the compressive force in the parabolic portion of depth 'a' of the stress block is given by

$$C_2 = \frac{2}{3} \{ 0.67 \frac{f_{cu}}{\gamma_m} b \, a \}$$

The centroid of C_2 is at a distance of 5a/8 from the neutral axis. The lever arm z_2 for C_2 from the centroid of steel area is

$$z_2 = d - x + 5a/8$$

Therefore taking moments about the centroid of steel area,

$$M = C_2 z_2 + C_1 z_1$$

For $f_{cu} = 30$ N/mm^2 and $\gamma_m = 1.5$,

$$\varepsilon_0 = 2.4 \times 10^{-4} \sqrt{\frac{f_{cu}}{\gamma_m}} = 0.00107$$

$$\varepsilon_0 / 0.0035 = 0.306$$
$$a = x \, (\varepsilon_0 / 0.0035) = 0.306x$$

$$C_1 = 0.31 \, f_{cu} \, bx$$
$$C_2 = 0.091 \, f_{cu} \, bx$$

$$z_1 = (d - 0.347x)$$
$$z_2 = d - 0.8088x$$

$$M = C_1 z_1 + C_2 z_2 = f_{cu} \, b \, d^2 \, \frac{x}{d} \{0.4014 - 0.1814 \frac{x}{d}\}$$

The corresponding equation for rectangular stress block assumption as derived in section 4.4.2 is

$$M = f_{cu} \, b \, d^2 \, \frac{x}{d} \{0.402 - 0.1809 \frac{x}{d}\}$$

The two equations differ by very little from each other. The Rectangular stress block assumption is therefore accurate for all practical calculations.

4.5 DOUBLY REINFORCED BEAMS

The normal design practice is to use singly reinforced sections. However if for any reason, for example headroom considerations, it is necessary to restrict the overall depth of a beam, then it becomes necessary to use steel in the compression zone as well because concrete alone cannot provide the necessary compression resistance.

4.5.1 Design Formulae Using the Simplified Stress Block

The formulae for the design of a doubly reinforced beam are derived using the rectangular stress block.

Let M be the design ultimate moment. As shown in section 4.4.2, a rectangular section as a *singly reinforced section* can resist a *maximum* value of the moment equal to

$$M_{sr} = 0.156 \, b \, d^2 \, f_{cu}$$

The corresponding neutral axis depth $x = 0.5 \, d$. The compressive force C_c in concrete is

$$C_c = 0.45 \, f_{cu} \, b \, 0.45 \, d = 0.2 \, f_{cu} \, b \, d$$

The lever arm z_c

$$z_c = 0.775d$$

If $M > M_{sr}$, then compression steel is required. The compressive force C_s due to compression steel of area A_s is

$$C_s = A_s^{'} f_s^{'}$$

where $f_s^{'}$ is the stress in compression steel.

As shown in Fig.4.12, the lever arm z_s for compression steel is

$$z_s = (d - d^{'})$$

The stress in the tensile steel is 0.95 f_y because the neutral axis depth is limited to 0.5d. However the stress $f_s^{'}$ in the compressive steel depends on the corresponding strain strain ε_{sc} in concrete at steel level. ε_{sc} is given by

$$\varepsilon_{sc} = 0.0035 \frac{(x - d^{'})}{x}$$

If the strain ε_{sc} is equal to or greater than the yield strain in steel, then steel yields and the stress $f_s^{'}$ in compression steel is equal to 0.95 f_y. Otherwise, the stress in compression steel is given by

$$f_s^{'} = E_s \, \varepsilon_{sc}.$$

If $f_y = 460$ N/mm^2 and $E_s = 200$ kN/mm^2, then the yield strain in steel is equal to

$$\varepsilon_{yield} = \frac{0.95 \, f_y}{E} = 0.0022$$

Therefore, steel will yield if

$$\varepsilon_{sc} = 0.0035 \frac{(x - d^{'})}{x} \geq (\frac{0.95 f_y}{E} = 0.0022)$$

$$\therefore \frac{d'}{x} \leq 0.37 \quad \text{If } x = 0.5d, \text{ then } \frac{d'}{x} \leq 0.186$$

If mild steel is used, then $f_y = 250$ N/mm^2. The above equations then become

$$\varepsilon_{yield} = \frac{0.95 \, f_y}{E} = 0.0012$$

Therefore, steel will yield if

$$\varepsilon_{sc} = 0.0035 \frac{(x - d^{'})}{x} \geq (\frac{0.95 f_y}{E} = 0.0012)$$

$$\therefore \frac{d'}{x} \leq 0.66 \quad \text{If } x = 0.5d, \text{ then } \frac{d'}{x} \leq 0.33$$

Taking moments about the tension steel,

$$M = C_c \, z_c + C_s \, z_s$$
$$M = 0.2 \, f_{cu} \, b \, d \, (0.775 \, d) + A_s^{'} \, f_s^{'} \, (d - d^{'})$$
$$M = 0.156 \, b \, d^2 \, f_{cu} + A_s^{'} \, f_s^{'} \, (d - d^{'})$$
$$A_s^{'} = (M - 0.156 \, b \, d^2 \, f_{cu}) / \{f_s^{'} \, (d - d^{'})\}$$

From equilibrium, the tensile force T is

$$T = A_s \, 0.95 \, f_y = C_c + C_s$$

One important point to remember is that to prevent steel bars in compression from buckling, it is necessary to restrain them using links. Clause 3.12.7 of the code says that links or ties at least one quarter of the size of the largest compression bar or 6 mm whichever is greater should be provided at a maximum spacing of 12 times the size of the smallest compression bar.

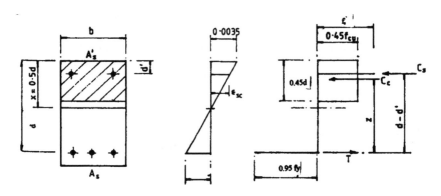

Fig.4.12 Doubly reinforced beam.

4.5.2 Examples of Rectangular Doubly Reinforced Concrete Beams

The use of the formulae developed in the previous section is illustrated by a few examples.

Example 1: A rectangular beam is simply supported over a span of 6 m and carries characteristic dead load including self-weight of 12.7 kN/m and characteristic imposed load of 6.0 kN/m. The beam is 200 mm wide by 300 mm effective depth and the inset of the compression steel is 40 mm. Design the steel for mid-span of the beam for grade C30 concrete and grade 460 reinforcement.

$$\text{design load} = (12.7 \times 1.4) + (6 \times 1.6) = 27.4 \text{ kN/m}$$

Required ultimate moment M:

$$M = 27.4 \times 6^2/8 = 123.3 \text{ kN m}$$

Maximum moment that the beam section can resist as a singly reinforced section is

$$M_{sr} = 0.156 \times 30 \times 200 \times 300^2 \times 10^{-6} = 84.24 \text{ kNm}$$

$M > M_{sr}$, Compression steel is required.

$$d'/x = 40/150 = 0.27 < 0.37$$

The compression steel yields. The stress f_s' in the compression steel is $0.95f_y$.

$$A_s' = \{ M - 0.156 \ b \ d^2 \ f_{cu} \}/[0.95 \ fy \ (d - d')]$$
$$A_s' = \{123.3 - 84.24\} \times 10^6/[0.95 \times 460 \times (300 - 40)] = 344 \text{ mm}^2$$

From equilibrium:

$$A_s \ 0.95 \ f_y = 0.2 \ b \ d \ f_{cu} + A_s' \ f_s'$$
$$A_s \ 0.95 \times 460 = 0.2 \times 200 \times 300 \times 30 + 344 \times 0.95 \times 460$$
$$A_s = 1168 \text{ mm}^2$$

For the tension steel (2T25 + 2T16) give A_s = 1383 mm². For the compression steel 2T16 give A_s' = 402 mm². The beam section and flexural reinforcement steel are shown in Fig.4.13.

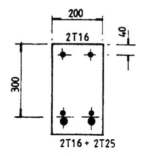

Fig.4.13 Doubly reinforced beam.

Example 2: Design the beam in Example 4.6 but with d' = 60 mm.
$$d'/x = 60/150 = 0.40 > 0.37$$
Compression steel does not yield. Strain in compression steel
$$\varepsilon_{sc} = 0.0035\frac{(x-d')}{x} = 0.0035\frac{(150-60)}{150} = 0.0021$$
Stress in compression steel is
$$f_s' = E_s\,\varepsilon_{sc} = 200 \times 10^3 \times 0.0021 = 420 \text{ N/mm}^2$$

$$A_s' = \{M - 0.156 \text{ b d}^2 f_{cu}\}/[420 (d - d')]$$
$$A_s' = \{123.3 - 84.24\} \times 10^6/[420 \times (300 - 40)] = 358 \text{ mm}^2$$

From equilibrium:
$$A_s\, 0.95\, f_y = 0.2 \text{ b d } f_{cu} + A_s'\, f_s'$$

$$A_s\, 0.95 \times 460 = 0.2 \times 200 \times 300 \times 30 + 358 \times 420, \; A_s = 1168 \text{ mm}^2$$

4.6 FLANGED BEAMS

4.6.1 General Considerations

In simple slab-beam system shown in Fig.4.14, the slab is designed to span between the beams. The beams span between external supports such as columns, walls, etc. The reactions from the slabs act as load on the beam. When a series of beams are used to support a concrete slab, because of the monolithic nature of concrete construction, the slab acts as the flange of the beams. The end beams become L-beams while the intermediate beams become T-beams. In designing the intermediate beams, it is assumed that the loads acting on half the slab on the two sides of the beam are carried by the beam. Because of the

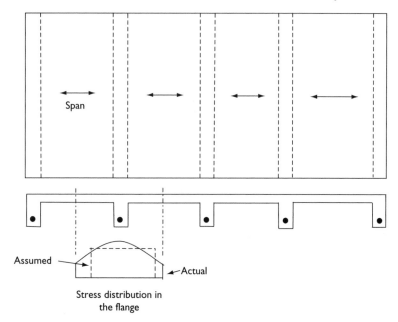

Fig.4.14 Beam-slab system.

comparatively small contact area at the junction of the flange and the rib of the beam, the distribution of the compressive stress in the flange is not uniform. It is higher at the junction and decreases away from the junction. This phenomenon is known as shear lag. For simplicity in design, it is assumed that only part of full physical flange width is considered to sustain compressive stress of uniform magnitude. This smaller width is known as effective breadth of the flange. Although the effective width actually varies even along the span as well, it is common to assume that the effective width remains constant over the entire span.

The effective breadth b of flanged beams (Fig.4.15) as given in BS 8110: Part 1, clause 3.4.1.5:

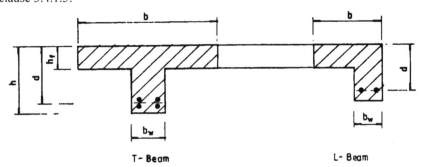

Fig.4.15: Cross section of flanged beams.

1. T-beams: b = {web width b_w + $\ell_z/5$} or the actual flange width if less;
2. L-beams: b = {web width b_w + $\ell_z/10$} or the actual flange width if less
where b_w is the breadth of the web of the beam and ℓ_z is the distance between points of zero moment in the beam. In simply supported beams it is the effective span where as in continuous beams ℓ_z may be taken as 0.7 times the effective span.

The design procedure for flanged beams depends on the depth of the stress block. Two possibilities need to be considered.

4.6.2 Stress Block within the Flange

If $0.9x \leq h_f$, the depth of the flange (same as the total depth of the slab) then all the concrete below the flange is cracked and the beam may be treated as a rectangular beam of breadth b and effective depth d and the methods set out in sections 4.4.6 and 4.4.7 above apply. The maximum moment of resistance when $0.9x = h_f$ is equal to

$$M_{flange} = 0.45\, f_{cu}\, b\, h_f (d - h_f/2)$$

Thus if the design moment $M \leq M_{flange}$, then design the beam as singly reinforced rectangular section $b \times d$.

4.6.3 Stress Block Extends into the Web

As shown in Fig.4.16, the compression forces are:
In the flange of width $(b - b_w)$, the compression force C_1 is
$$C_1 = 0.45\, f_{cu}\, (b - b_w)\, h_f$$
In the web, the compression force C_2 is
$$C_2 = 0.45\, f_{cu}\, b_w\, 0.9x$$
The corresponding lever arms about the tension steel are
$$z_1 = d - h_f/2$$
$$z_2 = (d - 0.9x/2)$$

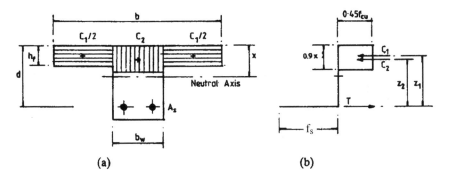

Fig.4.16: T-beam with the stress block extending into the web.

The moment of resistance M_R is given by

$$M_R = C_1 z_1 + C_2 z_2$$

$$M_R = 0.45 f_{cu} (b - b_w) h_f (d - h_f/2) + 0.45 f_{cu} b_w 0.9x (d - 0.9x/2)$$

$$\frac{M}{bd^2 f_{cu}} = 0.45(1 - \frac{b_w}{b})\frac{h_f}{d}(1 - \frac{h_f}{2d}) + 0.4\frac{b_w}{b}\frac{x}{d}(1 - 0.45\frac{x}{d})$$

From equilibrium,

$$T = A_s f_s = C_1 + C_2$$

If the amount of steel provided is sufficient to cause yielding of the steel, then $f_s = 0.95 f_y$. The maximum moment of resistance without any compression steel is when $x = 0.5d$. Substituting $x = 0.5d$ in the expression for M_R, the maximum moment of resistance is

$$M_{max} = 0.45 f_{cu} (b - b_w) h_f (d - h_f/2) + 0.156 f_{cu} b_w d^2$$

$$M_{max} = bd^2 f_{cu} [0.45(1 - \frac{b_w}{b})\frac{h_f}{d}(1 - \frac{h_f}{2d}) + 0.156\frac{b_w}{b}]$$

$$\text{or } M_{max} = \beta_f bd^2 f_{cu}$$

where

$$\beta_f = [0.45(1 - \frac{b_w}{b})\frac{h_f}{d}(1 - \frac{h_f}{2d}) + 0.156\frac{b_w}{b}]$$

Thus if the design moment $M_{flange} < M \leq M_{max}$, then determine the value of x from

$$\frac{M}{bd^2 f_{cu}} = 0.45(1 - \frac{b_w}{b})\frac{h_f}{d}(1 - \frac{h_f}{2d}) + 0.4\frac{b_w}{b}\frac{x}{d}(1 - 0.45\frac{x}{d})$$

where $x \leq 0.5d$ and the reinforcement required is obtained from the equilibrium condition,

$$A_s 0.95 f_y = C_1 + C_2$$

4.6.3.1 Code formula

As an alternative, a slightly conservative formula for calculating the steel area is given in clause 3.4.4.5 of the code. The equation in the code is derived using the simplified stress block with $x = 0.5d$ (Fig.4.16).

$$\text{depth of stress block} = 0.9x = 0.45d$$

The concrete forces in compression are

$$C_1 = 0.45 f_{cu} h_f (b - b_w)$$

$$C_2 = 0.45 f_{cu} \times 0.45 d b_w = 0.2 f_{cu} b_w d$$

The values of the lever arms for C_1 and C_2 from the steel force T are:

$$z_1 = d - 0.5h_f$$

$$z_2 = d - 0.5 \times 0.45d = 0.775d$$

The steel force in tension is
$$T = 0.95 \, f_y \, A_s$$
The moment of resistance of the section is found by taking moments about force C_1:
$$M = Tz_1 - C_2(z_1 - z_2)$$
$$M = 0.95 \, f_y \, A_s(d - 0.5h_f) - 0.2f_{cu} \, b_w \, d \, (0.225d - 0.5h_f)$$
$$M = 0.95 \, f_y \, A_s(d - 0.5h_f) - 0.1f_{cu} \, b_w \, d \, (0.45d - h_f)$$
from which
$$A_s = \frac{M + 0.1f_{cu} \, b_w \, d \, (0.45d - h_f)}{0.95 \, f_y \, (d - 0.5h_f)}$$

This is the expression given in the code. It gives conservative results for cases where x is less than 0.5d. The equation only applies when h_f is less than 0.45d, as otherwise the second term in the numerator becomes negative.

4.6.4 Steps in Reinforcement Calculation of a T- or an L-Beam

- Calculate the total design load (including self-weight) and the corresponding design moment M using appropriate load factors.
- Calculate the maximum moment M_{flange} that can be resisted, when the entire flange is in compression.
$$M_{flange} = 0.45 \, f_{cu} \, b \, h_f \, (d - h_f/2)$$
- Calculate the maximum moment that the section can withstand without requiring compression reinforcement.
$$M_{max} = 0.45 \, f_{cu} \, (b - b_w) \, h_f \, (d - h_f/2) + 0.156 \, f_{cu} \, b_w \, d^2$$
- If $M \le M_{flange}$, then design as a rectangular beam of dimensions, b × d.
- If $M_{flange} < M \le M_{max}$, then the required steel area can be determined to sufficient accuracy from the code formula
$$A_s = \frac{M + 0.1f_{cu} \, b_w \, d \, (0.45d - h_f)}{0.95 \, f_y \, (d - 0.5h_f)}$$
- If $M > M_{max}$, then compression steel is required or the section has to be revised. Compression steel is rarely required in the case of flanged beams.

4.6.5 Examples of Design of Flanged Beams

Example 1: A continuous slab 100 mm thick is carried on T-beams at 2 m centres. The overall depth of the beam is 350 mm and the breadth b_w of the web is 250 mm. The beams are 6 m span and are simply supported. The characteristic dead load including self-weight and finishes is 7.4kN/m^2 and the characteristic imposed load is 5 kN/m^2. Design the beam using the simplified stress block. The materials are grade C30 concrete and grade 460 reinforcement.

Since the beams are spaced at 2 m centres, the loads a the beam are:
$$\text{Dead load} = 7.4 \times 2 = 14.8 \text{ kN/m}$$

Live load = 5 × 2 = 10 kN/m

design load = (1.4 × 14.8) + (1.6 × 10) = 36.7 kN/m

ultimate moment at mid-span = 36.7 × 6^2/8 = 165 kN m

effective width b of flange: b = 250 + 6000/5 = 1450 mm

The beam section is shown in Fig.4.17. From BS8110: Part 1, Table 3.4, the nominal cover on the links is 25 mm for grade 30 concrete. If the links are 8 mm in diameter and the main bars are 25 mm in diameter, then

d = 350 − 25 − 8 − 12.5 = 304.5 mm, say 300 mm.

First of all check if the beam can be designed as a rectangular beam by calculating M_{flange}.

$$M_{flange} = 0.45 \ f_{cu} \ b \ h_f \ (d - h_f/2)$$
$$M_{flange} = 0.45 × 30 × 1450 × 100 \ (300 − 0.5 × 100) × 10^{-6} = 489.3 \ kNm$$

The design moment of 165 kNm is less than M_{flange}. The beam can be designed as a rectangular beam of size 1450 × 300. Using the code expressions in clause 3.4.4.4

$$k = M/ (b \ d^2 \ f_{cu}) = 165 × 10^6/ (1450 × 300^2 × 30) = 0.042$$
$$z/d = \{0.5 +\surd \ (0.25 − 0.042/0.9)\}= 0.95$$
$$z = 0.95d = 285 \ mm$$
$$A_s = 165 × 10^6/(0.95 × 460 × 285)= 1325 \ mm^2.$$

Provide 3T25; $A_s = 1472 \ mm^2$.

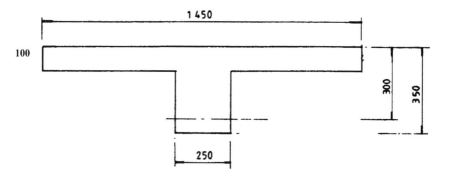

Fig.4.17 Cross section of T-beam.

Example 2: Determine the area of reinforcement required for the T-beam section shown in Fig.4.18 which is subjected to an ultimate moment of 260 kNm. The materials are grade C30 concrete and grade 460 reinforcement.

Calculate M_{flange} to check if the stress block is inside the flange or not.

$$M_{flange} = 0.45 × 30 × 600 × 100 \ (340 − 0.5 \ x \ 100) × 10^{-6} = 234.9 \ kNm$$

The design moment of 260 kNm is greater than M_{flange}. Therefore the stress block extends into the web.

Check if compression steel is required.

$$M_{max} = 0.45 \ f_{cu} \ (b − b_w) \ h_f \ (d − h_f/2) + 0.156 \ f_{cu} \ b_w \ d^2$$
$$M_{max} = \{0.45 × 30 × (600 − 250) ×100 × (340 − 100/2)$$
$$+ 0.156 ×30 × 250 × 340^2\}×10^{-6}$$

$$M_{max} = (137.0 + 135.3) = 272.3 \text{ kNm}$$
$$M_{max} > (M = 260 \text{ kNm})$$

The beam can be designed without any need for compression steel. Two approaches can be used for determining the area of tension steel required.

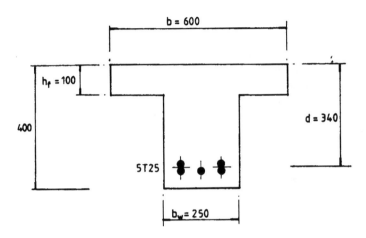

Fig.4.18 Cross section of T-beam.

(a) Exact approach

Determine the depth of the neutral axis from

$$\frac{M}{bd^2 f_{cu}} = 0.45(1 - \frac{b_w}{b})\frac{h_f}{d}(1 - \frac{h_f}{2d}) + 0.4\frac{b_w}{b}\frac{x}{d}(1 - 0.45\frac{x}{d})$$

$$\frac{260 \times 10^6}{600 \times 340^2 \times 30} = 0.45(1 - \frac{250}{600})\frac{100}{340}(1 - \frac{100}{2 \times 340}) + 0.4\frac{250}{600}\frac{x}{d}(1 - 0.45\frac{x}{d})$$

setting x/d = α

$$0.1250 = 0.0659 + 0.1667\,\alpha - 0.075\,\alpha^2$$

Simplifying

$$\alpha^2 - 2.22\,\alpha + 0.788 = 0$$

Solving the quadratic in α,

$$\alpha = x/d = (2.22 - 1.3328)/2 = 0.444 < 0.5$$
$$x = 0.444 \times 340 = 151 \text{ mm}$$
$$T = 0.95\, f_y\, A_s = C_1 + C_2$$
$$T = 0.45\, f_{cu}\, (b - b_w)\, h_f + 0.45\, f_{cu}\, b_w\, 0.9\, x$$
$$T = (0.45 \times 30 \times (600 - 250) \times 100 + 0.45 \times 30 \times 250 \times 0.9 \times 151) \times 10^{-3}$$
$$T = 0.95\, A_s = (472.5 + 458.7) = 931.2 \text{ kN}$$
$$A_s = 931.2 \times 10^3 /(0.95 \times 460) = 2131 \text{ mm}^2$$

(b) Code formula

Calculation of A_s using simplified code formula which assumes x/d = 0.5

$$A_s = \frac{260 \times 10^6 + 0.1 \times 30 \times 250 \times 340 \times (0.45 \times 340 - 100)}{0.95 \times 460 \times (340 - 0.5 \times 100)} = 2159\,\text{mm}^2$$

This is only 1% more than that calculated using the exact neutral axis depth!
Provide 5T25, A = 2454 mm^2

4.7 CHECKING EXISTING SECTIONS

In the previous sections methods have been described for designing rectangular
and flanged sections for a given moment. In practice it may be necessary to
calculate the ultimate moment capacity of a given section. This situation often
occurs when there is change of use in a building and the owner wants to see if the
structure will be suitable for the new use. Often moment capacity can be increased
either by

- Increasing the effective depth. This can be done by adding a well
  bonded layer of concrete at the top of the beam/slab
- Increasing the area of tension steel by bonding steel plates to the
  bottom of the beam.

4.7.1 Examples of Checking for Moment Capacity

Example 1: Calculate the moment of resistance of the singly reinforced beam
section shown in Fig.4.19(a). The materials are grade C30 concrete and grade 460
reinforcement. The tension reinforcement is 4T20 giving $A_s = 1256$ mm^2
Assuming that tension steel yields, total tensile force T is given by
$$T = 0.95\,f_y\,A_s = 0.95 \times 460 \times 1256 \times 10^{-3} = 548.7\,\text{kN}$$
If the neutral axis depth is x, then the compression force C is
$$C = 0.45 f_{cu}\,(0.9x \times b) = 0.45 \times 30 \times 0.9x \times 250 \times 10^{-3} = 3.0375x\,\text{ kN}$$
For equilibrium, T = C. Solving for x
$$x = 181\,\text{mm} < (0.5\,d = 200\,\text{mm})$$
Check the strain in steel
$$\varepsilon_s = \frac{0.0035}{x}(d-x) = \frac{0.0035}{181}(400-181) = 0.004 > (\text{yield strain} = 0.0022)$$
Steel yields. Therefore the initial assumption is valid.
$$z = d - 0.45x = 400 - 0.45 \times 181 = 310\,\text{mm}$$
$$z/d = 310/400 = 0.775 < 0.95$$
Moment of resistance M
$$M = T\,z = 548.7 \times 310 \times 10^{-3} = 169.9\,\text{kNm}$$

One can also use the Design Chart shown in Fig.4.10 to solve the problem.
$$100\frac{A_s}{bd}\frac{f_y}{f_{cu}} = 100\frac{1256}{250 \times 400}\frac{460}{30} = 19.26$$

From design chart Fig.4.10,

$$\frac{M}{bd^2 f_{cu}} = \frac{M \times 10^6}{250 \times 410^2 \times 30} = 0.145$$

$$M = 174.0 \text{ kNm}$$

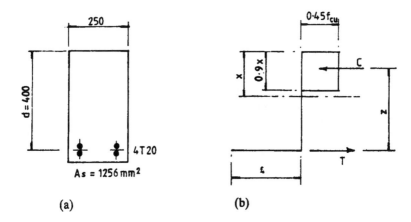

Fig 4.19 Cross section of rectangular beam.

Example 2: Determine the ultimate moment capacity of the beam in Fig.4.19, except, $A_s = 6T20 = 1885 \text{ mm}^2$

Proceeding as in Example 1, assume that steel yields and calculate

$$T = 0.95 \ f_y \ A_s = 0.95 \times 460 \times 1885 = 8.24 \times 10^5 \text{ N}$$
$$C = 0.45 \times f_{cu} \times 0.9 \times x \times b = 3037.5x \text{ N}$$

For equilibrium, $T = C$.

$$x = 271 \text{ mm, } x/d = 0.68 > 0.5$$

Check strain in steel to check the validity of the initial assumption.

$$\varepsilon_s = 0.0035\frac{(d-x)}{x} = 0.0035\frac{(400-271)}{271} = 0.0017 < (\text{yield strain} = 0.0022)$$

Since the strain in steel is less than yield strain, tension steel does not yield indicating that the initial assumption is wrong. Assume that the tension steel does not yield. For an assumed value of neutral axis depth x, strain ε_s in tension steel is

$$\varepsilon_s = 0.0035\frac{(d-x)}{x}$$

Since the steel is assumed not to yield, if Young's modulus for steel is $E_s = 200$ kN/mm^2, then stress f_s in tension steel is given by

$$f_s = E_s \times \varepsilon_s = 200 \times 10^3 \times 0.0035\frac{(d-x)}{x} = 700\frac{(d-x)}{x}$$

$$T = A_s \times f_s = \{1885 \times 700 \times (d-x)/x\} \times 10^{-3} = 1319.5 \ (400-x)/x \text{ kN}$$
$$C = 0.45 \ f_{cu} \ b \ 0.9x = \{0.45 \times 30 \times 250 \times 0.9x\} \times 10^{-3} = 3.0375 \ x \text{ kN}$$

For equilibrium, $T = C$

$$1319.5 \times (400 - x)/x = 3.0375x$$

Simplifying

$$x^2 + 434.40x - 173761.3 = 0$$

Solving the quadratic equation in x,

$$x = 253 \text{ mm}, x/d = 253/400 = 0.63 > 0.5$$

Calculate the strain in steel.

$$\varepsilon_s = 0.0035\frac{(d-x)}{x} = 0.0035\frac{(400-253)}{253} = 0.00204$$

$$f_s = E_s \times \varepsilon_s = 200 \times 10^3 \times 0.00204 = 408 \text{ N/mm}^2$$

$$z = d - 0.45x = 286 \text{ mm}$$

$$M = T \times z = 769 \times 286 \times 10^{-3} = 220 \text{ kNm}$$

Since x/d > 0.5, it is sensible to limit the permissible ultimate moment to a value less than 220 kNm. Assuming that x = 0.5d = 200 mm,

$$C = 0.45 f_{cu} b \, 0.9x = \{0.45 \times 30 \times 250 \times 0.9 \times 200\} \times 10^{-3} = 607.5 \text{ kN}$$

$$\text{Lever arm } z = 0.775 \, d = 0.775 \times 400 = 310 \text{ mm}$$

Taking moments about the steel centroid, M = C z = 188.3 kNm

Example 3: Calculate the moment of resistance of the beam section shown in Fig.4.20. The materials are grade C30 concrete and grade 460 reinforcement.

$$A_s = 4T25 = 1963 \text{ mm}^2, A_s^{'} = 2T20 + T16 = 829 \text{ mm}^2$$

Assume that both tension and compression steels yield and calculate the tension force T and compression force C_s in the steels.

$$T = 0.95 f_y A_s = 0.95 \times 460 \times 1963 \times 10^{-3} = 857.8 \text{ kN}$$

$$C_s = 0.95 f_y A_s^{'} = 0.95 \times 460 \times 829 \times 10^{-3} = 362.3 \text{ kN}$$

The compression force in concrete is

$$C_c = 0.45 f_{cu} (0.9x \times b) = 0.45 \times 30 \times 0.9x \times 250 \times 10^{-3} = 3.0375x \text{ kN}$$

For equilibrium,

$$C_c + C_s = T$$

$$3.0375x + 362.3 = 857.8.$$

Solving x = 163 mm, x/d = 0.47 < 0.5.

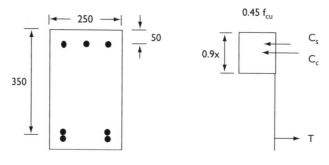

Fig.4.20 Cross section of doubly reinforced beam.

Calculate strain in tension and compression steels to verify the assumption.

$$\varepsilon_s = 0.0035 \frac{(d-x)}{x} = 0.0035 \frac{(350-163)}{163} = 0.004$$

$$\varepsilon_s' = 0.0035 \frac{(x-d')}{x} = 0.0035 \frac{(163-50)}{163} = 0.0024$$

Both strains are larger than yield strain of 0.0022. Therefore both steels yield and the initial assumption is correct.

$$C_c = 0.45 f_{cu} b \, 0.9x = 0.45 \times 30 \times 250 \times 0.9 \times 163 \times 10^{-3} = 495.1 \text{ kN}$$

Taking moments about the tension steel,

$$M = C_c (d - 0.45x) + C_s (d - d')$$

$$M = 495.1 (350 - 0.45 \times 163) \times 10^{-3} + 362.3 \times (350 - 50) \times 10^{-3} = 245.8 \text{ kNm}$$

4.7.2 Strain Compatibility Method

In the previous section, examples were given for calculating the moment of resistance of a given section. It required making initial assumptions about whether the steel yields or not. After calculating the neutral axis depth from equilibrium considerations, strains in tension and compression steels are calculated to validate the assumptions. The problem can become complicated if one steel yields while the other steel does not. A general approach in this case is the method of Strain Compatibility which has the advantage of avoiding the algebraic approach. The basic idea is to assume a neutral axis depth. From the assumed value of neutral axis depth, strains in steel in compression and tension are calculated. Thus

$$\varepsilon_s = \frac{0.0035}{x}(d-x), \ f_s = E\varepsilon_s \le 0.95 f_y$$

$$\varepsilon_s' = \frac{0.0035}{x}(x-d'), \ f_s' = E\varepsilon_s' \le 0.95 f_y$$

From the stresses, calculate the forces

$$T = A_s f_s, \ C_s = A_s' f_s', \ C_c = 0.45 f_{cu} b \, 0.9x, \ C = C_s + C_c$$

For equilibrium, $T = C$. If equilibrium is not satisfied, then adjust the value of x and repeat until equilibrium is established. Normally only two sets of calculations for neutral axis depth are required. Linear interpolation can be used to find the appropriate value of x to satisfy equilibrium. The following example illustrates the method.

4.7.2.1 Example of Strain-Compatibility Method

Example 1: Calculate the moment capacity of the section with b = 250 mm, d = 350 mm, d' = 50 mm,

$$A_s' = 3T20 = 942.5 \text{ mm}^2, \ A_s = 6T25 = 2945.2 \text{ mm}^2$$

Trial 1: Assume x = 220 mm
Strain ε_s' in compression steel is given by

$\varepsilon_s' = 0.0035(x - d')/x = 0.0035 \times (220 - 50)/220 = 0.0027 > 0.0022$

Therefore compression steel yields and the stress f_s' is equal to $0.95 f_y$

Similarly, strain ε_s in tension steel is given by

$\varepsilon_s = 0.0035(d - x)/x = 0.0035 \times (350 - 220)/220 = 0.00207 < 0.0022$

Therefore tension steel does not yield and the stress f_s is equal to

$$f_s = \varepsilon_s E_s = 0.00207 \times 200 \times 10^3 = 413.6 \text{ N/mm}^2$$
$$T = A_s \times f_s = 2945.2 \times 413.6 \times 10^{-3} = 1218.1 \text{ kN}$$
$$C = 0.45 f_{cu} \times b \times 0.9x + A_s' \times f_s'$$
$$C = \{0.45 \times 30 \times 250 \times 0.9 \times 220 + 942.5 \times 0.95 \times 460\} \times 10^{-3}$$
$$C = (668.25 + 411.87) = 1080.1 \text{ kN}$$
$$T - C = 138.0 \text{ kN}$$

Total tensile force T is greater than the total compressive force C. Therefore increase the value of x in order to increase the compression area of concrete and also reduce the strain in tension steel but increase the strain in compression steel.

Trial 2: Assume x = 240 mm say

Strain ε_s in compression steel is given by

$$\varepsilon_s' = 0.0035(x - d')/x = 0.00277 > 0.0022$$

Therefore compression steel yields and the stress f_s' is equal to $0.95 f_y$

Similarly, strain ε_s in tension steel is given by

$$\varepsilon_s = 0.0035(d - x)/x = 0.0016 < 0.0022$$

Therefore tension steel does not yield and the stress f_s is equal to

$$f_s = \varepsilon_s E = 0.001604 \times 200 \times 10^3 = 320.8 \text{ N/mm}^2$$
$$T = A_s \times f_s = 2945.2 \times 320.8 \times 10^{-3} = 944.8 \text{ kN}$$
$$C = 0.45 f_{cu} \times b \times 0.9x + A_s' \times f_s'$$
$$C = \{0.45 \times 30 \times 250 \times 0.9 \times 240 + 942.5 \times 0.95 \times 460\} \times 10^{-3}$$
$$C = (729.0 + 411.87) = 1140.9 \text{ kN}$$
$$T - C = -196.04 \text{ kN}$$

As shown in Fig.4.21, linearly interpolate between x = 220 and 240 to obtain the value of x giving T – C = 0.

$$x = 220 + (240 - 220) \times (138.0)/(138.0 + 196.04) = 228 \text{ mm}$$
$$x/d = 228/350 = 0.65 > 0.5$$

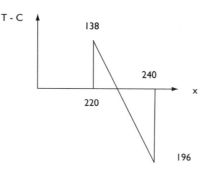

Fig.4.21 Linear interpolation

As a check calculate T and C for x = 228 mm
Strain $\varepsilon_s^{'}$ in compression steel is given by
$$\varepsilon_s^{'} = 0.0035(x - d^{`})/x = 0.0027 > 0.0022$$
Therefore compression steel yields and the stress f_s' is equal to 0.95 f_y
Similarly, strain ε_s in tension steel is given by
$$\varepsilon_s = 0.0035(d - x)/x = 0.00187 < 0.0022$$
Therefore tension steel does not yield and the stress f_s is equal to
$$f_s = \varepsilon_s E = 0.00187 \times 200 \times 10^3 = 374.6 \text{ N/mm}^2$$
$$T = A_s \times f_s = 2945.2 \times 374.6 \times 10^{-3} = 1103.2 \text{ kN}$$
$$C = 0.45 f_{cu} \times bx \, 0.9x + A_s^{'} \times f_s^{'} A_s^{'}$$
$$C = \{0.45 \times 30 \times 250 \times 0.9 \times 228 + 942.5 \times 0.95 \times 460\} \times 10^{-3}$$
$$C = (692.6 + 411.87) = 1104.4 \text{ kN}$$
$$T - C = -1.22 \text{ kN}$$
This is close enough to be zero.
Taking moments about the tension steel, the lever arm for compression force in concrete is (d – 0.45x) and for the compression force in steel it is (d – d').
$$M = \{692.6 \times (350 - 0.45 \times 228) + 411.87 \times (350 - 50)\} \times 10^{-3} = 294.9 \text{ kNm}$$
Since x/d > 0.5, it is sensible to limit the permissible ultimate moment to a value less than 294.9 kNm. Assuming that x = 0.5d = 175 mm,
$$C_c = 0.45 f_{cu} b \, 0.9x = \{0.45 \times 30 \times 250 \times 0.9 \times 175\} \times 10^{-3} = 531.6 \text{ kN}$$
$$\text{Lever arm } z_c = 0.775 \, d = 0.775 \times 350 = 271 \text{ mm}$$
Strain $\varepsilon_s^{'}$ in compression steel is given by $\varepsilon_s^{'} = 0.0035(x - d')/x = 0.0025 > 0.0022$
Therefore compression steel yields and the stress $f_s^{'}$ is equal to 0.95 f_y
$$C_s = \{942.5 \times 0.95 \times 460\} \times 10^{-3} = 411.9 \text{ kN, Lever arm } z_s = d - d^{`} = 300 \text{ mm}$$
Taking moments about the steel centroid, $M = C_c \, z_c + C_s \, z_s = 267.6$ kNm

5.1.2 Shear in a Reinforced Concrete Beam without Shear Reinforcement

(a) Shear failure
Shear in a reinforced concrete beam without shear reinforcement causes cracks on inclined planes near the support as shown in Fig.5.2.

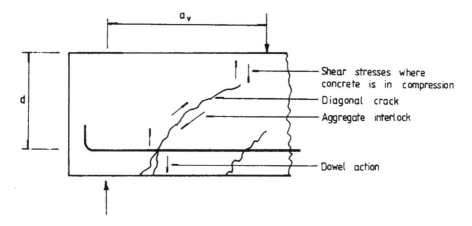

Fig.5.2 Different actions contributing to shear strength

The cracks are caused by the diagonal tensile stress mentioned above. The shear failure mechanism is complex and depends on the shear span a_v to effective depth d ratio (a_v/d). Shear span a_v is defined as the distance between the support and the major concentrated load acting on the span. When this ratio is large, the failure is as shown in Fig.5.2.

The following actions form the three mechanisms resisting shear in the beam:

1. shear stresses in the compression zone resisted by un-cracked concrete
2. aggregate interlock along the cracks: Although cracks exist in the web due to tensile stresses caused by shear stresses, the width of the cracks are not large enough prevent frictional forces between cracked surfaces. These frictional forces exist along the cracked surfaces and contribute to resisting shear force.
3. dowel action in the bars where the concrete between the cracks transmits shear forces to the bars

(b) Shear capacity
An accurate analysis for shear strength is not possible. The problem has been solved by testing beams of the type normally used in practice. The shear capacity is represented by the simple formula to calculate the notional shear stress given in BS 8110: Part 1, clause 3.4.5.2.

$$v = V/ (b_v\, d)$$

where b_v is the breadth of the section. For a flanged beam b_v is taken as the width of the web. V is the design shear force due to ultimate loads and d is the effective depth. The permissible shear stress in concrete v_c is used to determine the shear capacity of the concrete alone without any contribution from shear reinforcement. Value of v_c depends on several factors such as

- the percentage of flexural steel in the member: This affects the shear capacity by restraining the width of the cracks and thus enhancing the shear carried by the aggregate interlock along the cracks. It also naturally increases the shear capacity due to dowel action.
- the concrete grade: It affects by increasing the aggregate interlock capacity and also the shear capacity of the uncracked portion of the beam.
- Type of aggregate: This affects the shear resisted by aggregate interlock. For example, lightweight aggregate concrete has approximately 20% lower shear capacity compared to normal weight concrete, BS 8110, Part 2, clause 5.4).
- Effective depth: Tests indicate that deeper beams have proportionally lower shear capacity compared to shallow beams. The reason for this is not clear but it is thought it might have some thing to do with lower aggregate interlock capacity.

The design concrete shear stress is given by the following formula from Table 3.8 in the code:

$$v_c = \frac{0.79}{\gamma_m}\left(100\frac{A_s}{b_v\, d}\right)^{\frac{1}{3}}\left(\frac{400}{d}\right)^{\frac{1}{4}}\left(\frac{f_{cu}}{25}\right)^{\frac{1}{3}}$$

where:

γ_m = material safety factor is 1.25,
$$100A_s/(b_v\, d) \le 3.0$$
$$400/d \ge 1.0$$
$$f_{cu} \le 40 \text{ N/mm}^2$$

The code notes (clause 3.4.5.4) that for tension steel to be counted in calculating A_s it must continue for a distance at least of d past the section being considered. Anchorage requirements must be satisfied at supports (section 3.12.9.4 in the code).

Some values of v_c for grade C30 concrete are given in Table 5.1. v_c values for $f_{cu} = 40$ N/mm^2 are approximately 10% larger and the corresponding values for $f_{cu} = 25$ N/mm^2 are approximately 6% smaller than the values shown in Table 5.1.

(c) Enhanced shear capacity near supports

The code states (clause 3.4.5.8) that shear failure in beam sections without shear reinforcement normally occurs at about 30° to the horizontal. If the angle is steeper due to the load causing shear or because the section where the shear is to be checked is close to the support (Fig.5.3), the shear capacity is increased. The increase is because of the large vertical compressive stress in concrete due to the reaction and the load. The shear span ratio a_v/d is small in this case. The design

Table 5.1 Design concrete shear strength v_c for $f_{cu} = 30$ N/mm^2

100 (A$_s$/b$_v$d)	v_c (N/mm^2)							
	d = 125 mm	150	175	200	225	250	300	≥ 400
≤ 0.15	0.48	0.46	0.44	0.42	0.41	0.40	0.38	0.36
0.25	0.57	0.54	0.52	0.50	0.49	0.48	0.46	0.42
0.50	0.71	0.68	0.66	0.63	0.62	0.60	0.57	0.53
0.75	0.82	0.78	0.75	0.73	0.71	0.69	0.66	0.61
1.0	0.90	0.86	0.83	0.80	0.78	0.76	0.72	0.67
1.5	1.03	0.98	0.95	0.91	0.89	0.87	0.83	0.77
2.0	1.13	1.08	1.04	1.01	0.98	0.95	0.91	0.85
≥ 3.0	1.30	1.24	1.19	1.15	1.12	1.09	1.04	0.97

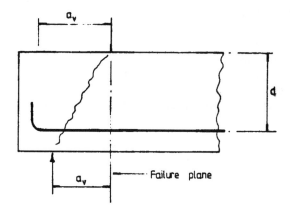

Fig.5.3 Shear failure close to the support

concrete shear can be increased from v_c as determined above to $2v_c d/a_v$ where a_v is the length of that part of a member traversed by a shear plane.

(d) Maximum shear stress
BS 8110: Part 1, clauses 3.4.5.2 and 3.4.5.8, state that the notional shear stress $v = V/(b_v d)$ must in no case exceed $0.8\sqrt{f_{cu}}$ or 5 N/mm^2, even if the beam is reinforced to resist shear. This upper limit prevents failure of the concrete in diagonal compression. If v is exceeds the specified maximum, the beam must be made larger.

5.1.3 Shear Reinforcement in the Form of Links

(a) Action of shear reinforcement
As stated in section 5.1.1 the complementary shear stresses give rise to diagonal tensile and compressive stresses as shown in Fig.5.1. Taking a simplified view, concrete is weak in tension, and so shear failure is caused by a failure in diagonal tension with cracks running at 45° to the beam axis. Shear reinforcement is

provided by bars which cross the cracks, and theoretically either vertical links or inclined bars will serve this purpose. In practice either vertical links alone or a combination of vertical links and bent-up bars are provided.

(b) Vertical links

As shown in Fig.5.4, a cracked beam essentially acts as a truss where the tension reinforcement acts as bottom chord, the stirrups act as the vertical members and the cracked concrete acts as diagonal compression members and the uncracked concrete at the top of the beam acting as the top chord.

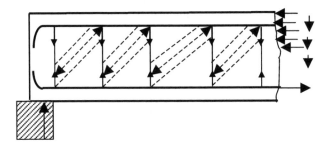

Fig.5.4 Analogous truss

If it is assumed that the distance between the top and bottom chords is approximately equal to d and that the cracks form at an angle of 45° to the neutral axis, then as shown in Fig.5.5, the *horizontal length* of the crack is approximately d. If the links are spaced at a distance s_v apart, then the number of links in a horizontal distance d is equal to d/s_v. Assuming that the stress in the links is equal to its yield stress, the shear force V_s resisted by the links is equal to

$$V_s = 0.95 \times f_{yv} \times A_{sv} \times \frac{d}{s_v}$$

where A_{sv} = Area of cross section of the legs of a 'single' link crossing the crack. Normally if single links are used, then A_{sv} = area of cross section of two legs of a link but in cases of heavy shear force, double links with four legs may be required. In designing the links, it is assumed that concrete resists shear force V_c equal to

$$V_c = v_c \, b_v \, d$$

The applied shear force V is resisted by a combination of concrete and steel links.

$$V = V_c + V_s$$

$$V = v b_v \, d = v_c \, b_v \, d + 0.95 \times f_{yv} \times A_{sv} \times \frac{d}{s_v}$$

Simplifying,

$$A_{sv} = \frac{b_v \, s_v \, (v - v_c)}{0.95 \, f_{yv}}$$

(ii) Calculate v
$$v = V/(b_v\, d) = 157.5 \times 10^3/(200 \times 375) = 2.1\ \text{N/mm}^2$$

(iii) Check maximum shear stress
$$v = 2.1 < (0.8\sqrt{f_{cu}} = 0.8\sqrt{30} = 4.4\ \text{N/mm}^2)$$

(iv) Calculate s_v
b_v = web width = 200 mm, d = 375 mm
$$A_{sv} = 157 = \frac{b_v\, s_v\,(v - v_c)}{0.95\, f_{yv}} = \frac{200 \times s_v \times (2.1 - 0.99)}{0.95 \times 250}\ ,\ s_v = 168\ \text{mm}$$
$$s_v = 168\ \text{mm} < (0.75\ d = 281\ \text{mm})$$

5.1.4 Shear Reinforcement Close to a Support

Clause 3.4.5.9 of the code deals with shear reinforcement for sections close to the support. The total area specified is
$$\Sigma A_{sv} = a_v\, b_v\,(v - 2d\, v_c/a_v)/(0.95\, f_{yv}) \geq 0.4\ a_v\, b_v\,/(0.95\, f_{yv})$$
Refer to Fig.5.3 where a_v is the distance from the support traversed by the failure plane. The term $2d\, v_c/a_v$ is the enhanced shear stress. The second expression ensures that at least minimum links are provided. This reinforcement should be provided within the middle three-quarters of a_v.

The code (clause 3.4.5.10) also gives a simplified approach for design taking enhanced shear strength into account. This applies to beams carrying uniform load or where the principal load is applied at more than 2d from the face of the support. The procedure given is as follows:
1. Calculate the notional shear stress v at *d* from the face of the support;
2. Check that v does not exceed the maximum permissible shear stress.
3. Determine v_c and the amount of shear reinforcement in the form of vertical links.
4. Provide this shear reinforcement between the section at *d* and the support. No further checks for shear reinforcement are required;
This approach is the most convenient to use in the majority of situations.

5.1.5 Examples of Design of Shear Reinforcement for Beams

Example 1: A simply supported T-beam of 6 m clear span (Fig.5.6) carries an ultimate load of 38 kN/m. The beam section dimensions, support particulars and tension reinforcement are shown in the figure. Design the shear reinforcement for the beam. The concrete is grade C30 and the shear reinforcement grade 250. The steps in design are as follows.

(a) Check the maximum shear stress at the face of the support
At the face of the support *d* = 400 mm.
$$V = 117.8 - 38.0 \times 0.2/2 = 114.0\ \text{kN}$$
$$v = 114 \times 10^3/(250 \times 400) = 1.14\ \text{N/mm}^2$$

This is not greater than $0.8\sqrt{f_{cu}} = 4.8$ N/mm^2 or 5 N/mm^2.
Section size is adequate.

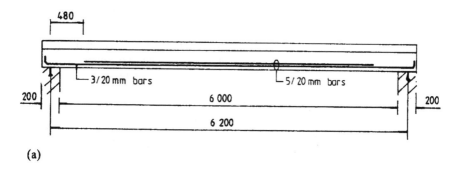

(a)

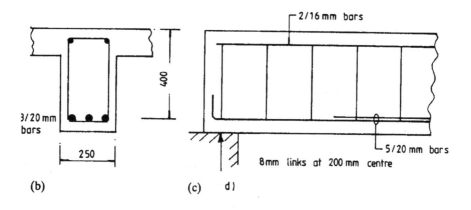

(b)   **(c)**   d)

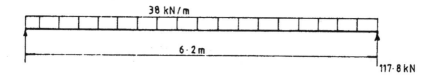

Fig.5.6 (a) Side elevation; (b) section at support; (c) shear reinforcement; (d) load.

(b) Determine the length in the centre of the beam over which nominal links are required

The shear resistance of concrete where 5T20 mm diameter bars form tension reinforcement is to be determined.

$$A_s = 5T20 = 1570 \text{ mm}^2$$

With 5T20,

$$d = 400 - 20 = 380\text{mm}$$
$$100 A_s/ (b_v d) = 100 \times 1570 (250 \times 380) = 1.65 < 3.0$$
$$400/d = 400/380 = 1.05 > 1.0$$

$$f_{cu} = 30 < 40 \text{ N/mm}^2$$
$$v_c = 0.79 \times \{(1.65)^{1/3} \times (1.05)^{1/4} (30/25)^{1/3}\}/1.25 = 0.80 \text{ N/mm}^2$$

The shear resistance of the concrete is

$$V_c = v_c \times b_v \times d = 0.80 \times 380 \times 250 \times 10^{-3} = 76.0 \text{ kN}$$

Using the simplified approach described in section 5.1.4, this value of v_c is valid from a distance of $480 + (d = 380) = 860$ mm from the centre of supports.

Adopt nominal 8 mm diameter links where the area of two legs is 100 mm². The spacing of the links is calculated from

$$A_{sv} = [0.4 \ b_v \ s_v]/ 0.95 \ f_{yv}$$
$$100 = 0.4 \times 250 \times s_v/ (0.95 \times 250), s_v = 238 \text{ mm}$$
$$0.75d = 0.75 \times 380 = 285 \text{ mm} > 238 \text{ mm}; \text{ space links at 200 mm c/c.}$$

With nominal links:

$$v - v_c = 0.4 \text{ and with } v_c = 0.80 \text{ N/mm}^2, v = 1.2 \text{ N/mm}^2.$$
$$V = v \times b_v \times d = 1.2 \times 250 \times 380 \times 10^{-3} = 114.0 \text{ kN}$$

Shear force at centre of support is

$$V_{Support} = 38 \times 6.2/2 = 117.8 \text{ kN}$$

Shear force V at x = 860 mm from the support is

$$V = 117.8 - 38 \times 0.860 = 85.12 \text{ kN}$$

This is less than 114.0 kN, the shear capacity with minimum area of links. Provide 8 mm diameter links at 200 mm centres over the centre 4.48 m length of beam, i.e. at 0.76 m from the *face* of each support.

(c) Shear reinforcement
The shear reinforcement is shown in Fig.5.6(c). Two 16 mm diameter bars are provided at the top of the beam to carry the links. Note that the top bars are not designed as compression steel. Thus the requirements of clause 3.12.7.1 set out in section 5.1.3(b) do not apply.

Example 2: Design the shear reinforcement for a rectangular beam with the dimensions, the design ultimate load and moment steel shown in Fig.5.7. The concrete is grade C30 and the shear reinforcement grade 250.

(a) Maximum shear at the face of the support
$$V = 330 - 165 \times 0.15 = 305.25 \text{ kN}$$
$$v = V/ (b_v \ d) = 305.25 \times 10^3 / (300 \times 450) = 2.26 \text{ N/mm}^2$$
$$2.26 \text{ N/mm}^2 < (0.8 \ \sqrt{f_{cu}} = 4.38 \text{ N/mm}^2)$$

Section size is satisfactory.

(b) Minimum links
The top layer 3T20 at centre stop at 320 mm from the centre of the support. For all the six bars to be considered as effective, they have to extend a distance d from the section. Therefore all bars are effective from a distance of

$$320 + (d = 425) = 745 \text{ mm from the centre of support}$$
$$A_s = 3T25 + 3T20 = 2415 \text{ mm}^2$$
$$d = 425 \text{ mm}$$

$$100 \ A_s / \ (b_v \ d) = 100 \times 2415 / \ (300 \times 425) = 1.89 < 3.0$$
$$400/d = 400/425 = 0.94 < 1.0. \ \text{Therefore assume } 1.0.$$
$$f_{cu} = 30 \ \text{N/mm}^2 < 40 \ \text{N/mm}^2$$
$$v_c = 0.79 \times [1.89^{\ 1/3} \times (1.0)^{1/4} \times (30/25)^{1/3}]/1.25 = 0.83 \ \text{N/mm}^2$$
$$V_c = v_c \ b_v \ d = 0.83 \times 300 \times 425 \times 10^{-3} = 105.8 \ \text{kN}$$

Adopt 10 mm diameter links; $A_{sv} = 157 \ \text{mm}^2$. Spacing dictated by minimum shear reinforcement. Therefore

$$(v - v_c) = 0.4 \ \text{N/mm}^2$$
$$A_{sv} = 157 = \frac{0.4 \, b_v \, s_v}{0.95 \, f_{yv}} = \frac{0.4 \times 300 \times s_v}{0.95 \times 250}, s_v = 311 \, \text{mm}$$

$$s_v = 311 \ \text{mm} < (0.75 \ d = 319 \ \text{mm})$$

Top bars (2T20) are designed as compression steel:

$$s_v \leq (12 \times \text{bar diameter} = 12 \times 20) = 240 \ \text{mm}$$
$$v - v_c = 0.4, \ v = 0.83 + 0.4 = 1.23 \ \text{N/mm}^2$$
$$V = v \ b_v \ d = 1.23 \times 300 \times 425 \times 10^{-3} = 156.83 \ \text{kN}$$

Shear force at centre of support is

$$V_{support} = 165 \times 4.0/2 = 330 \ \text{kN}$$

The distance a from the support where the shear is equal to 156.83 kN is given by

$$156.83 = 330 - 165a, \ a = 1.05 \text{m}$$

The face of support is at 150 mm from the centre of support. Provide eleven number 10 mm links at 190 mm centres in the centre at $(1.05 - 0.15) = 0.9$ m from the faces of the supports (Fig.5.7(c)).

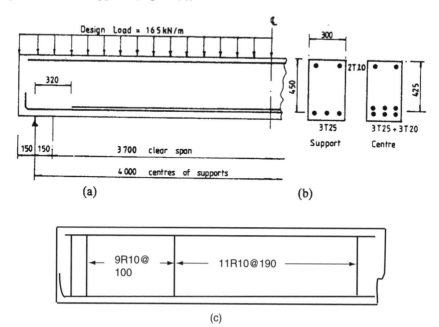

Fig.5.7 (a) Side elevation; (b) section and moment steel; (c) links.

(c) Section between d = 450 mm and 900 mm from the face of the support

$$A_s = 3T25 = 1473 \text{ mm}^2,$$
$$d = 450 \text{ mm}$$
$$100 \, A_s / (b_v \, d) = 100 \times 1473 / (300 \times 450) = 1.09 < 3.0$$
$$400/d = 400/450 < 1.0, \text{ take as } 1.0.$$
$$f_{cu} = 30 \text{ N/mm}^2 < 40 \text{ N/mm}^2$$
$$v_c = 0.79(1.09)^{1/3} \, (1.0)^{1/4} \, (30/25)^{1/3}/1.25 = 0.69 \text{ N/mm}^2$$

V at d from the face of support is

$$V = 330 - 165 \times (0.45 + 0.15) = 234 \text{ KN}$$
$$v = 234.0 \times 10^3 / (300 \times 450) = 1.73 \text{ N/mm}^2$$
$$v - v_c = 1.73 - 0.69 = 1.04 \text{ N/mm}^2 > 0.4 \text{ N/mm}^2$$

Provide 10mm diameter links:

$$A_{sv} = 157 \text{ mm}^2$$

The spacing required is

$$A_{sv} = 157 = \frac{b_v \, s_v \, (v - v_c)}{0.95 \, f_{yv}} = \frac{300 \times s_v \times (1.73 - 0.69)}{0.95 \times 250}, s_v = 120 \text{ mm}$$

As the bending moment is low in this section, the compression stress in the top bars will be very low. Therefore there is no need to restrain them.

$$s_v < (0.75 \, d = 0.75 \times 450 = 338 \text{ mm})$$

Provide nine number 100 mm diameter links at a spacing of 100 mm centres to 900 mm from the support and then change to minimum links at 190 mm centres.

(d) Shear reinforcement

This is shown in Fig.5.7(c).

5.1.6 Shear Reinforcement in The Form of Bent-Up Bars

Although use of links to resist shear is the preferred option, because bending moment decreases towards the support, instead of curtailing the flexural steel bars, they can be bent up at approximately 45° to cross the tension cracks which also form at approximately 45°. BS 8110: Part 1, clause 3.4.5.6, states that the design shear resistance of a system of bent-up bars may be calculated by assuming that the bent-up bars form the tension members of one or more single systems of trusses in which the concrete forms the compression members as shown in Fig.5.8(a).

The following terms are as defined:

s_b spacing of the bent-up bars

A_{sb} cross-sectional area of a **pair** of bent-up bars

f_{yv} characteristic strength of the bent-up bars

α inclination of bent-up bars

β inclination of the crack

The truss is to be arranged so that the angles α and β are greater than or equal to 45° giving a maximum value of $s_b = 1.5d$

A single truss system is shown in Fig.5.8(a), where the spacing of the bent-up bars is equal to

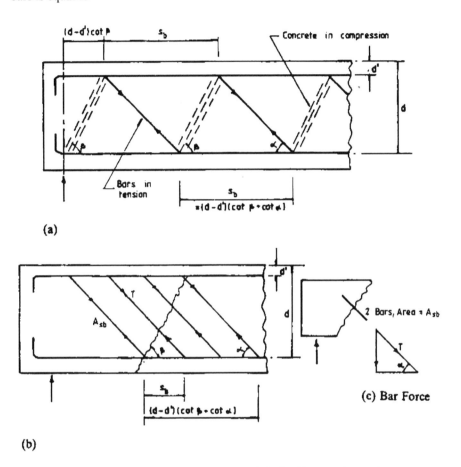

(a)

(b)

(c) Bar Force

Fig.5.8 (a) Equivalent single-truss system; (b) inclined bars crossing crack.

$$s_b = (d - d') (\cot \beta + \cot \alpha)$$

In the single truss system, at any vertical section over a distance s_b, the vertical component of either of the bent-up bar or of the concrete strut resists shear. The force T in a pair of bent-up bars is

$$T = 0.95 f_{yv} A_{sb}$$

The shear force V_b resisted by the bent-up bars is equal to the vertical component of the forces in the bars. Therefore

$$V_b = T \sin \alpha = 0.95 f_{yv} A_{sb} \sin \alpha$$

If $\alpha = 45°$ and $s_b = (d - d') (\cot \beta + \cot \alpha) \approx 1.5$ d, then $\cot \beta = 0.5$ or $\beta = 63°$
If the spacing $s_b < (d - d') (\cot \beta + \cot \alpha)$ we get a multiple truss system as shown in Fig.5.8(b). The number of bars inclined at an angle α crossing a crack inclined at an angle β is $(d - d') (\cot \beta + \cot \alpha)/s_b$. The design shear strength of the bent-up

5.1.7 Shear Resistance of Solid Slabs

Slab design is treated in Chapter 8. One-way and two-way solid slabs are designed on the basis of a strip of unit width of 1 m. The shear resistance of solid slabs is set out in BS8110: Part 1, section 3.5.5.

The design shear stress is given by

$$v = V/ (b\ d)$$

where b is the breadth of slab considered, generally 1 m. The form and area of shear reinforcement is given in Table 3.16 of the code. When v is less than the design concrete shear stress v_c given by the formula in Table 3.8, no shear reinforcement is required, not even minimum links.

Slabs carrying moderate loads such as floor slabs in office buildings and apartments do not normally require shear reinforcement. It is not desirable to have shear reinforcement in slabs with an effective depth of less than 200 mm. Where shear reinforcement is required, reference should be made to Table 3.16 of the code. The approach and equations are similar to those for rectangular beams discussed earlier in this chapter.

5.1.8 Shear Due to Concentrated Loads on Slabs

Shear in slabs under concentrated loads is set out in BS 8110: Part 1, section 3.7.7. Fig.5.10 shows situations where a slab is subjected to concentrated forces such as when the concentrated load is caused by a column reaction in a flat slab or in a pad footing or due to a concentrated wheel load on slabs in bridge decks.

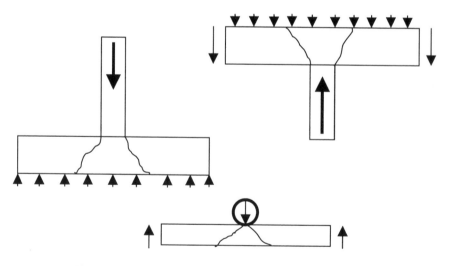

Fig.5.10 Punching shear (a) pad footing; (b) flat slab-column junction; (c) wheel load on bride deck.

A concentrated load causes punching failure which occurs on inclined faces of a truncated cone or pyramid depending on the shape of the loaded area. The clause states that it is satisfactory to consider rectangular failure perimeters. The main rules for design for punching are as follows (Fig.5.11).

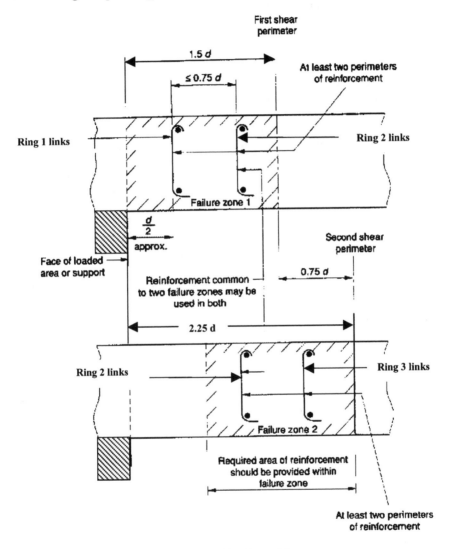

Fig.5.11 Zones for punching shear reinforcement.

(a) Maximum design shear capacity
The maximum design shear stress is
$$v_{max} = V / (u_o\, d)$$

where u_o is the effective length of the perimeter which touches a loaded area. This is the perimeter of the column in Fig.5.11. The shear stress v_{max} is not to exceed $0.8\sqrt{f_{cu}}$ or 5 N/mm².

(b) Design shear stress for a failure zone

The notional shear stress in a failure zone is given by

$$v = V/(u\,d)$$

where u is the effective length of the perimeter of the failure zone.

If v is less than v_c (Table 3.8 of the code) no shear reinforcement is required. The code also states that enhancement of v_c may not be applied to the shear strength of a perimeter at a distance of 1.5d or more from the face of the loaded area. For a perimeter less than 1.5d, v_c can be enhanced by the factor $1.5d/a_v$ where a_v is the distance from the face of the loaded area to the perimeter considered. Fig. 3.17 of the code shows the location of successive shear perimeters on which shear stresses are checked and the division of the slab into shear zones for design of shear reinforcement. Shear perimeters and zones are shown in Fig.5.11. The design procedure is set out in clause 3.7.7.6 of the code.

Zone 1: It has its shear perimeter at 1.5d from the loaded area, is checked first. The clause states that if this zone does not require reinforcement, i.e. the shear stress v is less than v_c, then no further checks are required.

Design of shear reinforcement is set out in clause 3.7.7.5 of the code for slabs over 200 mm thick. If $v_c < v \le 2\ v_c$, shear reinforcement is required and is calculated as follows.

(i) $v_c < v \le 1.6\ v_c$,

$$\Sigma A_{sv} = \frac{(v - v_c)u\,d}{0.95\,f_{yv}}$$

(ii) $1.6v_c < v \le 2\ v_c$,

$$\Sigma A_{sv} = \frac{5(0.7v - v_c)u\,d}{0.95\,f_{yv}}$$

(iii) Minimum

$$\Sigma A_{sv} = \frac{0.4u\,d}{0.95\,f_{yv}}$$

The design procedure involves checking the shear stress on a series of perimeters 0.75d apart.

Shear steel of area ΣA_{sv} is divided between two 'rings'. 'Ring–1' at a distance of approximately 0.5d from the face of the column and contains at least 40% of ΣA_{sv} and 'Ring–2' at a distance of approximately $(0.5 + 0.75)\ d = 1.25\ d$ from the face of the column and contains the rest of ΣA_{sv}.

Zone 2: The second shear perimeter at a distance of $(1.5d + 0.75d) = 2.25$ d from the face of the column. The corresponding value of $\sum A_{sv}$ is calculated. Value of steel equal to $\{\sum A_{sv}$ – steel provided on 'Ring-2'$\}$ is provide on 'Ring-3' at a distance of approximately 0.75 d from 'Ring 2' i.e. at a distance of $(1.25d + 0.75d)$ = 2.0 d from the face of the column.

The procedure is continued on further perimeters spaced at 0.75d from the previous perimeters until $v \le v_c$. The links should not be spaced at a distance greater than 1.5d.

5.1.8.1 Example of Punching Shear Design

Design the shear reinforcement around the column of a flat slab. The flat slab is supported by 400×600 mm columns spaced at 7.5 m in both directions. The slab is 400 mm thick and is reinforced with 20 mm bars at 150 mm c/c in both directions with 30 mm cover. Assume $f_{cu} = 35$ N/mm^2, $f_{yv} = 460$ N/mm^2 and shear links are T8 *single* leg.
The nominal (unfactored) loads on the slab are: Live load = 15 kN/m^2
Dead load including self weight, screed, partitions, etc = 12 kN/m^2

(i) Effective depths
$$\text{In x-direction} = 400 - 30 - 20/2 = 360 \text{ mm}$$
$$\text{In y-direction} = 400 - 30 - 20 - 20/2 = 340 \text{ mm}$$
$$\text{Average d} = 350 \text{ mm}$$

(ii) Steel percentage

$$A_s = 20 \text{ mm bars at 150 mm c/c} = 2094 \text{ mm}^2/\text{m}$$
$$100 \, A_s/(bd) = 100 \times 2094/(1000 \times 350) = 0.60 < 3.0$$
$$400/d = 400/350 = 1.14 > 1.0$$
$$f_{cu} = 35 < 40$$

(iii) Calculate v_c

$$v_c = \frac{0.79}{1.25}(0.6)^{0.33} \, (1.14)^{0.25} \, (\frac{35}{25})^{0.33} = 0.62 \text{ N/mm}^2$$
$$1.6 \, v_c = 0.99 \text{ N/mm}^2, \, 2 \, v_c = 1.24 \text{ N/mm}^2$$

(iv) Column reaction

Design load on slab:
$$q = 1.4 \times 12 + 1.6 \times 15 = 40.8 \text{ kN/m}^2$$
Column reaction $V = q \times$ spacing in x-direction $\times$ spacing in y-direction
$$V = 40.8 \times 7.5 \times 7.5 = 2295 \text{ kN}$$

(v) Check for maximum shear around the column perimeter

$$u_0 = \text{Column perimeter} = 2(400 + 600) = 2000 \text{ mm}$$
$$v_{max} = V/(u_0\, d) = 2295 \times 10^3/(2000 \times 350) = 3.28 < (0.8\sqrt{f_{cu}} = 4.7) \text{ N/mm}^2$$

The slab thickness is therefore adequate.

(vi) Calculate shear stress at the first shear perimeter at 1.5 d from the column face:

$$u = 2\{(400 + 3d) + (600 + 3d)\} = 6200 \text{ mm}$$

The load acting within the perimeter is equal to

$$(400 + 3d) \times (600 + 3d) \times 40.8 = 97.6 \text{ kN}$$
$$V = 2295 - 97.6 = 2197.4 \text{ kN}$$
$$v = V/(u \times d) = 2197.4 \times 10^3/(6200 \times 350) = 1.01 \text{ N/mm}^2$$

(vii) Range of v

$$1.6\, v_c < v < 2\, v_c$$
$$5(0.7v - v_c) = 0.44 \geq 0.40$$

(viii) Shear reinforcement on first two rings

$$\Sigma A_{sv} = \frac{5(0.7v - v_c)u\, d}{0.95\, f_{yv}} = \frac{0.44 \times 6200 \times 350}{0.95 \times 460} = 2185 \text{ mm}^2$$

Area of a single leg of an 8 mm diameter link = 50.3 mm^2.
Number of links = 2185/(50.3) = 44 links.

(ix) Distribution of shear steel on two rings

(a) Ring–1
Number and distribution of *single leg links* on Ring–1:
Number of links = 40% of 44 = 18 links.
Ring–1 is at 0.5 d from the column face. Lengths of the sides of 'Ring–1' are
$$(400 + d) = 750 \text{ mm}, (600 + d) = 950 \text{ mm}$$
On each of the two shorter sides place 4 *single leg links* at a spacing of 250 mm and on each of the two longer sides place an additional 5 *single leg links* at an equal spacing of 158 mm. The maximum spacing does not exceed 1.5 d = 525 mm.

(b) Ring–2
Number of *single leg links* on Ring–2:
Number of *single leg links* = 44 – 18 = 26 links.
Ring–2 is approximately 1.25 d from 'Ring–1'. Lengths of the sides of 'Ring–2' are

$$(400 + 2.5d) = 1275, (600 + 2.5\, d) = 1475 \text{ mm}$$

On each of the two shorter sides place 6 *single leg links* at a spacing of 255 mm and place on each of the two longer sides place an additional 7 *single leg links* at equal spacing of 184 mm. The maximum spacing does not exceed 1.5 d = 525 mm.

(x) Calculate shear stress at the second shear perimeter at 2.25 d from the column face

$$u = 2\{(400 + 4.5\ d) + (600 + 4.5d)\} = 8300 \text{ mm}$$

The load inside the perimeter is

$$(400 + 4.5d) \times (600 + 4.5d) \times 40.8 = 175.3 \text{ kN}$$
$$V = 2295 - 175.3 = 2119.7 \text{ kN}$$
$$v = V/(u \times d) = 2119.7 \times 10^3 / (8300 \times 350) = 0.73 \text{ N/mm}^2$$

(xi) Range of v

$$v_c < v < 1.6\ v_c$$
$$(v - v_c) = (0.73 - 0.62) = 0.11 < 0.4$$
$$(v - v_c) = 0.4$$

(xii) Shear reinforcement on second and third rings

Total shear reinforcement area:

$$\sum A_{sv} = \frac{(v - v_c)u\ d}{0.95\ f_{yv}} = \frac{0.4 \times 8300 \times 350}{0.95 \times 460} = 2659 \text{ mm}^2$$

Number of 8 mm single leg links = 2659/ (50.3) = 53 links.

(xiii) Distribution of shear steel on two rings
(c) Ring 3

Number and distribution of links on Ring 3:

Number of *single leg links* = 53 – links on ring 2
Number of *single leg links* = 53 – 26 = 27

Ring 3 is at 2.0 d from the column face. Lengths of the sides of 'Ring3' are

$$(400 + 4d) = 1800, (600 + 4d) = 2000$$

On each of the two shorter sides place 7 links at a spacing of 300 mm and place on each of the two longer sides place additional 7 links at equal spacing of 250 mm. The maximum spacing does not exceed 1.5 d = 525 mm.

(xiv) Calculate shear stress at the third shear perimeter at 3.0 d from the column face:

$$u = 2\{(400 + 6.0\ d) + (600 + 6.0d)\} = 10400 \text{ mm}$$

The load inside the perimeter is equal to

$$(400 + 6.0\ d) \times (600 + 6.0d) \times 40.8 = 275.4 \text{ kN}$$
$$V = 2295 - 275.4 = 2019.6 \text{ kN}$$
$$v = V/(u \times d) = 2019.6 \times 10^3 / (10400 \times 350) = 0.56 \text{ N/mm}^2$$
$$v < (v_c = 0.62 \text{ N.mm}^2)$$

No further checks are required. Design is complete.

5.2 BOND, LAPS AND BEARING STRESSES IN BENDS

Bond is the grip due to adhesion or mechanical interlock and bearing in deformed bars between the reinforcement and the concrete. Anchorage is the embedment of a bar in concrete so that it can carry load through bond between the steel and concrete. A pull-out test on a bar is shown in Fig.5.12(a). If the anchorage length is sufficient, then the full strength of the bar can be developed by bond. The area over which the bond stress acts is the anchorage length multiplied by the perimeter of the bar. Anchorages for bars in a beam to external column joints and in a column base are shown in 5.12(b) and 5.12(c) respectively.

Clause 3.12.8.1 of the code states that the embedment length in the concrete is to be sufficient to develop the design force in the bar. Clause 3.12.8.3 of the code states that the design anchorage bond stress is assumed to be constant over the anchorage length. It is given by

$$f_b = \frac{F_s}{\pi \phi_c \ell}$$

where F_s is the force in the bar or group of bars, ℓ is the anchorage length and ϕ_c is the nominal diameter for a single bar or the diameter of a bar of equal total area for a group of bars.

$$F_s = \text{stress in the bar} \times (\pi/4)\phi_c^2$$

$$f_b = \frac{F_s}{\pi \phi_c \ell} = \frac{\text{stress in the bar} \times \phi_c}{4 \times \ell}$$

The anchorage length ℓ should be such that the bond stress does not exceed the design ultimate anchorage bond stress given by

$$f_{bu} = \beta \sqrt{f_{cu}}$$

where β is a bond coefficient that depends on the bar type. Values of β are given in Table 3.26 of the code from which the following is extracted:

Type 2 deformed bars are rolled with transverse ribs. The values of β apply in slabs and beams with minimum links. End bearing is taken into account in bars in compression and this gives a higher ultimate bond stress. The values include a partial safety factor $\gamma_m = 1.4$.

Table 5.1 Bond stresses

Type of bar	tension	compression
Plain bars	$\beta = 0.28$	$\beta = 0.35$
Type 2 deformed bars	$\beta = 0.50$	$\beta = 0.63$

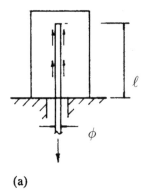

(a)

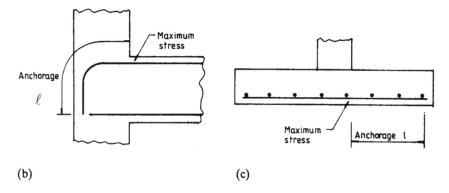

(b) **(c)**

Fig.5.12 (a) Pull-out test; (b) beam to external column; (c) column base.

5.2.1 Example of Calculation of Anchorage Lengths

Calculate the anchorage lengths in tension and compression for a grade 460 type 2 deformed bar of diameter ϕ in grade C30 concrete.

(a) Tension anchorage
The ultimate anchorage bond stress is

$$f_{bu} = \beta \sqrt{f_{cu}} = 0.5 \sqrt{30} = 2.74 \text{ N/mm}^2$$

Equate the anchorage bond resistance to the ultimate strength of the bar. Stress in the bar = 0.95 f_y.

$$f_b = 2.74 = \frac{stress\ in\ the\ bar \times \phi}{4 \times \ell} = \frac{0.95 \times 460 \times \phi}{4 \times \ell}$$

$$\ell = 40\phi$$

(b) Compression anchorage

The ultimate anchorage bond stress is

$$f_{bu} = \beta\sqrt{f_{cu}} = 0.63 \sqrt{30} = 3.45 \text{N/mm}^2$$

$$f_b = 3.45 = \frac{stress\ in\ the\ bar \times \phi}{4 \times \ell} = \frac{0.95 \times 460 \times \phi}{4 \times \ell} \quad \text{giving } \ell = 32\phi$$

Ultimate anchorage bond lengths as a multiple of a integral number of bar diameters are given in BS8110: Part 1, Table 3.27.

5.2.2 Hooks and Bends

Hooks and bends are used to shorten the length required for anchorage. Clause 3.12.8.23 of the code states that the effective length of a hook or bend is the length of a straight bar which has the same anchorage value as that part of the bar between the start of the bend and a point four bar diameters past the end of the bend.

The effective anchorage lengths given in the code are as follows.

(i) 180° hook: (see Fig.5.13(a))

The effective anchorage length is the *greater* of

(a) Eight times the *internal radius* r of the hook but not greater than 24 times the bar diameter

If the length of the bar *beyond the end of the bend* is L, then from the start of the bend to the end of the bar

Effective anchorage length = 8 r ($\leq$ 24ϕ) + L − 4 ϕ

or

(b) the actual length of the bar from the start of the bend

Effective anchorage length = π r + L

(ii) 90° bend: (see Fig.5.13(ba))

The effective anchorage length is the greater of

(a) four times the internal radius of the bend but not greater than 12 times the bar diameter. If the length of the bar *beyond the end of the bend* is L, then from the *start* of the bend to the end of the bar

Effective anchorage length = 4 r ($\leq$ 12ϕ) + L − 4 ϕ

or

(b) the actual length of the bar from the start of the bend

Effective anchorage length = 0.5π r + L

The radius of bend should not be less than twice the radius of the bend guaranteed by the manufacturer. A radius of two bar diameters is generally used for mild steel and three bar diameters for high yield steel. The hook and bend are shown in Fig.5.13.

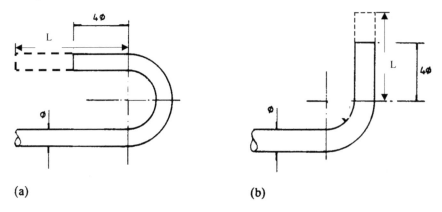

Fig.5.13 (a) 180° hook; (b) 90° bend.

5.2.2.1 Examples of anchorage length calculation

Example 1: Calculate the effective anchorage length of a 90° bend with an internal radius of 70 mm in a 16 mm diameter bar. The straight length of the bar beyond the end of the curve is 100 mm.

The effective anchorage length is the greater of

$$4\,r\,(\leq 12\phi) + L - 4\,\phi = 4 \times 70\,(\leq 12 \times 16) + 100 - 4 \times 16$$
$$= 192 + 100 - 64 = 228 \text{ mm}$$
$$0.5\pi\,r + L = 0.5 \times \pi \times 70 + 100 = 210 \text{ mm}$$

Effective anchorage length = 228 mm = 14.25 ϕ.

Therefore for a full anchorage length of 40 ϕ, the straight length of the bar *before* the start of the bend must be (40 – 14.25) ϕ = 412 mm.

Example 2: Calculate the effective anchorage length of an 180° bend with an internal radius of 80 mm in a 16 mm diameter bar. The straight length of the bar beyond the end of the curve is 100 mm.

The effective anchorage length is the *greater* of

$$8\,r\,(\leq 24\phi) + L - 4\,\phi = 8 \times 80\,(\leq 24 \times 16) + 100 - 4 \times 16 = 384 + 100 - 64 = 420 \text{ mm}$$
$$\pi\,r + L = \pi \times 80 + 100 = 351 \text{ mm}$$
$$\text{Effective anchorage length} = 420 \text{ mm} = 26.25 \; \phi$$

Therefore for a full anchorage length of 40 ϕ, the straight length of the bar *before* the start of the bend must be (40 – 26.25) ϕ = 220 mm.

5.2.2.2 Curtailment and anchorage of bars

The minimum anchorage length is the effective depth d of the section or 12 ϕ, which ever is larger. In the tension zone of a flexural member, the bar should be taken a full anchorage length beyond a point where it is no longer required or to a point where the remaining bars continuing beyond provide a moment of resistance twice the bending moment at the point. Curtailment or stopping of the bars when they are no longer needed should always be staggered.

5.2.3 Laps and Joints

Lengths of reinforcing bars are joined by lapping, by mechanical couplers or by butt or lap welded joints. Only lapping which is the usual way of joining bars is discussed here.

The minimum lap length specified in clause 3.12.8.11 is not to be less than 15 ϕ or 300 mm whichever is greater. From clause 3.12.8.13 the requirements for tension laps are the following:
1. The lap length is not to be less than the tension anchorage length;
2. If the lap is at the top of the section and the cover is less than two bar diameters the lap length is to be increased by a factor of 1.4;
3. If the lap is at the corner of a section and the cover is less than two bar diameters the lap length is to be increased by a factor of 1.4;
4. If conditions 2 and 3 both apply the lap length is to be doubled.
The length of compression laps should be 1.25 times the length of compression anchorage.

Note that all lap lengths are based on the smaller bar diameter. The code gives values for lap lengths in Table 3.27. It also sets out requirements for mechanical couplers in clause 3.12.8.16.2 and for the welding of reinforcing bars in clauses 3.12.8.17 and 3.12.8.18.

5.2.4 Bearing Stresses Inside Bends

It is often necessary to anchor a bar by extending it around a bend in a stressed state, as shown in Fig.5.14(a). It may also be necessary to take a stressed bar through a bend as shown in Fig.5.14(b).

In BS 8110: Part 1, clause 3.12.8.25.1, it is stated that if the bar does not extend or is not assumed to be stressed beyond a point four times the bar diameter past the end of the bend no check need be made. If it is assumed to be stressed beyond this point, the bearing stress inside the bend must be checked using the equation:

$$Bearing\ stress = \frac{F_{bt}}{r\,\phi} \le \frac{2\,f_{cu}}{[1+2\dfrac{\phi}{a_b}]}$$

where F_{bt} is the tensile force due to ultimate loads in the bar or group of bars, r is the internal radius of the bend, ϕ is the bar diameter or, for a group, the size of bar of equivalent area and a_b, for a bar or group of bars in contact, is the centre-to-centre distance between the bars or groups perpendicular to the bend; for a bar or group of bars adjacent to the face of a member a_b is taken as the cover plus ϕ.

5.2.4.1 Example of design of anchorage at beam support

Referring to Fig.5.14(a), three 20 mm diameter grade 460 deformed type 2 bars are to be anchored past the face of the column. The concrete is grade C30. From Table 3.3 of the code the nominal cover to 10 mm diameter links for mild exposure is 25 mm. The area of tension reinforcement *required* in the design is 810 mm^2. Design the anchorage required for the bars.

The ultimate anchorage bond stress is
$$f_{bu} = 0.5 \times \sqrt{30} = 2.74 \text{ N/mm}^2$$
Area provided $= 3T20 = 943$ mm^2,
$$\text{Stress in the bars} = 0.95 \, f_y \text{ (Area required/area provided)}$$
$$= 0.95 \times 460 \times (810/943) = 376 \text{ N/mm}^2$$
The anchorage length (see section 5.2) is
$$\ell = \text{Stress in bar} \times \phi / (4 \times f_{bu}) = 376 \times 20/ (4 \times 2.74) = 686 \text{ mm}$$
The internal radius of the bends is taken to be 100 mm and the cover on the main bars is 35 mm. The arrangement for the anchorage is shown in Fig.5.13(a). Referring to section 5.2.3 the anchorage length for the *lower* 90° bend is the *greater* of
$$4 \, r \, (\leq 12\phi) + L - 4 \, \phi = 4 \times 100 \, (\leq 12 \times 20) + 80 - 4 \times 20$$
$$= 240 + 80 - 80 = 240 \text{ mm}$$
$$0.5\pi \, r + L = 0.5 \times \pi \times 100 + 80 = 237 \text{ mm}$$
Thus the total anchorage provided past the face of the column is
$$145 + 173 + 140 + 253 = 711 \text{ mm}$$
$$\text{Actual bond stress} = \text{stress in bar} \times \phi / (4 \times \text{bond length})$$
$$= 376 \times 20/ (4 \times 711) = 2.64 \text{ N/mm}^2$$
From the figure the bars are at 80 mm centres. The ultimate bearing stress is
$$2 \, f_{cu}/ [1 + 2\phi/a_b] = 2 \times 30/ [1 + 2 \times 20/80] = 40 \text{ N/mm}^2$$
$$\text{The tensile force in a bar} = 376 \times \{(\pi/4) \, 20^2\} \times 10^{-3} = 118.1 \text{ kN}$$
Allow for the reduction in the force in the bar due to bond stress: Anchorage length from the point where the stress is a maximum to the centre of the top bend is
$$\text{anchorage length} = 145 + 173/2 = 232 \text{ mm}$$
$$F_{bt} \text{ at the top of the bend} = \text{force in the bar} - \text{bond force over 232 mm}$$
$$= 118.1 - 2.64 \times 232 \times \pi \times \phi \times 10^{-3} = 118.1 - 38.5 = 79.6 \text{ kN}$$
$$\text{Bearing stress} = F_{bt}/(r \, \phi)$$
$$= 79.6 \times 10^3/ (100 \times 20) = 39.8 \text{ N/mm}^2 < 40 \text{ N/mm}^2$$
The arrangement is satisfactory.

Full anchorage length of 40 $\phi = 40 \times 20 = 800$ mm could be provided by increasing the length past the lower bend from 80 to 170 mm. n in Fig.5.25.

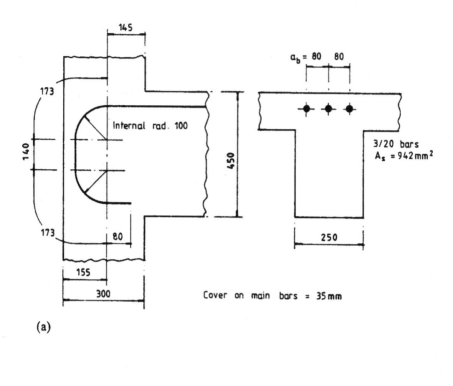

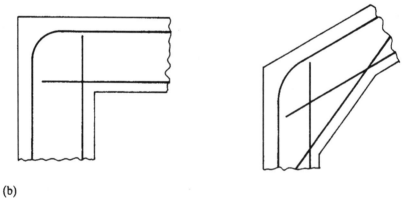

Fig.5.14 (a) Anchorage at end of beam; (b) stressed bars carried around bends.

5.3 TORSION

5.3.1 Occurrence and Analysis of Torsion

BS8110: Part 1, clause 3.4.5.13, states that in normal slab-beam or framed construction specific calculations for torsion are not usually necessary. Shear reinforcement will control cracking due to torsion adequately. However when the design relies on torsional resistance, specific design for torsion is required. Such a case is the overhanging slab shown in Fig.5.15 (a).

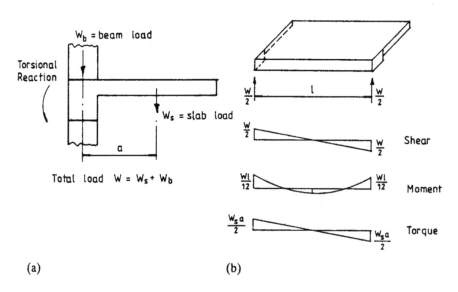

(a)   (b)

Fig.5.15 (a) Overhanging slab; (b) design actions.

5.3.2 Structural Analysis Including Torsion

Rigid–jointed frame buildings, although three dimensional, are generally analysed as a series of plane frames. This is a valid simplification because the torsional stiffness is much less than the bending stiffness. Figure 5.16(a) shows where bending in the beams in the transverse frames causes torsion in the longitudinal side beams, where only the end frame beams are loaded. In Fig.5.16(b) the loading on the intermediate floor beam causes torsion in the support beams. Analysis of the building as a space frame for various arrangements of loading would be necessary to determine maximum design conditions including torsion for all members.

If torsion is to be taken into account in structural analysis BS 8110: Part 2, clause 2.4.3, specifies that

$$\text{torsional rigidity} = GC, \text{ shear modulus } G = 0.42E_c$$

where E_c is the modulus of elasticity of the concrete and C is the torsional constant equal to *one-half* of the St Venant value for the plain concrete section.

The code states that the St Venant torsional stiffness may be calculated from the equation

$$C = \beta \, h_{min}^3 \, h_{max}$$

where β depends on the ratio *h/b* of the overall depth divided by the breadth. Values of β are given in Table 2.2 in the code. If the section is square $\beta = 1.4$. Also, h_{max} is the larger dimension of a rectangular section and h_{min} the smaller dimension. The code also gives a procedure for dealing with non–rectangular sections.

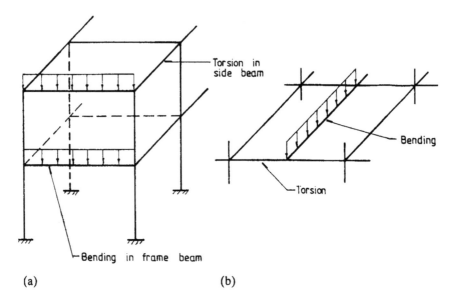

Fig.5.16 (a) Three-dimensional frame; (b) floor system.

5.3.3 Torsional Shear Stress in a Concrete Section

Fig.5.17 shows a rectangular box beam whose wall thickness can be considered as small compared to other cross sectional dimensions. It is shown in books on Strength of Materials that when the box section is subjected to a torsional moment T, the shear flow defined as the product of shear stress in the wall and its thickness is a constant. The walls of the box are in a state of pure shear.

The shear stress v_t is given by

$$v_t = \frac{T}{2abt}$$

where a, b are the centre line dimensions of the sides of the box and t is wall thickness.

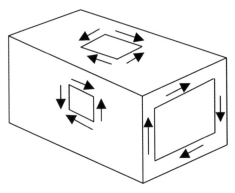

Fig.5.17 Stresses in a thin walled box beam under torsion.

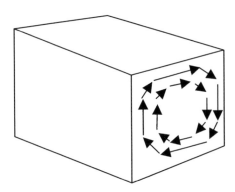

Fig 5.18 Torsional stress distribution in a solid rectangular section.

Fig 5.18 shows the elastic stress distribution in a solid rectangular section subjected to a torsional moment. The shear stresses due to torsion are tangential to the sides and in an elastic material, the maximum shear stress occurs in the middle of the longer side of a rectangular section. The stress is zero at the centroid and increases in a non–linear manner towards the edges. If the material is ductile, then at the ultimate or plastic state, the stress is the same everywhere.

The constant state of stress is represented by slope a heap of sand would take when it is poured over a plate of the same cross section as the beam under torsion. This is called as Sand Heap analogy. The heap is conical for a circular section, pyramid shaped for a square and pitched–roof–shape for the rectangular section as shown

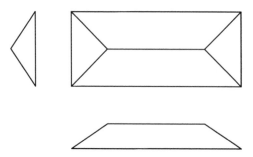

Fig.5.19 Sand heap for a rectangular section.

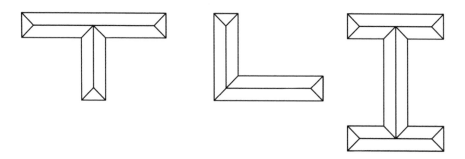

Fig.5.20 Sand heap for T, L and I sections.

in Fig.5.19. Fig.5.20 shows the plan for the sand heap in the case of T-, L- and I-sections

The slope of the sand heap is proportional to the shear stress v_t and twice the volume is proportional to the corresponding plastic torsional resistance. The expression for the torsional shear stress and ultimate torque given in BS 8110: Part 2, clause 2.4.4, can be derived for the sand heap shape as follows, where h_{max}, h_{min} are section dimensions, a is the height of the sand heap, the torsional shear stress v_t is equal to the slope of the sand heap, i.e. $v_t = 2a/h_{min}$, and the ultimate torque T is twice the volume of the sand heap, i.e.

$$T = 2[\frac{1}{3}h_{min}^2 a + \frac{1}{2}h_{min}\, a\,(h_{max} - h_{min})]$$

Substituting $a = 0.5\ v_t\ h_{min}$,

$$T = 0.5\ v_t\ h_{min}^2\ (h_{max} - h_{min}/3)$$

$$v_t = \frac{2T}{h_{min}^2\ (h_{max} - \dfrac{h_{min}}{3})}$$

The code states that T- and L- or I-sections are to be treated by dividing them into component rectangles. The division is to be such as to maximize the function

$\Sigma(h_{min}^3 h_{max})$. This will be achieved if the widest rectangle is made as long as possible. The torque resisted by *each* component rectangle is to be taken as

$$T \frac{h_{min}^3 h_{max}}{\Sigma(h_{min}^3 h_{max})}$$

If the torsional shear stress v_t exceeds the value of $v_{t\ min}$, reinforcement must be provided.

$$v_{t,min} = 0.067\sqrt{f_{cu}} \leq 0.4 \text{ N/mm}^2$$

The sum $(v + v_t)$ of the shear stresses from direct shear and torsion must not exceed the value of v_{tu}.

$$v_{tu} = 0.8\sqrt{f_{cu}} \leq 5.0 \text{ N/mm}^2$$

In addition, for small sections where $y_1 < 550$ mm,

$$v_t \leq v_{tu} y_1/550$$

where y_1 is the larger dimension of the link in the cross section (See Fig.5.21(c)). This restriction is to prevent concrete breaking away at the corners of small sections.

5.3.4 Torsional Reinforcement

A concrete beam subjected to torsion fails in diagonal tension on each face to form cracks running in a spiral around the beam, as shown in Fig.5.21(a). The torque may be replaced by the shear forces V on each face. The action on each face is similar to vertical shear in a beam. Reinforcement to resist torsion is provided in the form of closed links and longitudinal bars. This steel together with diagonal bands of concrete in compression can be considered to form a space truss which resists torsion.

This is illustrated in Fig.5.21(b). Fig.5.21(c) shows the torsional reinforcement. Let:

$\qquad x_1$ smaller dimension of the link
$\qquad y_1$ larger dimension of the link
$\qquad A_{sv}$ area of two legs of the link
$\qquad f_{yv}$ characteristic strength of the link
$\qquad s_v$ longitudinal spacing of the links

Assuming cracks at 45°, the number of links crossing the cracks is y_1/s_v on the sides and x_1/s_v on the top and bottom faces. The force in one link due to torsion is $0.95 f_{yv} A_{sv}/2$. The torsional T resistance of all links crossing the cracks is

$$T = 0.95 f_{yv} \frac{A_{sv}}{2} [\frac{x_1 y_1}{s_v} + \frac{y_1 x_1}{s_v}] = 0.95 f_{yv} A_{sv} [\frac{x_1 y_1}{s_v}]$$

The expression given in the BS 8110: Part 2. Clause 2.4.7. is

$$\frac{A_{sv}}{s_v} > \frac{T}{[0.8 x_1 y_1 0.95 f_{yv}]}$$

A safety factor of 0.8 (equal to reciprocal of 1.25) has been introduced into the value of resistance torque T.

The links and longitudinal bars should fail together. This is achieved by making the steel volume multiplied by the characteristic strength the same for each set of bars. This gives

$$A_{sv} (x_1 + y_1) f_{yv} = A_s s_v f_y$$

$$A_s = A_{sv} [\frac{f_{yv}}{f_y}] \{\frac{(x_1 + y_1)}{s_v}\}$$

where A_s is the area of longitudinal reinforcement and f_y is the characteristic strength of the longitudinal reinforcement. This is the expression given in the code.

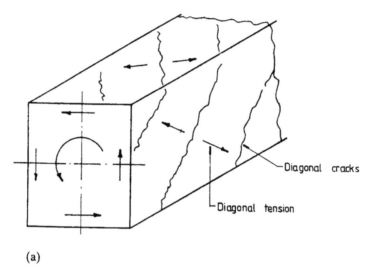

(a)

Fig.5.21(a) Diagonal cracking pattern.

The code also states that the spacing of the links is not to exceed x_1, $y_1/2$ or 200 mm. The links are to be of the closed type as shown in Fig.5.22(a). The longitudinal reinforcement is to be distributed evenly around the inside perimeter of the links. The clear distance between these bars should not exceed 300 mm and at least four bars, one in each corner, are required.

The torsion reinforcement is in addition to that required for moment and shear. In design, the longitudinal steel areas for moment and torsion and the link size and spacing for shear and torsion are calculated separately and combined.

BS8110: Part 2, clause 2.4.10, states that the link cages should interlock in T– and L–sections and tie the component rectangles together as shown in Fig.5.22(b). If the torsional shear stress in a minor component rectangle does not exceed $v_{t.min}$ then no torsional shear reinforcement need be provided in that rectangle.

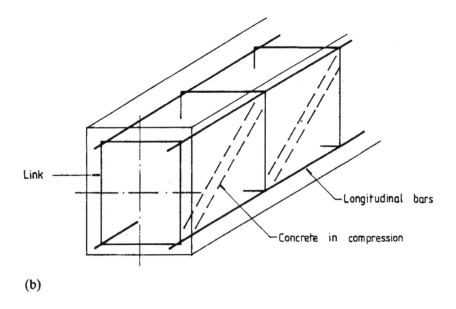

Link

Longitudinal bars

Concrete in compression

(b)

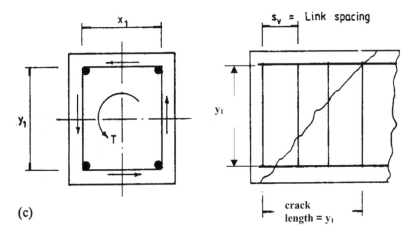

x_1

y_1

s_v = Link spacing

y_1

T

crack
length = y_1

(c)

Fig.5.21 (b) space truss; (c) torsion resistance.

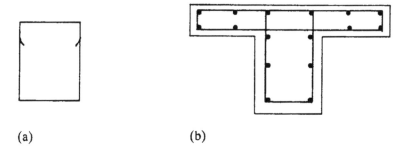

(a)

(b)

Fig. 5:22 (a) Closed link; (b) torque reinforcement for a T-beam.

5.3.4.1 *Example of Design of Torsion Steel for Rectangular Beam*

A rectangular beam section has an overall depth of 500 mm and a breadth of 300 mm. It is subjected at ultimate to a vertical hogging moment of 387.6 kNm, a vertical shear of 205 kN and a torque of 13 kN m. Design the longitudinal steel and links required at the section. The materials are grade C30 concrete, grade 460 reinforcement for the main bars and grade 250 for the links.

The section is shown in Fig.5.23(a) with the internal dimensions for locations of longitudinal bars and links taken for design. These dimensions are based on 25mm cover, 12 mm diameter links and 25 mm diameter main bars at the top in vertical pairs and 20 mm bars at the bottom.

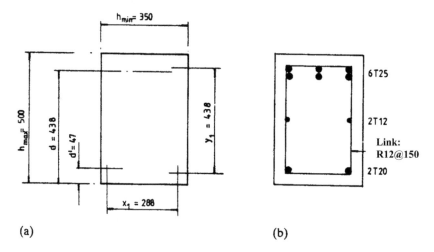

(a)   (b)

Fig.5.23 (a) Section and design dimensions; (b) reinforcement.

(a) Bending moment in the vertical plane:
Calculate the moment that the beam can sustain without any compression steel:
$$M_{SR} = 0.156 \, b \, d^2 \, f_{cu} = 0.156 \times 350 \times 438^2 \times 30 \times 10^{-6} = 314.2 \text{ kNm}$$
$$d - d' = 438 - 47 = 391 \text{ mm}$$
$$d'/d = 47/438 = 0.107 < 0.186, \text{ compression steel yields.}$$
Moment to be resisted by compression steel:
$$\text{Applied moment} - M_{Rc} = 387.6 - 314.2 = 73.4 \text{ kNm}$$
$$A_s' = 73.4 \times 10^6 / [(d - d') \, 0.95 \, f_y] = 73.4 \times 10^6 / [391 \times 0.95 \times 460] = 430 \text{ mm}^2$$
$$0.95 \, f_y \, A_s = 0.45 \, f_{cu} \, b \, 0.45d + 0.95 \, f_y \, A_s'$$
$$0.95 \times 460 \times A_s = 0.45 \times 30 \times 350 \times 0.45 \times 438 + 0.95 \times 460 \times 430$$
$$A_s = 2561 \text{ mm}^2$$

(b) Shear reinforcement for direct shear
Calculate v_c:
$$100 \, A_s / (b_v \, d) = 100 \times 2561 / (350 \times 438) = 1.67 < 3.0$$
$$400/d = 400/438 = 0.91 < 1.0, \text{ use } 1.0$$

$$v_c = 0.79 \, (1.67)^{1/3} \, (1.0)^{1/4} \, (30/25)^{1/3} / 1.25 = 0.80 \text{ N/mm}^2$$
$$v = V/ (b_v \, d) = 205 \times 10^3 / (350 \times 438) = 1.34 \text{ N/mm}^2$$
$$v - v_c = 0.54 > 0.4. \text{ Needs designed links.}$$

$$A_{sv} = \frac{b_v \, s_v \, (v - v_c)}{0.95 \, f_{yv}}$$

Assuming a spacing of 150 mm for the links,
$$A_{sv} = 350 \times 150 \times 0.54/ (0.95 \times 250) = 119.4 \text{ mm}^2.$$
The spacing selected is less than $(0.75d = 0.75 \times 438 = 323 \text{ mm})$. The link steel area will be added to that required for torsion below.

(c) Torsion reinforcement

$$v_t = \frac{2T}{h_{min}^2 \, (h_{max} - \dfrac{h_{min}}{3})} = \frac{2 \times 13 \times 10^6}{350^2 \, \{500 - \dfrac{350}{3}\}} = 0.55 \text{ N/mm}^2$$

$$v_{t.min} = 0.067 \sqrt{f_{cu}} = 0.37 \text{ N/mm}^2$$
$$(v_t = 0.55) > (v_{t,min} = 0.37)$$

Shear reinforcement is necessary for torsion.
Calculate the sum of direct and torsional shear stresses.
$$v + v_t = 1.34 + 0.55 = 1.89 \text{ N/mm}^2$$
$$v_{tu} = 0.8 \sqrt{f_{cu}} = 4.38 \text{ N/mm}^2$$
$$(v + v_t = 1.89) < (v_{tu} = 4.38)$$

y_1 is the larger centre to centre dimension of the link
$$y_1 = 500 - 2 \times 25 - 12 = 438 \text{ mm}$$

Check also that
$$(v_t = 0.55) < \{ v_{tu} \frac{y_1}{550} = 4.38 \frac{438}{550} = 3.49 \}$$

Taking $s_v = 150$ mm, torsional steel is calculated from
$$\frac{A_{sv}}{s_v} > \frac{T}{[0.8 \, x_1 \, y_1 \, 0.95 \, f_{yv}]}$$

$$A_{sv} = 13 \times 10^6 \times 150/ (0.8 \times 288 \times 438 \times 0.95 \times 250) = 81.4 \text{ mm}^2$$

Since A_{sv} corresponds to area of two legs of a link, the total area of *one leg* of a link is
$$0.5 \, (A_{sv} \text{ for torsion} + A_{sv} \text{ for direct shear}) = 0.5 \times (81.4 + 119.4) = 100.4 \text{ mm}^2$$
Links 12 mm in diameter with an area of 113 mm^2 are required.
The spacing must not exceed $x_1 = 288$ mm, $y_1/2 = 438/2 = 219$ mm or 200 mm.
The spacing of 150 mm is satisfactory.
The area of longitudinal reinforcement
$$A_s = A_{sv} \frac{f_{yv}}{f_y} \frac{x_1 + y_1}{s_v}$$

$$A_s = 81.4 \times (250/460) \times (288 + 438)/150 = 214 \text{ mm}^2$$
This area is to be distributed equally around the perimeter.

(d) Arrangement of reinforcement
The clear distance between longitudinal bars required to resist torsion is not to exceed 300 mm. Six bars with a theoretical area of 214/6 = 35.7 mm^2 per bar are required.
For the bottom steel:
$$A_s' = 430 + 2 \times 35.7 = 501 \text{ mm}^2$$
Provide 2T20 of area 628 mm^2.
For the top steel:
$$A_s = 2561 + 2 \times 35.7 = 2637 \text{ mm}^2$$
Provide 6T25 in two rows of 3 each, giving an area of 2945 mm^2.
For the centre bars provide two 12 mm diameter bars of area 113 mm^2 per bar. The reinforcement is shown in Fig.5.23(b).

5.3.4.2 Example of T-beam Design for Torsion Steel

The T-beam shown in Fig.5.24(a) spans 8 m. The ends of the beam are simply supported for vertical load and restrained against torsion. The beam carries an ultimate distributed vertical load of 24 kN/m. A column is supported on one flange at the centre of the beam and transmits an ultimate load of 50 kN to the beam as shown. Design the reinforcement at the centre of the beam for the T-beam section only. A transverse stiffening beam would be provided at the centre of the beam but design for this is not part of the exercise. The materials are grade C30 concrete, grade 460 for the longitudinal reinforcement and grade 250 for the links.

(a) The ultimate beam actions:
The maximum vertical shear is
$$V_A = (0.5 \times 24 \times 8) + (0.5 \times 50) = 121 \text{ kN}$$
The maximum moment in the vertical plane is
$$M = 24 \times 8^2/8 + 50 \times 8/4 = 292 \text{ kNm}$$
The torque is
$$T = 50 \times 0.7 \times 0.5 = 17.5 \text{ kNm}$$
The load, shear force, bending moment and torque diagrams are shown in Fig.5.24(c). Cover has been taken as 25mm, the links are 12 mm in diameter and the main bars 25 mm diameter. The dimensions adopted for design as shown in Fig.5.22(b) are: d = 438 mm, y_1 = 438 mm, x_1 = 238 mm

(b) Moment in the vertical plane:
$$M = 292 \text{ kNm}$$
Check if the beam needs to be designed as a rectangular beam or a T-beam.
$$M_{flange} = 0.45 \, f_{cu} \times b \times h_f \times (d - h_f/2)$$
$$= 0.45 \times 30 \times 1600 \times 150 \times (438 - 150/2) \times 10^{-6} = 1176 \text{ kNm.}$$
$$M_{flange} > M.$$
Therefore the stress block lies in the flange and the beam can be designed as a rectangular beam 1600 × 438.
$$k = M/(b \, d^2 \, f_{cu}) = 292 \times 10^6/(1600 \times 438^2 \times 30) = 0.032 < 0.156.$$

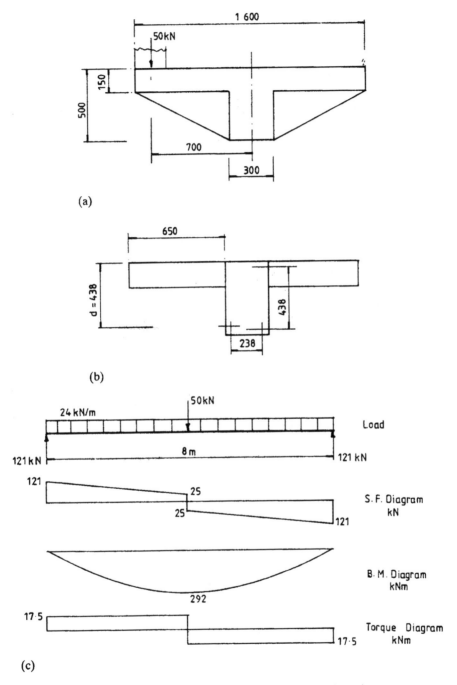

Fig.5.24 (a) section; (b) dimensions for design; (c) beam design actions.

Therefore no compression steel is required.
$$z/d = 0.5 + \sqrt{(0.25 - k/0.9)} = 0.96 > 0.95. \text{ Therefore } z/d = 0.95$$
$$A_s = M/ [0.95 \, f_y \, z] = 292 \times 10^6 / (0.95 \times 460 \times 0.95 \times 438) = 1606 \text{ mm}^2$$

(c) Vertical shear at the centre of the beam
Calculate v_c:
$$100 \, A_s/ (b_v \, d) = 100 \times 1606 / (300 \times 438) = 1.22 < 3.0$$
$$400/d = 400/438 = 0.91 < 1.0, \text{ use } 1.0$$
$$v_c = 0.79 \, (1.22)^{1/3} \, (1.0)^{1/4} \, (30/25)^{1/3} \, /1.25 = 0.72 \text{ N/mm}^2$$
$$v = V/ (b_v \, d) = 25 \times 10^3/ (300 \times 438) = 0.19 \text{ N/mm}^2 < v_c$$
Beam needs only nominal links.

$$A_{sv} = \frac{b_v \, s_v \, 0.4}{0.95 \, f_{yv}}$$

Assuming a spacing of 175 mm for the links,
$$A_{sv} = 300 \times 175 \times 0.4/ (0.95 \times 250) = 88.4 \text{ mm}^2$$
The spacing selected is less than $(0.75d = 0.75 \times 438 = 323 \text{ mm})$. The steel area will be added to that required for torsion below.

(d) Torsion reinforcement
The T-section is split into component rectangles such that $\Sigma(h_{min}^3 h_{max})$ is a maximum. Check the following two alternatives.
(a) flange (1600×150) + rib (350×300)
$$\Sigma(h_{min}^3 h_{max}) = (150^3 \times 1600) + (300^3 \times 350) = 5.4 \times 10^9 + 9.45 \times 10^9 = 14.85 \times 10^9$$
(b) rib (500×300) + two flanges (650×150)
$$\Sigma(h_{min}^3 h_{max}) = (300^3 \times 500) + (2 \times 150^3 \times 650) = 13.5 \times 10^9 + 4.39 \times 10^9$$
$$= 17.89 \times 10^9$$
Arrangement (b) is adopted where the widest rectangle has been made as long as possible. Torque taken by the flange and the ribs are:
$$\text{Rib}: T \times 13.5 \times 10^9/\{17.89 \times 10^9\} = 17.5 \times 0.755 = 13.21 \text{ kNm}$$
$$\text{Two flange}: T \times 4.39 \times 10^9/ \{17.89 \times 10^9\} = 17.5 \times 0.245 = 4.29 \text{ kNm}$$

$$v_t = \frac{2T}{h_{min}^2 \, (h_{max} - \dfrac{h_{min}}{3})}$$

$$\text{Rib}: v_t = 2 \times 13.21 \times 10^6 / [300^2(500 - 300/3) = 0.73 \text{ N/mm}^2$$
$$\text{Flanges}: v_t = 2 \times 0.5 \times 4.29 \times 10^6 / \{[150^2(650 - 150/3)]\} = 0.32 \text{ N/mm}^2$$
$$v_{t.min} = 0.067\sqrt{f_{cu}} = 0.37 \text{ N/mm}^2$$
$$\text{Rib}: v + v_t = 0.19 + 0.73 = 0.92 \text{ N/mm}^2$$
This is less than $v_{tu} = 0.8\sqrt{f_{cu}} = 4.38 \text{ N/mm}^2$
$$v_t < [v_{tu} \, (y_1/550) = 4.38 \times 438/550 = 3.49 \text{ N/mm}^2]$$
Taking $s_v = 175$ mm, torsional steel is calculated from
$$\frac{A_{sv}}{s_v} > \frac{T}{[0.8 x_1 \, y_1 \, 0.95 \, f_{yv}]}$$

$A_{sv} = 13.21 \times 106 \times 175/ (0.8 \times 238 \times 438 \times 0.95 \times 250) = 116.7 \text{ mm}^2$

Since A_{sv} corresponds to area of two legs of a link, the total area of <u>one leg</u> of a link is

$0.5 \ (A_{sv} \text{ for torsion} + A_{sv} \text{ for direct shear}) = 0.5 \times (116.7 + 88.4) = 102.6 \text{ mm}^2$

Links 12 mm in diameter with an area of 113 mm^2 are required. The spacing must not exceed [$x_1 = 238$ mm, ($y_1/2 = 438/2 = 219$ mm) or 200 mm]. The spacing of 175 mm is satisfactory.

The area of longitudinal reinforcement

$$A_s = A_{sv} \frac{f_{yv}}{f_y} \frac{x_1 + y_1}{s_v}$$

$A_s = 116.7 \times (250/460) \times (238 + 438)/175 = 245 \text{ mm}^2$

This area is to be distributed equally around the perimeter. Using 6 bars, each bar has an area of 40.8 mm^2. In the flange $v_t = 0.32 \text{ N/mm}^2$ is less than $v_{t.min} = 0.37$ N/mm^2 and therefore no torsional steel is required.

(e) Arrangement of reinforcement

For the bottom steel

$$A_s = 1606 + 2 \times 40.8 = 1688 \text{ mm}^2.$$

Provide 4T25 of area 1963 mm^2.

For the top and centre of the rib, provide two 12 mm diameter bars at each location. The distance between the longitudinal bars is not to exceed 300 mm. Reinforcement would have to be provided to support the load on the flange. The moment, direct shear and torsion reinforcement for the rib is shown in Fig.5.25.

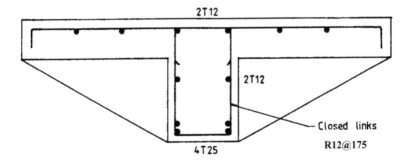

Fig.5.25 Reinforcement in centre rib.

CHAPTER 6

SERVICEABILITY LIMIT STATE CHECKS

6.1 SERVICEABILITY LIMIT STATE

In chapter 4 and chapter 5, design procedure for the ultimate limit state (ULS) in bending, shear and torsion were described. It is necessary in practice to ensure that the structure can not only withstand the forces at the ultimate limit state but also that it behaves satisfactorily at working loads. The main aspects to be satisfied at serviceability limit state (SLS) are that of deflection and cracking. In this chapter checks that are normally used to ensure satisfactory behaviour under SLS conditions without detailed calculations are considered. These are known as 'deemed to satisfy' clauses. Methods requiring detailed calculations are discussed in Chapter 19.

6.2 DEFLECTION

6.2.1 Deflection Limits and Checks

Limits for the serviceability limit state of deflection are set out in BS 8110: Part 2, clause 3.2.1. It is stated in this clause that the deflection is noticeable if it exceeds $L/250$ where L is the span of a beam or length of a cantilever. Deflection due to dead load can be offset by pre-cambering.

The code also states that damage to partitions, cladding and finishes will generally occur if the deflection exceeds

1. L/500 or 20 mm whichever is the lesser for brittle finishes
2. L/350 or 20 mm whichever is the lesser for non-brittle finishes

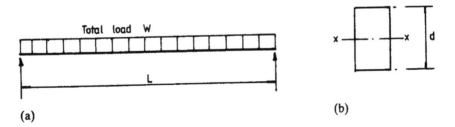

Fig. 6.1 (a) Beam load; (b) section.

Design can be made such as to accommodate the deflection of structural members without causing damage to partitions or finishes. Two methods are given in BS 8110: Part 1 for checking that deflection is not excessive:
1. limiting the span–to–effective depth ratio using the procedure set out in clause 3.4.6. This method should be used in all normal cases
2. calculation of deflection from curvatures set out in BS 8110: Part 2, sections 3.6 and 3.7 . This will be discussed in Chapter 19.

6.2.2 Span–to–Effective Depth Ratio

In a homogeneous elastic beam of span L, if the maximum stress is limited to an allowable value σ and the deflection Δ is limited to span/q, then for a given load a unique value of span–to–depth ratio L/d can be determined to limit stress and deflection to their allowable values simultaneously. Thus for the simply supported beam with a uniform load shown in Fig. 6.1

$$\text{Maximum bending moment} = W\,L/8$$

where W = total load on the beam

$$\text{maximum stress } \sigma = M\,y/I = \frac{W\,L}{8}\frac{y}{I} = \frac{W\,L}{8}\frac{0.5d}{I}$$

where I is the second moment of area of the beam section. d is the depth of the beam and L is the span.

The allowable deflection Δ is

$$\Delta = \frac{L}{q} = \frac{5}{384}\frac{WL^3}{EI} = \frac{WL\,0.5d}{8I}\frac{5L^2}{24E\,d} = \sigma\,\frac{5L^2}{24E\,d}$$

$$\frac{span}{depth} = \frac{L}{d} = \frac{4.8E}{\sigma\,q}$$

where E is Young's modulus.

Similar reasoning may be used to establish span–to–effective depth ratios for reinforced concrete beams to control deflection. The method in the code is based on calculation and confirmed by tests. The main factors affecting the deflection of the beam are taken into account. The allowable value for the span–to–effective depth ratio calculated using the procedure given in clause 3.4.6 of the code for normal cases depends on
1. the basic span–to–effective depth ratio for rectangular or flanged beams and the support conditions
2. the amount of tension steel and its stress
3. the amount of compression steel
These considerations are discussed briefly below.

(a) Basic span–to–effective depth ratios
The code states that the basic span–to–effective depth ratios given in Table 3.9 for rectangular and flanged beams are so determined as to limit the total deflection to span/250. This ensures that deflection occurring after construction is limited to

span/350 or 20 mm whichever is the less. The support conditions have also to be taken into account.

The basic span–to–effective depth ratios from Table 3.9 of the code are given in Table 6.1. The values in the table apply to beams with spans up to 10 m. Refer to clause 3.4.6.4 of the code for beams of longer span.

If $b_w/b > 0.3$, linear interpolation between values for $b_w/b = 0.3$ and $b_w/b = 1$ (Rectangular beam) can be used.

The allowable L/d is lower for flanged beams because in the flanged beam there is not as much concrete in the tension zone as in a rectangular beam and the stiffness of the beam is therefore reduced.

Table 6.1 Basic span–to–effective depth ratios

Support conditions	Span to effective depth ratio	
	Rectangular beam	Flanged beam, $b_w/b \leq 0.3$
Cantilever	7	5.6
Simply supported beam	20	16.0
Continuous beam	26	20.8

(b) Tension reinforcement

The deflection is influenced by the amount of tension reinforcement and the value of the stress at service loads at the centre of the span for beams or at the support for cantilevers. According to clause 3.4.6.5 of the code the basic span–to–effective depth ratio from Table 3.9 of the code is multiplied by the modification factor from Table 3.10. The modification factor is given by the formula in the code:

$$\text{modification factor} = 0.55 + \frac{477 - f_s}{120\{0.9 + \dfrac{M}{bd^2}\}} \leq 2.0$$

Note that the amount of tension reinforcement present is measured by the $M/(bd^2)$ term.

The service stress f_s is estimated from the equation

$$f_s = \frac{2}{3} f_y \frac{A_{s.req}}{A_{s.prov}} \frac{1}{\beta_b}$$

where

$A_{s.req}$ is the area of tension steel required at mid–span to support ultimate loads (at the support for a cantilever)

$A_{s.prov}$ is the area of tension steel provided at mid–span (at the support for a cantilever)

$\beta = 1.0$ for statically indeterminate beams. Values of β_b in the case of statically indeterminate structures will be discussed in Chapter 13.

The following comments are made concerning the expression for service stress:
1. The stress due to service loads is given by (2/3) f_y. This takes account of partial factors of safety for loads and materials used in design for the ultimate limit state;

2. If more steel is provided than required the service stress is reduced by the ratio $A_{s.req}/A_{s.prov}$

It can be noted from the equation for modification factor, that for a given section with the reinforcement at a given service stress the allowable *span/d* ratio is lower when the section contains a larger amount of steel. This is because the steel stress as measured by $M/(bd^2)$, is increased

1. the depth to the neutral axis is increased and therefore the curvature for a given steel stress increases (see calculation of deflections in Chapter 19.)

2. there is a larger area of concrete in compression, which leads to larger deflections due to creep

3. the smaller portion of concrete in the tension zone reduces the stiffness of the beam

Providing more steel than required reduces the service stress and this increases the allowable *span/d* ratio for the beam.

(c) Compression reinforcement

All reinforcement in the compression zone reduces concrete shrinkage and creep and therefore the curvature. This effect decreases the deflection. The modification factors for compression reinforcement are given in BS 8110: Part 1, Table 3.11. The modification factor is given by the formula

$$=1.0+\frac{\dfrac{100\,A'_{s.prov}}{bd}}{3.0+\dfrac{100\,A'_{s.prov}}{bd}} \leq 1.5$$

where $A'_{s.prov}$ is the area of compression reinforcement provided.

(d) Deflection check

The allowable span–to–effective depth ratio is the basic ratio multiplied by the modification factor for tension reinforcement multiplied by the modification factor for compression reinforcement. This value should be greater than the actual *span/d* ratio for the beam to be satisfactory with respect to deflection.

(e) Deflection checks for slabs

The deflection checks applied to slabs are discussed under design of the various types of slab in Chapter 8.

6.2.2.1 Example of Deflection Check for T–Beam

The section at mid–span designed for a simply supported T–beam of 6 m span is shown in Fig. 6.2. The design moment is 165 kNm. The calculated area of tension reinforcement was 1447mm^2 and three 25 mm diameter bars of area 1472 mm^2 were provided. To carry the links, two 16 mm diameter bars have been provided at the top of the beam. Using the rules set out above, check whether the beam is satisfactory for deflection. The materials used are concrete grade 30 and reinforcement grade 460.

From BS 8110: Part 1, Table 3.9:

web width/ effective flange width = $b_w/b = 250/1450 = 0.17 < 0.3$

The basic span–to–effective depth ratio is 16.

$$M/(bd^2) = 165 \times 10^6 / (1450 \times 300^2) = 1.26$$

The service stress f_s is

$$f_s = \frac{2}{3} f_y \frac{A_{s.req}}{A_{s.prov}} \frac{1}{\beta_b} = \frac{2}{3} \times 460 \times \frac{1447}{1472} 1.0 = 302 \text{ N/mm}^2$$

The modification factor for tension reinforcement using the formula given in Table 3.10 in the code is

$$0.55 + \frac{477 - f_s}{120\{0.9 + \dfrac{M}{bd^2}\}} = 0.55 + \frac{477 - 302}{120\{0.9 + 1.26\}} = 1.23 < 2.0$$

For the modification factor for compression reinforcement
$A'_{s.prov} = 2T12 = 226 \text{ mm}^2$

$$100 \frac{A_{s.prov}}{bd} = 100 \frac{226}{1450 \times 300} = 0.052$$

The modification factor for compression steel is

$$1.0 + \frac{0.052}{\{3.0 + 0.052\}} = 1.017 < 1.5$$

Allowable span/depth = $16 \times 1.23 \times 1.017 = 20.0$

Actual span/depth = $6000/300 = 20.0$

The beam is only just satisfactory with respect to deflection.

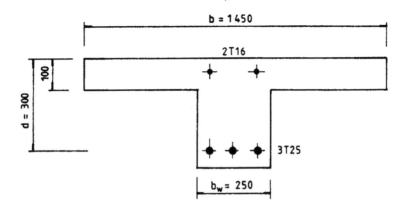

Fig. 6.2 Doubly reinforced T–beam.

6.3 CRACKING

6.3.1 Cracking Limits and Controls

Any prominent crack in reinforced concrete greatly detracts from the appearance. Excessive cracking and wide deep cracks affect durability and can lead to corrosion of reinforcement. BS 8110: Part 1, clause 2.2.3.4.1, states that for reinforced concrete cracking should be kept within reasonable bounds. The clause points to two methods for crack control:

1. in normal cases a set of rules for limiting the maximum bar spacing in the tension zone of members
2. in special cases use of a formula given in BS 8110: Part 2, section 3.8, for assessing the design crack width

In this Chapter only rules for the normal case are considered. Rules for special cases are discussed in Chapter 19.

6.3.2 Bar Spacing Controls in Beams

Cracking is controlled by specifying the maximum distance between bars in tension. The spacing limits are specified in clause 3.12.11.2. The clause indicates that in normal conditions of internal or external exposure, the bar spacings given will limit crack widths to 0.3 mm. Calculations of crack widths can often be made to justify larger spacings. The rules are as follows.

1. Bars of diameter less than 0.45 of the largest bar in the section should be ignored except when considering bars in the side faces of beams.
2. The clear horizontal distance S_1 between bars or groups near the tension face of a beam should not be greater than the values given in Table 3.28 of the code which are given by the expression (Fig. 6.3)

$$clear\ spacings\ \leq \frac{70000}{f_y}\beta_b \leq 300$$

$\beta = 1.0$ for statically indeterminate beams. Value of β in the case of statically indeterminate structures will be discussed in Chapter 13.

The maximum clear distance depends on the grade of reinforcement and a smaller spacing is required with high yield bars to control cracking because stresses and strains are higher than with mild steel bars.

3. As an alternative the clear spacing between bars can be found from the formula (Clause 3.12.11.2.4 of the Code)

$$clear\ spacings\ \leq \frac{47000}{f_s} \leq 300$$

$$where\ f_s = \frac{2}{3}f_y\frac{A_{s.req}}{A_{s.prov}}\frac{1}{\beta_b}$$

4. The clear distance s_2 from the corner of a beam to the surface of the nearest horizontal bar should not exceed one–half of the values given in BS 8110: Part 1, Table 3.28 or the alternative formula.

5. If the overall depth of the beam exceeds 750 mm, longitudinal bars should be provided at a spacing of s_3 not exceeding 250 mm over a distance of two–thirds of the overall depth from the tension face. The size of bar should not be less than $\sqrt{(S_b\, b/f_y)}$ where S_b is the bar spacing and b is the breadth of the beam (see clause 3.12.5.4 of the Code).

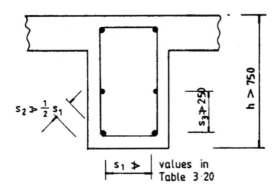

Fig.6.3 Rules for crack width limitation.

6.3.2.1 Examples of maximum bar spacings in beams

Example 1: Fig. 6.3 shows a T–beam. Apply 'Deemed to satisfy' rules to check the maximum bar spacings.

Dimensions of beam: b = 600 mm, b_w = 250 mm, h_f = 100 mm, h = 410 mm, cover = 30 mm, links 8 mm diameter, main steel 25 mm bars, effective depth = 348 mm. The beam has been designed for an ultimate moment of 260 kNm using f_{cu} = 30 N/mm² and f_y = 460 N/mm².

Using the code formula for designing the steel in T–beam,

$$A_{s\,Required} = \frac{260\times10^6 + 0.1\times30\times250\times348\times(0.45\times348-100)}{0.95\times460\times(348-0.5\times100)} = 2110\,\text{mm}^2$$

Using 5T25, $A_{s\,provided}$ = 2454 mm².

(i) Clear distance s_1 between bars:
$$s_1 = \{250 - 2\times30 - 2\times8 - 2\times25 - 25)/2 = 49.5\,\text{mm}$$
Allowable value:

$$clear\ spacings \leq \{\frac{70000}{f_y}\beta_b = \frac{70000}{460} = 152\} \leq 300$$

The spacing s_1 is with in the permissible limit.

Using the alternative formula

$$f_s = \frac{2}{3} f_y \frac{A_{s.req}}{A_{s.prov}} \frac{1}{\beta_b} = \frac{2}{3} 460 \frac{2110}{2454} = 264\,\text{N/mm}^2$$

$$\text{clear spacings} \leq \{\frac{47000}{f_s} = \frac{47000}{264} = 178\} \leq 300$$

The alternative spacing s_1 is with in the permissible limit.

(ii) Check s_2

Distance from the side and bottom to the centre of the bottom left bar is
30 (cover) + 10 (Link) + 25/2 = 52.5 mm

$$s_2 = \sqrt{\{52.5^2 + 52.5^2\}} - 25/2 = 62\,\text{mm}$$

Allowable value = 0.5 s_1 = 0.5 × 163 = 82 mm.
The maximum spacing is satisfactory.

(iii) Minimum reinforcement in the side faces

Beam is less than 750 mm deep. Additional steel on side faces not required.

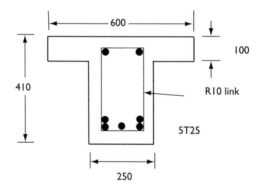

Fig. 6.3 T−beam.

Example 2: Fig. 6.4 shows a rectangular beam. Apply 'Deemed to satisfy' rules to check the maximum bar spacings.

Dimensions of beam: breadth b = 400 mm, overall depth h = 800 mm, cover = 30 mm, links 10 mm diameter, main steel 25 mm bars, effective depth = 735 mm.
   The beam has been designed for an ultimate moment of 900 kNm using f_{cu} = 30 N/mm² and f_y = 460 N/mm².

$$k = 900 \times 10^6 / (400 \times 735^2 \times 30) = 0.139 < 0.156$$
$$z/d = 0.5 + \sqrt{(0.25 - 0.139/0.9)} = 0.81 < 0.95$$
$$A_{s\,required} = 900 \times 10^6 / (0.81 \times 735 \times 0.95 \times 460) = 3459\,\text{mm}^2$$
$$A_{s\,Provided} = 8T25 = 3927\,\text{mm}^2$$

(i) Clear distance s_1 between bars:
$$s_1 = \{400 - 2 \times 30 - 2 \times 10 - 25)/3 - 25 = 73 \text{ mm}$$

Allowable value:
$$\text{clear spacings} \leq \{\frac{70000}{f_y} \beta_b = \frac{70000}{460} = 152\} \leq 300$$

Using the alternative formula
$$f_s = \frac{2}{3} f_y \frac{A_{s.req}}{A_{s.prov}} \frac{1}{\beta_b} = \frac{2}{3} 460 \frac{3459}{3927} = 270 \text{ N/mm}^2$$

$$\text{clear spacings} \leq \{\frac{47000}{f_s} = \frac{47000}{270} = 174\} \leq 300$$

The spacing is with in the permissible limit.

(ii) Check s_2:
Distance from the side and bottom to the centre of the bottom left bar is
30 (cover) + 10 (Link) + 25/2 = 52.5 mm
$$s_2 = \sqrt{\{52.5^2 + 52.5^2\}} - 25/2 = 62 \text{ mm}$$
Allowable value = 0.5 s_1 = 0.5 × 163 = 82 mm

(iii) Minimum reinforcement in the side faces
Beam is more than 750 mm deep. Additional steel on side faces is required.
Assume a bar spacing S_b = 250 mm

$$\text{Minimum diameter of bar} = \sqrt{\frac{S_b b}{f_y}} = \sqrt{\frac{250 \times 400}{460}} = 14.7 \text{ say } 16 \text{ mm}$$

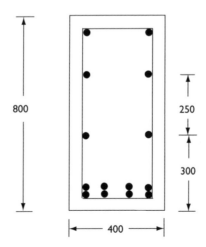

Fig. 6.4 Deep rectangular beam.

Provide 16 mm bars at a spacing of 250 mm for the $(2/3) \times 800 = 533$ mm from the bottom. Provide two layers of bars, first pair of 16 mm bars at 300 mm from the bottom and the second pair at 550 mm from bottom.
All maximum spacings are satisfactory.

6.3.3 Bar Spacing Controls in Slabs

The maximum clear spacing between bars in slabs is given in BS 8110: Part 1, clause 3.12.11.2.7. This clause states that the absolute clear distance between bars should not exceed three times the effective depth or 750 mm. It also states that no further checks are required if any of (a), (b) 0r (c) are satisfied, where:
(a) grade 250 steel is used and the slab depth does not exceed 250 mm
(b) grade 460 steel is used and the slab depth does not exceed 200 mm
(c) the reinforcement percentage $100A_s/(bd)$ is less than 0.3, b is the breadth of the slab considered and d is the effective depth
Refer to clauses 3.12.11.2.7 and 3.12.11.2.8 for other requirements regarding crack control in slabs.

6.3.3.1 Example of maximum bar spacings in slabs

Example 1: A 350 mm deep slab has been designed for an ultimate moment of 200 kNm/m using $f_{cu} = 30$ N/mm^2 and $f_y = 460$ N/mm^2. Apply 'Deemed to satisfy' rules to check the maximum bar spacings.
    Dimensions of slab: $b = 1000$ mm, $h = 350$ mm, cover = 30 mm, main steel 16 mm bars,

$$\text{effective depth} = 350 - 30 - 16/2 = 312 \text{ mm.}$$
$$k = 200 \times 10^6/ (1000 \times 312^2 \times 30) = 0.069 < 0.156$$
$$z/d = 0.5 + \sqrt{(0.25 - 0.069/0.9)} = 0.92 < 0.95$$
$$A_{s\ required} = 200 \times 10^6/ (0.92 \times 312 \times 0.95 \times 460) = 1595 \text{ mm}^2/\text{m}$$
$$A_{s\ Provided} = T16@125 \text{ mm} = 1609 \text{ mm}^2$$

(i) $f_y = 460$ N/mm^2, $h = 350$ mm > 200 mm,
$$100\ A_s/ (bd) = 100 \times 1609/ (1000 \times 312) = 0.52 > 0.3$$
$$1.0 > \{100\ A_s/ (bd) = 0.52\} > 0.3$$

Further checks on bar spacing are necessary.
Maximum spacings allowed
$$= \{155 \text{ mm from Table 3.28 of code}\}/(\% \text{ steel area})$$
$$= 155/ (0.52) = 298 \text{ mm}$$

Actual spacing of 125 mm is less than permitted maximum spacing. Design is satisfactory.

CHAPTER 7

SIMPLY SUPPORTED BEAMS

The aim in this chapter is to put together the design procedures developed in Chapters 4, 5 and 6 to make a complete design of a reinforced concrete beam. Beams carry lateral loads in roofs, floors etc. and resist the loading in bending, shear and bond. The design must comply with the ultimate and serviceability limit states.

7.1 SIMPLY SUPPORTED BEAMS

Simply supported beams do not occur as frequently as continuous beams in *in-situ* concrete construction, but are an important element in pre-cast concrete construction.

The effective span of a simply supported beam is defined in BS 8110: Part 1, clause 3.4.1.2. This should be taken as the smaller of
1. the distance between centres of bearings or
2. the clear distance between the faces supports plus the effective depth

7.1.1 Steps in Beam Design

Although the steps in beam design as shown in (a) to (i) below are presented in a sequential order, it is important to appreciate that design is an iterative process. Initial assumptions about size of the member, diameter of reinforcement bars are made and after calculations it might be necessary to revise the initial assumptions and start from the beginning. Experience built over some years helps to speed up the time taken to arrive at the final design. The two examples in this chapter do not show this iterative aspect of design.

(a) Preliminary size of beam
The size of beam required depends on the moment and shear that the beam carries. The maximum reinforcement provided must not exceed 4% of the cross sectional area of concrete and the minimum percentage must comply with the values in Table 3.25 of the code.

A general guide to the size of beam required may be obtained from the basic span/effective depth ratio from Table 3.9 of the code. The values are:

Rectangular sections: 20.0

Flanged sections with $b_w/b \le 0.3$, it is 16.0.

The following values are generally found to be suitable.

overall depth $\approx$ span/15

breadth $\approx$ (0.4 to 0.6) $\times$ depth

The breadth may have to be very much greater in some cases. The size is generally chosen from experience. Many design guides are available which assist in design.

(b) Estimation of loads:
The loads should include an allowance for self–weight which will be based on experience or calculated from the assumed dimensions for the beam. The original
The loads should include an allowance for self–weight which will be based on experience or calculated from the assumed dimensions for the beam. The original estimate may require checking after the final design is complete. The estimation of loads should also include the weight of screed, finish, partitions, ceiling and services if applicable.

The following values are often used:

$$screed: 1.8 \text{ kN/m}^2$$
$$ceiling \text{ and service load: } 0.5 \text{ kN/m}^2$$
$$demountable \text{ light weight partitions: } 1.0 \text{ kN/m}^2$$
$$block–work \text{ partitions: } 2.5 \text{ kN/m}^2$$

The imposed loading, depending on the type of occupancy, is taken from
BS 6399: Part 1.1996. *Code of practice for dead and imposed loads.*

(c) Analysis
The ultimate design loads are calculated using appropriate partial factors of safety from BS 8110: Part 1, Table 2.1. The load factors are

Dead load: Adverse = 1.4, Beneficial = 1.0
Imposed load: Adverse = 1.6, Beneficial = 0

The ultimate reactions, shears and moments are determined and the corresponding shear force and bending moment diagrams are drawn.

(d) Design of moment reinforcement
The flexural reinforcement is designed at the point of maximum moment. Refer to Chapter 4 for the steps involved.

(e) Curtailment and end anchorage
A sketch of the beam in elevation is made and the cut–off point for part of the tension reinforcement is determined. The end anchorage for bars continuing to the end of the beam is set out to comply with code requirements in clause 3.12.9 of the code.

(f) Design for shear
Design ultimate shear stresses are checked and shear reinforcement is designed using the procedures set out in BS 8110: Part 1, section 3.4.5. Refer to this and Chapter 5.

Note that except for minor beams such as lintels all beams must be provided with at least minimum links as shear reinforcement. Small diameter bars are required in the top of the beam to carry and anchor the links.

(g) Deflection

Deflection is checked using the rules from BS 8110: Part 1, section 3.4.6. Refer to Chapter 6.

(h) Cracking

The maximum clear distance between bars on the tension face is checked against the limits given in BS 8110: Part 1, clause 3.12.11 and Table 3.28. Refer to Chapter 6.

(i) Design sketch

Design sketches of the beam with elevation and sections are needed to show all information for the draughtsperson.

7.1.2 Curtailment and Anchorage of Bars

General and simplified rules for curtailment of bars in beams are set out in BS 8110: Part 1, section 3.12.9. The same section also sets out requirements for anchorage of bars at a simply supported end of a beam. These provisions are set out below.

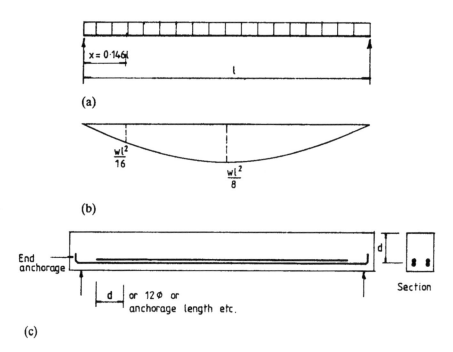

Fig.7.1 (a) Load; (b) bending moment diagram; (c) beam and moment reinforcement.

(a) General rules for curtailment of bars
Clause 3.12.9.1 of the code states that except at end supports every bar should extend beyond the point at which it is theoretically no longer required to resist moment by a distance equal to the greater of
    i. the effective depth of the beam
    ii. twelve times the bar size
In addition, where a bar is stopped off in the tension zone, one of the following conditions must be satisfied:

iii. The bar must extend an anchorage length past the theoretical cut–off point; or
iv. The bar must extend to the point where the shear capacity is twice the design shear force; or
v. The bars continuing past the actual cut–off point provide double the area to resist moment at that point.

These requirements are set out in Fig.7.1 for the case of a simply supported beam with uniform load. The section at the centre has four bars of equal area. The theoretical cut–off point or the point at which two of the bars are no longer required to resist the moment is found from the equation

$$\frac{1}{2}\frac{w\ell^2}{8} = \frac{w\ell x}{2} - \frac{wx^2}{2}$$

$$x^2 - \ell x + \frac{\ell^2}{8} = 0$$

$$x = 0.146\ell \text{ and } 0.854\ell$$

In a particular case calculations can be made to check that one only of the three conditions above is satisfied. *Extending a bar a full anchorage length beyond the point at which it is no longer required is the easiest way of complying with the requirements.*

(b) Anchorage of bars at a simply supported end of a beam
BS 8110: Part 1, clause 3.12.9.4, states that at the ends of simply supported beams the tension bars should have an anchorage equal to one of the following lengths:
i. Twelve bar diameters beyond the centre of the support; no hook or bend should begin before the centre of the support.
ii. Twelve bar diameters plus one–half the effective depth from the face of the support; no hook or bend should begin before $d/2$ from the face of the support.

(c) Simplified rules for curtailment of bars in beams
The simplified rules for curtailment of bars in simply supported beams and cantilevers are given in clause 3.12.10.2 and Figs. 3.24(b) and 3.24(c) of the code. As these code figures are rather difficult to interpret, the rules for simply supported beam and cantilever are shown much more clearly in Fig.7.2. The clause states that the beams are to be designed for predominantly uniformly distributed loads

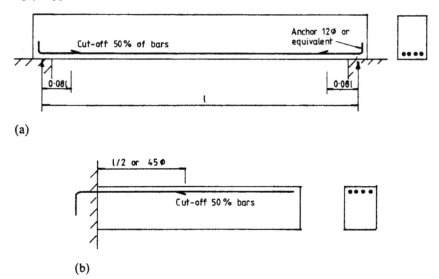

(a)

(b)

Fig.7.2 (a) Simply supported bean; (b) Cantilever

7.1.3 Example of Design of a Simply Supported L–Beam in a Footbridge

(a) Specification
The section through a simply supported reinforced concrete footbridge of 7 m span is shown in Fig.7.3(a). The characteristic imposed load is 5 kN/m² and the materials to be used are grade C30 concrete and grade 460 reinforcement. Design the L–beams that support the bridge. Concrete weighs 23.5 kN/m³, and the unit mass of the handrails is 16 kg/m per side.

(b) Loads, shear force and bending moment diagram
The total load is carried by two L–beams. All the load acting on 0.8 m width acts on an L–beam.
The dead load carried by each L–beam is
$$[(0.12 \text{ slab} + 0.03 \text{ screed}) \times 0.8 + 0.2 \times (0.4 - 0.12) \text{ rib}] \times 23.5$$
$$+ 16 \times 9.81 \text{x } 10^{-3} \text{ hand rails} = 4.3 \text{ kN/m}$$
The total live load acting on each beam is $0.8 \times 5 = 4.0$ kN/m
The design load at ultimate limit state is
$$(1.4 \times 4.3) + (1.6 \times 4) = 12.42 \text{ kN/m}$$
The ultimate moment at the centre of the beam is
$$12.42 \times 7^2/8 = 76.1 \text{ kN m}$$
The load, shear force and bending moment diagrams are shown in Figs 7.3(b), 7.3(c) and 7.3(d) respectively.

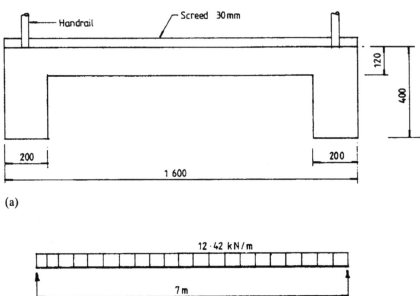

(a)

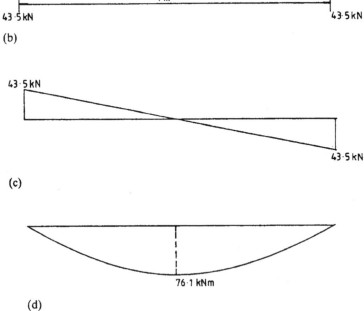

(b)

(c)

(d)

Fig 7.3 (a) Section through the footbridge; (b) design load; (c) shear force diagram; (d) bending moment diagram.

(c) Design of moment reinforcement
The effective width b of the flange of the L−beam is given by the lesser of
$$\text{(i) the actual width, 800 mm, or}$$
$$\text{(ii) } 200 + 8000/10 = 1000 \text{ mm}$$
$$b = 800 \text{ mm}$$
From BS 8110: Part 1, Table 3.3, the cover for moderate exposure is 35 mm. Assuming 25 mm bars, the effective depth d is estimated as
$$d = 400 - 35 \text{ (cover)} - 8 \text{ (link diameter)} - 25/2 = 344.5 \text{ mm, say } 340 \text{ mm}$$
The L−beam is shown in Fig.7.4(a).
Check for the depth of the stress block:
The moment of resistance of the section when the stress block is equal to the slab depth $h_f = 120$ mm is
$$M_{Flange} = 0.45 \; f_{cu} \times b \times h_f \times (d - h_f/2)$$
$$M_{Flange} = 0.45 \times 30 \times 800 \times 120 \times (340 - 0.5 \times 120) \times 10^{-6} = 362.9 \text{ kNm}$$
$$(M = 76.1) < (M_{Flange} = 362.9)$$
The stress block is inside the slab and the beam can be designed as a rectangular section.
$$k = M/ (bd^2 \; f_{cu}) = 76.1 \times 10^6/ (800 \times 340^2 \times 30) = 0.027$$
$$z/d = 0.5 + \sqrt{(0.25 - k/0.9)} = 0.5 + \sqrt{(0.25 - 0.027/0.9)} = 0.97 > 0.95$$
$$z = 0.95 \times 340 = 323 \text{ mm}$$
$$A_s = M/ (0.95 \; f_y \; z) = 76.1 \times 10^6/ (0.95 \times 460 \times 323) = 539 \text{ mm}^2$$
Provide 4T16, $A_s = 804$ mm^2.
From Fig.7.4(a), b = 800 mm, $b_w = 200$ mm, h = 400 mm, $b_w/b = 0.25$, 100 $A_s/(b_w \; h) = 1.005 > 0.18$ for minimum steel requirement.

Using the simplified rules for curtailment of bars (Fig.7.2(a)) two bars are cut off as shown in Fig.7.4(c) at 0.08 of the span from each end.

(d) Design of shear reinforcement
The enhancement of shear strength near the support using the simplified approach given in clause 3.4.5.10 is taken into account in the design for shear. The maximum shear stress at the support is,
$$V = V/ (b_v \; d) = 43.5 \times 10^3/ (200 \times 340) = 0.64 \text{ N/mm}^2$$

This is less than $0.8 \times \sqrt{30} = 4.38$ N/mm^2 or 5 N/mm^2.
The width of supports from Fig.7.4(b) is 200 mm. The shear at $d = 340$ mm from face of the support is
$$V = 43.5 - 12.42 \times (0.34 + 0.2/2) = 38.0 \text{ kN}$$
$$v = 38.0 \times 10^3/ (200 \times 340) = 0.56 \text{ N/mm}^2$$
The effective area of steel at d from the face of support is 2T16 of area 402 mm^2
$$100 \; A_s / (b_v \; d) = 100 \times 402/ (200 \times 340) = 0.59 < 3.0$$
$$400/d = 400/340 = 1.18 > 1.0$$
$$v_c = 0.79 \times (0.59)^{1/3}(1.18)^{1/4}(30/25)^{1/3}/1.25 = 0.59 \text{ N/mm}^2$$
$$v < v_c$$
Therefore only nominal links required. Provide 8 mm diameter two−leg vertical links, $A_{sv} = 100$ mm^2, in grade 250 reinforcement. The spacing required is determined using the formula

$$A_{sv} \geq 0.4 \, b_v \, s_v \, / \, (0.95 \, f_{yv})$$
$$100 \geq 0.4 \times 200 \times s_v / (0.95 \times 250), \, s_v \leq 297 \text{ mm}$$
$$0.75 \, d = 0.75 \times 340 = 255 \text{ mm}$$

8 mm links will be spaced at 250 mm throughout the beam. Two 12 mm diameter bars are provided to carry the links at the top of the beam. The shear reinforcement is shown in Figs 7.4(b) and 7.4(c).

(e) End anchorage

The anchorage of the bars at the supports must comply with BS 8110: Part 1, clause 3.12.9.4. The bars are to be anchored 12 bar diameters equal to 192 mm past the centre of the support. This will be provided by a 90° bend with an internal radius of three bar diameters equal to 48 mm. From clause 3.12.8.23, the anchorage length is the greater of

(i) 4 × internal radius = 4 × 48 = 192 mm but not greater than 12 × ϕ = 192 mm
(ii) the actual length of the bar $(4 \times \phi) + (\pi/2) \, (r + \phi /2) = 152$ mm
The anchorage needed is 192 mm.

(f) Deflection check

The deflection of the beam is checked using the rules given in BS 8110: Part 1. clause 3.4.6. Referring to Table 3.9 of the code,

$$b_w/b = 200/800 = 0 \, 25 < 0 \, 3$$

(i) The basic span−to−effective depth ratio is 16.

(ii) The modification factor for tension reinforcement

$$M/ (bd^2) = 76.1 \times 10^6/ (800 \times 340^2) = 0.82$$

The service stress f_s is

$$f_s = (2/3) \, f_y \, (A_{s.prov}/A_{s.Req}) = (2/3) \times 460 \times (539/804) = 206 \text{ N/mm}^2$$

Using the formula in Table 3.10 in the code is

$$0.55 + \frac{(477 - f_s)}{120(0.9 + \dfrac{M}{bd^2})} \leq 2.0$$

$$0.55 + \frac{(477 - 206)}{120(0.9 + 0.82)} = 1.86 < 2.0$$

(iii) For the modification factor for compression reinforcement

$$A_{s.prov}' = 2T12 = 226 \text{ mm}^2$$
$$100 \, A_{s.prov}' / (bd) = 100 \times 226/ (800 \times 340) = 0.083$$

The modification factor for compression steel from the formula in Table 3.11 is

$$1.0 + \frac{\dfrac{100 \times A_{s.prov}'}{bd}}{3 + \dfrac{100 \times A_{s.prov}'}{bd}}$$

$$= 1.0 + \frac{0.083}{(3.0 + 0.083)} = 1.03 < 1.5$$

(iv) Allowable span/d

$$\text{Span/d} = 16 \times 1.86 \times 1.03 = 30.65$$
$$\text{Actual span/d} = 7000/340 = 20.5$$

Hence the beam is very satisfactory with respect to deflection.

(g) Check for cracking

The clear distance between bars on the tension face is

$$200 - 2 \times 35 - 2 \times 8 - 2 \times 16 = 82 \text{ mm}$$

This does not exceed 155 mm as per Table 3.28 of the code.
The distance from the side or bottom to the nearest longitudinal bar is

$$35 + 8 + 16/2 = 51 \text{ mm}$$

The distance from the corner to the nearest longitudinal bar is

$$\sqrt{(51^2 + 51^2)} - 16/2 = 64 \text{ mm} < (155/2 = 77.5 \text{ mm})$$

The beam is satisfactory with regard to cracking.

(h) End bearing

No particular design is required in this case for the end bearing. With the arrangement shown in Fig.7.4(b) the average bearing stress is 1.09 N/mm². The ultimate bearing capacity of concrete is $0.35f_{cu} = 10.5$ N/mm².

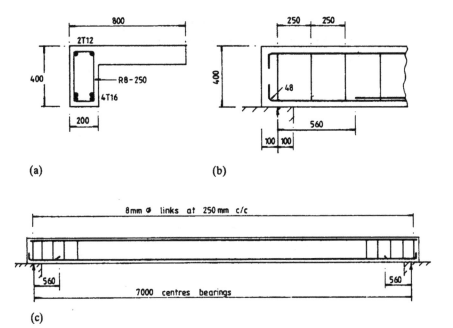

(a)   (b)

(c)

Fig.7.4 (a) Beam section; (b) beam support; (c) beam elevation

(i) Beam reinforcement

The reinforcement for each L−beam is shown in Fig.7.4. Note that the slab reinforcement also provides reinforcement across the flange of the L−beam.

7.1.4 Example of Design of Simply Supported Doubly Reinforced Rectangular Beam

(a) Specification

A rectangular beam is 300 mm wide by 520 mm overall depth with inset to the compression steel of 55 mm. The beam is simply supported and spans 8 m. The characteristic dead load including an allowance for self−weight is 20 kN/m and the characteristic imposed load is 11 kN/m. The materials to be used are grade 30 concrete and grade 460 reinforcement. Design the beam.

(b) Loads and shear force and bending moment diagrams

$$\text{design load} = (1.4 \times 20) + (1.6 \times 11) = 45.6 \text{ kN/m}$$
$$\text{ultimate moment} = 45.6 \times 8^2/8 = 364.8 \text{ kN m}$$

The loads and shear force and bending moment diagrams are shown in Fig.7.5.

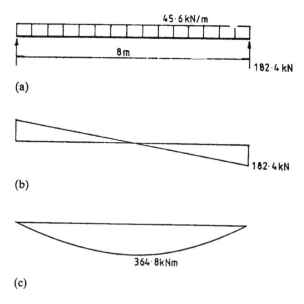

Fig.7.5 (a) Design loading; (b) ultimate shear force diagram; (c) ultimate bending moment diagram.

(c) Design of the moment reinforcement

Calculate the effective depth d:

Assuming 25 mm bars for reinforcement in two layers, 8 mm diameter for links and cover to the reinforcement is taken as 35 mm for moderate exposure, effective depth d is

$$d = 520 - 35 - 10 - 25 = 450.$$

The maximum moment of resistance of a singly reinforced rectangular beam is

$$0.156 \ b \ d^2 \ f_{cu} = 0.156 \times 300 \times 450^2 \times 30 \times 10^{-6} = 284.3 \text{ kNm} < 364.8 \text{ kNm}$$

Compression reinforcement is required.

$$d'/ x = 55/(0.5 \times 450) = 0.24 < 0.43$$

The compression steel yields. Stress in the compression steel is $0.95 \ f_y$.

The area of compression steel is

$$A'_s = (M - 0.156 \ bd^2 \ f_{cu}) / [(d - d') \times 0.95 \ f_y]$$
$$= (364.8 - 283.3)10^6 / [(450 - 55) \times 0.95 \times 460] = 472 \text{ mm}^2$$

Provide 2T20, $A'_s = 628 \text{ mm}^2$.

Equate total tensile and compressive forces:

$$0.45 \ f_{cu} \times b \times 0.45d + A'_s \times 0.95 \ f_y = A_s \times 0.95 \ f_y$$
$$(0.45 \times 30 \times 300 \times 0.45 \times 450) + (0.95 \times 460 \times 472) = A_s \times 0.95 \times 460$$
$$A_s = 2349 \text{ mm}^2$$

Provide 6T25, $A_s = 2945 \text{ mm}^2$. The reinforcement is shown in Fig.7.6(a). In accordance with the simplified rules for curtailment, 3T25 tension bars will be cut off at $(0.08 \ L = 0.08 \times 8000 = 640 \text{ mm})$ from each support. The compression bars will be carried through to the ends of the beam to anchor the links. The end section of the beam is shown in Fig.7.6(b) and the side elevation in 7.6(c).

(d) Design of shear reinforcement

The design for shear is made using the simplified approach in clause 3.4.5.10.

(i) Maximum shear stress at support:

$V = 182.4$ kN, $v = V/ (b_v \ d) = 182.4 \times 10^3 / (300 \times 462.5) = 1.31 \text{ N/mm}^2$
This is less than $0.8 \times \sqrt{30} = 4.38 \text{ N/mm}^2$ or 5 N/mm^2.
Section size is adequate.

(ii) v_c with $A_s = 3T25$:

The top layer bars are curtailed at 640 mm from centre of support. The calculation with 3T25 is valid up to $[640 + (d = 450 \text{ for 6T25})] = 1090$ mm from centres of supports.

$$d = 520 - 35 - 10 - 25/2 = 462.5 \text{ mm}$$
$$A_s \text{ for 3T25} = 1472 \text{ mm}^2.$$
$$100 \ A_s/ (b_v d) = 100 \times 1472/ (300 \times 462.5) = 1.06 < 3.0$$
$$400/d = 0.87 < 1.0. \text{ Therefore take as 1.0.}$$
$$v_c = 0.79 \times (1.06)^{1/3} (1.0)^{1/4} (30/25)^{1/3} /1.25 = 0.69 \text{ N/mm}^2$$

Shear force that can be supported with minimum links is

$$v - v_c = 0.4, \ v = 0.4 + 0.69 = 1.09 \text{ N/mm}^2$$
$$V = v \ b_v \ d = 1.09 \times 300 \times 462.5 \times 10^{-3} = 151.24 \text{ kN}$$

Using 10 mm two leg links, spacing required is

$$A_{sv} \geq [0.4 \ b_v \ s_v] / (0.95 \ f_{yv})$$
$$157 \geq 300 \times s_v \times 0.40 / (0.95 \times 250), \ s_v \leq 311 \ mm,$$

Because of the presence of compression reinforcement, maximum spacing is limited to twelve times the size of the smallest compression bar (clause 3.12.7.1 of code).

Maximum spacing is $12 \times 20 = 240$ mm $< (0.75 \ d = 0.75 \times 462.5 = 347$ mm).

Shear force that can be supported by 10 mm diameter two leg links spaced at 240 mm is

$$A_{sv} \geq [b_v \ s_v \ (v - v_c)] / (0.95 \ f_{yv})$$
$$157 \geq 300 \times 240 \times (v - 0.69) / (0.95 \times 250), \ v \leq 1.21 \ N/mm^2$$
$$V = 300 \times 462.5 \times 1.21 \times 10^{-3} = 167.9 \ kN$$

Taking the width of support as 250 mm, shear at $d = 462.5$ mm from the face of support is

$$V = 182.4 - 45.6 \times (462.5 + 250/2) \times 10^{-3} = 155.6 \ kN < 167.9 \ kN$$

(iii) v_c with $A_s = 6T25$:

Valid over the central distance $(8000 - 1090 - 1090) = 5820$ mm.
$$d = 450 \ mm, \ A_s \ for \ 6T25 = 2945 \ mm^2.$$
$$100 \ A_s / (b_v d) = 100 \times 2945 / (300 \times 450) = 2.18 < 3.0$$
$$400/d = 0.89 < 1.0. \ Therefore \ take \ as \ 1.0.$$
$$v_c = 0.79 \times (2.18)^{1/3} \ (1.0)^{1/4} \ (30/25)^{1/3} / 1.25 = 0.87 \ N/mm^2$$

Shear force that can be supported with minimum links is

$$v - v_c = 0.4, \ v = 0.4 + 0.87 = 1.27 \ N/mm^2$$
$$V = v \ b_v \ d = 1.27 \times 300 \times 450 \times 10^{-3} = 171.45 \ kN > 155.6 \ kN$$

(iv) Rationalization of link spacing

Since minimum spacing is required over the entire span, provide 34R10 links at 235 centres.

(e) End anchorage

As in Example in section 7.1.3, the tension bars are anchored 12 bar diameters past the centre of the support. The end anchorage is shown in Fig.7.6(c) where a 90° bend with an internal radius of three bar diameters is provided.

(f) Deflection check

The deflection of the beam is checked using the rules given in BS 8110: Part 1, clause 3.4.6.

(i) The basic span/d ratio

From Table 3.9 of the code the ratio is 20 for a simply supported rectangular beam.

(ii) Calculate the modification factor due to tension steel

$$M / (bd^2) = 364.8 \times 10^6 / (300 \times 450^2) = 6.0$$

The service stress f_s is

$$f_s = (2/3) \ f_y \ (A_{s.prov}/A_{s.Req}) = (2/3) \times 460 \times (2350/2945) = 248 \ N/mm^2$$

The modification factor for tension reinforcement using the formula in Table 3.11 in the code is

$$= 0.55 + \frac{(477 - f_s)}{120(0.9 + \frac{M}{bd^2})} \leq 2.0$$

$$= 0.55 + \frac{(477 - 248)}{120(0.9 + 6.0)} = 0.83 < 2.0$$

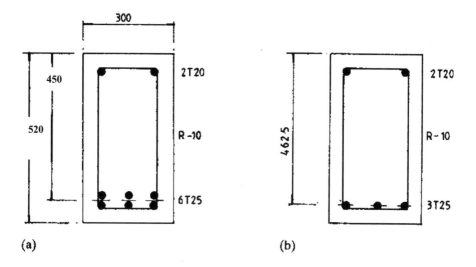

(a)

(b)

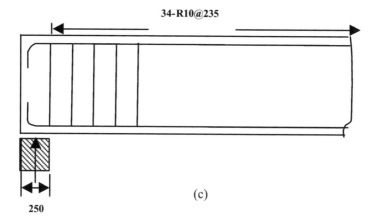

34-R10@235

(c)

250

Fig.7.6 (a) Section at mid–span; (b) section at support; (c) part side elevation.

(iii) For the modification factor for compression reinforcement
$$A'_{\text{s.prov}} = 2T20 = 628 \text{ mm}^2,$$
$$100\, A'_{\text{s.prov}} / (bd) = 100 \times 628/ (300 \times 450) = 0.465$$
The modification factor from the formula in Table 3.11 is

$$1.0 + \dfrac{\dfrac{100 \times A'_{s.prov}}{bd}}{3 + \dfrac{100 \times A'_{s.prov}}{bd}} = 1.0 + \dfrac{0.465}{(3.0 + 0.465)} = 1.13 < 1.5$$

(iv) Allowable span/depth ratio
$$\text{Allowable span/}d = 20 \times 0.83 \times 1.13 = 18.76$$
$$\text{Actual span/}d = 8000/450 = 17.78.$$
The beam is satisfactory with respect to deflection.

(g) Check for cracking
The clear distance between bars on the tension face is
$$(300 - 2 \times 35 - 2 \times 10 - 3 \times 25)/2 = 68 \text{ mm}$$
This does not exceed 155 mm as per Table 3.28 of the code.
The distance from the side or bottom to the nearest longitudinal bar is
$$35 + 10 + 25/2 = 58 \text{ mm}$$
The distance from the corner to the nearest longitudinal bar is
$$\sqrt{(58^2 + 58^2)} - 25/2 = 70 \text{ mm} < (155/2 = 77.5 \text{ mm})$$
The beam is satisfactory with regard to cracking.

(h) Beam reinforcement
The beam reinforcement is shown in Fig.7.6.

7.2 REFERENCES

Manual for the design of reinforced concrete building structures, (Institution of Structural Engineers, London, 2002)
Ray, S.S. 1995, *Reinforced Concrete*, (Blackwell Science).
Goodchild, C.H. 1997, *Economic concrete frame elements*, (Reinforced Concrete Council).

CHAPTER 8

REINFORCED CONCRETE SLABS

8.1 TYPES OF SLAB AND DESIGN METHODS

Slabs are plate elements forming floors and roofs in buildings, which normally carry uniformly, distributed loads acting normal to the plane of the slab. In general at a slab section, bending moments in two orthogonal directions and twisting moments are present. Slabs may be simply supported or continuous over one or more supports and are classified according to the method of support as follows:

1. spanning one way between beams or walls
2. spanning two ways between the support beams or walls
3. flat slabs carried on columns and edge beams or walls with no interior beams

Slabs may be solid of uniform thickness or ribbed with ribs running in one or two directions. Slabs with varying depth are generally not used. Stairs with various support conditions form a special case of sloping slabs.
Slabs may be analysed using the following methods.
1. Elastic analysis covers two techniques:
(a) idealization into strips or beams spanning one way or a grid with the strips spanning two ways
(b) elastic plate analysis using analytical methods for many cases of rectangular slabs and finite element analysis for irregularly shaped slabs or slabs with non-uniform loads.
2. Using design coefficients for moment and shear coefficients given in the code, which have been obtained from yield line analysis
3. The yield line and Hillerborg strip methods.

8.2 ONE-WAY SPANNING SOLID SLABS

8.2.1 Idealization for Design

(a) Uniformly loaded slabs
One-way slabs carrying predominantly uniform load are designed on the assumption that they consist of a series of rectangular beams 1 m wide spanning between supporting beams or walls. The sections through a simply supported slab and a continuous one-way slab are shown in Fig.8.1 (a) and Fig.8.1 (b) respectively.

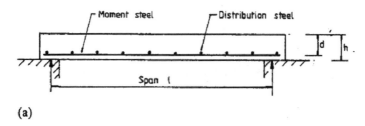

(a)

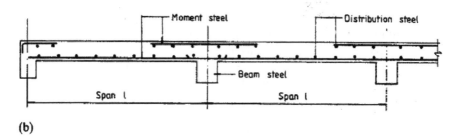

(b)

Fig.8.1 (a) Simply supported slab; (b) Continuous one-way slab

(b) Concentrated loads on a solid slab

BS 8110: Part 1 specifies in clause 3.5.2.2 that, for a slab simply supported on two edges carrying a concentrated load, the effective width of slab resisting the load may be taken as

$$w = \text{width of load} + 2.4 \times (1 - x/l)$$

where x is the distance of the load from the nearer support and l is the span of the slab.

If the load is near an unsupported edge the effective width should not exceed
1. w as defined above, or
2. $0.5w$ + distance of the centre of the load from the unsupported edge

The effective widths of slab supporting a load in the interior of the slab and near an unsupported edge are shown in Figs 8.2(a) and 8.2(b) respectively.
Refer to the code for provisions regarding other types of slabs.

8.2.2 Effective Span, Loading and Analysis

(a) Effective span

The effective span for one-way slabs is the same as that set out for beams in section 7.1. Refer to BS8110: Part 1, clauses 3.4.1.2 and 3.4.1.3. The effective spans are
i. Simply supported slabs:
    Effective span = The smaller of the centres of bearings or the clear span + d

ii. Continuous slabs:

Effective span = distance between centres of supports

(b) Arrangement of loads

The code states in clause 3.5.2.3 that in principle the slab should be designed to resist the most unfavourable arrangement of loads. However, normally it is only necessary to design for the single load case of maximum design load on all spans or panels. This is permitted subject to the following conditions:

1. The area of each bay exceeds 30 m². In this connection bay is defined as a strip across the full *width* of a structure bounded on the other two sides by a line of supports. In effect the area of a bay is equal to the product of the building width times column spacing normal to the width.

2. The ratio of characteristic imposed load to characteristic dead load does not exceed 1.25.

3. The characteristic imposed load does not exceed 5 kN/m² excluding partitions.

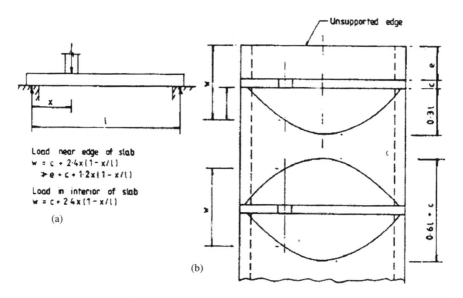

Load near edge of slab
$w = c + 2\cdot4 \times (1 - x/l)$
$\not> e + c + 1\cdot2 \times (1 - x/l)$

Load in interior of slab
$w = c + 2\cdot4 \times (1 - x/l)$

(a)

(b)

Fig 8.2 The effective widths of slab: (a). interior of slab; (b) near an unsupported edge.

(c) Analysis and redistribution of moments

Clause 3.5.2.3 referred to above states that if the analysis is carried out for the single load case of all spans loaded, the support moments except at supports of cantilevers should be reduced by 20%. This gives an increase in the span moments. The moment envelope should satisfy all the provisions of clause 3.2.2.1 in the code regarding redistribution of moments (See Chapter 13). No further redistribution is to be carried out.

The case of a slab with cantilever overhang is also discussed in clause 3.5.2.3 of the code. In this case if the cantilever length is of significant length compared to the span adjacent to the cantilever, the load case of $(1.4G_k + 1.6Q_k)$ on the cantilever and $1.0G_k$ on the span should be considered, where G_k is the characteristic dead load and Q_k is the characteristic imposed load.

(d) Analysis using moment coefficients
The code states in clause 3.5.2.4 that where the spans of the slab are approximately equal and conditions set out in clause 3.5.2.3 discussed above are met, the moments and shears for design may be taken from code Table 3.12. This table allows for 20% redistribution and is reproduced here as Table 8.1.

Table 8.1 Ultimate moments and shears in one-way spanning slabs

	At outer support	Near middle of end span	At first interior support	At middle of interim support	At interior supports
Moment	0 $(-0.04\,Fl^*)$	$0.086\,Fl$ $(0.075\,Fl^*)$	$-0.086\,Fl$	$0.063\,Fl$	$-0.063\,Fl$
Shear	0.4 F $(0.46\,F^*)$	-	0.6 F	-	0.5F

F = Total design load on span, l = span,
*refer to case where the end support is fixed.

8.2.3 Section Design and Slab Reinforcement Curtailment and Cover

(a) Cover
The amount of cover required for durability and fire protection is taken from Tables 3.3 and 3.4 of the code. For grade C30 concrete the cover is 25 mm for mild exposure and this will give 2 h of fire protection in a continuous slab.

(b) Main moment steel
The main moment steel spans between supports and over the interior supports of continuous slabs as shown in Fig. 8.1. The slab sections are designed as rectangular beam sections 1 m wide.
   The minimum area of main reinforcement is given in Table 3.25 of the code. For rectangular sections and solid slabs this is

Mild steel: f_y = 250 N/mm^2, 100A$_s$/A$_c$ = 0.24
High yield steel: f_y = 460 N/mm^2, 100A$_s$/A$_c$ = 0.13

where A$_s$ is the minimum area of reinforcement and A_c is the total area of concrete.

(c) Distribution steel
The distribution, transverse or secondary steel runs at right angles to the main moment steel and serves the purpose of tying the slab together and distributing non-uniform loads through the slab. The area of this secondary reinforcement is the same as the minimum area for main reinforcement set out in (a) above. Note

that distribution steel is required at the top parallel to the supports of continuous slabs. The main steel is placed nearest to the surface to give the greatest effective depth.

Table 8.2 Fabric types

Fabric reference	Longitudinal wire			Cross wire		
	Wire size (mm)	Pitch (mm)	Area (mm²/m)	Wire size (mm)	Pitch (mm)	Area (mm²/m)
Square mesh						
A393	10	200	393	10	200	393
A252	8	200	252	8	200	252
A193	7	200	193	7	200	193
A142	6	200	142	6	200	142
A98	5	200	98	5	200	98
Structural mesh						
B1131	12	100	1131	8	200	252
B785	10	100	785	8	200	252
B503	8	100	503	8	200	252
B385	7	100	385	7	200	193
B285	6	100	285	7	200	193
B196	5	100	196	7	200	193
Long mesh						
C785	10	100	785	6	400	70.8
C636	9	100	636	6	400	70.8
C503	8	100	503	5	400	49.1
C385	7	100	385	5	400	49.1
C283	6	100	283	5	400	49.1
Wrapping mesh						
D98	5	200	98	5	200	98
D49	2.5	100	49	2.5	100	49

(d) Slab reinforcement

Slab reinforcement is a mesh and may be formed from two sets of bars placed at right angles. Table 4.4 gives bar spacing data in the form of areas of steel per metre width for various bar diameters and spacings. Alternatively cross-welded wire fabric to BS 4483 can be used. This is produced from cold reduced steel wire with a characteristic strength of 460 N/mm². The particulars of fabric used are given in Table 8.2, taken from BS 4483: 1985

(e) Curtailment of bars in slabs

The general recommendations given in clause 3.12.9.1 for curtailment of bars apply. The code sets out simplified rules for slabs in clause 3.12.10.3 and Fig. 3.25 in the code. These rules may be used subject to the following provisions:

1. The slabs are designed for predominantly uniformly distributed loads;
2. In continuous slabs the design has been made for the single load case of maximum design load on all spans.

The simplified rules for simply supported, cantilever and continuous slabs are as shown in Figs 8.3(a), 8.3(b) and 8.3(c) respectively. It should be noted that while Figs 8.3(a) and 8.3(c) specify 40% of the midspan steel to be continued to supports, often it is much more convenient to continue 50% of the steel. This is accomplished by stopping off every alternate bar at midspan provided the maximum permitted spacing between bars and also minimum steel requirement are not violated.

The code states in clause 3.12.10.3.2 that while the supports of simply supported slabs or the end support of a continuous slab cast integral with an L-beam have been taken as simple supports for analysis, the end of the slab might not be permitted to rotate freely as assumed. Hence negative moments may arise and cause cracking. To control this, bars are to be provided in the top of the slab of area equal to 50% of the steel at mid-span but not less than the minimum area specified in Table 3.25 in the code. This is accomplished by turning up every alternate bar at midspan provided the maximum permitted spacing between bars and also minimum steel requirement are not violated. The bars are to extend not less than $0.15l$ or 45 bar diameters into the span. These requirements are shown in Figs 8.3(a) and 8.3(c).

Bottom bars at a simply supported end are generally anchored 12 bar diameters past the centreline of the support as shown for the right hand support in Fig.8.3 (a). However, these bars may be stopped at the line of the effective support where the slab is cast integral with the edge beam as shown in left hand support in Fig.8.3(a) (but see design for shear below).

Note that where a one-way slab ends in edge beams or is continuous across beams parallel to the span some two-way action with negative moments occurs at the top of the slab. Reinforcement in the top of the slab, of the same area as that provided in the direction of the span at the discontinuous edge should be provided to control cracking. This is shown in Fig.8.4(c).

8.2.4 Shear

Under normal loads shear stresses are not critical and shear reinforcement is not required. Shear reinforcement is provided in heavily loaded thick slabs but should not be used in slabs less than 200 mm thick. The shear resistance is checked in accordance with BS8110: Part 1, section 3.5.5.

The shear stress is given by

$$v = V/ (bd)$$

where V is the shear force due to ultimate loads. If v is less than the value of v_c given in Table 3.8 in the code no shear reinforcement is required. Enhancement in design shear strength close to supports can be taken into account. This was discussed in section 5.1.2. The form and area of shear reinforcement in solid slabs is set out in Table 3.16 in the code. The design is similar to that set out for beams in section 5.1.3.

The shear resistance at the end support, which is integral with the edge beam and where the slab has been taken as simply supported in the analysis, depends on the detailing.

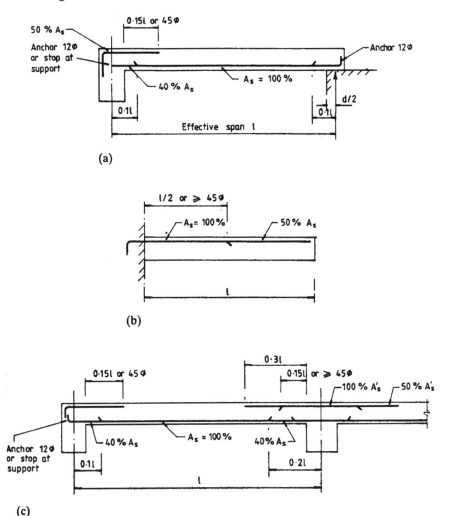

Fig.8.3 (a) Simply supported span. Left and right hand support details are two possible methods of detailing; (b) cantilever; (c) continuous slab.

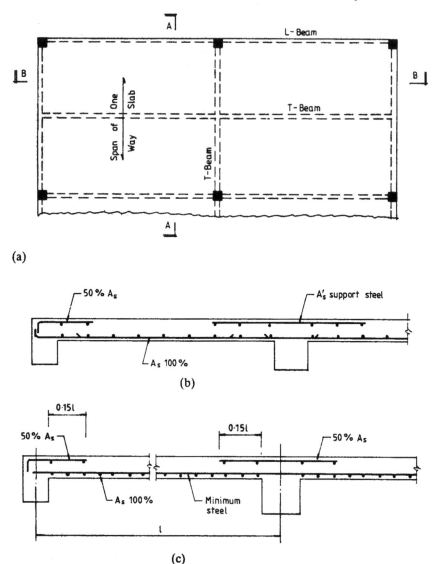

Fig.8.4 (a) Part floor plan; (b) section AA; (c) section BB.

The following procedures are specified in clauses 3.12.10.3.2 of the code.
1. If the bottom tension bars are anchored 12 diameters past the centre line of the support, the shear resistance is based on the steel percentage of bottom bars.

2. If the tension bars are stopped at the line of effective support, the shear resistance is based on the steel percentage top bars.

8.2.5 Deflection

The check for deflection is a very important consideration in slab design and usually controls the slab depth. The deflection of slabs is discussed in BS8110: Part 1, section 3.5.7. In normal cases a strip of slab 1 m wide is checked against span-to-effective depth ratios including the modification for tension reinforcement set out in section 3.4.6 of the code. Only the tension steel at the centre of the span is taken into account.

8.2.6 Crack Control

To control cracking in slabs, maximum values for clear spacing between bars are set out in BS8110: Part 1, clause 3.12.11.2.7. These were discussed in Chapter 6, section 6.2.3

8.3 EXAMPLE OF DESIGN OF CONTINUOUS ONE-WAY SLAB

(a) Specification
A continuous one-way slab has three equal spans of 3.5 m each. The slab depth is assumed to be 140 mm.
The loading is as follows:
Dead loads due to self-weight, screed, finish, partitions, ceiling: 5.2 kN/m^2
Imposed load: 3.0 kN/m^2
The construction materials are grade C30 concrete and grade 460 reinforcement. The condition of exposure is mild and the cover required is 25 mm. Design the slab and show the reinforcement on a sketch of the cross-section.

(b) Design loads

Consider a strip 1 m wide.
design ultimate load = $(1.4 \times 5.2) + (1.6 \times 3) = 12.08$ kN/m
design load F per span = $12.08 \times 3.5 = 42.28$ kN
The single load case of maximum design loads on all spans is shown in Fig.8.5 where the critical points for shear and moment are also indicated.

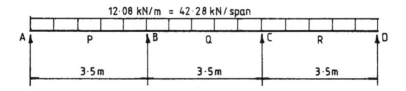

Fig 8.5 Continuous one-way slab.

(c) Shear forces and bending moments in the slab

The shear forces and moments in the slab are calculated using BS8110: Part 1 Table 3.12. The values are shown in Table 8.3. The redistribution is 20%.

Table 8.3 Design ultimate shears and moments

Position	Shear (kN/m)	Moment (kNm/m)
A	$0.4 \times 42.28 = 16.91$	
P		$+0.086 \times 42.28 \times 3.5 = 12.73$
B	$0.6 \times 42.28 = 25.37$	$-0.086 \times 42.28 \times 3.5 = -12.73$
Q		$+0.063 \times 42.28 \times 3.5 = 9.32$

(d) Design of moment steel

Assume 10 mm diameter bars with 25 mm cover. The effective depth is
$d = 140 - 25 - 10/2 = 110$ mm
The calculations for steel areas are set out below. Reference is made to clause 3.4.4.4 in the code for the section design.

(i) Section at support B

$$M = 12.73 \text{ kN m/m (hogging)}$$

When values from Table 3.12 are used, redistribution is 20% ($\beta_b = 0.8$),
$$K' = 0.402\,(\beta_b - 0.40) - 0.18\,(\beta_b - 0.40)^2$$
$$K' = 0.402\,(0.8 - 0.40) - 0.18\,(0.8 - 0.40)^2 = 0.132$$
$$k = M/\,(bd^2\,f_{cu}) = 12.73 \times 10^6 /\,(1000 \times 110^2 \times 30) = 0.035 < K'$$
Therefore no compression steel is required.
$$z/d = 0.5 + \sqrt{(0.25 - k/0.9)} = 0.5 + \sqrt{(0.25 - 0.035/0.9)} = 0.96 > 0.95$$
$$z = 0.95d = 105 \text{ mm}$$
$$A_s = M/\,(0.95\,f_y\,z) = 12.73 \times 10^6/\,(0.95 \times 460 \times 105) = 279 \text{ mm}^2/\text{m}$$
Provide in the top surface 8 mm bars at 175 mm centres to give an area of 288 mm^2/m.

(ii) Section in span at P

$$M = 12.73 \text{ kN m (sagging)}$$

Redistribution of support moment increases the span moment. Therefore take maximum k = 0.156.
$$k = M/\,(bd^2\,f_{cu}) = 12.73 \times 10^6 /\,(1000 \times 110^2 \times 30) = 0035 < 0.156$$
Therefore no compression steel is required.
$$z/d = 0.5 + \sqrt{(0.25 - k/0.9)} = 0.5 + \sqrt{(0.25 - 0.035/0.9)} = 0.96 > 0.95$$
$$z = 0.95d = 105 \text{ mm}$$
$$A_s = M/\,(0.95\,f_y\,z) = 12.73 \times 10^6/\,(0.95 \times 460 \times 104.5) = 279 \text{ mm}^2/\text{m}$$
Provide in the top surface 8 mm bars at 175 mm centres to give an area of 288 mm^2/m.

(iii) *Section* Q
Details as at P except that
$$M = 9.32 \text{ kNm/m, } z = 0.95d, A_s = 204 \text{ mm}^2/\text{m}$$
Provide 8 mm bars at 225 mm centres to give an area of 226 mm^2/m.

(iv) The minimum area of reinforcement
$$0.13\% \times A_c = (0.13/100) \times 1000 \times 140 = 182 \text{ mm}^2/\text{m} < 226 \text{ mm}^2/\text{m}.$$
The moment reinforcement is shown in Fig.8.6. Curtailment of bars has not been made because one-half of the calculated steel areas would fall below the minimum area of steel permitted.

At the end support A, top steel equal in area to 50% of the mid-span steel, i.e. 140 mm^2/m, but not less than the minimum area of 182 mm^2/m has to be provided. The clear spacing between bars is not to exceed $3d = 330$ mm. Provide 8 mm bars at 250 mm centres to give 201 mm^2/m. The tension bars in the bottom of the slab at support A are stopped off at the line of support.

(e) Distribution steel
The minimum area of reinforcement (182 mm^2/m) has to be provided. The spacing is not to exceed $3d = 330$ mm. Provide 8 mm bars at 250 mm centres to give an area of 201 mm^2/m.

(f) Shear resistance
Assume that enhancement in design strength close to the support has not been taken into account.

(i) End support
$$V = 16.91 \text{ kN.}$$
$$v = 16.91 \times 10^3/ (1000 \times 110) = 0.15 \text{ N/mm}^2$$
The shear resistance is based on top bars, T8-250 = 201 mm^2/m.
$$100 \ A_s/ (b_v d) = 100 \times 201/ (1000 \times 110) = 0.18 < 3.0$$
$$400/d = 400/110 = 3.64 > 1.0$$
$$v_c = 0.79 \times (0.18)^{1/3} (3.64)^{1/4} (30/25)^{1/3} /1.25 = 0.52 \text{ N/mm}^2$$
$$v < v_c$$
No shear reinforcement is required.

(ii) Interior Support
$$V = 25.37 \text{ kN}$$
$$v = 25.37 \times 10^3/ (1000 \times 110) = 0.23 \text{ N/mm}^2$$
The shear resistance is based on top bars, T8-175 = 288 mm^2/m.
$$100 \ A_s/ (b_v d) = 100 \times 288/ (1000 \times 110) = 0.26 < 3.0$$
$$400/d = 400/110 = 3.64 > 1.0$$
$$v_c = 0.79 \times (0.26)^{1/3} (3.64)^{1/4} (30/25)^{1/3} /1.25 = 0.59 \text{ N/mm}^2$$
$$v < v_c$$
No shear reinforcement is required.

(g) Deflection

The slab is checked for deflection using the rules from section 3.4.6 of the code. The end span is checked because it is continuous on one side only and will therefore deflect more than the interior span. The basic span-to-effective depth ratio is 26 for the continuous slab.

Modification factor for tension steel: Using the moment at P,

$$M = 12.73 \text{ kNm/m}$$
$$M/(bd^2) = 12.73 \times 10^6/(1000 \times 110^2) = 1.05$$
$$f_s = (2/3)\, 460 \times (279/288) = 297 \text{ N/mm}^2$$

The modification factor for tension steel is

$$0.55 + (477 - 297)/[120(0.9 + 1.05)] = 1.32$$
$$\text{allowable } span/d \text{ ratio} = 1.32 \times 26 = 34.3$$
$$\text{actual } span/d \text{ ratio} = 3500/110 = 31.8$$

The slab is satisfactory with respect to deflection.

Note that basic L/d ratio for continuous slab refers to case where both supports are continuous. In the case of only one support being continuous, it is perhaps sensible to use L/d = 23 based on average of values for simply supported slab (L/d = 20) and continuous slab (L/d = 26).

(h) Crack control

Because the steel grade is 460, the slab depth is less than 200 mm and the clear spacing does not exceed 3d = 330 mm, the slab is satisfactory with respect to cracking. Refer to BS8110: Part 1, clause 3.12.11.2.7.

(i) Sketch of cross-section of slab

A sketch of the cross-section of the slab with reinforcement is shown in Fig.8.6.

8.4 ONE-WAY SPANNING RIBBED SLABS

8.4.1 Design Considerations

When spans are long (perhaps over 5 m) but the live loads are relatively moderate or light, it is advantageous to reduce the dead weight of the slab. By having a series of ribs (beams) connected by structural topping as shown in Fig.8.7, the weight of the slab in between the ribs is considerably reduced.

Ribbed slabs may be constructed in a variety of ways as discussed in BS8110: Part 1, section 3.6. Two principal methods of construction are:

i. ribbed slabs without permanent blocks
ii. ribbed slabs with permanent hollow or solid blocks

These two types are shown in Fig.8.7. The topping or concrete floor panels between ribs may or may not be considered to contribute to the strength of the slab. The hollow or solid blocks may also be counted in assessing the strength using rules given in the code. The design of slabs with topping taken into account but without permanent blocks is discussed.

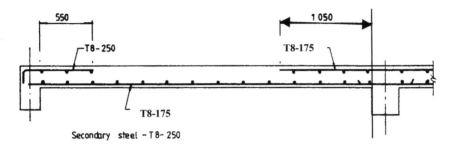

Fig.8.6 Reinforcement details in one-way continuous slab.

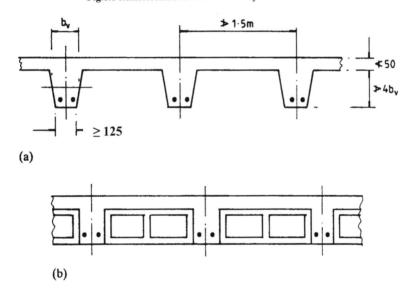

(a)

(b)

Fig.8.7 (a) Ribbed slab; (b) ribbed slab with hollow blocks.

8.4.2 Ribbed slab proportions

Proportions for ribbed slabs without permanent blocks are set out in section 3.6 of the code. The main requirements are as follows:

1. The centres of ribs should not exceed 1.5 m.
2. The depth of ribs *excluding topping* should not exceed four times their average width.

3. The minimum rib width should be determined by consideration of cover, bar spacing and fire resistance. Referring to Fig. 3.2 in the code, the minimum rib width is 125 mm.

4. The thickness of structural topping or flange should not be less than 50 mm or one-tenth of the clear distance between ribs (Table 3.17 in the code).

Note that, to meet a specified fire resistance period, non-combustible finish, e.g. screed on top or sprayed protection, can be included to give the minimum thickness for slabs set out in Fig. 3.2 in the code. See also Part 2, section 4.2, of the code. For example, a slab thickness of 125 mm is required to give a fire resistance period of 2 h. The requirements are shown in Fig.8.7 (a).

8.4.3 Design Procedure and Reinforcement

(a) Shear forces and moments

Shear forces and moments for continuous slabs can be obtained by analysis as set out for solid slabs in section 8.2 or by using Table 3.12 in the code.

(b) Design for moment and moment reinforcement

The mid-span section is designed as a T-beam with flange width equal to the distance between ribs. The support section is designed as a rectangular beam. The slab may be made solid near the support to increase shear resistance.

Moment reinforcement consisting of one or more bars is provided in the top and bottom of the ribs. If appropriate, bars can be curtailed in a similar way to bars in solid slabs.

(c) Shear resistance and shear reinforcement

The design shear stress is given by

$$v = V/ (b_v d)$$

where V is the ultimate shear force on a width of slab equal to the distance between ribs, b_v is the average width of a rib and d is the effective depth. In no case should the maximum shear stress v exceed $0.8\sqrt{f_{cu}}$ or 5 N/mm². No shear reinforcement is required when v is less than the value of v_c given in Table 3.8 of the code. Shear reinforcement is required when v exceeds v_c. Clause 3.6.6.3 states that if the rib contains two or more bars, links are recommended for constructional purposes, except in waffle slabs. The spacing of links can generally be of the order of 1 m to 1.5 m depending on the size of the main bars.

(d) Reinforcement in the topping

The code states in clause 3.6.6.2 that fabric with a cross-sectional area of not less than 0.12% of the area of the topping should be provided in each direction. The spacing of wires should not exceed one-half the centre-to-centre distance of the ribs. The mesh is placed in the centre of the topping and requirements for cover given in section 3.3.7 of the code should be satisfied. If the ribs are widely spaced the topping may need to be designed for moment and shear as a continuous one-way slab between ribs.

8.4.4 Deflection

The deflection can be checked using the span-to-effective depth rules given in section 3.4.6 of the code.

8.4.5 Example of One-way Ribbed Slab

(a) Specification
A ribbed slab is continuous over four equal spans each of 6 m. The characteristic dead loading including self-weight, finishes, partitions etc. is 4.7 kN/m² and the characteristic imposed load is 2.5 kN/m². The construction materials are grade C30 concrete and grade 460 reinforcement. Design the end span of the slab.

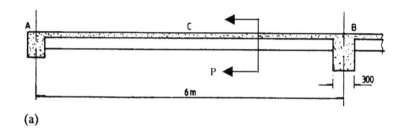

(a)

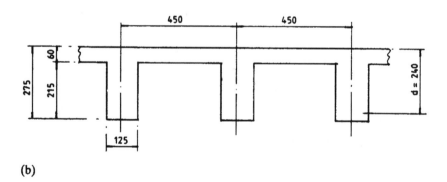

(b)

Fig.8.8 (a) Section through floor; (b) section PP through slab.

(b) Trial section
A cross-section through the floor and a trial section for the slab are shown in Fig.8.8. The thickness of topping is made 60 mm and the minimum width of a rib is 125 mm. The deflection check will show whether the depth selected is satisfactory. The cover for mild exposure is 25 mm. For 12 mm diameter bar effective depth d is

$$d = 275 - 25 - 12/2 = 244, \text{ say } 240 \text{ mm}$$

Note: Minimum thickness of topping = 50 mm or Clear distance between ribs/10 whichever is greater. In this case topping is 60 mm which is greater than 50 mm or $(450 – 125)/10 = 32$ mm.

(c) Shear and moments in the rib

Consider 0.45 m width of floor. The design load per span is
$$0.45[(1.4 \times 4.7) + (1.6 \times 2.5)] = 4.76 \text{ kN/m}$$
$$F = 4.76 \times 6 = 28.57 \text{ kN}$$
The design shears and moments taken from Table 3.12 in the code are as follows:
$$\text{shear at A} = 0.4 \times 28.57 = 11.43 \text{ kN}$$
$$\text{shear at B} = 0.6 \times 28.57 = 17.14 \text{ kN}$$
$$\text{moment at C} = + 0.086 \times 28.57 \times 6 = 14.74 \text{ kN m (sagging)}$$
$$\text{moment at B} = - 0.086 \times 28.57 \times 6 = - 14.74 \text{ kN m (hogging)}$$

(d) Design of moment reinforcement

(i) Mid-span T-section
The flange breadth is 450 mm, $h_f = 60$ mm, $d = 240$ mm
$$\text{Effective width} = 125 + (0.7 \times 6000)/5 = 965 > 450 \text{ mm}$$
$$b = 450 \text{ mm}$$
Check if the stress block is inside the flange.
$$M_{flange} = 0.45 \times 30 \times 450 \times 60 \times (240 – 60/2) \times 10^{-6} = 76.6 > 14.74 \text{ kNm}.$$
The neutral axis lies in the flange. The beam is designed as a rectangular beam.
$$k = 14.74 \times 10^6 / [30 \times 450 \times 240^2] = 0.019 < 0.156$$
$$z/d = 0.5 + \sqrt{(0.25 – 0.019/0.9)} = 0.98 > 0.95$$
$$z = 0.95 \text{ d} = 0.95 \times 240 = 228 \text{ mm}$$
$$A_s = 14.74 \times 10^6 / (0.95 \times 460 \times 228) = 149 \text{ mm}^2$$
Provide 2T10 at top of the rib. $A_s = 157$ mm^2.

(ii) Section at support
This is a rectangular section 125 mm wide, because the flange is in tension.
The redistribution is 20%, $\beta_b = 0.8$. From Clause 3.4.4.4 of the code
$$K^{'} = 0.402 \times (0.8 – 0.40) – 0.18 \times (0.8 – 0.40)^2 = 0.132$$
$$k = M/ (bd^2 f_{cu}) = 14.74 \times 10^6 / (125 \times 240^2 \times 30) = 0.068 < K^{'}$$
Therefore no compression steel is required.
$$z/d = 0.5 + \sqrt{(0.25 – k/0.9)} = 0.5 + \sqrt{(0.25 – 0.068/0.9)} = 0.92 < 0.95$$
$$z = 0.92d = 220 \text{ mm}$$
$$A_s = M/ (0.95 \, f_y \, z) = 14.74 \times 10^6 / (0.95 \times 460 \times 220) = 153 \text{ mm}^2$$
Provide 2T10 at bottom of the rib. $A_s = 157$ mm^2.

(e) Shear resistance
No account will be taken of enhancement to shear strength.
At support B:
$$V = 17.14 \text{ kN}.$$
$$v = 17.14 \times 10^3 / (125 \times 240) = 0.57 \text{ N/mm}^2$$
$$100 A_s / (b_v d) = 100 \times 157/(125 \times 240) = 0.52 < 3.0$$

$$400/d = 400/240 = 1.67 > 1.0$$
$$v_c = 0.79 \, (0.52)^{1/3} \, (1.67)^{1/4} \, (30/25)^{1/3}/1.25 = 0.58 \text{ N/mm}^2$$
$$v < v_c$$

Links not required. However for construction purposes provide 6 mm links in 250 grade steel at 1000 mm c/c.

At the simple support A, the bottom bars are to be anchored 12 bar diameters past the centre of the support.

(f) Deflection

$$b_w / b = 125/450 = 0.27 < 0.3$$

From Table 3.10 of the code, the basic span/d ratio is 20.8
Modification factor for tension steel:

$$M/ (bd^2) = 14.74 \times 10^6/ (450 \times 240^2) = 0.57$$
$$f_s = (2/3) \, 460 \, (153/157) = 299 \text{ N/mm}^2$$

The amount of redistribution at mid-span is not known, but the redistributed moment is greater than the elastic ultimate moment. Take β_b equal to 1.0. The modification factor is

$$0.55 + (477 - 299)/[120(0.9 + 0.57)] = 1.56 < 2.0.$$

Limit the modification factor to 2.0.

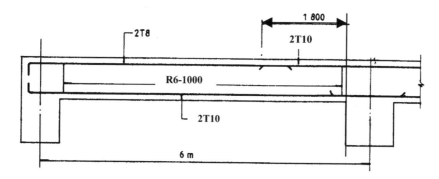

Fig.8.9 Reinforcement detail in the ribs of a ribbed slab.

allowable *span/d* ratio = 20.8 × 1.56 = 32.45
actual *span/d* ratio = 6000/240 = 25

The slab is satisfactory with respect to deflection.

(g) Arrangement of reinforcement in ribs

The arrangement of moment and shear reinforcement in the rib is shown in Fig.8.9.

(h) Reinforcement in topping

The area required per metre width is

$$(0.12/100) \times 60 \times 1000 = 72 \text{ mm}^2/\text{m}$$

The spacing of wires is not to be greater than one-half the centre-to-centre distance of the ribs, i.e. 225 mm. Refer to Table 8.2. Provide D98 wrapping mesh with an area of 98 mm^2/m and wire spacing 200 mm both ways in the centre of the topping.

8.5 TWO-WAY SPANNING SOLID SLABS

8.5.1 Slab Action, Analysis and Design

When floor slabs are supported on four sides, two-way spanning action occurs as shown in Fig.8.10 (a). In a square slab the action is equal in each direction. In long narrow slabs where the length is greater than twice the breadth the action is effectively one way. However, the end beams always carry some slab load.

Slabs may be classified according to the edge conditions. In the following the word continuous over supports also includes the case where the slab is built in at the supports. They can be defined as follows:

1. simply supported one-panel slabs where the corners can lift away from the supports
2. a one panel slab held down on four sides by integral edge beams (the stiffness of the edge beam affects the slab design)
3. slabs with all edges continuous over supports
4. slab with one, two or three edges continuous over supports. The discontinuous edge(s) may be simply supported or held down by integral edge beams

Elastic analysis of rectangular and circular slabs using analytical solutions for standard cases are given in textbooks on the theory of plates. Irregularly shaped slabs, slabs with openings or slabs carrying non-uniform or concentrated loads, slabs with edge beams can be analysed to give solutions based on finite element analysis.

Commonly occurring cases in slab construction in buildings are discussed. The design is based on shear and moment coefficients and the procedures and provisions set out in BS8110: Part 1, section 3.5.3. The slabs are square or rectangular in shape and support predominantly subjected to uniformly distributed load.

8.5.2 Rectangular Slabs Simply Supported on All Four Edges

The design of simply supported slabs that do not have adequate provision either to resist torsion at the corners or to prevent the cornets from lifting may be made in accordance with BS 8110: Part 1, clause 3.5.3.3. This clause gives the following equations for the maximum bending moments m_{sx} and m_{sy} at mid-span on strips of unit width for spans l_x and l_y respectively:

$$m_{sx} = \alpha_{sx}\, n\, l_x^2 \quad \text{(Code equation 10)}$$
$$m_{sy} = \alpha_{sy}\, n\, l_x^2 \quad \text{(Code equation 11)}$$

where

l_x is the length of the *shorter* span,

l_y is the length of the longer span,

$n = 1.4G_k + 1.6Q_k$ is the total ultimate load per unit area

α_{sx}, α_{sy} are moment coefficients from Table 3.14 in the code. .

G_k and Q_k respectively are unfactored dead and imposed loads.

The centre strips and locations of the maximum moments are shown in Fig.8.11 (a). A simple support for a slab on a steel beam is shown in 8.11(b). Alternatively the slab support might be a wall. The expressions for α_{sx} and α_{sy} are derived as follows.

If n is the load applied to the slab, as shown in Fig.8.11(a), let load n_x be carried by a simply supported strip in direction l_x and n_y be carried a simply supported strip in direction l_y. Therefore

$$n_x + n_y = n$$

For equal deflection at mid-span of the strips,

$$n_x\, l_x^{\,4} = n_y\, l_y^{\,4}$$

Solving for n_x and n_y,

$$n_x = n\, r^4 / (1+r^4)$$
$$n_y = n / (1+r^4)$$
$$r = (l_y / l_x), \quad l_y \geq l_x$$

The bending moments at the middle of the strips are

$$m_{sx} = n_x\, l_x^2/8 = \alpha_{sx}\, n\, l_x^2, \quad \alpha_{sx} = r^4/\{8(1+r^4)\} \quad \text{(Code equation 12)}$$
$$m_{sy} = n_y\, l_y^2/8 = \alpha_{sy}\, n\, l_x^2, \quad \alpha_{sy} = r^2/\{8(1+r^4)\} \quad \text{(Code equation 13)}$$

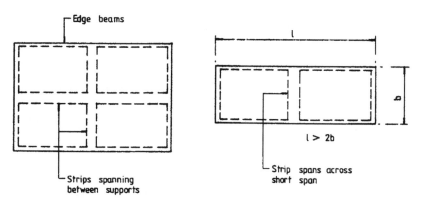

(a) Two-Way Action   (b) One-Way Action

Fig.8.10 One-way and two-way action in slabs

Some values of the coefficients from BS8110: Part 1, Table 3.13, are

$$l_y/l_x = 1.0;\ \alpha_{sx} = 0.062,\ \alpha_{sy} = 0.062$$
$$l_y/l_x = 1.5;\ \alpha_{sx} = 0.104,\ \alpha_{sy} = 0.046$$

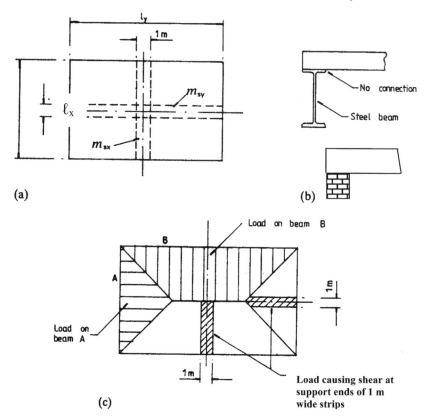

Fig.8.11 (a) Centre strips; (b) end supports; (c) loads on edge beam/wall supports and slab shears.

As the ratio of sides l_y/l_x increases, the shorter span supports an increasing share of the load.

The tension reinforcement can be designed using the formulae for rectangular beams in clause 3.4.4.4 of the code. The area must exceed the values for minimum reinforcement for solid slabs given in Table 3.25 of the code. The simplified rules for curtailment given in Fig.3.25 in the code and shown in Fig.8.3 here apply. These rules state that 40% of the mid-span reinforcement should extend to the support and be anchored 12 bar diameters past the centre of the support. The other 60% of bars are stopped off at 0.1 of the span from the support.

The generally assumed load distribution to the beams and the loads causing shear on strips 1 m wide are shown in Fig.8.11(c). The shear forces in both strips have the same value. The shear resistance is checked using formulae given in clause 3.5.5 and Table 3.16 of the code.

The deflection of solid slabs is discussed in BS 8110: Part 1, clause 3.5.7, where it is stated that in normal cases it is sufficient to check the span-to-effective depth ratio of the unit strip spanning in the shorter direction against the requirements given in section 3.4.6 of the code. The amount of steel in the direction of the shorter span is used in the calculation.

Crack control is dealt with in clause 3.5.8 of the code which states that the bar spacing rules given in clause 3.12.11 are to be applied.

8.5.3 Example of a Simply Supported Two-way Slab

(a) Specification

A slab in an office building measuring 5 m × 7.5 m is simply supported at the edges with no provision to resist torsion at the corners or to hold the corners down. The slab is assumed initially to be 200 mm thick. The total characteristic dead load including self-weight, screed, finishes, partitions, services etc. is 6.2 kN/m². The characteristic imposed load is 2.5 kN/m². Design the slab using grade C30 concrete and grade 250 reinforcement.

(b) Design of the moment reinforcement

Consider centre strips in each direction 1 m wide. The design load is

$$n = (1.4 \times 6.2) + (1.6 \times 2.5) = 12.68 \text{ kN/m}^2$$
$$l_y/l_x = 7.5/5 = 1.5$$

From BS8110: Part 1, Table 3.13 (or from equations 12 and 13 of the code), the moment coefficients are

$$\alpha_{sx} = 0.104, \ \alpha_{sy} = 0.046$$

For cover of 25 mm and 16 mm diameter bars the effective depths are as follows:

for short span bars in the bottom layer: $d_x = 200 - 25 - 8 = 167$ mm
for long span bars in the top layer: $d_y = 200 - 25 - 16 - 8 = 151$ mm

(i) *Short span*

$$m_{sx} = 0.104 \times 12.68 \times 5^2 = 32.97 \text{ kN m/m}$$
$$k = 32.97 \times 10^6 / (1000 \times 167^2 \times 30) = 0.039 < 0.156$$
$$z/d_x = 0.5 + \sqrt{(0.25 - 0.039/0.9)} = 0.954 > 0.95$$
$$z = 0.95 \times 167 = 159 \text{ mm}$$
$$A_s = 32.97 \times 10^6 / (0.95 \times 250 \times 159) = 873 \text{ mm}^2/\text{m}$$

Provide 16 mm diameter bars at 200 mm centres to give an area of 1005 mm²/m.

$$\text{Steel percentage} = 100 \times 1005/(1000 \times 200) = 0.5$$

Curtailing 50% of bars gives a steel percentage of 0.25 > 0.24 (minimum). Therefore provide 16 mm diameter bars at 200 mm centres but curtail alternate bars at 0.1 of span from the supports.

(ii) *Long span*

$$m_{sy} = 0.046 \times 12.68 \times 5^2 = 14.58 \text{ kN m/m}$$
$$k = 14.58 \times 10^6 / (1000 \times 151^2 \times 30) = 0.021 < 0.156$$
$$z/d_y = 0.5 + \sqrt{(0.25 - 0.021/0.9)} = 0.98 > 0.95$$
$$z = 0.95 \times 151 = 144 \text{ mm}$$
$$A_s = 14.58 \times 10^6 / (0.95 \times 250 \times 144) = 426 \text{ mm}^2/\text{m}$$

Provide 12 mm diameter bars at 225 mm centres to give an area of 503 mm²/m.

$$\text{Steel percentage} = 100 \times 502/ (1000 \times 200) = 0.25$$

Curtailing 50% of bars gives a steel percentage of 0.125 < 0.24 (minimum). Therefore no curtailment is possible. All bars must be anchored 12 diameters past the centre of the support.

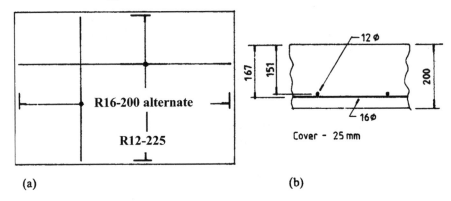

Fig 8.12 Slab steel (a) plan; (b) part section.

(c) Shear resistance
Check the shear stress on the long span. This will have the greatest value because d_y is less than on the short span. The design concrete shear stress will also be lower for the long span because the steel area is less than in the short span.
Referring to Fig.8.11(c) the maximum shear at the support is given by
$$V = 12.68 \times 2.5 = 31.7 \text{ kN}$$
$$v = 31.7 \times 10^3 / (1000 \times 151) = 0.21 \text{ N/mm}^2$$
$$100 \, A_s/ (b_v \, d) = 100 \times 503/(1000 \times 151) = 0.33 < 3.0$$
$$400/151 = 2.65 > 1.0$$
$$v_c = 0.79 \times (0.33)^{1/3} (2.65)^{1/4} (30/25)^{1/3}/1.25 = 0.59 \text{ N/mm}^2$$
$$v < v_c$$
No shear reinforcement is necessary and the slab is satisfactory with respect to shear.

(d) Deflection
The slab is checked for deflection across the short span as this carries the major part of the load.
From Table 3.9 in the code the basic span-to-effective ratio is 20.
Modification factor for tension steel:
$$m_{sx} / (bd^2) = 32.97 \times 10^6/ (1000 \times 167^2) = 1.18$$
$$f_s = (2/3) \, 250 \, (873/1005) = 145 \text{ N/mm}^2$$
The modification factor *for* tension steel is
$$0.55 + (477 - 145)/ [120 \times (0.9 + 1.18)] = 1.88 < 2.0$$
$$\text{allowable span/d ratio} = 20 \times 1.88 = 37.6$$

actual span/d ratio = 5000/167 = 29.9

The slab is very satisfactory with respect to deflections.

If high yield reinforcement is used, then

$$A_s = 32.97 \times 10^6/ (0.95 \times 460 \times 159) = 475 \text{ mm}^2/\text{m}$$

$$\text{Provide T12 at 200 mm c/c, } A_s = 565 \text{ mm}^2/\text{m}$$

$$f_s = (2/3) \times 460 \times (475/565) = 258 \text{ N/mm}^2$$

The modification factor *for* tension steel is

$$0.55 + (477 - 258)/ [120 \times (0.9 + 1.18)] = 1.43 < 2.0$$

$$\text{Allowable span/d ratio} = 20 \times 1.43 = 28.55 < 29.9$$

A thicker slab is needed to comply with the deflection limit.

(e) Cracking

Referring to BS8110: Part 1, clause 3.12.11.2.7, using average value of d in x and y-directions, d = (151 + 167)/2 = 160 mm. The clear spacing is not to exceed $3d = 480$ mm. The maximum actual spacing is only 225 mm. In addition the slab depth does not exceed 250 mm for grade 250 reinforcement. No further checks are required.

(f) Slab reinforcement

The slab reinforcement is shown in Fig.8.12.

8.6 RESTRAINED SOLID SLABS

8.6.1 Design and Arrangement of Reinforcement

The design method for restrained slabs is given in BS8110: Part 1, clause 3.5.3.4. In these slabs the corners are prevented from lifting and provision is made for resisting torsion near the corners. The maximum moments at mid-span on strips of unit width for spans l_x and l_y are given by

$$m_{sx} = \beta_{sx} \, n \, l_x^2 \quad \text{(Code equation 14)}$$

$$m_{sy} = \beta_{sy} \, n \, l_x^2 \quad \text{(Code equation 15)}$$

The clause states that these equations may be used for continuous slabs when the following provisions are satisfied:

1. The characteristic dead and imposed loads are approximately the same on adjacent panels as on the panel being considered;
2. The spans of adjacent panels in the direction perpendicular to the line of the common support are approximately the same as that of the panel considered in that direction.

The moment coefficients β_{sx} and β_{sy} in the equations above are given in BS8110: Part 1, Table 3.14. The coefficients have been derived using Yield Line analysis and will be discussed in section 8.9.16. Nine slab support arrangements are covered, the first four of which are shown in Fig.8.13. The moment coefficients are given both for support moment (hogging) and span moment (sagging) in both the short span and long span directions. The design rules for slabs are as follows.

1. The slabs are divided in each direction into middle and edge strips as shown in Fig.8.14.
2. The maximum moments defined above apply to the middle strips. The moment reinforcement is designed for 1m wide strips using formulae in Chapter 4. The amount of reinforcement provided must not be less than the minimum area given in BS8110: Part 1, Table 3.25. The bars are spaced at the calculated spacing uniformly across the middle strip.
3. The reinforcement is to be detailed in accordance with the simplified rules for curtailment of bars in slabs given in clause 3.12.10.3 and shown in Fig. 3.25 of the code. At the discontinuous edge, top steel of one-half the area of the bottom steel at mid-span is to be provided as specified in clause 3.12.10.3.2 to control cracking.
Provisions are given in the same clause regarding shear resistance at the end support. This depends on the detailing of the bottom reinforcement and was discussed in section 8.2.4 above.
4. The minimum tension reinforcement specified in Table 3.25 of the code is to be provided in the edge strips together with the torsion reinforcement specified in rule5 below. The edge strips occupy a width equal to total width/8 parallel to the supports as shown in Fig.3.9 of the code and here in Fig.8.14.

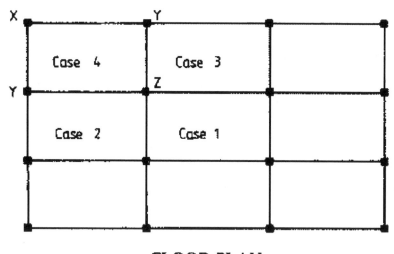

FLOOR PLAN

Fig.8.13 (a) Slab arrangement, floor plan: case 1: interior panel; case 2: one short edge discontinuous; case 3: one long edge discontinuous; case 4: two adjacent edges discontinuous.

5. Torsion reinforcement is to be provided at corners where the slab is simply supported on *both edges meeting at the corners*. Corners X and Y shown in Fig.8.13 require torsion reinforcement. This is to consist of a top and bottom mesh

with bars parallel to the sides of the slab and extending from the edges a distance of one-fifth of the shorter span. The area of bars in each of the four layers should be, at X, three-quarters of the area of bars required for the maximum mid-span moment and at Y, one-half of the area of the bars required at corner X. Note that no torsion reinforcement is required at the internal corners Z shown in Fig.8.13.

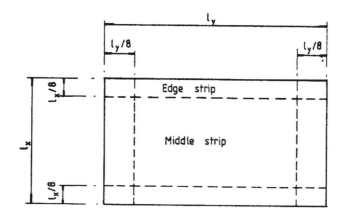

Fig.8.14 (b) Middle and edge strips.

8.6.2 Adjacent Panels with Markedly Different Support Moments

The moment coefficients in Table 3.14 of the code apply to slabs with similar spans and loads giving similar support moments. If the support moments for adjacent panels differ significantly, the adjustment procedure set out in clause 3.5.3.6 of the code must be used. This case is not discussed further.

8.6.3 Shear Forces and Shear Resistance

(a) Shear forces
Shear force coefficients β_{vx} and β_{vy} for various support cases for continuous slab strips are given in Table 3.15 of the code. The maximum shear force per unit width in the slab are given by

$$V_{sx} = \beta_{vx}\, n\, l_x \quad \text{(Code equation 19)}$$
$$V_{sy} = \beta_{vy}\, n\, l_x \quad \text{(Code equation 20)}$$

These are numerically the same as the design loads on supporting beams per unit length over the middle three quarters of the span as shown in Fig. 3.10 in the code which is also shown here in Fig.8.15.

The coefficients have been derived using Yield Line analysis and will be discussed in section 8.9.16.

(b) Shear resistance
The shear resistance for solid slabs is covered in BS 8110: Part 1, section 3.5.5, and the form and area of shear reinforcement are given in Table 3.16. Shear resistance was discussed in section 8.2.4 above.

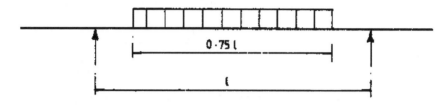

Fig.8.15 Distribution of load on support beam.

8.6.4 Deflection

Deflection is checked in accordance with section 3.4.6 of the code by comparing the actual span-to-effective depth ratio with the corresponding allowable ratio. Clause 3.5.7 states that the ratio is to be based on the shorter span and its amount of tension reinforcement in that direction.

8.6.5 Cracking

Crack control is discussed in BS 8110: Part 1, clause 3.5.8. This states that the bar spacing rules given in clause 3.12.11 are the best means of controlling flexural cracking in the slabs.

8.6.6 Example of Design of Two-way Restrained Solid Slab

(a) Specification
The part floor plan for an office building is shown in Fig.8.16. It consists of restrained slabs poured monolithically with the downstand edge beams. The slab is 175 mm thick and the loading is as follows:

$$\text{total characteristic dead load } G_k = 6.2 \text{ kN/m}^2$$
$$\text{characteristic imposed load } Q_k = 2.5 \text{ kN/m}^2$$

Design the comer slab using grade C35 concrete and grade 460 reinforcement. Show the reinforcement on sketches.

(b) Slab division, moments and reinforcement
The corner slab is divided into middle and edge strips as shown in Fig.8.16 (a). The moment coefficients are taken from BS 8110: Part 1, Table 3.14 for a square slab for the case with two adjacent edges discontinuous. The values of the coefficients and locations of moments are shown in Fig.8.16 (b).

$$\text{design ultimate load} = (1.4 \times 6.2) + (1.6 \times 2.5) = 12.68 \text{ kN/m}^2$$

Assuming 10 mm diameter bars and 20 mm cover from Table 3.3 in the code, the effective depth of the outer layer to be used in the design for moments in the short span direction is

$$d = 180 - 20 - 5 = 155 \text{ mm}$$

The effective depth of the inner layer to be used in the design for moments in the other span direction is

$$d = 180 - 20 - 5 - 10 = 145 \text{ mm}$$

The moments and steel areas for the middle strips are calculated. Because the slab is square only one direction need be considered.

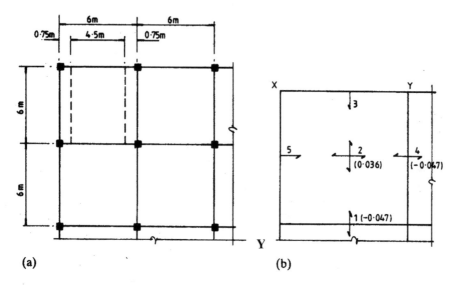

Fig.8.16 (a) Part floor plan; (b) symmetric moment coefficients.

(i) *Positions 1 and 4* (Over supports)
$$d = 155 \text{ mm}$$
$$m_{sx} = -0.047 \times 12.68 \times 6^2 = -21.45 \text{ kN m/m}$$
$$k = 21.45 \times 10^6 / (35 \times 1000 \times 155^2) = 0.026 < 0.156$$
$$z/d = [0.5 + \sqrt{(0.25 - 0.026/0.9)}] = 0.97 > 0.95$$
$$z = 0.95 \times 155 = 147 \text{ mm}$$
$$A_s = 21.45 \times 10^6 / (0.95 \times 460 \times 147) = 334 \text{ mm}^2/\text{m}$$
Provide 10 mm diameter bars at 200 mm centres to give an area of 392 mm²/m.

(ii) *Position 2* (mid-span)
Use the smaller value of *d*.
$$d = 145 \text{ mm}$$
$$m_{sx} = 0.036 \times 12.68 \times 6^2 = 16.43 \text{ kN m/m}$$
$$k = 16.43 \times 10^6 / (35 \times 1000 \times 145^2) = 0.022 < 0.156$$
$$z/d = [0.5 + \sqrt{(0.25 - 0.022/0.9)}] = 0.98 > 0.95$$

$$z = 0.95 \times 145 = 138 \text{ mm}$$
$$A_s = 16.43 \times 10^6 / (0.95 \times 460 \times 138) = 272 \text{ mm}^2/\text{m}$$

Provide 8 mm diameter bars at 160mm centres to give an area of 314 mm^2/m.

(iii) Minimum steel

Check the minimum area of steel in tension from BS8110: Part 1, Table 3.25:

$$0.13 \times 1000 \times 180/100 = 234 \text{ mm}^2/\text{m}$$

(iv) *Positions* 3 and 5 (Discontinuous edges)

According to clause 3.12.10.3, top steel one half of the area of steel at mid-span to be provided.
$$A_s = 0.5 \times 314 = 157 \text{ mm}^2/\text{m} < 234 \text{ mm}^2/\text{m (minimum steel)}$$
Provide 8 mm diameter bars at 200 mm centres to give an area of 251 mm^2/m.
In detailing, the moment steel will not be curtailed because both negative and positive steel would fall below the minimum area if 50% of the bars were stopped off.

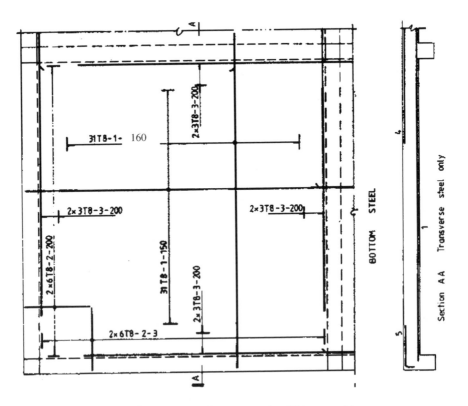

Fig.8.17 (a) Bottom steel arrangement in solid two-way slab.

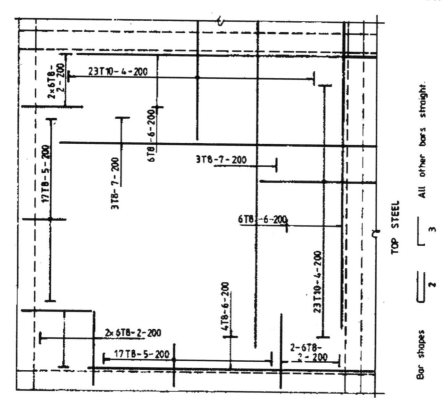

Fig.8.17 (b) Top steel arrangement in solid two-way slab.

(c) Shear forces and shear resistance

(i) *Positions* 1 and 4

$$d = 155 \text{ mm}$$
$$V_{sx} = 0.4 \times 12.68 \times 6 = 30.43 \text{ kN/m}$$
$$v = 30.43 \times 10^3 / (1000 \times 155) = 0.20 \text{ N/mm}^2$$
$$100 A_s/ (b_v d) = 100 \times 392/ (1000 \times 155) = 0.25 < 3.0$$
$$400 /d = 400/155 = 2.58 > 1.0$$
$$v_c = 0.79 \times (0.25)^{1/3}(2.58)^{1/4}(35/25)^{1/3}/1.25 = 0.57 \text{ N/mm}^2$$
$$v < v_c$$

No shear reinforcement is required.

(ii) *Positions* 3 and 5

$$d = 145 \text{ mm}$$

The bottom tension bars are to be stopped at the centre of the support. The shear resistance is based on the top steel with $A_s = 251 \text{ mm}^2/\text{m}$.

$$V_{sx} = 0.26 \times 12.68 \times 6 = 19.78 \text{ kN/m}$$
$$v = 19.78 \times 10^3 / (1000 \times 145) = 0.14 \text{ N/mm}^2$$

$$100 \, A_s/ \, (b_v d) = 100 \times 251/ \, (1000 \times 145) = 0.17 < 3.0$$
$$400 \, /d = 400/145 = 2.76 > 1.0$$
$$v_c = 0.79 \times (0.17)^{1/3}(2.76)^{1/4}(35/25)^{1/3}/1.25 = 0.51 \text{ N/mm}^2$$
$$v < v_c$$

No shear reinforcement is required.

(d) Torsion steel

Torsion steel of length equal to $1/5^{th}$ of shorter span = 6/5 = 1.2 m is to be provided in the top and bottom of the slab at the three external corners marked X and Y in Fig.8.16 (b).

(i) Corner X

The area of torsion steel is 0.75 × (Required steel at maximum *mid-span* moment)
$$A_s = 0.75 \times 272 = 204 \text{ mm}^2/\text{m}$$
This will be provided by the minimum steel of 8 mm diameter bars at 200 mm centres giving a steel area of 251 mm^2/m.

(ii) Corner Y

The area of torsion steel is one half of that at corner X.
$$A_s = 0.5 \times 204 = 102 \text{ mm}^2/\text{m}.$$
Again provide minimum 8 mm diameter bars at 200 mm centres giving a steel area of 251 mm^2/m.

(e) Edge strips

Provide minimum reinforcement, 8 mm diameter bars at 200 mm centres, in the edge strips both at top and bottom.

(f) Deflection

Check using steel at mid-span with d = 145 mm.
$$\text{basic } span \, d \text{ ratio} = 26 \text{ (BS8110: Part 1, Table 3.9)}$$

Modification factor for tension steel:
$$m_{sx} / (bd^2) = 16.43 \times 10^6 / (1000 \times 145^2) = 0.78$$
$$f_s = (2/3) \times 460 \times (272/314) = 266 \text{ N/mm}^2$$
The modification factor due to tension steel is
$$0.55 + (477 - 266)/ \, [120(0.9 + 0.78)] = 1.60 < 2.0$$

$$\text{allowable span/d ratio} = 1.60 \times 26 = 41.60$$
$$\text{actual span/d ratio} = 6000/145 = 41.37$$
The slab can be considered to be just satisfactory.

(g) Cracking

The bar spacing does not exceed $3d = 3 \times 145 = 435$ mm and in addition for grade 460 steel the depth is less than 200 mm. No further checks are required as stated in clause 3.12.11.2.7 of the code.

(h) Sketch of slab
The arrangements of reinforcement are shown in Fig.8.17 (a) and Fig.8.17 (b). The top and bottom bars are shown separately for clarity. The moment steel in the bottom of the slab is stopped at the supports at the outside edges and lapped with steel in the next bays at the continuous edges. Secondary steel is provided in the top of the slab at the continuous edges to tie in the moment steel.

8.7 WAFFLE SLABS

8.7.1 Design Procedure

Two-way spanning *ribbed* slabs are termed waffle slabs. The general provisions for construction and design procedure are given in BS 8110: Part 1, section 3.6. These conditions are set out in section 8.3 above dealing with one-way ribbed slabs. Moments for design may be taken from Table 3.13 of the code for slabs simply supported on four sides or from Table 3.14 for panels supported on four sides with provision for torsion at the corners. Slabs may be made solid near supports to increase moment and shear resistance and provide flanges for support beams. In edge slabs, solid areas are required to contain the torsion steel.

8.7.2 Example of Design of a Waffle Slab

(a) Specification
Design a waffle slab for an internal panel of a floor system that is constructed on an 8 m square module. The total characteristic dead load is 6.5 kN/m^2 and the characteristic imposed load is 2.5 kN/m^2. The materials for construction are grade C30 concrete and grade 460 reinforcement.

(b) Arrangement of slab
A plan of the slab arrangement is shown in Fig.8.18 (a). The slab is made solid for 500 mm from each support. The proposed section through the slab is shown in 8.18(b). The proportions chosen for rib width, rib depth, depth of topping and rib spacing meet various requirements set out in BS 8110: Part 1, section 3.6. The rib width is the minimum specified for fire resistance given in Fig. 3.2 of the code. From Table 3.3 the cover required for mild exposure is 25 mm.

(c) Reinforcement
$$\text{design ultimate load} = (1.4 \times 6.5) + (1.6 \times 2.5) = 13.1 \text{ kN/m}^2$$
The middle strip moments for an interior square panel are, from Table 3.14,
$$\text{Support } m_{sx} = -0.031 \times 13.1 \times 8^2 = -26.00 \text{ kNm/m}$$
$$\text{Mid-span } m_{sx} = 0.024 \times 13.1 \times 8^2 = 20.12 \text{ kNm/m}$$
Slab width supported by one rib = 500 mm

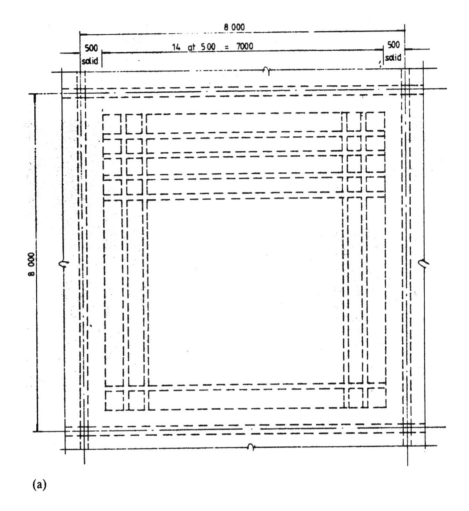

(a)

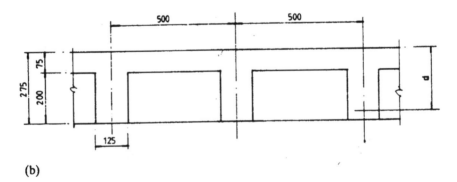

(b)

Fig 8.18 (a) Plan of waffle slab; (b) section through the slab

The moment per rib is therefore

$$\text{Support } m_{sx} = -26.0 \times 0.5 = -13.00 \text{ kNm}$$
$$\text{Mid-span } m_{sx} = 20.12 \times 0.5 = 10.06 \text{ kNm}$$

The effective depths assuming 12 mm diameter main bars and 6 mm diameter links are as follows:

$$\text{Outer layer } d = 275 - 25 - 6 - 12/2 = 238 \text{ mm}$$
$$\text{Inner layer } d = 275 - 25 - 6 - 12 - 12/2 = 226 \text{ mm}$$

(i) Support-solid section 500 mm wide

As the flooring dimensions and supports are symmetrical, it is convenient to have the steel arrangement also symmetrical. Section design is based on the smaller value of d equal to 226 mm.

$$k = 13.0 \times 10^6 / (500 \times 226^2 \times 30) = 0.017 < 0.156$$
$$z/d = [0.5 + \sqrt{(0.25 - 0.017/0.9)}] = 0.98 > 0.95$$
$$z = 0.95 \, d = 215 \text{ mm}$$
$$A_s = 13.0 \times 10^6 / (0.95 \times 460 \times 215) = 138 \text{ mm}^2$$

Provide two 10 mm diameter bars to give a steel area of 157 mm².

At the end of the solid section, the maximum moment of resistance of the concrete ribs with width 125 mm is given by

$$M = 0.156 \times 30 \times 125 \times 226^2 \times 10^{-6} = 29.88 \text{ kNm}$$

This exceeds the applied moment at the support and so the ribs are able to resist the applied moment without compression steel. The applied moment at 500 mm from the support will be somewhat less than the support moment.

(ii) Centre of span. T-beam, d = 226 mm

The flange breadth b is 500 mm and $h_f = 75$ mm.

$$M_{flange} = 0.45 \times 30 \times 500 \times 75(226 - 75/2) \times 10^{-6} = 95.4 \text{ kNm} > 10.06 \text{ kNm}$$

Hence the neutral axis lies in the flange and the beam is designed as a rectangular beam.

$$k = 10.06 \times 10^6 / (500 \times 226^2 \times 30) = 0.013$$
$$z/d = [0.5 + \sqrt{(0.25 - 0.013/0.9)}] = 0.99 > 0.95$$
$$z = 0.95 \, d = 215 \text{ mm}$$
$$A_s = 10.06 \times 10^6 / (0.95 \times 460 \times 215) = 107 \text{ mm}^2$$

Provide two 10 mm diameter bars with area 157 mm².

(d) Shear resistance

The shear force coefficient is taken from BS8110: Part 1, Table 3.15. The shear at the support (Table 3.15) is

$$V_{sx} = 0.33 \times 13.1 \times 8 = 34.58 \text{ kN/m}$$

The shear at the support for the width of 500 mm supported by one rib is

$$V_{sx} = 34.58 \times 0.5 = 17.29 \text{kN}$$

The shear on the ribs at 500 mm from support is

$$V = 17.29 - 0.5 \times 13.1 \times 0.5 = 14.02 \text{ kN}$$
$$v = 14.02 \times 10^3 / (125 \times 238) = 0.47 \text{ N/mm}^2$$
$$100 \, A_s / (b_v \, d) = 100 \times 157 / (125 \times 238) = 0.53 < 3.0$$
$$400/d = 400/238 = 1.68 > 1.0$$

$$v_c = 0.79 \times (0.53)^{1/3}(1.68)^{1/4}(30/25)^{1/3}/1.25 = 0.62 \text{ N/mm}^2$$
$$0.5\, v_c < v < v_c$$

Provide 6 mm diameter in grade 250 reinforcement nominal links and two bars in the rib to anchor the links. Area of two legs of 6 mm diameter links $A_{sv} = 57 \text{ mm}^2$. Calculate the spacing.

$$57 > 0.4 \times 125 \times s_v / (0.95 \times 250)$$
$$s_v = 271 \text{ mm}$$
$$0.75d = 0.75 \times 226 = 170 \text{ mm}$$

Space the links at 160 mm along the rib.

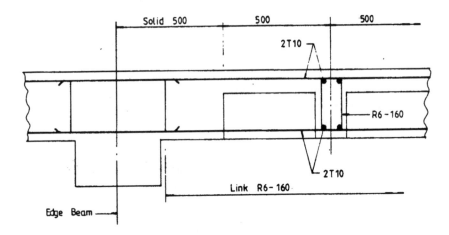

Fig 8.19 Reinforcement detail in the rib including shear reinforcement.

(e) Deflection

$$b_w/b = 125/500 = 0.25 < 0.3 \text{ (Table 3.9 of code)}$$

Basic span/depth ratio = 20.8 (Continuous over all supports)
Modification factor for tension steel:

$$M/(bd^2) = 10.06 \times 10^6/(500 \times 226^2) = 0.394$$
$$f_s = (2/3) \times 460 \times (107/157) = 209 \text{ N/mm}^2$$

The modification factor for tension steel is

$$0.55 + (477 - 209)/[120(0.9 + 0.394) = 2.28 > 2.0$$
$$\text{allowable (span/d)} = 20.8 \times 2 = 41.6$$
$$\text{actual span/d ratio} = 8000/226 = 35.4$$

The slab is satisfactory with respect to deflection.

(f) Reinforcement in topping

According to clause 3.6.5.2, for a topping 75 mm thick the area required per metre width is

$$0.12 \times 75 \times 1000/100 = 90 \text{ mm}^2/\text{m}$$

The spacing of the wires is not to be greater than one-half the centre-to-centre distance of the ribs, i.e. 250 mm. Refer to Table 8.2. Provide D98 wrapping mesh with an area 98 mm²/m and wire spacing of 200 mm in the centre of the topping.

(g) Arrangement of the reinforcement

The arrangement of the reinforcement and shear reinforcement in the rib is shown in Fig.8.19.

8.8 FLAT SLABS

8.8.1 Definition and Construction

The flat slab is defined in BS8110: Part 1, clause 1.3.2.1, as a slab with or without drops, supported generally without beams by columns with or without column heads. The code states that the slab may be solid or have recesses formed on the soffit to give a waffle slab. Here only solid slabs will be discussed.

Flat slab construction is shown in Fig.8.20 for a building with circular internal columns, square edge columns and drop panels. The slab is thicker than that required in T-beam floor slab construction but the omission of beams gives a smaller storey height for a given clear height and simplification in construction and formwork. Various column supports for the slab either without or with drop panels are shown in Fig.8.21. The effective column head is defined in the code.

8.8.2 General Code Provisions

The design of slabs is covered in BS 8110: Part 1, section 3.7. General requirements are given in clause 3.7.1, as follows.

1. The ratio of the longer to the shorter span should not exceed 2.

2. Design moments may be obtained by
   (a) equivalent frame method
   (b) simplified method
   (c) finite element analysis

3. The effective dimension l_h of the column head is taken as the lesser of
   (a) the actual dimension l_{ho} or
   (b) $l_{h\,max} = l_c + 2(d_h - 40)$

where l_c is the column dimension measured in the same direction as l_h. For a flared head l_{ho} is measured 40 mm below the slab or drop. Column head dimensions and the effective dimension for some cases are shown in Fig.8.22 (see also BS8110: Part 1, Fig. 3.11).

4. The effective diameter of a column or column head is as follows:
(a) For a column, the diameter of a circle whose area equals the area of the column
(b) for a column head, the area of the column head based on the effective dimensions defined in requirement 3 above.

The effective diameter of the column or column head must not be greater than one-quarter of the shorter span framing into the column.

5. Drop panels only influence the distribution of moments if the smaller dimension of the drop is at least equal to one-third of the smaller panel dimension. Smaller drops provide resistance to punching shear.

6. The panel thickness is generally controlled by deflection. The thickness should not be less than 125 mm.

8.8.3 Analysis

The code states that normally it is sufficient to consider only the single load case of maximum design load, (1.4 × dead load + 1.6 × imposed load) on all spans. The following two methods of analysis are set out in section 3.7.2 of the code to obtain the moments and shears for design.

(a) Frame analysis method
The structure is divided longitudinally and transversely into frames consisting of columns and strips of slab. Either the entire frame or sub-frames can be analysed by frame analysis programs. This method is not considered further.

Table 8.4 Moments and shear forces for flat slabs for internal panels

	At first interior support	*Middle of interior span*	*Interior supports*
Moment	−0.086F*l*	+0.063F*l*	−0.063F*l*
Shear	0.6F		0.5F

l = full panel length in the direction of the span; *F*, total design load on the strip of slab between adjacent columns due to 1.4 times the dead load plus 1.6 times the imposed load.

(b) Simplified method
In this method, for structures where lateral stability does not depend on slab-column connections, moments and shears are taken from Table 3.12 of the code for one-way spanning continuous slabs. The total moment across the full width of the panel is calculated and the proportion resisted by the column strip and middle strip are taken from Table 3.18 of the code. The design moments and shears for internal panels from Table 3.12 of the code are given in Table 8.4. Refer to the code for the complete table. Table 8.5 reproduces figures from Table 3.18 of the code.

Table 8.5 Distribution of moments in flat slabs

	Distribution between column and middle strip as percentage of total negative or positive moment	
	Column strip	Middle strip
Negative	75	25
Positive	55	45

The following provisions apply:

1. Design is based on the single load case mentioned above;
2. The structure has at least three rows of panels of approximately equal span in the direction considered.
3. Moments at supports from Table 3.12 may be reduced by 0.15 F h_c, where F = Total design load, h_c = Effective diameter of column head.

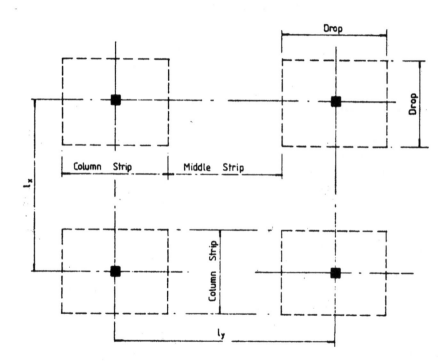

Fig.8.23 Division of a panel into column and middle strips.

8.8.4 Division of Panels and Moments

The code rules have been derived on the basis of extensive analytical studies of plate problems. For a full discussion see Reference 2 at the end of the chapter.

(a) Panel division

Flat slab panels are divided into column and middle strips as shown in Fig. 3.12 of the code. The division is shown in Fig.8.23 for a slab with drop panels.

(b) Moment division

The design moments obtained from Table 3.12 of the code are divided between column and middle strips in accordance with Table 3.18 of the code. The proportions are given in Table 8.5. Refer to the code for modifications to the table for the case where the middle strip is increased in width.

8.8.5 Design of Internal Panels and Reinforcement Details

The slab reinforcement is designed to resist moments derived from Tables 3.12 and 3.18 of the code. The code states in clause 3.7.3.1 for an internal panel that two-thirds of the amount of reinforcement required to resist negative moment in the column strip should be placed in a central zone of width one-half of the column strip.

Reinforcement can be detailed in accordance with the simplified rules given in clause 3.12.10.3.1 and Fig. 3.25 of the code (section 8.2.3(d) above).

8.8.6 Design of Edge Panels

Design of edge panels is not discussed. Reference should be made to the code for design requirements. The design is similar to that for an interior panel. The moments are given in Table 3.12 of the code. The column strip is much narrower than for an internal panel (Fig. 3.13 of the code). The slab must also be designed for large shear forces as shown in Fig. 3.15 of the code.

8.8.7 Shear Force and Shear Resistance

The code states in clause 3.7.6.1 that punching shear around the column is the critical consideration in flat slabs. Rules are given for calculating the ultimate design shear force and checking shear stresses.

(a) Shear forces

Equations are given in the code for calculating the design effective shear force V_{eff} at a shear perimeter in terms of the design shear V_t transferred to the column. The equations for V_{eff} include an allowance for moment transfer, i.e. the design moment transferred from the slab to the column.

The code states that in the absence of calculations it is satisfactory to take

$$V_{eff} = 1.15V_t$$

for internal columns in braced structures with approximately equal spans. To calculate V_t all panels adjacent to the column are loaded with the maximum design load.

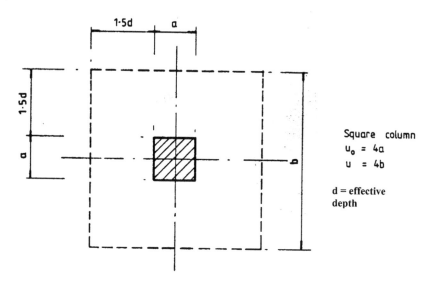

Fig.8.24 Punching shear perimeter

(b) Shear resistance
Guidance on shear due to concentrated loads on slabs is given in BS 8110: Part 1, section 3.7.7 (Refer to section 5.1.9, Chapter 5). The checks are as follows.

(i) Maximum shear stress at the face of the column
$$v_{max} = V/ (u_0\, d) \leq 0.8\sqrt{f_{cu}} \text{ or } 5 \text{ N/mm}^2$$
where u_o is the perimeter of the column (Fig.8.24) and V is the design ultimate value of the concentrated load promoting punching.

(ii) Shear stress on a failure zone 1.5d from the face of the column
$$v = V/ (u\, d)$$
where u is the perimeter of the failure zone 1.5*d* from the face of the column (Fig.8.24). If *v* is less than the design concrete shear stress given in Table 3.8 of the code, no shear reinforcement is required. If the failure zone mentioned above does not require shear reinforcement, no further checks are required. As conventional shear reinforcement in the form of links greatly complicates and slows down the steel fixing process, it is not desirable to have shear reinforcement in light or moderately loaded slabs. However in the last ten years some prefabricated proprietary shear reinforcement have become available which considerably simplify the provision of shear reinforcement. Another form of shear reinforcement used is Stud rails which consist of headed shear studs welded to a steel plate.

8.8.8 Deflection

The code states in clause 3.7.8 that for slabs with drops, if the width of drop at least equal to one-third of the span, the rules limiting span-to-effective depth ratios given in section 3.4.6 of the code can be applied directly. In other cases span-to-effective depth ratios are to be multiplied by 0.9. The check is to be carried out for the most critical direction, i.e. for the longest span.

8.8.9 Crack Control

The bar spacing rules for slabs given in clause 3.12.11.2.7 of the code apply.

8.8.10 Example of Design for an Internal Panel of a Flat Slab Floor

(a) Specification
The floor of a building constructed of flat slabs is 30 m × 24 m. The column centres are 6 m in both directions and the building is braced with shear walls. The panels are to have drops of 3 m × 3 m. The depth of the drops is 250 mm and the slab depth is 200 mm. The internal columns are 450 mm square and the column heads are 900 mm square. The depth of the column head is 600 mm.
The loading is as follows:

$$\text{Screed, floor finishes, partitions and ceiling} = 2.5 \text{ kN/m}^2$$
$$\text{Imposed load} = 3.5 \text{ kN/m}^2$$

The materials are grade C30 concrete and grade 250 reinforcement.
Design an internal panel next to an edge panel on two sides and show the reinforcement on a sketch.

(b) Slab and column details and design dimensions
A part floor plan and column head, drop and slab details are shown in Fig.8.25. The drop panels are made one-half of the panel dimension.
The column head dimension l_{h0}, 40 mm below the soffit of the drop panel, is 870 mm. The effective dimension l_h of the column head is the *lesser* of

$$l_{h0} = 450 + (900 - 450) \times \{(600 - 40)/600\} = 870 \text{ mm, and}$$
$$l_{hmax} = 450 + 2(600 - 40) = 1570 \text{ mm}$$

Therefore $l_h = 870$ mm.
The effective span l is 6000 mm
The effective diameter of the column head is

$$h_c = \sqrt{(4 \times 870^2/\pi)} = 982 \text{ mm} \le (\tfrac{1}{4}) \times 6000 = 1500 \text{ mm}$$
$$h_c = 982 \text{ mm}$$

The column and middle strips are shown in Fig.8.25(a).

(c) Design loads and moments
Taking unit weight of concrete as 23.6 kN/m³, the average load due to the weight of the slabs and drops is

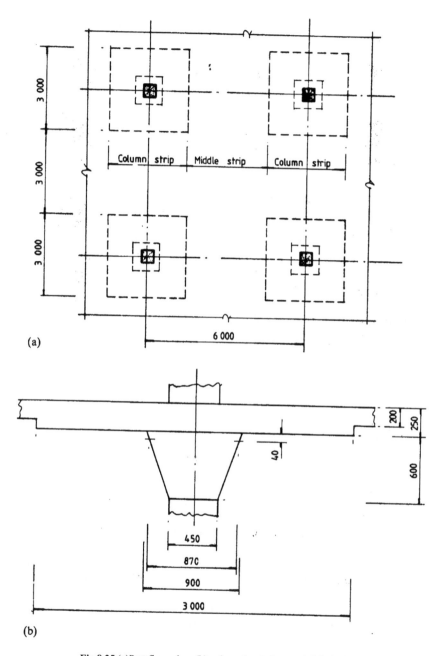

Fig 8.25 (a)Part floor plan; (b) column head, drop and slab details.

$[(3 \times 3 \times 0.25)\ (\text{Drops}) + (6 \times 6 - 3 \times 3) \times 0.2)] \times 23.6/6^2 = 5.02\ \text{kN/m}^2$

Note: The area of drops in a square panel 3 m × 3 m is 9 m²

The design ultimate load is
$$n = (5.02 + 2.5) \times 1.4 + (3.5 \times 1.6) = 16.13\ \text{kN/m}^2$$

The total design load on 1 m strip of slab is
$$F = 16.13 \times 6 = 96.78\ \text{kN/m}$$
$$Fl = 96.78 \times 6 = 580.68\ \text{kNm/m}$$

The moments in the flat slab are calculated using coefficients from Table 3.12 of the code and the distribution of the design moments in the panels of the flat slab is made in accordance with Table 3.18. The moments in the flat slab are as follows.

(i) Negative moment at first interior support
$$M = -086 \times 580.68 = -49.94\ \text{kNm/m}$$

This moment can be reduced by 0.15 Fh_c.
$$0.15\ F\ h_c = 0.15 \times 96.78 \times 0.982 = 14.26\ \text{kNm/m}.$$
$$M = -49.94 + 14.26 = -35.68\ \text{kNm/m}$$

Total negative moment across the whole panel width of 6 m
$$M = -35.68 \times 6 = -214.08\ \text{kNm}$$

(ii) Positive moment at centre of the interior span
$$M = +0.063 \times 580.68 = 36.58\ \text{kNm/m}$$

(iii) Total positive moment across the whole panel width of 6 m
$$M = 36.58 \times 6 = 219.48\ \text{kNm}$$

The distribution in the panels between the column and middle strips is as follows.

(iv) Column strip
$$\text{negative moment} = -0.75 \times 214.08 = -160.6\ \text{kN m}$$
$$\text{positive moment} = 0.55 \times 219.48 = 120.7\ \text{kN m}$$

(v) Middle strip
$$\text{negative moment} = -0.25 \times 214.6 = -53.5\ \text{kNm}$$
$$\text{positive moment} = 0.45 \times 219.5 = 98.8\ \text{kNm}$$

Note: There is 20% redistribution assumed in the above values.

(d) Design of moment reinforcement
The cover is 25 mm and 16 mm diameter bars in two layers are assumed. At the drop the effective depth for the inner layer is
$$d = 250 - 25 - 16 - 8 = 201\ \text{mm}$$

In the slab the effective depth of the inner layer is
$$d = 200 - 25 - 16 - 8 = 151\ \text{mm}$$

The design calculations for the reinforcement in the column and middle strip are made with width $b = 3000$ mm.

(i) Column strip negative moment reinforcement (Steel at top)
With 20% redistribution, $\beta_b = 0.8$.

$$K^{'} = 0.402 \, (\beta_b - 0.4) - 0.18 \, (\beta_b - 0.4)^2 = 0.132$$
$$k = 160.6 \times 10^6 / (3000 \times 201^2 \times 30) = 0.044 < 0.132$$
$$z/d = [0.5 + \sqrt{(0.25 - 0.044/0.9)}] = 0.95$$
$$z = 0.95 \, d = 191 \text{ mm}$$
$$A_s = 160.6 \times 10^6 / (0.95 \times 250 \times 191) = 3540 \text{ mm}^2$$

The total width of the column strip is 3000 mm.

Two thirds of 3540 mm^2 equal to 2360 mm^2 is placed in the centre half of the column strip of width of 1500 mm. Provide 12T16-150 mm c/c giving steel area of 2413 mm^2. The remaining one third of 3540 mm^2 viz. 1180 mm^2 are placed in each of the two outer 750 mm strips. The required steel area is half that in the central strip and is satisfied by providing 3T-16 at 250 mm spacing in *each* outer half giving a total steel area of 1206 mm^2. Although according to the simplified detailing rules in Fig. 3.25 of the code, all the bars need to continue only up to $0.15l = 900$ mm from the face of the column on either side of the column and then only 50% of that steel needs to continue a further 900 mm, it is convenient to continue all the bars to a distance of $0.3l = 1800$ mm on either side of the column face.

(ii) Column strip positive moment reinforcement (Steel at bottom)
Redistribution increases the span moment. Therefore $K^{'} = 0.156$.
$$k = 120.7 \times 10^6 / (3000 \times 151^2 \times 30) = 0.059 < 0.156$$
$$z/d = [0.5 + \sqrt{(0.25 - 0.06/0.9)}] = 0.93 < 0.95$$
$$z = 0.93 \, d = 140 \text{ mm}$$
$$A_s = 120.7 \times 10^6 / (0.95 \times 250 \times 140) = 3630 \text{ mm}^2$$

This is over a width of 3 m. Steel area required is 1210 mm^2/m, which is satisfied by providing 19T16-150 mm spacing giving a total steel area of 3819 mm^2. Only 40% of the steel need go over to the support. This can be achieved by curtailing alternate bars at $0.2l$ from the centre of the column.

(iii) Middle strip negative moment reinforcement (Steel at top)
With 20% redistribution, $\beta_b = 0.8$, $K^{'} = 0.132$
$$k = 53.5 \times 10^6 / (3000 \times 151^2 \times 30) = 0.026 < 0.132$$
$$z/d = [0.5 + \sqrt{(0.25 - 0.026/0.9)}] = 0.97 > 0.95$$
$$z = 0.95 \, d = 144 \text{ mm}$$
$$A_s = 53.5 \times 10^6 / (0.95 \times 250 \times 144) = 1564 \text{ mm}^2$$

This steel is placed uniformly over a width of 3 m. The steel area required is therefore 521 mm^2/m. Provide fourteen 12 mm bars at 200 mm c/c giving a steel area of 1582 mm^2. As in the case of the steel in the column strip, it is convenient to continue all the bars to a distance of $0.3l = 1800$ mm on either side of the column.

(iv) Middle strip positive moment reinforcement (Steel at bottom)
Redistribution increases the span moment. Therefore $K^{'} = 0.156$.
$$k = 98.8 \times 10^6 / (3000 \times 151^2 \times 30) = 0.048 < 0.156$$
$$z/d = [0.5 + \sqrt{(0.25 - 0.048/0.9)}] = 0.94 < 0.95$$
$$z = 0.94 \, d = 142 \text{ mm}$$

$$A_s = 98.8 \times 10^6 / (0.95 \times 250 \times 142) = 2930 \text{ mm}^2$$

This is over a width of 3 m. Steel area required is 977 mm²/m, which is satisfied by providing 15T16 at 200 mm spacing giving a total steel area of 3015 mm². Only 40% of the steel need go over to the support. This can be achieved by curtailing alternate bars at $0.2l$ from the centre of the column.

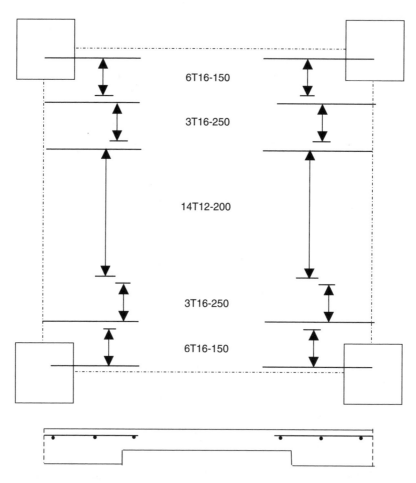

Fig.8.26 (a) Top steel in x-direction.

(e) Shear resistance

(i) At the column face 40 mm below the soffit

$$V_t = 1.15 \times 16.13 \times (6^2 - 0.87^2) = 653.7 \text{kN}$$
$$u_0 = 4 \times 870 = 3480 \text{ mm}$$
$$v_{max} = 653.7 \times 10^3 / (3480 \times 201) = 0.94 \text{ N/mm}^2 < (0.8\sqrt{30} = 4.38 \text{ N/mm}^2)$$

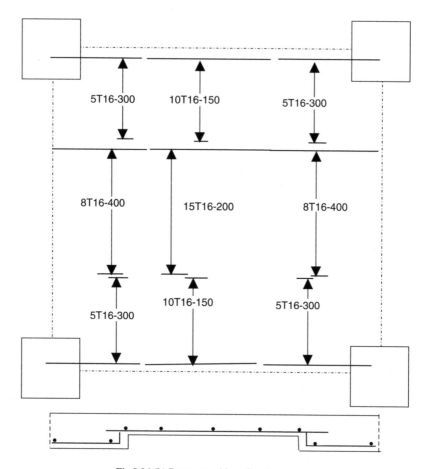

Fig.8.26 (b) Bottom steel in x-direction.

The maximum shear stress is satisfactory.
(ii) At 1.5 d from the column face
Side of shear perimeter
$$= 870 + 2 \times 1.5 \times d = 870 + 3 \times 201 = 1473 \text{ mm}$$
$$u = 4 \times 1473 = 5892 \text{ mm}$$
$$V_{eff} = 1.15 \times 16.13 \times (6^2 - 1.473^2) = 627.5 \text{ kN}$$
$$v = 627.5 \times 10^3 / (5892 \times 201) = 0.53 \text{ N/mm}^2$$
In the centre half of the column strip 16 mm diameter bars are spaced at 150 mm centres giving an area of 1340 mm^2/m.
$$100 A_s / (bd) = 100 \times 1340 / (1000 \times 201) = 0.67 < 3.0$$
$$400/d = 400/201 = 2.0 > 1.0$$
The design concrete shear stress is
$$v_c = 0.79 \times (0.67)^{1/3} (2.0)^{1/4} (30/25)^{1/3} / 1.25 = 0.70 \text{ N/mm}^2$$
$$v < v_c$$
The shear stress is satisfactory and no shear reinforcement is required.

(f) Deflection
The basic span/d ratio is 26 (Table 3.9 of the code).
Modification factor for tension steel:
$$M/(bd^2) = 219.5 \times 10^6/(6000 \times 151^2) = 1.61$$
The calculations are made for the middle strip using the total moment at mid-span
and the average of the column and middle strip tension steel.
$$\text{Required steel} = \text{Average of } (3630 + 2930) = 3280 \text{ mm}^2$$
$$\text{Provided steel} = \text{Average of } (3819 + 3015) = 3417 \text{ mm}^2$$
$$f_s = (2/3) \times 250 \times (3280/3417) = 160 \text{ N/mm}^2$$
The modification factor for tension steel is
$$0.55 + (477 - 160)/[120 \times (0.9 + 1.61)] = 1.60$$
$$\text{allowable span/d ratio} = 1.60 \times 26 = 41.7$$
$$\text{actual span/d ratio} = 6000/151 = 39.7$$
Hence the slab is satisfactory with respect to deflection.

(g) Cracking
The bar spacing does not exceed $3d$, i.e. 603 mm for the drop panel and 453 mm
for the slab. In accordance with BS8110: Part 1, clause 3.12.11.2.7, for grade 250
reinforcement the drop panel depth does not exceed 250 mm and so no further
checks are required.

(h) Arrangement of reinforcement
The arrangement of the reinforcement is shown in Fig.8.26(a) and Fig.8.26(b). For
clarity, only the steel in x-direction is shown separately for top and bottom steel.
The steel arrangement is identical in both the x and y-directions. Note that
although in the diagrams steel shown arrangement is shown confined to the
individual panel, in reality steel extends into adjacent panels.

8.9 YIELD LINE METHOD

8.9.1 Outline of Theory

The yield line method developed by Johansen is applicable to collapse by yielding
of under-reinforced concrete slab. It is based on the Upper bound theorem (also
known as the Kinematics theorem) of the classical Theory of Plasticity. According
to this theorem, for any assumed collapse mechanism, if the collapse load is
calculated by equating the energy dissipation at the plastic 'hinges' to the work
done by the external load, then the load so calculated is equal to or greater than the
true collapse load. The Yield line method applied to slabs is analogous of the
calculation of ultimate load of frames by the formation of plastic hinges in the
members of the frame. The collapse mechanism of a frame consists of a set
of rigid members connected at plastic hinges. The only difference between a frame

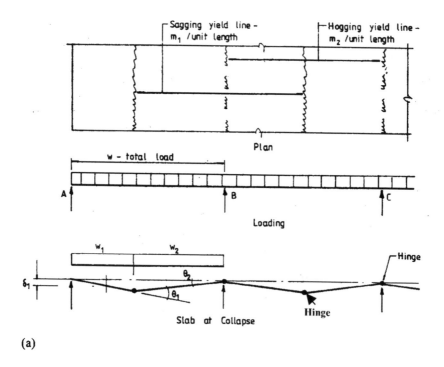

Plan

w - total load

A   B   C

Loading

Slab at Collapse

(a)

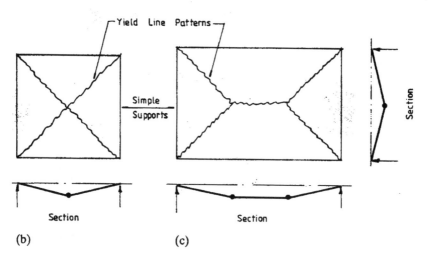

Yield Line Patterns

Simple
Supports

Section

Section   Section

(b)   (c)

Fig 8.27 (a) Continuous one-way slab; (b) square slab; (c) rectangular slab.

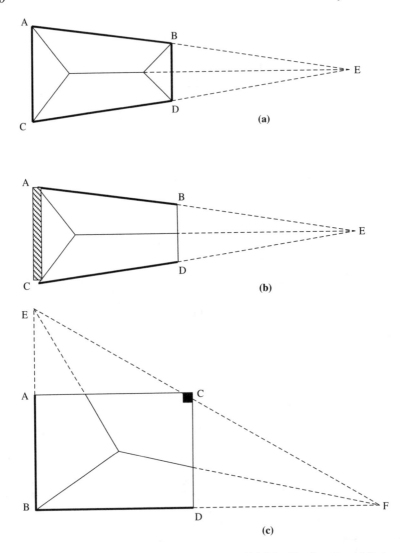

Fig.8.28 (a) Simply supported trapezoidal slab; (b) Trapezoidal slab with a free edge; (c) Rectangular slab with a column support.

of hinges, referred to as yield lines. All deformations are assumed to take place at the yield lines and the fractured slab at collapse consists of rigid portions held together by the yielded reinforcement at the yield lines. It is important to appreciate that the method assumes ductile behaviour at yield lines and does not consider the possibility of shear failure. Another important point to bear in mind is that because the method gives an upper bound solution to the true collapse load, various yield line patterns must be examined so as to determine which gives the minimum collapse load. Fig.8.27 shows some yield line patterns.

In the one-way continuous slab shown in Fig.8.27(a), straight yield lines form with a sagging yield line at the bottom of the slab near mid-span and hogging yield lines over the supports. The yield line patterns for a square and a rectangular simply supported two-way slab subjected to a uniform load are shown in Fig.8.27(b) and Fig.8.27(c) respectively. The deformed shape of the square slab is a inverted pyramid and that of the rectangular slab is an inverted roof shape.

8.9.1.1 Properties of yield lines

The following properties of the yield lines will be found useful in proposing possible collapse mechanisms.
(i) Yield lines are generally straight and they must end at a slab boundary.
(ii) A yield line between two rigid regions must pass through the intersection of the axes of rotation of the two rigid regions. Edge supports act as axes of rotation.
(iii) Axes of rotation lie along the line of supports. They can pass over a column at any angle.
   Fig 8.28 shows some yield line patterns, which illustrate the above properties.
(i) Fig.8.28 (a) shows a trapezoidal slab simply supported on all four edges. The yield line between the two trapezoidal rigid regions passes through E where the axes of rotations AB and CD meet. The yield line between the trapezoidal rigid region rotating about AB and the triangular region rotating about BD meets at B, the intersection point of the two axes of rotation.
(ii) Fig.8.28 (b) shows a trapezoidal slab simply supported on two opposite edges AB and CD, while edge AC is fixed against rotation while edge BD is free. The yield line between the two trapezoidal rigid regions passes through E where the axes of rotations AB and CD meet. The yield line ends at the free edge. The yield line between the trapezoidal rigid region rotating about AB and the triangular region rotating about AC meets at A, the intersection point of the two axes of rotation.
(iii) Fig.8.28(c) shows a rectangular slab simply supported on edges AB and BD and supported on a column at C. The axes of rotations are AB, BD and EF. The yield lines terminating at a free edge intersect the intersection of the two axes of rotations. The axis of rotation ECF passes over the column.

8.9.2 Johansen's Stepped Yield Criterion

As remarked earlier, the slab yields only at yield lines. Yielding is governed by Johansen's Stepped yield criterion which assumes that yielding takes place when the applied moment normal to the yield line is equal to the moment of resistance provided by the reinforcement crossing the yield line. It assumes that all reinforcement crossing a yield line yield and that the reinforcement bars stay in their original directions.
   As shown in Fig.8.29, let the two sets of reinforcement in the x and y directions respectively have ultimate moment of resistance such that for a yield line parallel

to the x–axis the normal moment of resistance is m_x and this resistance is provided by flexural steel in the *y-direction*. Similarly for a yield line parallel to the y–axis, the normal moment of resistance is m_y and this is provided by flexural steel in the *x-direction*.

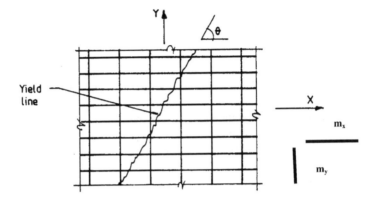

Fig.8.29 Yield line in an orthogonally reinforced slab.

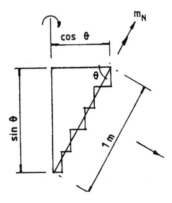

Fig 8.30 Resistant moments on an inclined yield line.

If a yield line forms at an angle θ to the x–axis, then as shown in Fig 8.30, the yield line can be imagined to be made up of a series of steps parallel to the reinforcement directions. For a unit length of yield line, the lengths of the horizontal and vertical steps are respectively $\cos\theta$ and $\sin\theta$. The moment of resistance on the horizontal step is $m_x \cos\theta$ and on the vertical step it is $m_y \sin\theta$. The components of these moment of resistance parallel to the yield line are $m_x \cos^2\theta$ and $m_y \sin^2\theta$ respectively. Thus the normal moment of resistance along the yield line is

$$m_n = m_x \cos^2\theta + m_y \sin^2\theta$$

Note that if $\theta = 0$, then the yield line is perpendicular to the reinforcement in the y-direction and hence $m_n = m_x$. Similarly if $\theta = 90°$, then the yield line is perpendicular to the reinforcement in the x-direction and hence $m_n = m_y$. If $m_x = m_y = m$, a case of isotropic reinforcement, then $m_n = m$ irrespective of the direction of the yield line.

8.9.3 Energy Dissipated in a Yield Line

Consider a slab ABCD, simply supported on the two adjacent edges AB and BC and free on the other two edges AD and CD as shown in Fig.8.31(a). Let a yield line BD form between the two rigid regions ABD and CBD as shown in Fig.8.31(b). Let the dimensions of the slab be as follows:
AF = 1.5, BF = 6.0, FD = 2.0, BG = 4.0, CG = 0.5
From geometry, the values for the following angles can be calculated.

$$\text{Angle ABF} = \tan^{-1}(1.5/6.0) = 14.04°,$$
$$\text{Angle FBD} = \tan^{-1}(2.0/6.0) = 18.44°$$
$$\text{Angle CBG} = \tan^{-1}(0.5/4.0) = 7.13°,$$
$$\text{Angle DBG} = 90° - \text{Angle FBD} = 71.56°$$
$$\text{The length L of the yield line BD} = \sqrt{(2^2 + 6^2)} = 6.325$$

The energy dissipated at a yield line is given by the equation
$$E = m_n L \, \theta_n$$
where m_n = normal moment on the yield line, L = length of the yield line, θ_n = rotation at the yield line.

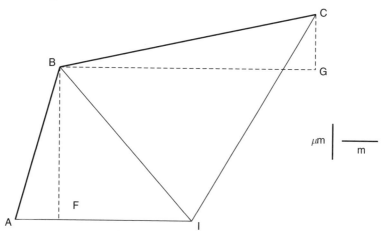

Fig.8.31 (a) Slab supported on two edges only.

If the moment of resistance due to steel in the x and y-directions respectively are μm and m, then the value of m_n on a yield line inclined at an angle ϕ to the x-axis is given by

$$m_n = m \cos^2 \phi + \mu m \sin^2 \phi$$

The energy dissipation at a yield line can be calculated by any of the three methods as follows.

Method 1: This is the most general and direct method but is not always the most convenient method to use. The inclination of the yield line to the horizontal is
$$\phi = \text{Angle FDB} = 90° - 18.44° = 71.56°.$$
The length L of the yield lines is L = 6.325 and the moment of resistance normal to the yield line is
$$m_n = m \cos^2\phi + \mu m \sin^2\phi = 0.1m + 0.9\mu m$$
In order to calculate θ_n, draw a line JK perpendicular to the yield line BD as shown in Fig.8.31(b). From geometry,
$$\text{Angle (JBD)} = 14.04° + 18.11° = 32.48°$$
$$\text{BD} = 6.325, \quad \text{JD} = \text{BD tan (JBD)} = 4.063$$
$$\text{Angle(DBK)} = 71.56° + 7.13° = 78.69°$$
$$\text{KD} = \text{BD tan (DBK)} = 31.625$$
If point D deflects vertically by Δ, then
$$\theta_n = \Delta/\text{JD} + \Delta/\text{KD} = 0.2777\ \Delta$$
Energy dissipated in the yield line is
$$m_n\ L\ \theta_n = (0.1m + 0.9\ \mu m)\ (6.325)(\ 0.2777\ \Delta)$$
$$= (1.5812\ \mu m + 0.1757\ m)\ \Delta$$

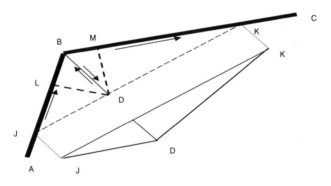

Fig.8.31(b) Deformation along the yield line BD.

Method 2: If the yield line is inclined at an angle ϕ to the x–axis, then the energy dissipated can be expressed as
$$m_n\ L\ \theta_n = (m_x \cos^2\phi + m_y \sin^2\phi)\ L\ \theta_n$$
$$m_n\ L\ \theta_n = m_x\ (L \cos\phi)(\theta_n \cos\phi) + m_y\ (L \sin\phi)(\ \theta_n \sin\phi)$$
$$L_x = L \cos\phi, \quad L_y = L \sin\phi,$$
$$\theta_x = \theta_n \cos\phi, \quad \theta_y = \theta_n \sin\phi$$
$$m_n\ L\ \theta_n = m_x\ L_x\ \theta_x + m_y\ L_y\ \theta_y$$
This formulation avoids having to calculate m_n and L and can be useful in some instances. Referring to Fig.8.31 (a),
$$L_x = \text{FD} = 2.0, L_y = \text{FB} = 6.00,$$
$$m_x = m, m_y = \mu m,$$

From Method 1,

$$\theta_n = 0.2777\ \Delta,$$
$$\phi = \text{Angle FDB} = 71.56°$$
$$\theta_x = \theta_n \cos \phi = 0.0879\ \Delta,$$
$$\theta_y = \theta_n \sin \phi = 0.2635\ \Delta$$
$$m_n\ L\ \theta_n = m\ (2.0)(0.0879\ \Delta) + \mu m(6.0)(0.2635\ \Delta)$$
$$= (1.5812\ \mu m + 0.1757\ m)\ \Delta$$

Method 3: This is the best approach if the axes of rotation of the rigid regions lie along the coordinate axes and the steel is orthogonal and the *reinforcement directions coincide with the coordinate axes*. The method is based on the fact that θ_n is the sum of the rotation of the two rigid regions. In Fig.8.31(b),

$$\theta_n = \text{Angle DJK} + \text{Angle DKJ}.$$
$$m_n\ L\ \theta_n = m_x\ L_x\ \theta_x + m_y\ L_y\ \theta_y$$
$$= (m_x\ L_x\ \theta_{x1} + m_y\ L_y\ \theta_{y1}) + (m_x\ L_x\ \theta_{x2} + m_y\ L_y\ \theta_{y2})$$

Where $(\theta_{x1}, \theta_{y1})$ and $(\theta_{x2}, \theta_{y2})$ refer respectively to the x and y components of the rotation at the yield line due to rigid regions ABD (i.e. Angle DJK) and CBD (i.e. Angle DKJ).

In order to use this method, it is important to use a consistent notation for the moment and rotation vectors. For the **rotation vector**, it is assumed that it is positive if the right hand's thumb points along the positive direction of the rotation vector, then the slab rotates in the **clockwise** direction. For the **moment vector**, it is assumed that it is positive if the right hand's thumb points along the positive direction of the moment vector, then the moment acts in the **anticlockwise** direction.

In the example considered, the rigid portion ABD rotates about the support AB in a clockwise direction. Therefore the rotation vector points in the direction from A to B. Similarly, the rigid portion DBC rotates about the support BC in a clockwise direction. Therefore the rotation vector points in the direction from B to C.

The normal moment on the yield line BD causes tension on the bottom side. Therefore in the rigid portion ABD, the moment vector points in the direction from D to B while in the rigid portion DBC the moment vector points in the opposite direction from B to D.

Rotation of the rigid region ABD about the axis AB is

$$\theta_1 = \Delta/DL, \text{ where DL is perpendicular to AB.}$$
$$DL = BD \sin ABD = 6.325 \sin (14.04 + 18.44 = 32.48) = 3.397$$
$$\theta_1 = \Delta/3.397 = 0.2944\ \Delta$$
$$\text{Angle FAB} = 90 - 14.04° = 75.96°$$
$$\theta_{x1} = \theta_1 \cos (FAB) = 0.2944\ \Delta \times 0.2425 = 0.0714\ \Delta$$
$$\theta_{y1} = \theta_1 \sin (FAB) = 0.2944\ \Delta \times 0.9701 = 0.2856\ \Delta$$
$$m_x = -m, \quad m_y = \mu\ m, \quad L_x = 2.0, \quad L_y = 6.0$$

Note that the sign of m_x is negative because the horizontal component of the moment vector points in a direction opposite to that of the corresponding rotation component.

Rotation of the rigid region DBC about the axis BC is

$$\theta_2 = \Delta/DM, \text{ where DM is perpendicular to AB}$$

$$DM = BD \sin DBM = 6.325 \sin (90 - 18.44 + 7.13 = 78.69) = 6.202$$
$$\theta_2 = \Delta/6.202 = 0.1612 \, \Delta$$
$$\theta_{x2} = \theta_2 \cos (GBC) = 0.1612 \, \Delta \cos(7.13) = 0.16 \, \Delta$$
$$\theta_{y2} = \theta_2 \sin (GBC) = 0.1612 \, \Delta \sin(7.13) = 0.02 \, \Delta$$
$$m_x = m, \quad m_y = -\mu \, m, \quad L_x = 2.0, \quad L_y = 6.0$$

Note that the sign of m_y is negative because the vertical component of the moment vector points in the direction opposite to that of the corresponding rotation component.

Energy E dissipated on the yield line is
$$E = - \, m(2.0)(\, 0.0714 \, \Delta) + \mu \, m \, (6.0)(\, 0.2856 \, \Delta)$$
$$+ \, m(2.0)(\, 0.16 \, \Delta) - \mu \, m \, (6.0)(\, 0.02\Delta)$$
$$E = (1.5936 \, \mu \, m + 0.1772 \, m) \, \Delta$$

Although Method 3 appears to be more complicated than Method 1, in most cases of rectangular slabs where the axes of rotation coincide with the coordinate axes, Method 3 will be found to be the ideal method to use.

8.9.4 Work Done by External Loads

If a rigid region carries a uniformly distributed load q and rotates by θ about an axis AB as shown in Fig.8.32, then the work done by q is given by
$$\text{Work done} = \int q \, dA \, r \, \theta$$
where dA = an element of area, r = perpendicular distance to the element of area from the axis of rotation.
Since q and θ are constant, work done $= q \, \theta \int r \, dA$
But $\int r \, dA$ = first moment of area about the axis of rotation.
$$W = q \, \theta \, \{\text{First moment of area about the axis of rotation}\}$$
$$= q \, \theta \times \text{Area} \times \text{Distance to the centroid of area from the axis of rotation}$$
$$= q \times \text{Area} \times \text{Deflection at the centroid}$$

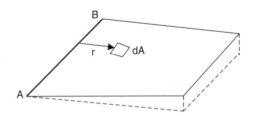

Fig.8.32 External work done by loads on a slab.

8.9.5 Example of a Continuous One-way Slab

Consider a strip of slab 1 m wide where the mid-span positive reinforcement has a moment of resistance of *m* per metre and the support negative reinforcement has a moment of resistance of m' per metre. The slab with ultimate load W per span is shown in Fig.8.33(a).

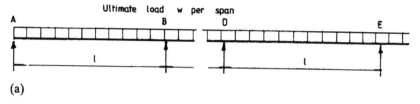

(a)

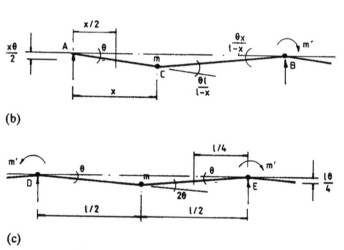

(b)

(c)

Fig.8.33 (a) Continuous one-way slab; (b) end span; (c) internal span.

(a) End span AB

The yield line in the span forms at point C at x from A. The rotation at A is θ. The deflection Δ at the hinge in the span is θx. If the rotation at the hinge over the support B is ϕ, then

$$\phi(l-x) = \Delta = \theta x$$
$$\phi = \theta x / (l-x)$$

The net rotation ψ at the hinge C is

$$\psi = (\theta + \phi) = \theta l / (l-x)$$

The rotations at A, C and B are shown in Fig.8.33(b).
The work done by the loads is $W(x\,\theta)/2$.
The energy dissipated E in the yield lines is

$$E = (m\,\psi + m'\,\phi)$$
$$E = m\,\theta l/(l-x) + m'\,\theta x/(l-x)$$
$$E = \theta l/(l-x)\,\{m\,l + m'\,x\}$$

Equating the work done by the loads to energy dissipated at the hinges,

$$W(x\,\theta)/2 = \theta l/(l-x)\,\{m\,l + m'\,x\}$$
$$W = 2\,\{m\,l + m'\,x\}\,/[x\,(l-x)]$$

The position x of the yield line in the span is determined so that the collapse load is minimum. Differentiating W with respect to x,

$$dW/dx = x(l-x)\,m' - (m\,l + m'\,x)\,(l-2x) = 0$$
$$m'\,x^2 - m\,l\,(l-2x) = 0$$

This equation can be solved for x for a given value of the ratio m'/m. Clause 3.5.2.1 of the code states that values of the ratio between support and span moments should be similar to those obtained by elastic theory. Values of m'/m should normally lie in the range 1.0 to 1.5. This limitation ensures that excessive cracking does not occur over the support B.

For the special case where $m = m'$, the equation $dW/dx = 0$ reduces to

$$x^2 + 2\,l\,x - l^2 = 0, \quad x = 0.414\,l$$

Substitute in the work equation x = 0.414 l to obtain the value of m:

$$m = m' = 0.086\,Wl$$

Note that the moment value 0.086 Wl is same that in Table 3.12 of the code. Since the maximum moment in span is at x = 0.414 l, the contra-flexure point is at a distance of 2x from support A. Therefore the theoretical cut-off point for the top reinforcement is at 2x = 0.828l from the support A or at 0.172l from support B.

(b) Internal span DE

The hinge is at mid-span and the rotations are shown in Fig.8.31(c). The work equation is

$$W\,(0.5\,l\,\theta)/2 = m'\,\theta + m\,2\,\theta + m'\,\theta$$

For the case where $m = m'$

$$m = Wl/16 = 0.063Wl$$

Note that the moment value 0.063 Wl is same that in Table 3.12 of the code. The theoretical cut–off points for the top bars are at 0.147 l from each support.

8.9.6 Simply Supported Rectangular Two-Way Slab

The slab and yield line pattern are shown in Fig.8.34. The ultimate loading is w per square metre. As shown in Fig.8.34, steel in the shorter y-direction provides a moment of resistance of m per unit length and the steel in the longer x-direction provides a moment of resistance of μm per unit length. The yield line pattern is defined by one parameter, β.

1. Work done by external loads

The work done by the loads can be calculated by assuming that points E and F deflect by Δ,

(i) Triangles ACE and BFD

Area = 0.5 b βa, deflection at the centroid = $\Delta/3$.
Work done by external loads is

$$W_1 = 2[q\ 0.5\,b\,\beta a\,\frac{\Delta}{3}] = qab\,\frac{\beta}{3}\,\Delta$$

(ii) Trapeziums CEFD and AEFB

Dividing the trapezium into two triangles and a rectangle,

Triangle: area = 0.5 b/2 βa, deflection at the centroid = Δ/3
Rectangle: area = b/2 (a – 2 βa), deflection at the centroid = Δ/2

Work done by external loads is

$$W_2 = 2[2\{q\ 0.5\frac{b}{2}\beta a\frac{\Delta}{3}\} + q\frac{b}{2}(a-2\beta a)\frac{\Delta}{2}] = qab\{\frac{(3-4\beta)}{6}\}\Delta$$

Total work done by the loads W = W_1 + W_2 = $q\dfrac{ab}{6}(3-2\beta)\Delta$

2. Energy dissipated at the yield lines

The energy dissipated at the yield lines can be calculated using Method 3.

(i) Yield line in triangles ACE and BFD: The triangles rotate only about y–axis.

$$l_y = b, \quad m_y = \mu m, \quad \theta_y = \frac{\Delta}{\beta a}$$

Hence the energy dissipated E_1 on the yield lines in triangles ACE and BFC is

$$E_1 = 2[\ell_x m_x \theta_x + \ell_y m_y \theta_y]$$

$$E_1 = 2[2\beta a\ m\ 0 + b\ \mu m\ \frac{\Delta}{\beta a}] = 2m\frac{b}{a}\frac{\mu}{\beta}\Delta$$

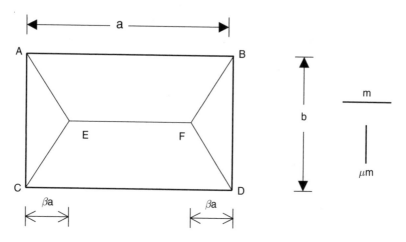

Fig.8.34 Collapse mode for a simply supported slab.

(ii) Yield line in trapeziums AEFB and CEFD: The trapeziums rotates only about x–axis.

$$l_x = a\ , m_x = m,\ \theta_x = \frac{\Delta}{0.5b}$$

Hence the energy dissipated E_2 on the yield lines in trapeziums AEFB and CEFD is

$$E_2 = 2[a\,m\,\frac{\Delta}{0.5b} + b\,\mu m\,0] = 4m\frac{a}{b}\Delta$$

The total energy dissipated E is therefore

$$E = E_1 + E_2 = \{2m\frac{b}{a}\frac{\mu}{\beta} + 4m\frac{a}{b}\}\Delta$$

3. Calculation of moment of resistance:

Equating the work done by the external loads to the energy dissipated at the yield lines,

$$E = 2m\frac{b}{a}\frac{\mu}{\beta} + 4m\frac{a}{b} = W = q\frac{ab}{6}(3-2\beta)$$

Solving for m,

$$m = q\frac{b^2}{12}\frac{(3\beta - 2\beta^2)}{(\mu\frac{b^2}{a^2} + 2\beta)}$$

In order to calculate the maximum value of m required, set dm/dβ = 0

$$(\mu\frac{b^2}{a^2} + 2\beta)(3 - 4\beta) - (3\beta - 2\beta^2)(2) = 0$$

Simplifying

$$4\beta^2 + 4\mu\frac{b^2}{a^2}\beta - 3\mu\frac{b^2}{a^2} = 0$$

Solving the quadratic in β,

$$\beta = \frac{1}{2}\frac{b^2}{a^2}\{-\mu + \sqrt{(\mu^2 + 3\mu\frac{a^2}{b^2})}\}$$

8.9.6.1 Example of yield line analysis of a simply supported rectangular slab

A simply supported rectangular slab 4.5 m long by 3 m wide carries an ultimate load of 16 kN/m². Determine the design moments for the case when the value of μ = 0.5.

Substituting a = 4.5 m, b = 3.0 m, b/a = 0.667, μ = 0.5, in the formula for β,

$$\beta = 0.312$$

Substituting β = 0.312 and q = 16.0 in the equation for m,

$$m = 10.51 \text{ kNm/m}, \mu m = 5.26 \text{ kNm/m}$$

It is usual in designs based on the yield line analysis for the reinforcement to remain uniform in each direction. It is evident from the collapse mechanism, Yield Line analysis provides no information on where the reinforcement can be curtailed, nor does it give any information on the shear force distribution in the slab.

8.9.7 Rectangular Two-Way Slab Continuous Over Supports

The solution derived in section 8.9.6 can be extended to the case of a continuous slab. The slab shown in Fig.8.35 has a continuous hogging yield line around the supports. The negative moment of resistance of the slab at the supports has a value of $\gamma\, m'$ per unit length in the shorter direction and m' per unit length in the longer direction.

The basic yield line pattern will be as for the simply supported slab shown in Fig.8.34 except that negative yield lines (tension at the top face) form parallel and close to the supports as shown in Fig.8.35 by dotted lines. The work done by external loads and the energy dissipated at the positive yield lines (tension at the bottom face) remain as for the simply supported slab. The extra aspect to be considered is the energy dissipated at the negative yield lines.

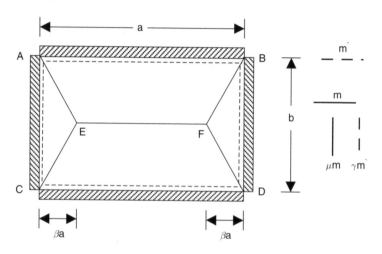

Fig.8.35 Collapse mode for a clamped slab.

(i) For the *two* negative yield lines parallel to the *shorter* sides

$$l_y = b,\; m_y = \gamma m', \; \theta_y = \frac{\Delta}{\beta a}$$

Hence the energy dissipated by the *two* negative yield lines parallel to the *shorter* side is

$$E_3 = 2\{b\,\gamma m'\,\frac{\Delta}{\beta a}\} = 2m'\frac{b}{a}\frac{\gamma}{\beta}\Delta$$

(ii) For the *two* negative yield lines parallel to the *longer* sides

$$l_x = a\,,\; m_x = m', \; \theta_x = \frac{\Delta}{0.5b}$$

Hence the energy dissipated by the *two* negative yield lines parallel to the longer sides is

$$E_4 = 2\{a\,m'\,\frac{\Delta}{0.5b}\} = 4m'\frac{a}{b}\Delta$$

Total energy dissipated at the negative yield lines is

$$E_{negative\,yield\,lines} = E_3 + E_4 = 2m'\{\frac{b}{a}\frac{\gamma}{\beta} + 2\frac{a}{b}\}\Delta$$

Total energy dissipated at the positive and negative yield lines is

$$E = [2m'\{\frac{b}{a}\frac{\gamma}{\beta} + 2\frac{a}{b}\} + 2m\{\frac{b}{a}\frac{\mu}{\beta} + 2\frac{a}{b}\}]\Delta$$

Equating the work done by the external loads to the energy dissipated at all the yield lines,

$$m = q\frac{b^2}{12}\frac{(3\beta - 2\beta^2)}{(\mu\dfrac{b^2}{a^2} + 2\beta) + \dfrac{m'}{m}(\gamma\dfrac{b^2}{a^2} + 2\beta)}$$

In order to calculate the maximum value of m required, set dm/dβ = 0

$$\{(\mu + \gamma\frac{m'}{m})\frac{b^2}{a^2} + 2\beta(1 + \frac{m'}{m})\}(3 - 4\beta) - (3\beta - 2\beta^2)(2 + 2\frac{m'}{m}) = 0$$

Simplifying

$$4(1 + \frac{m'}{m})\beta^2 + 4(\mu + \gamma\frac{m'}{m})\frac{b^2}{a^2}\beta - 3(\mu + \gamma\frac{m'}{m})\frac{b^2}{a^2} = 0$$

Solving the quadratic in β,

$$\beta = \frac{1}{2}\frac{b^2}{a^2}[\frac{-(\mu + \gamma\frac{m'}{m}) + \sqrt{\{(\mu + \gamma\frac{m'}{m})^2 + 3(\mu + \gamma\frac{m'}{m})(1 + \frac{m'}{m})\frac{a^2}{b^2}\}}}{(1 + \frac{m'}{m})}]$$

8.9.7.1 Example of Yield line analysis of a clamped rectangular slab

A clamped rectangular slab 4.5 m long by 3 m wide carries an ultimate load of 16 kN/m^2. Determine the design moments for the case where

$$m'/m = 1.3, \quad \mu = 0.6, \quad \gamma = 0.6$$

as obtained from *average moment ratios* from elastic analysis.
Substituting in the formula for β,

$$a = 4.5\,m, \quad b = 3.0\,m, \quad b/a = 0.667, \quad m'/m = 1.3, \quad \mu = 0.6, \quad \gamma = 0.6$$
$$\beta = 0.77 > 0.5$$

As the value of β cannot be greater than 0.5, substituting β = 0.5 and
q = 16.0 kN/m^2, m = 4.1 kNm/m, μ m = 2.47 kNm/m, m$'$ = 5.33 kNm/m,
γm$'$ = 3.2 kNm/m
   It is usual in designs based on the yield line analysis for the reinforcement to remain uniform in each direction.   However, the extent of the negative reinforcement required can be determined by finding the dimensions of a simply

supported central rectangular region of dimensions $\alpha a \times \alpha b$ which has a collapse moment in the yield lines of m = 4.1 kNm/m, μ m = 2.47 kNm/m and q = 16.0.

Since at the cut–off of top bars, the hogging yield line has zero strength, this simulates simple support. The bars must be anchored beyond the theoretical cut–off lines.

Substituting μ = 0.6, b/a = 3/ (4.5) = 0.67 in the equation for β in section 8.9.6, gives β = 0.33

Substituting for m = 4.1 kN/m, b = 3 α, a = 4.5 α, μ = 0.6, β = 0.33, q = 16kN/m² in the equation for m in section 8.9.6, where

$$m = q\frac{(\alpha b)^2}{12}\frac{(3\beta - 2\beta^2)}{(\mu\frac{b^2}{a^2} + 2\beta)}$$

gives α = 0.64. The theoretical cut–off lengths are therefore $(1 - \alpha)/2 = 0.18$ of the side dimensions.

8.9.8 Clamped Rectangular Slab with One Long Edge Free

A rectangular slab continuous on three edges and free on a long edge has two distinct modes of collapse as shown in Fig.8.36 and Fig.8.37. The slab shown in Fig.8.36 and Fig.8.37 has a continuous hogging yield line around the supports in addition to positive yield lines. As shown in Fig.8.36, the positive and negative moment of resistance of the slab have values of μm and $\gamma m'$ per unit length respectively in yield lines parallel to the y-direction and m and m' per unit length respectively in yield lines parallel to the x-direction.

In the case of simply supported and clamped rectangular slabs, there was only one mode of collapse defined by a single parameter β. However when one of the edges is free, there are two different modes of collapse possible as shown in Fig.8.36 and Fig.8.37. Calculations have to be done for both modes of collapse to determine either the minimum collapse load or the maximum moment of resistance required.

8.9.8.1 Calculations for collapse Mode 1

The mode of collapse is shown in Fig.8.36. Assume that EF deflects by Δ.

(1) Energy dissipated at the yield lines:
(a) Trapeziums ACFE and BDFE: The trapeziums rotate only about the y–axis.

(i) For the negative yield line

$$l_y = b, \quad m_y = \gamma m', \quad \theta_y = \frac{\Delta}{0.5a}$$

Total energy dissipation for the two regions is

$$E_1 = 2\{b\ \gamma m'\ \frac{\Delta}{0.5a}\} = 4m'\gamma\frac{b}{a}\Delta$$

(ii) For the positive yield line

$$l_y = b, \quad m_y = \mu m, \quad \theta_y = \frac{\Delta}{0.5a}$$

Total energy dissipation for the two regions is

$$E_2 = 2\{b\ \mu m\ \frac{\Delta}{0.5a}\} = 4m\frac{b}{a}\mu\Delta$$

(b) Triangle CFD: The triangle rotates only about the x–axis.
(i) For the negative yield line

$$l_x = a, \quad m_x = m', \quad \theta_x = \frac{\Delta}{\beta b}$$

Energy dissipation is

$$E_3 = a\ m'\ \frac{\Delta}{\beta b} = m'\frac{a}{b}\frac{1}{\beta}\Delta$$

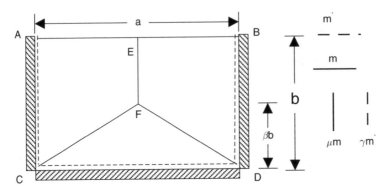

Fig.8.36 Clamped slab with a free edge; collapse mode 1.

(ii) For the positive yield line

$$l_x = a, \quad m_x = m, \quad \theta_x = \frac{\Delta}{\beta b}$$

Energy dissipation is

$$E_4 = a\ m\ \frac{\Delta}{\beta b} = m\frac{a}{b}\frac{1}{\beta}\Delta$$

Total energy dissipated at the positive and negative yield lines is

$$E = E_1 + E_2 + E_3 + E_4 = \{4\frac{b}{a}(\gamma m' + \mu m) + \frac{a}{b}(m' + m)\frac{1}{\beta}\}\Delta$$

Simplifying

$$E = \frac{1}{\beta ab}\{4(\gamma m' + \mu m)\beta b^2 + (m' + m)a^2\}\Delta$$

(2) Work done by external loads

(a) Triangle CFD

Area = 0.5 a βb, deflection at the centroid = $\Delta/3$

Work done by external loads is

$$W_1 = q \ 0.5 \ a \ \beta b \frac{\Delta}{3}$$

(b) Trapeziums ACFE and BDFE

Dividing it into a triangle and a rectangle

Triangle: area = 0.5 a/2 βb, deflection at the centroid = $\Delta/3$
Rectangle: area = a/2 (b – βb), deflection at the centroid = $\Delta/2$

Work done by the external loads is

$$W_2 = 2[\{q \ 0.5 \frac{a}{2}\beta b\frac{\Delta}{3}\} + q\frac{a}{2}(b - \beta b)\frac{\Delta}{2}] = q\frac{ab}{6}(3 - 2\beta)\Delta$$

Total work W done by the loads = $W_1 + W_2$

$$W = q\frac{ab}{6}(3 - \beta)\Delta$$

(3) Calculation of m: Equating the work done by the external loads to the energy dissipated at all the yield lines,

$$m = q\frac{b^2}{6}\frac{(3\beta - \beta^2)}{\{4(\gamma\frac{m'}{m} + \mu)\beta\frac{b^2}{a^2} + (1 + \frac{m'}{m})\}}$$

In order to calculate the maximum value of m required, set dm/dβ = 0

$$\{4(\gamma\frac{m'}{m} + \mu)\beta\frac{b^2}{a^2} + (1 + \frac{m'}{m})\}(3 - 2\beta) - (3\beta - 2\beta^2)4(\gamma\frac{m'}{m} + \mu)\beta\frac{b^2}{a^2} = 0$$

Simplifying

$$4(\gamma\frac{m'}{m} + \mu)\frac{b^2}{a^2}\beta^2 + 2(1 + \frac{m'}{m})\beta - 3(1 + \frac{m'}{m}) = 0$$

8.9.8.2 Calculations for Collapse Mode 2

The mode of collapse is shown in Fig.8.37. Assume that E F deflects by Δ.

1 Energy dissipated at the yield lines

(a) Triangles ACE and BDF: Rotation of the triangles is about y–axis only.

(i) For the negative yield line

$$l_y = b, \quad m_y = \gamma m', \quad \theta_y = \frac{\Delta}{\beta a}$$

Total energy dissipation for the two triangles is

$$E_1 = 2\{b\,\gamma m'\,\frac{\Delta}{\beta a}\} = 2m'\gamma\frac{b}{a}\frac{1}{\beta}\Delta$$

(ii) For the positive yield line

$$l_y = b, \quad m_y = \mu m, \quad \theta_y = \frac{\Delta}{\beta a}$$

Total energy dissipation for the two triangles is

$$E_2 = 2\{b\,\mu m\,\frac{\Delta}{\beta a}\} = 2m\mu\frac{b}{a}\frac{\Delta}{\beta}$$

(b) Trapezium CEFD: The trapezium rotates only about the x–axis.

(i) For the negative yield line

$$l_x = a, \quad m_x = m', \quad \theta_x = \frac{\Delta}{b}$$

Energy dissipation is

$$E_3 = a\,m'\,\frac{\Delta}{b}$$

(ii) For the positive yield lines

$$l_x = 2(\beta a), \quad m_x = m, \quad \theta_x = \frac{\Delta}{b}$$

Energy dissipation is

$$E_4 = 2\beta a\,m\,\frac{\Delta}{b}$$

Total energy dissipated at the positive and negative yield lines is

$$E = E_1 + E_2 + E_3 + E_4 = \{2\frac{b}{\beta a}(\gamma m' + \mu m) + \frac{a}{b}(m' + 2\beta m)\}\Delta$$

Simplifying

$$E = \frac{1}{\beta ab}\{2(\gamma m' + \mu m)b^2 + (m'\beta + 2\beta^2 m)a^2\}\Delta$$

2. Work done by external loads

(i) Triangles ACE and BDF:
Area = 0.5 b βa, deflection at the centroid = Δ/3

$$W_1 = 2\{q\frac{1}{2}b\,\beta a\frac{\Delta}{3}\} = q\frac{ab}{3}\beta\Delta$$

ii. Trapezium CEFD
Dividing it into two triangles and a rectangle

Two triangles: area = 0.5 b βa, deflection at the centroid = Δ/3
Rectangle: area = b (a – 2βa), deflection at the centroid = Δ/2

$$W_2 = 2\{q\frac{1}{2}b\,\beta a\frac{\Delta}{3}\} + qb(a-2\beta a)\frac{\Delta}{2} = q\frac{ab}{6}(3-4\beta)\Delta$$

Total work W done by the loads = $W_1 + W_2$

$$W = q\frac{ab}{6}(3-2\beta)\Delta$$

Equating the work done by the external loads to the energy dissipated at all the yield lines,

$$m = q\frac{b^2}{6}\frac{(3\beta - 2\beta^2)}{\{2(\gamma\frac{m'}{m}+\mu)\frac{b^2}{a^2}+(2\beta^2 + \frac{m'}{m}\beta)\}}$$

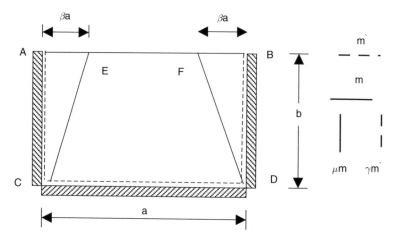

Fig.8.37 Clamped slab with a free edge; collapse mode 2

In order to calculate the maximum value of m required, set dm/dβ = 0

$$\{2(\gamma\frac{m'}{m}+\mu)\frac{b^2}{a^2}+2\beta^2+\frac{m'}{m}\beta)\}(3-4\beta)-(3\beta-2\beta^2)(4\beta+\frac{m'}{m}) = 0$$

Simplifying

$$(\frac{m'}{m}-3)\beta^2 - 2\{\frac{m'}{m}+4(\gamma\frac{m'}{m}+\mu)\frac{b^2}{a^2}\}\beta + 3(\gamma\frac{m'}{m}+\mu)\frac{b^2}{a^2} = 0$$

Solving the quadratic in β,

$$\beta = \frac{\{\frac{m'}{m} + 4(\gamma\frac{m'}{m} + \mu)\frac{b^2}{a^2}\} + \sqrt{[\{\frac{m'}{m} + 4(\gamma\frac{m'}{m} + \mu)\frac{b^2}{a^2}\}^2 - 3(\gamma\frac{m'}{m} + \mu)\frac{b^2}{a^2}(\frac{m'}{m} - 3)]}}{(\frac{m'}{m} - 3)}$$

8.9.8.3 Example of yield line analysis of a clamped rectangular slab with one free long edge

A clamped rectangular slab with one long edge free is 4.5 m long by 3 m wide and carries an ultimate load of 16 kN/m². Determine the design moments for the case when

$$m'/m = 5.0, \quad \mu = 3.0, \quad \gamma = 1.5$$

as obtained from maximum moment ratios from elastic finite element analysis.

When a slab has more than one distinct mode of failure, it is necessary to investigate both modes of failure and accept the *larger* of the two moments as the design moment.

Mode 1: Substituting the values of the parameters in the formula for β,

$$\beta = 0.712 < 1.0$$

Using this value of β,

$$m = 2.03 \text{ kNm/m}$$

Mode 2: Using the same parameters as for mode 1, calculate the value of β. The smaller root for β is

$$\beta = 0.3 < 0.5$$

Using this value of β,

$$m = 1.57 \text{ kNm/m}.$$

For design the larger value for m is obtained from mode 1. Therefore
$m = 2.03$ kNm/m, $\mu m = 6.1$ kNm/m, $m' = 10.2$ kNm/m, $\gamma m' = 15.2$ kNm/m

8.9.9 Trapezoidal Slab Continuous Over Three Supports and Free on a Long Edge

Fig.8.38 shows a uniformly loaded trapezoidal slab with three edges clamped and one edge free. Normal moment of resistance per unit length on positive and negative yield lines parallel to x and y–axes are respectively $(m, \mu m)$ and $(m', \gamma m')$ respectively.

In the previous examples, the rotations of the rigid regions took place about edges which were parallel to x or y–axis. In this example only one axis of rotation is parallel to the coordinate axes. One possible mode of collapse is shown by the positive yield lines CE and DF and negative yield lines parallel to the supports.

Assume that EF deflects by Δ. As shown in Fig.8.39, let the yield line CE be inclined to the vertical by ϕ and the support CA is inclined to the vertical by ψ. Let EG be perpendicular to support AC. From geometry

$$\text{Angle ACH} = \psi, \quad \tan\psi = \alpha\frac{a}{b}, \quad \cos\psi = \frac{h}{AE} = \frac{h}{a}\frac{1}{(\alpha+\beta)}$$

$$\text{Angle ECH} = \phi, \quad \tan\phi = \beta\frac{a}{b}$$

$$\text{Angle CAH} = 90 - \psi$$
$$GE = h, \quad AE = (\alpha+\beta)\,a$$

Triangle ACE rotates clockwise about the support AC by θ. The rotational components of θ are

$$\theta = \frac{\Delta}{h}, \quad \theta_x = \theta\sin\psi, \quad \theta_y = \theta\cos\psi$$

Substituting for $\tan\psi$ and $\cos\psi$

$$\theta_x = \frac{\alpha}{(\alpha+\beta)}\frac{\Delta}{b}, \quad \theta_y = \frac{1}{(\alpha+\beta)}\frac{\Delta}{a}$$

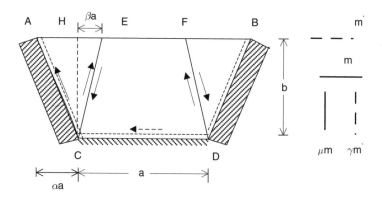

Fig.8.38 Trapezoidal clamped slab with a free edge.

1. Energy dissipated in yield lines

(i) Negative yield line in triangles ACE and BDF
The triangles rotate about an axis inclined to both x and y–axes.

$$l_x = \alpha\,a, \quad m_x = m\dot{}, \quad \theta_x = \frac{\alpha}{(\alpha+\beta)}\frac{\Delta}{b}$$

$$l_y = b, \quad m_y = \gamma m\dot{}, \quad \theta_y = \frac{1}{(\alpha+\beta)}\frac{\Delta}{a}$$

Hence the energy dissipated on the negative yield line in the two triangles is

$$E_1 = 2\{\ell_x\,m_x\,\theta_x + \ell_y\,m_y\,\theta_y\}$$

$$E_1 = 2\{\alpha a\, m'\,\frac{\alpha}{(\alpha+\beta)}\frac{\Delta}{b} + b\,\gamma m'\,\frac{1}{(\alpha+\beta)}\frac{\Delta}{a}\}$$

$$E_1 = 2\frac{m'}{(\alpha+\beta)}[\alpha^2\frac{a}{b}+\gamma\frac{b}{a}]\Delta$$

Note that because both the moment vector and the rotation vector act in the same direction, the energy dissipated in both the x and y–components are positive.

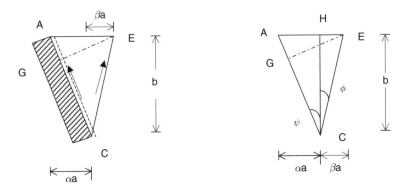

Fig.8.39 Rotation and moment vectors.

(ii) Positive yield line in triangles ACE and BDF

$$l_x = \beta\, a, \quad m_x = m, \quad \theta_x = \frac{\alpha}{(\alpha+\beta)}\frac{\Delta}{b}$$

$$l_y = b, \quad m_y = \mu m, \quad \theta_y = \frac{1}{(\alpha+\beta)}\frac{\Delta}{a}$$

Note that x–components of the moment vector and rotation vector point in opposite directions. Therefore the energy dissipated on the positive yield line in triangle ACE is

$$E_2 = 2\{\ell_x m_x \theta_x + \ell_y m_y \theta_y\}$$

$$E_2 = 2\{-\beta a\, m\,\frac{\alpha}{(\alpha+\beta)}\frac{\Delta}{b} + b\,\mu m\,\frac{1}{(\alpha+\beta)}\frac{\Delta}{a}\}$$

$$E_2 = 2\frac{m}{(\alpha+\beta)}[-\alpha\beta\frac{a}{b}+\mu\frac{b}{a}]\Delta$$

(iii) Negative yield line in the trapezium ECDF: The trapezium rotates only about the x–axis.

$$l_x = a, \quad m_x = m', \quad \theta_x = \frac{\Delta}{b}$$

$$E_3 = \ell_x m_x \theta_x = a\, m'\,\frac{\Delta}{b}$$

(iv) Positive yield line in the trapezium ECDF

$$l_x = 2(\beta\, a), \quad m_x = m, \quad \theta_x = \frac{\Delta}{b}$$

$$E_4 = \ell_x\, m_x\, \theta_x = 2\beta a\, m\, \frac{\Delta}{b}$$

Therefore total energy dissipation is

$$E = E_1 + E_2 + E_3 + E_4$$

$$E = \frac{1}{(\alpha+\beta)}\frac{1}{ab}[m'\{(2\alpha^2+\alpha+\beta)a^2+2\gamma b^2\}+2m\{\beta^2 a^2+\mu b^2\}]\Delta$$

2. Work done by external loads

(i) Triangles ACE and BDF

Area = 0.5 $(\alpha + \beta)$ a b, deflection of the centroid = $\Delta/3$

$W_1 = 2$ q $\{0.5\ (\alpha + \beta)$ a b $\Delta/3\}$

(ii) Trapezium ECDF

Divide into two triangles and a rectangle.
For each triangle:

area = 0.5 β a b, deflection of the centroid = $\Delta/3$

For the rectangle:

Area = $(1 - 2\beta)$ a b, deflection of the centroid = $\Delta/2$

Work done is

$$W_2 = 2\text{ q } \{0.5\text{ a b }\beta\}\ \Delta/3 + \text{q a b }(1 - 2\ \beta)\ \Delta/2$$

Total work W done is $W = W_1 + W_2$

$$W = \text{q ab } \{3 + 2\ (\alpha - \beta)\}\ \Delta/6$$

Equating W = E and simplifying,

$$m = q b^2 \frac{1}{6} \frac{3(\alpha+\beta)+2(\alpha^2-\beta^2)}{[\dfrac{m'}{m}\{(2\alpha^2+\alpha+\beta)+2\gamma\dfrac{b^2}{a^2}\}+2\{\beta^2+\mu\dfrac{b^2}{a^2}\}]}$$

If $\alpha = 0$, then the equation will be same as for Mode 2 collapse of a clamped rectangular slab with a free edge.
Assuming:

a = 4.5 m, b = 3.0 m, α a = 1.0, m'/m = 5.0, μ = 3.0, γ = 1.5, q = 16 kN/m^2
giving, β = 0.5, m = 3.04 kNm/m.

8.9.10 Slab with Hole

Fig.8.40 shows a simply supported rectangular slab of dimensions a × b with a central rectangular hole of dimensions αa × αb. There are three distinct modes of collapse which have to be analysed in calculating the minimum collapse load.

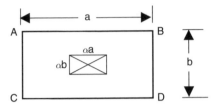

Fig.8.40 Slab with a hole.

8.9.10.1 Calculations for collapse Mode 1

Fig.8.41 shows the collapse mode 1. Let the deflection at the apex of the triangle be Δ.

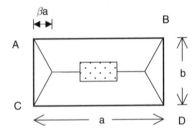

Fig 8.41 Collapse Mode 1.

1. Work done by external loads
(i) Two side triangles
Area = 0.5 b βa,   deflection at the centroid = Δ/3

$$W_1 = 2\{q\,0.5b\,\beta a\frac{\Delta}{3}\} = qab\frac{\beta}{3}\Delta$$

(ii) Two Trapeziums
Dividing each into two triangles, two rectangles and a rectangle adjoining the side of the hole.
(a) Triangle
Area = 0.5 b/2 βa,   deflection at the centroid = Δ/3
(b) Rectangle
Area = b/2 (a – αa – 2 βa),   deflection at the centroid = Δ/2
(c) Rectangle adjoining the hole
Note that the deflection at the edge of the hole is not Δ but (1 – α) Δ
Area = {b(1 – α)/2} αa ,   deflection at the centroid = (1 – α)Δ/2

$$W_2 = 2[2q\{\frac{1}{2}\frac{b}{2}\beta a\frac{\Delta}{3}\} + q\frac{b}{2}(a - \alpha a - 2\beta a)\frac{\Delta}{2} + qb\frac{(1-\alpha)}{2}\alpha a(1-\alpha)\frac{\Delta}{2}]$$

$$W_2 = q\frac{ab}{6}\{3 - 4\beta - 6\alpha^2 + 3\alpha^3\}\Delta$$

Total work W done by the loads = $W_1 + W_2$

$$W = q\frac{ab}{6}(3 - 2\beta - 6\alpha^2 + 3\alpha^3)\Delta$$

2. The energy dissipated at the yield lines
(i) Yield lines in the two side triangles

$$l_y = b, \quad m_y = \mu m, \quad \theta_y = \frac{\Delta}{\beta a}$$

$$E_1 = 2\{b\ \mu m\ \frac{\Delta}{\beta a}\} = 2m\frac{b}{a}\frac{\mu}{\beta}\Delta$$

(ii) Yield lines in the two trapeziums

$$l_x = (a - \alpha a), \quad m_x = m, \quad \theta_x = \frac{\Delta}{0.5b}$$

$$E_2 = 2\{a(1 - \alpha)\ m\ \frac{\Delta}{0.5b}\} = 4m(1 - \alpha)\frac{a}{b}\Delta$$

The total energy dissipated is therefore

$$E_1 + E_2 = E = \{2m\frac{b}{a}\frac{\mu}{\beta} + 4m(1 - \alpha)\frac{a}{b}\}\Delta$$

Equating the work done by the external loads to the energy dissipated at the yield lines,

$$E = 2m\frac{b}{a}\frac{\mu}{\beta} + 4m(1 - \alpha)\frac{a}{b}\Delta = W = q\frac{ab}{6}(3 - 2\beta - 6\alpha^2 + 3\alpha^3)\Delta$$

Solving for m,

$$m = q\frac{b^2}{12}\frac{\{(3 - 6\alpha^2 + 3\alpha^3)\beta - 2\beta^2\}}{\{\mu\frac{b^2}{a^2} + 2(1 - \alpha)\beta\}}$$

In order to calculate the maximum value of m required, set dm/dβ = 0

$$\{\mu\frac{b^2}{a^2} + 2(1 - \alpha)\beta\}(3 - 6\alpha^2 + 3\alpha^3 - 4\beta) - \{(3 - 6\alpha^2 + 3\alpha^3)\beta - 2\beta^2\}2(1 - \alpha) = 0$$

The resulting quadratic equation in β can be solved numerically for specific values of the parameters α and μ and the corresponding value of m can be determined. Note that from geometry, the above equations are valid for $0 \le \beta \le (1 - \alpha)/2$.

8.9.10.2 *Calculations for collapse Mode 2*

Fig.8.42 shows the collapse mode 2. Assume that the deflection at the apex of side triangles is Δ.

1. Work done by external loads

(i) Side trapeziums, two in number

Dividing each into two triangles and a rectangle

(a) Triangle

$$\text{Area} = 0.5 \{(1 - \alpha)a \,/2\}\, \beta b, \quad \text{deflection at the centroid} = \Delta/3$$

(b) Rectangle:

$$\text{Area} = (1 - 2\beta)\, b \,(1 - \alpha)\, a/2, \quad \text{deflection at the centroid} = \Delta/2$$

$$W_1 = 2[2q\{\frac{1}{2}\frac{(1-\alpha)a}{2}\,\beta b\,\frac{\Delta}{3}\} + q(1-2\beta)b\frac{(1-\alpha)a}{2}\frac{\Delta}{2}\,]$$

$$W_1 = q\frac{ab}{6}\{3 - 3\alpha - 4\beta(1-\alpha)\}\Delta$$

(ii) Top and bottom trapeziums two in number

Dividing it into two triangles and a rectangle which is part of the hole.

(a) Triangle

$$\text{Area} = 0.5 \{(1 - \alpha)\, a \,/2\}\, \beta b, \quad \text{deflection at the centroid} = \Delta/3$$

(b) Rectangle:

Note that the deflection at the edge of the hole is not Δ but $(1 - \alpha)\Delta$

$$\text{Area} = \alpha a \,(1 - \alpha)\, b/2, \quad \text{deflection at the centroid} = (1 - \alpha)\Delta/2$$

$$W_2 = 2[2q\{\frac{1}{2}\frac{(1-\alpha)}{2}\,a\,\beta b\,\frac{\Delta}{3}\} + q\{\alpha a\frac{(1-\alpha)b}{2}\,(1-\alpha)\frac{\Delta}{2}\}]$$

$$W_2 = q\frac{ab}{6}(1-\alpha)\{2\beta - 3\alpha^2 + 3\alpha\}\Delta$$

Total work W done by the loads

$$W = W_1 + W_2 = q\frac{ab}{6}(1-\alpha)(3 - 2\beta - 3\alpha^2 + 3\alpha)\Delta$$

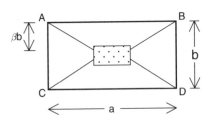

Fig 8.42 Collapse Mode 2.

2. The energy dissipated at the yield lines

(i) Yield lines in the two side trapeziums

$$l_y = 2(\beta b), \quad m_y = \mu m, \quad \theta_y = \frac{\Delta}{0.5(1-\alpha)a}$$

$$E_1 = 2\{2\beta b\,\mu m\,\frac{\Delta}{0.5(1-\alpha)a}\} = 8m\frac{b}{a}\frac{\beta\mu}{(1-\alpha)}\Delta$$

(ii) Yield line in the top and bottom trapeziums

$$l_x = 2(1-\alpha)\,a/2, \quad m_x = m, \quad \theta_x = \frac{\Delta}{\beta b}$$

$$E_2 = 2\{(1-\alpha)a\,m\,\frac{\Delta}{\beta b}\} = 2m(1-\alpha)\frac{1}{\beta}\frac{a}{b}\Delta$$

The total energy dissipated is therefore

$$E = E_1 + E_2 = \{8m\frac{b}{a}\frac{\beta}{(1-\alpha)}\mu + 2m\frac{(1-\alpha)}{\beta}\frac{a}{b}\}\Delta$$

Equating the work done by the external loads to the energy dissipated at the yield lines,

$$\{8m\frac{b}{a}\frac{\beta}{(1-\alpha)}\mu + 2m\frac{(1-\alpha)}{\beta}\frac{a}{b}\}\Delta = q\frac{ab}{6}(1-\alpha)(3-2\beta-3\alpha^2+3\alpha)\Delta$$

Solving for m,

$$m = q\frac{b^2}{12}\frac{(1-\alpha)^2\{(3-3\alpha^2+3\alpha)\beta - 2\beta^2\}}{\{4\mu\dfrac{b^2}{a^2}\beta^2 + (1-\alpha)^2\}}$$

In order to calculate the maximum value of m required, setting $dm/d\beta = 0$, gives

$$\{4\mu\frac{b^2}{a^2}\beta^2 + (1-\alpha)^2\}(3-3\alpha^2+3\alpha-4\beta) - \{(3-3\alpha^2+3\alpha)\beta - 2\beta^2\}8\mu\frac{b^2}{a^2}\beta = 0$$

The resulting quadratic equation in β can be solved numerically for specific value of the parameters and the corresponding value of m can be determined. The above equations are valid only for $0.5(1-\alpha) \le \beta \le 0.5$.

8.9.10.3 Calculations for collapse Mode 3

Fig.8.43 shows the collapse mode 3. Assume that the deflection at the longer sides of the hole is Δ.

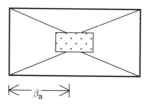

Fig 8.43 Collapse Mode 3.

1. Work done by external loads

(i) Side trapeziums, two in number: Divide each into two triangles and a rectangle.

(a) Triangle

Area = 0.5 $\{(1 - \alpha)$b /2$\}$ βa,   deflection at the centroid = Δ/3

(b) Rectangle: Note that the deflection at the edge of the hole is not Δ but $\{0.5(1 - \alpha)/ \beta\}$ Δ

Area = αb (1 - α)a/2, deflection at the centroid = $\{0.5(1 - \alpha)/ \beta\}\Delta$/2

$$W_1 = 2[2q\{\frac{1}{2}\frac{(1-\alpha)b}{2} \beta a \frac{\Delta}{3}\} + q\,\alpha b (1-\alpha)\frac{a}{2}\frac{(1-\alpha)}{2\beta}\frac{\Delta}{2}\ ]$$

$$W_1 = q\frac{ab}{12}\frac{(1-\alpha)}{\beta}\{4\beta^2 + 3\alpha(1-\alpha)\}\Delta$$

(ii) Top and bottom trapeziums two in number: Divide it into two triangles and a rectangle which is part of the hole.

(a) Triangle

Area = 0.5 $\{(1 - \alpha)$b /2$\}$ βa,   deflection at the centroid = Δ/3

(b) Rectangle

Area = $(1 - 2\beta)$a $(1 - \alpha)$b/2,   deflection at the centroid = Δ/2

$$W_2 = 2[2q\{\frac{1}{2}\frac{(1-\alpha)}{2} b \,\beta a \frac{\Delta}{3}\} + q\{(1-2\beta)a\frac{(1-\alpha)b}{2}\frac{\Delta}{2}\}]$$

$$W_2 = q\frac{ab}{6}(1-\alpha)\{3-4\beta\}\Delta$$

Total work W done by the loads = $W_1 + W_2$

$$W = q\frac{ab}{12\beta}(1-\alpha)(6\beta - 4\beta^2 - 3\alpha^2 + 3\alpha)\Delta$$

2. The energy dissipated at the yield lines
(i) Yield lines in the two side trapeziums

$$l_y = 2(1 - \alpha)b/2, \quad m_y = \mu m, \quad \theta_y = \frac{\Delta}{\beta a}$$

$$E_1 = 2\{2(1-\alpha)\frac{b}{2} \,\mu m \frac{\Delta}{\beta a}\} = 2m\frac{b}{a}\frac{(1-\alpha)\mu}{\beta}\Delta$$

(ii) Yield line in the top and bottom trapeziums

$$l_x = 2(\beta a), \quad m_x = m, \quad \theta_x = \frac{\Delta}{0.5(1-\alpha)b}$$

$$E_2 = 2\{2\beta a \, m \frac{\Delta}{0.5(1-\alpha)b}\} = 8m\frac{\beta}{(1-\alpha)}\frac{a}{b}\Delta$$

The total energy dissipated is $E = E_1 + E_2$

$$E = \{ 2m\frac{(1-\alpha)}{\beta}\mu\frac{b}{a} + 8m\frac{a}{b}\frac{\beta}{(1-\alpha)}\}\Delta$$

Equating the work done by the external loads to the energy dissipated at the yield lines,

$$\{2m\frac{(1-\alpha)}{\beta}\mu\frac{b}{a}+8m\frac{a}{b}\frac{\beta}{(1-\alpha)}\}\Delta = q\frac{ab}{12\beta}(1-\alpha)(6\beta-4\beta^2-3\alpha^2+3\alpha)\Delta$$

Solving for m,

$$m = q\frac{b^2}{24}\frac{(1-\alpha)^2\{3\alpha(1-\alpha)\beta+6\beta^2-4\beta^3\}}{\{4\beta^2+(1-\alpha)^2\mu\frac{b^2}{a^2}\}}$$

In order to calculate the maximum value of m required, set dm/dβ = 0

$$\{4\beta^2+(1-\alpha)^2\mu\frac{b^2}{a^2}\}\{3\alpha(1-\alpha)+12\beta-12\beta^2\}-\{3\alpha(1-\alpha)\beta+6\beta^2-4\beta^3\}8\beta=0$$

The resulting quadratic equation in β can be solved numerically for specific values of the parameters and the corresponding value of m can be determined. The above equations are valid only for $0.5(1-\alpha)\le\beta\le0.5$.

8.9.10.4 Calculation of moment of resistance

For calculating the required ultimate moment, ultimate moment from all the three modes are calculated and the largest value is chosen. Results of calculations for a/b = 1.5, $\mu = 0.5$ for a range of $0 \le \alpha \le 0.9$ are shown in Table 8.6.

Table 8.6 Collapse load for a simply supported slab with a hole

α	Mode 1		Mode 2		Mode 3	
	β	m/ (qb²)	β	m/ (qb²)	β	m/ (qb²)
0	0.312	**0.0730**	N/A	-	N/A	-
0.05	0.317	**0.0753**	0.5*	N/A	0.5*	N/A
0.10	0.320	**0.0770**	0.5*	0.0716	0.5*	0.0336
0.20	0.322	**0.0779**	0.5*	0.0742	0.5*	0.0325
0.30	0.317	**0.0760**	0.4984	0.0767	0.5*	0.0290
0.40	0.300*	0.0701	0.4549	0.0754	0.5*	0.0242
0.50	0.250*	0.0607	0.4601	**0.0705**	0.5*	0.0189
0.6	0.20*	0.0474	0.3383	**0.0627**	0.2*	0.0136
0.7	0.15*	0.0316	0.2673	**0.0525**	0.15*	0.0120
0.8	0.10*	0.0158	0.1875	**0.0404**	0.10*	0.0074
0.9	0.05*	0.0041	0.10*	**0.0272**	0.05*	0.0036

*Not a stationary minimum.

In some cases the stationary minimum value of β is obtained in the non–valid region. In such cases the minimum value of m/(qb²) has been calculated by limiting the value of β to the valid region. This is indicated in the table by *. It is noticed that up to α ≈ 0.25, Mode 1 governs and afterwards mode 2 governs. It appears that mode 3 never governs.

8.9.11 Slab–Beam Systems

Combined slab–beam systems are commonly met in practice. Fig.8.44 shows a typical case of a slab supported on beams cast integral with slabs, and which in turn are supported on columns at the corners of the rectangle. In considering this type of system, it is important to investigate yield line collapse modes involving independent collapse of the slab only and *combined* slab–beam collapse.

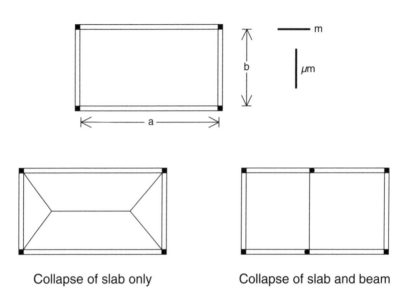

Collapse of slab only   Collapse of slab and beam

Fig.8.44 Collapse of beam–slab systems.

If the torsional strength of the supporting beams is ignored, then the slab can be assumed to be simply supported on beams and the collapse of the slab only is treated as the collapse of a simply supported slab as discussed in section 8.9.6. The calculated moment of resistance is the minimum that should be provided in the slab.

In considering the combined slab–beam collapse, if the deflection at the plastic hinge is Δ, then the total rotation at the plastic hinge is
$$\theta_n = 2(\Delta/0.5a) = 4\Delta/a$$
The work done by external loads is
$$W = qb\,(\Delta a/2) = 0.5q\,a\,b\,\Delta$$
If the moment capacity of the beams is M_b, then the energy dissipated at the plastic hinge due to slab and beam is
$$E = (\mu m\,b + 2M_b)\,\theta_n = (m\,b + 2M_b)\,4\Delta/a$$
Equating $W = E$,
$$qa^2b/8 = \mu m\,b + 2M_b$$
Knowing m,
$$M_b = qa^2b/16 - \mu m\,b/2$$

It is possible to increase the value of m to a value larger than the minimum for collapse of slab only and provide lighter beams. For example using the slab designed in section 8.8.6.1, $q = 16$ kN/m^2, $a = 4.5$ m, $b = 3.0$ m, $\mu = 0.5$, $m = 10.51$ kNm/m, $\mu\, m = 5.26$ kNm/m, $M_b = 52.86$ kNm.

If it is decided to decrease the moment capacity of the beams towards the supports, then other possible collapse modes such as that shown in Fig.8.45 need to be investigated.

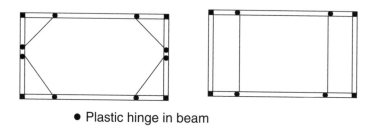

● Plastic hinge in beam

Fig.8.45 Collapse of beam–slab systems.

8.9.12 Corner Levers

In sections 8.9.6 and 8.9.7 the yield lines for both simply supported and continuous slabs were assumed to run directly into the corners (Fig.8.34 and Fig.8.35). This situation will develop only if there is sufficient top steel at the corner region and the corner is held down. However if the corners are not held down then the yield line will divide to form a corner lever as shown in Fig.8.46. Two possible situations occur.

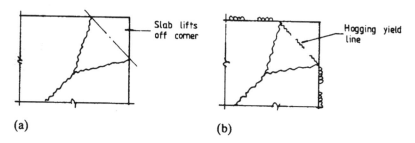

(a)   **(b)**

Fig.8.46 (a) Corner not held down; (b) corner held down.

1. Simply supported corner not held down: In this case the slab lifts off the corner and the sagging yield line divides as shown in Fig.8.46(a) and the triangular portion rotates about the chain–dotted line.

2. Simply supported corner held down: In this case the sagging yield line divides and a hogging yield line forms as shown in Fig.8.46(b).

Solutions have been obtained for these cases which show that for a 90° corner, the corner lever mechanism decreases the overall strength of the slab by about 10%. In the case of slabs with acute corners the reduction in the calculated ultimate load due to corner levers is much larger. The reinforcement should be increased accordingly when the simplified solution is used. The top reinforcement commonly known as torsional reinforcement will prevent cracking in continuous slabs on the corner lever hogging yield line.

8.9.13 Collapse Mechanisms with More than One Independent Variable

The collapse mechanisms considered previously were governed by a single variable β. Unfortunately this is not always the case. Fig.8.47 shows a case of a slab clamped on two adjacent edges and the other two edges simply supported. In this case the collapse mechanism is defined by three independent variables β_1, β_2 and β_3. The problem of finding the maximum (or minimum) value of a function of several variables is not a trivial task. Fortunately computer programs are available for solving such problems.

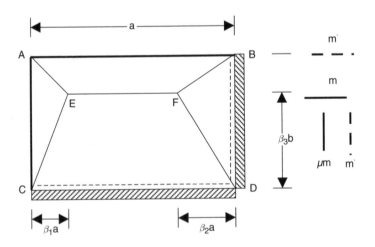

Fig.8.47 Collapse mode for a slab with two adjacent edges discontinuous.

8.9.14 Circular Fans

When concentrated loads act, flexural failure modes are likely to involve concentration of yield lines around the loaded area. This generally involves curved negative moment yield lines with radial positive moment yield lines as shown in Fig.8.48.

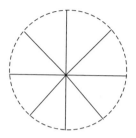

Fig.8.48 Yield lines in a circular fan.

If the moment of resistance is same in both directions and the radius of the fan is r, then the energy dissipated can be calculated by assuming that the deflection at the centre is Δ.

1. Energy dissipated at yield line

(i) Negative yield line

The total length L of the negative yield line is $2\pi r$
rotation θ_n at the yield line is Δ/r
moment of resistance is m'
Therefore energy dissipated is

$$E_1 = m' (2\pi r) \Delta/r = 2\pi \, m' \, \Delta$$

(ii) Positive yield lines

As the reinforcement in each direction is same, the slab is isotropically reinforced. Therefore the x and y axes for each triangular segment can be different. Assuming that the x and y–axes coincide with the radial and tangential directions of the circle, each segment rotates about the tangent only. The projection of the yield lines of each segment on the tangent is equal to the arc length corresponding to that segment.

The total projected length L of all positive yield line is $2\pi r$
The tangential rotation θ_n at the yield line is Δ/r
moment of resistance is m.
Therefore energy dissipated is

$$E_2 = m(2\pi r) \Delta/r = 2\pi \, m \, \Delta$$

The total energy dissipation is

$$E = E_1 + E_2 = 2\pi (m' + m) \Delta$$

2. Work done by external loads

Let q be the uniformly distributed load due to self weight and other externally applied loads and P is the concentrated load at the centre of the circle. The concentrated load could be an external load or a reaction from a column as in flat slab construction. The work done by the external uniformly distributed load q is calculated by noting that at the centroid of each triangular segment, deflection is $\Delta/3$ and the total load is $q(\pi\, r^2)$. Therefore

$$W = q(\pi\, r^2)\, \Delta/3 + P\, \Delta$$

Equating E and W,

$$(m' + m) = q\, r^2\, /6 + P/\, (2\, \pi)$$

8.9.14.1 Collapse mechanism for a flat slab floor

Fig.8.49 shows a flat slab floor with columns spaced at L_x and L_y in the x and y-directions respectively. If a collapse mechanism is postulated where the entire floor deflects by Δ with circular fans around columns as shown in Fig.8.49, then in any one panel

(i) Energy dissipated at yield lines
From section 8.9.14 above,

$$E = 2\pi\, (m' + m)\, \Delta$$

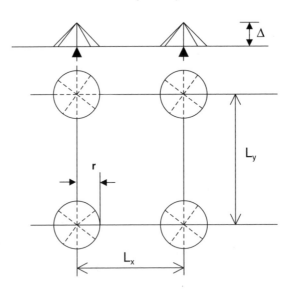

Fig.8.49 Collapse of a flat slab floor

(ii) Work done by external load

(a) Uniformly distributed load outside the circular fans
$$\text{Area} = L_x\, L_y - \pi\, r^2,\quad \text{deflection} = \Delta$$
(b) Uniformly distributed load inside the circular fans
$$\text{Area} = \pi\, r^2,\quad \text{deflection} = 2\Delta/3$$

Total work done is

$$W = q(L_x L_y - \pi r^2)\Delta + q \pi r^2 (2\Delta/3)$$
$$W = q(L_x L_y - \pi r^2/3)\Delta$$

Equating W and E,

$$m + m' = \frac{q}{2\pi} \{L_x L_y - \frac{\pi r^2}{3}\}$$

8.9.15 Design of a Corner Panel of Floor Slab Using Yield Line Analysis

A square corner panel of a floor slab simply supported on the outer edges on steel beams and continuous over the interior beams is shown in Fig.8.50. The design ultimate load is 12.4 kN/m². The slab is to be 175mm thick and reinforced equally in both directions. The moment of resistance in the hogging and sagging yield lines is to be the same. The materials are grade C30 concrete and grade 250 reinforcement. Design the slab using the yield line method.

The yield line pattern, which is symmetric about the diagonal, depends on one variable β. Assuming the deflection at the meeting point of the sagging yield lines as Δ,

1. Energy dissipated at yield lines

(i) Energy dissipated by the positive yield lines in the left and bottom triangles
(a) Bottom triangle: Rotates about x–axis only.
$$\theta_x = \Delta/ (6\beta), \quad m_x = m, \quad l_x = 6$$
(b) Left triangle: Rotates about y–axis only.
$$\theta_y = \Delta/ (6\beta), \quad m_y = m, \quad l_y = 6$$
Total energy dissipation E_1 is
$$E_1 = 2\{m \, 6 \, \Delta/ (6\beta)\} = 2m \, \Delta/\beta$$

(ii) Energy dissipated by the positive and negative yield lines in the top and right triangles
Note that $m_y = (m+m')$ accounts for both positive and negative yield lines in the triangle.
(a) Right triangle Rotates about y–axis only.
$$\theta_y = \Delta/ (6 - 6\,\beta), \quad m_y = (m + m'), \quad l_y = 6$$
(b) Top triangle: Rotates about x–axis only.
$$\theta_x = \Delta/ (6 - 6\,\beta), \quad m_x = (m + m'), \quad l_x = 6$$
Total energy dissipation E_2 is
$$E_2 = 2(m + m') \Delta/ (1 - \beta)$$
The total energy dissipated E by all yield lines is
$$E = E_1 + E_2 = 2\{(m + m')/ (1 - \beta) + m/ \beta\} \Delta$$
If $m = m'$,
$$E = 2m\{2/ (1 - \beta) + 1/ \beta\} \Delta$$

2. External work done by the loads
(i) Bottom and left triangles

$$\text{Area} = 0.5 \times 6 \times (6\ \beta), \quad \text{deflection} = \Delta/3$$
$$W_1 = 2[q\{0.5 \times 6 \times (6\ \beta)\}\ \Delta/3] = 12q\beta\ \Delta$$

(ii) Top and right triangles

$$\text{Area} = 0.5 \times 6 \times (6 - 6\ \beta), \quad \text{deflection} = \Delta/3$$
$$W_2 = 2[q\{0.5 \times 6 \times (6 - 6\ \beta)\}\ \Delta/3] = 12q(1 - \beta)\ \Delta$$

The total work W done is

$$W = W_1 + W_2 = 12\ q\ \Delta$$

3. Calculation of moment capacity required

Equating E and W, and solving for m

$$m = 6q\ \frac{\beta - \beta^2}{(1 + \beta)}$$

For maximum m, $dm/d\beta = 0$,

$$(1 + \beta)\ (1 - 2\ \beta) - (\beta - \beta^2) = 0$$

Simplifying,

$$\beta^2 + 2\beta - 1 = 0, \beta = (\sqrt{2} - 1) = 0.4142$$

Substituting for β, $m = 1.03q$. If $q = 12.4$ kN/m^2, $m = 12.76$ kNm/m

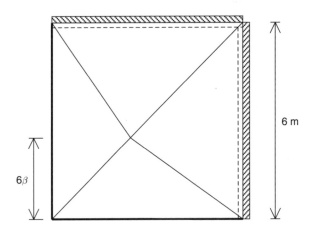

Fig.8.50 Corner slab.

4. Design for flexure

Assuming 10 mm diameter bars and 25 mm cover, the effective depth d of the inner layer is

$$d = 175 - 25 - 10 - 5 = 135 \text{ mm}$$
$$k = m/(bd^2\ f_{cu}) = 12.76 \times 10^6/ (1000 \times 135^2 \times 30) = 0.023 < 0.156$$
$$z/d = 0.5 + \sqrt{(0.25 - 0.023/0.9)} = 0.97 > 0.95$$
$$z = 0.95 \times 135 = 128 \text{ mm}$$
$$A_s = 12.76 \times 10^6/ (0.95 \times 250 \times 128) = 420 \text{ mm}^2/\text{m}$$

Increase the steel area by 10% to 462 mm²/m to allow for the formation of corner levers.

Provide 10 mm diameter bars at 150 mm centres to give a steel area of 523 mm²/m. The minimum area of reinforcement is

$A_s = (0.24/100) \times 175 \times 1000 = 420$ mm²/m

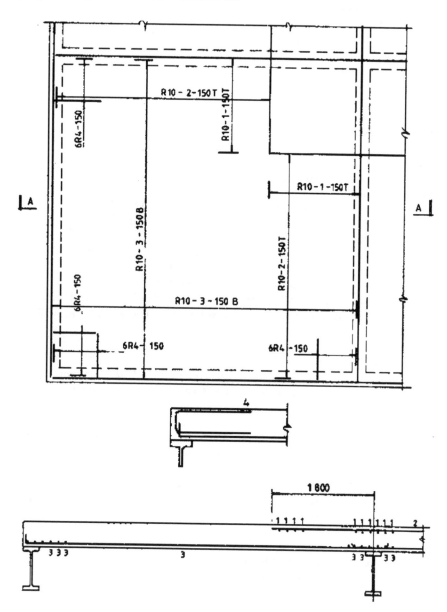

Fig.8.51 (a) Plan; (b) corner detail; (c) section AA.

Using the simplified rules for curtailment of bars in slabs from Fig. 3.25 of the code, the negative steel is to continue for 0.3 of the span, i.e. 1800 mm past the interior supports. The positive steel is to run into and be continuous through the supports.

5. Check shear capacity
The shear at the continuous edge is
$$V = 12.4 \times (6/2) + 12.76/6 = 39.3 \text{kN} /\text{m}$$
$$v = 39.3 \times 10^3 / (1000 \times 135) = 0.29 \text{ N/mm}^2$$
$$100 A_s/ (bd) = 100 \times 523/ (1000 \times 135) = 0.39 < 3.0$$
$$400/135 = 2.96 > 1.0$$
$$v_c = 0.79 \times (0.39)^{1/3}(2.96)^{1/4}(30/25)^{1/3}/1.25 = 0.64 \text{ N/mm}^2$$
$$v < v_c$$

The shear stress is satisfactory.

6. Deflection
From Table 3.9 of the code, the basic span/d ratio is 26.
Modification factor for tension steel:
$$m/ (bd^2) = 12.76 \times 10^6/ (1000 \times 135^2) = 0.7$$
$$f_s = (2/3) \times 250 \times (462/523) = 147 \text{ N/mm}^2$$
The modification factor is
$$0.55 + (477 - 147)/ \{120 \times (0.9 + 0.7)\} = 2.27 > 2.0$$
$$\text{Allowable span/d ratio} = 26 \times 2.0 = 52$$
$$\text{Actual span/d ratio} = 6000/135 = 44$$
The slab is satisfactory with respect to deflection.   Note that if grade 460 reinforcement is used, then a deeper slab would be required to comply with limit on deflection.

7. Cracking
The minimum clear distance between bars is not to exceed $3d$ = 405 mm. The slab depth 175 mm does not exceed 250 mm and so no further checks are required.

8. Reinforcement details
The reinforcement is shown in Fig.8.51. Note that U–bars are provided at the corners to act as torsion reinforcement. This design should be compared with the example in section 8.6.6.
   If the slab was supported on reinforced concrete L-beams on the outer edges, a value for the ultimate negative resistance moment at these edges could be assumed and used in the analysis.

**8.9.16 Derivation of BS 8110 Moment and Shear Coefficients
      for the Design of Restrained Slabs**

Bending moment and shear force coefficients in Tables 3.14 and 3.15 of the code for the design of two-way restrained slabs with corners held down and with provision for resisting torsion are derived on the basis of Yield Line analysis. The ratio of negative moment to positive moment is kept constant at 1.33. The 'long span' moments derived for a square slab are assumed to hold good for other values of the aspect ratio. Yield line analysis assumes that reinforcement in each direction is uniformly distributed over the width but the code recommends that the main steel is provided only in the middle strip which is 3/4 times the relevant width and only minimum steel in the edge strips. Therefore the value obtained from the Yield line analysis is multiplied by 4/3. The shear in the slab is calculated by assuming that the total load on the support is uniformly distributed over the middle three quarters of the beam span (see Fig.8.15)

8.9.16.1 Simply supported slab (case 9 in BS Table 3.14)

Using the formulae derived in section 8.9.6,

$$2m\frac{b}{a}\frac{\mu}{\beta} + 4m\frac{a}{b} = q\frac{ab}{6}(3-2\beta)$$

Multiplying through by (b/a) and simplifying, the above equation becomes

$$m[\frac{b}{a}]^2\frac{\mu}{\beta} + 2m = q\frac{b^2}{12}(3-2\beta)$$

(i) Square slab

$$a/b = 1, \quad \mu m = m, \quad \beta = 0.5, \quad m = 0.0417 \, qb^2.$$

Multiplying this value for m by 4/3, $m = 0.056 \, qb^2.$, $\beta_{sx} = 0.056$.

(ii) Rectangular slab

$$a/b > 1.0.$$

Keep $\mu m = 0.0417 \, qb^2$ as constant for all values of a/b. Substituting this value in the above equation for m and simplifying,

$$m = q \, b^2 \{0.125 - 0.0833\beta - 0.0209(\frac{b}{a})^2\frac{1}{\beta}\}$$

For a maximum value of m, dm/dβ = 0.

$$-0.0833 + 0.0209(\frac{b}{a})^2\frac{1}{\beta^2} = 0, \beta = 0.5\frac{b}{a}$$

Substituting this value of β in the equation for m

$$m = q \, b^2 \{0.125 - 0.0833(\frac{b}{a})\}$$

Multiplying this value for m by 4/3,

$$4/3 \, m = q \, b^2 \{0.1667 - 0.1111(\frac{b}{a})\}, \beta_{sx} = \{0.1667 - 0.1111(\frac{b}{a})\}$$

$$\mu m = 0.0417 \, qb^2$$

Multiplying this value for μm by 4/3,

$$\mu m = 0.0556 \, qb^2, \beta_{sy} = 0.056$$

(iii) Shear coefficients

$$\beta = 0.50 \text{ b/a}$$

Short beam

$$\text{Load} = 0.5 \times q \times b \times \beta a.$$

Spreading this uniformly over a length of 0.75 b,

$$v_x = qb\{0.6667\beta\frac{a}{b}\} = 0.333qb, \beta_{vy} = 0.3333$$

Long beam

$$\text{Load} = 0.5 \times q \times 0.5b \times (2 - 2\beta) \text{ a.}$$

Spreading this uniformly over a length of 0.75 a,

$$v_y = qb\{0.6667(1-\beta)\} = qb\{0.6667(1-0.5\frac{b}{a})\}, \beta_{vx} = \{0.6667(1-0.5\frac{b}{a})$$

8.9.16.2 Clamped slab (case 1 in BS Table 3.14)

Using the formula derived in section 8.9.7,

$$2(\frac{b}{a}\frac{\mu m}{\beta} + 2\frac{a}{b}m) + 2m'(\frac{b}{a}\frac{\gamma m'}{\beta} + 2\frac{a}{b}m') = q\frac{ab}{6}(3-2\beta)$$

Multiplying through by b/a and simplifying, the above equation becomes

$$[\frac{b}{a}]^2\frac{(\mu m + \gamma m')}{\beta} + 2(m+m') = q\frac{b^2}{12}(3-2\beta)$$

(i) Square slab

$$a/b = 1, \quad \mu m = m, \quad \gamma m' = m' = 1.33 \text{ m}, \quad \beta = 0.5$$
$$m = 0.0179 \text{ qb}^2, \quad m' = 0.024 \text{ qb}^2.$$

Multiplying these value for μm by 4/3,

$$m = 0.024 \text{ qb}^2, \quad \beta_{sx} = 0.024 \text{ and m}' = 0.032 \text{ qb}^2, \quad \beta_{sy} = 0.032.$$

(ii) Rectangular slab: a/b > 1.0

Keep $\mu m = 0.0179 \text{ qb}^2$, $\gamma m' = 0.024 \text{ qb}^2$ and $m' = 1.33m$ as constant for all values of a/b. Substituting these values in the above equation for m and simplifying,

$$m = qb^2\{0.0536 - 0.0357\beta - 0.009(\frac{b}{a})^2\frac{1}{\beta}\}$$

For a maximum value of m, dm/dβ = 0.

$$-0.0357 + 0.009(\frac{b}{a})^2\frac{1}{\beta^2} = 0, \beta = 0.502\frac{b}{a}$$

Substituting this value of β in the equation for m

$$m = qb^2\{0.0536 - 0.0359(\frac{b}{a})\}$$

Multiplying this value for μm by 4/3,

$$m = q\,b^2\{0.0715 - 0.0479(\frac{b}{a})\}$$

$$m^{'} = 1.33m = q\,b^2\{0.0953 - 0.0639(\frac{b}{a})\}$$

For positive moment at mid-span in short span

$$\beta_{sx} = \{0.0715 - 0.0479(\frac{b}{a})\}$$

Negative moment at short edge

$$\beta_{sx} = \{0.0953 - 0.0639(\frac{b}{a})\}$$

$$\mu m = 0.0179\ qb^2,\ \gamma m^{'} = 0.024\ qb^2$$

Multiplying the above values by 4/3

$$\mu m = 0.0239\ qb^2,\ \gamma m^{'} = 0.032\ qb^2$$

In the long span direction, coefficients for positive and negative moments are respectively

$$\beta_{sy} = 0.0239\ \text{and}\ \beta_{sy} = 0.032$$

(iii) Shear coefficients

As for the simply supported slab in section 8.9.16.1

8.9.16.3 Slab with two short edges discontinuous (case 5 in BS Table 3.14)

Fig.8.52 shows a slab with two short edges discontinuous.

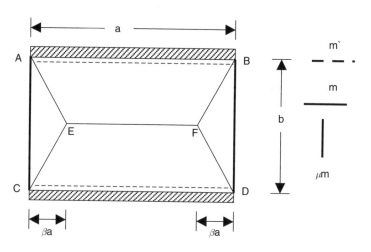

Fig.8.52 Collapse mode for a slab with two short edges discontinuous.

Using the formulae derived in section 8.8.16.2 but with $\gamma m' = 0$,

$$[\frac{b}{a}]^2 \frac{\mu m}{\beta} + 2(m+m') = q\frac{b^2}{12}(3-2\beta)$$

(i) Square slab

$$a/b = 1, \quad \mu m = m, \quad m' = 1.33 \text{ m}.$$

For maximum m, $dm/d\beta = 0$

$$\beta^2 + 0.4286\beta - 0.3214 = 0, \quad \beta = 0.3918.$$

Using $\beta = 0.3918$, $m = 0.026 \text{ qb}^2$, $m' = 0.034 \text{ qb}^2$.

Multiplying these values by 4/3,

$$m = 0.034 \text{ qb}^2, \quad \beta_{sx} = 0.034 \text{ and } m' = 0.046 \text{ qb}^2, \quad \beta_{sy} = 0.046.$$

(ii) Rectangular slab: a/b > 1.0

Keep $\mu m = 0.026 \text{ qb}^2$ and $m' = 1.33 \text{ m}$ as constant for all values of a/b.
Substituting these values in the equation for m and simplifying,

$$m = q b^2 \{0.0536 - 0.0357\beta - 0.0056(\frac{b}{a})^2 \frac{1}{\beta}\}$$

For a maximum value of m, $dm/d\beta = 0$

$$-0.0357 + 0.0056(\frac{b}{a})^2 \frac{1}{\beta^2} = 0, \quad \beta = 0.3950\frac{b}{a}$$

Substituting this value of β in the equation for m

$$m = q b^2 \{0.0536 - 0.0282(\frac{b}{a})\}$$

Multiplying the above value by 4/3,

$$m = q b^2 \{0.0715 - 0.0376(\frac{b}{a})\}, \quad m' = 1.333 m = q b^2 \{0.0953 - 0.0501(\frac{b}{a})\}$$

Short span
(a) For positive moment

$$\beta_{sx} = \{0.0715 - 0.0376(\frac{b}{a})\}$$

(b) Negative moment at short edge

$$\beta_{sx} = \{0.0953 - 0.0501(\frac{b}{a})\}$$

$$\mu m = 0.0260 \text{ qb}^2$$

Multiplying the above values by 4/3, $\mu m = 0.0347 \text{ qb}^2$

(c) For positive moment at mid-span in long span

$$\beta_{sy} = 0.0347$$

(iii) Shear coefficients

As for the simply supported slab but use:

$$\beta = 0.3918, \text{ a/b} = 1$$
$$\beta = 0.3950 \text{ b/a}, \text{ a/b} > 1.0.$$

$$v_x = qb\{0.6667\beta\frac{a}{b}\}$$

$$v_x = 0.2633\,qb, \beta_{vy} = 0.2633, b/a < 1.0$$

$$v_x = 0.2612\,qb, \beta_{vy} = 0.2612, b/a = 1.0$$

$$v_y = qb\{0.6667(1-\beta)\}$$

$$v_y = qb\{0.6667(1-0.3950\frac{b}{a})\},\ \beta_{vy} = \{0.6667(1-0.3950\frac{b}{a}), b/a < 1.0$$

$$v_y = qb\{0.6667(1-0.3918)\},\ \beta_{vy} = 0.4055, b/a = 1.0$$

8.9.16.4 Slab with two long edges discontinuous (case 6 in BS Table 3.14)

Fig.8.53 shows a slab with two long edges discontinuous. Using the formulae derived in section 8.9.16.2 and substituting m` = 0,

$$[\frac{b}{a}]^2 \frac{(\mu m + \gamma m`)}{\beta} + 2m = q\frac{b^2}{12}(3-2\beta)$$

(i) Square slab

$$a/b = 1, \mu m = m, \gamma m` = 1.33\ m.$$

For maximum, dm/ $\beta = 0$.

$$\beta^2 + 2.333\beta - 1.75 = 0, \beta = 0.5972 > 0.5$$

Restrict $\beta = 0.5$ as this is the maximum value permissible. Using $\beta = 0.5$,
m = 0.025 qb^2, γm` = 0.033 qb^2.
Multiplying these values by 4/3,

$$m = \mu m = 0.033\ qb^2,\ \beta_{sx} = 0.033\ \text{and}\ \gamma m` = 0.044\ qb^2,\ \beta_{sy} = 0.044.$$

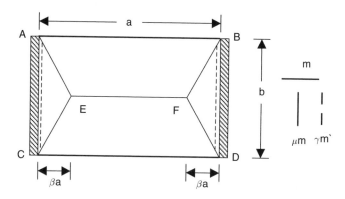

Fig.8.53 Collapse mode for slab with two long edges discontinuous.

(ii) Rectangular slab: a/b > 1.0

Keep $\mu m = 0.025\ qb^2$ and $\gamma m` = 0.033\ qb^2$ as constant for all values of a/b.
Substituting these values in the equation for m and simplifying,

$$m = q\,b^2\{0.125 - 0.0833\beta - 0.029(\frac{b}{a})^2 \frac{1}{\beta}\}$$

For a maximum value of m, dm/d β = 0.

$$-0.0833 + 0.029(\frac{b}{a})^2\frac{1}{\beta^2} = 0, \ \beta = 0.59\frac{b}{a}$$

Substituting this value of β in the equation for m and simplifying,

$$m = qb^2\{0.125 - 0.098(\frac{b}{a})\}$$

Multiplying the above value by 4/3, $m = qb^2\{0.1667 - 0.1307(\frac{b}{a})\}$

(a) For positive moment at mid-span in short span

$$\beta_{sx} = \{0.1667 - 0.1307(\frac{b}{a})\}, b/a<1.0, \ \beta_{sx} = 0.033, b/a = 1.0$$

(b) Long span direction

$$\mu m = 0.033 \ qb^2, \gamma m' = 0.044 \ qb^2$$

Positive moment

$$\mu m = 0.033 \ qb^2, \ \beta_{sy} = 0.033$$

Negative moment

$$\gamma m' = 0.044 \ qb^2, \ \beta_{sy} = 0.044$$

(iii) Shear coefficients
As for the simply supported slab but use:
$\beta = 0.5$ for a/b = 1 and for a/b > 1.0, $\beta = 0.59$ b/a

$$v_x = qb\{0.6667\beta\frac{a}{b}\}$$

$$v_x = 0.3933 \ qb, \beta_{vy} = 0.3933, b/a<1.0$$

$$v_x = 0.3333 \ qb, \beta_{vy} = 0.3333, b/a = 1.0$$

$$v_y = qb\{0.6667(1-\beta)\}$$

$$v_y = qb\{0.6667(1-0.59\frac{b}{a})\}, \ \beta_{vy} = \{0.6667(1-0.59\frac{b}{a}), b/a<1.0$$

$$v_y = qb\{0.6667(1-0.5)\}, \ \beta_{vy} = 0.3333, b/a = 1.0$$

8.9.16.5 Slab with one long edge discontinuous (case 3 in BS Table 3.14)

Fig.8.54 shows the collapse mode which is governed by two parameters β_1 and β_2.
It can be shown that the basic equation for solving the problem is

$$2[\frac{b}{a}]^2\frac{(\mu m+\gamma m')}{\beta_1} + \frac{(m+m')}{\beta_2} + \frac{m}{(1-\beta_2)} = q\frac{b^2}{6}(3-2\beta_1)$$

(i) Square slab
a/b = 1, $\mu m = m$, $\gamma m' = m' = 1.33$ m.

$$m\{\frac{4.6667}{\beta_1} + \frac{2.333}{\beta_2} + \frac{1}{(1-\beta_2)}\} = q\frac{b^2}{6}(3-2\beta_1)$$

For maximum m, $dm/d\beta_1 = 0$ and $dm/d\beta_2 = 0$.

$dm/d\beta_2 = 0$ leads to $\beta_2^2 - 3.5\beta_2 + 1.75 = 0$, $\beta_2 = 0.604 < 1.0$

Value of $\beta_2 = 0.604$ is independent of β_1. Using this value of β_2,

$dm/d\beta_1 = 0$ leads to $\beta_1^2 + 1.4612\beta_1 - 1.0959 = 0$, $\beta_1 = 0.546 > 0.5$

Using $\beta_1 = 0.5$ and $\beta_2 = 0.604$, $m = \mu m = 0.020\ qb^2$, $m' = \gamma m' = 0.027\ qb^2$

Multiplying these values by 4/3,

$m = \gamma m' = 0.027\ qb^2$, $\beta_{sx} = 0.027$ and $m' = \gamma m' = 0.036\ qb^2$, $\beta_{sy} = 0.036$.

(ii) Rectangular slab: a/b > 1.0

Keep $\mu m = 0.020\ qb^2$ and $\gamma m' = 0.027\ qb^2$ as constant for all values of a/b and $m' = 1.33\ m$. Substituting these values in the equation for m and simplifying,

$$m\{\frac{2.333}{\beta_2} + \frac{1}{(1-\beta_2)}\} = qb^2\{0.5 - 0.333\beta_1 - \frac{0.094}{\beta_1}(\frac{b}{a})^2\}$$

$dm/d\beta_2 = 0$ leads to $\beta_2^2 - 3.5\beta_2 + 1.75 = 0$, $\beta_2 = 0.604 < 1.0$

$dm/d\beta_1 = 0$ leads to $-0.333 + 0.094(\frac{b}{a})^2\frac{1}{\beta_1^2} = 0$, $\beta_1 = 0.531\frac{b}{a}$

Substituting these values of β_1 and β_2 in the equation for m and simplifying,

$$m = qb^2\{0.0783 - 0.0554(\frac{b}{a})\}$$

Multiplying the above value by 4/3,

$$m = qb^2\{0.1044 - 0.0739(\frac{b}{a})\}, \quad m' = 1.33\ m = qb^2\{0.1392 - 0.0985(\frac{b}{a})\}$$

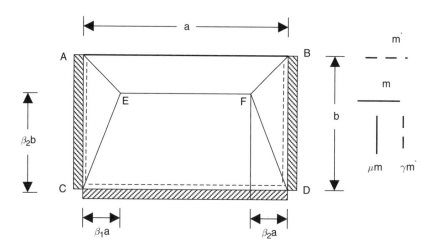

Fig.8.54 Collapse mode for a slab with one long edge discontinuous.

(i) Short span
(a) For positive moment

$$m = qb^2\{0.1044 - 0.0739(\tfrac{b}{a})\}, \quad \beta_{sx} = \{0.1044 - 0.0739(\tfrac{b}{a})\}, b/a < 1$$

$$m = qb^2 0.027, \quad \beta_{sx} = 0.027, b/a = 1$$

(b) Negative moment at short edge

$$m' = 1.33m = qb^2\{0.1392 - 0.0985(\tfrac{b}{a})\}, \quad \beta_{sx} = \{0.1392 - 0.0985(\tfrac{b}{a})\}, b/a < 1$$

$$m' = qb^2 0.036, \quad \beta_{sx} = 0.036, b/a = 1$$

(ii) Long span

$$\mu m = 0.027\, qb^2 \text{ and } \gamma m' = 0.036\, qb^2$$

For positive and negative moments, the moment coefficients β_{sy} are 0.027 and 0.036 respectively.

(iii) Shear coefficients
Use and $\beta_1 = 0.5$ for a/b = 1 and for a/b > 1.0, use $\beta_1 = 0.531\, b/a$. $\beta_2 = 0.604$ for all aspect ratios.

(a) Short beam: Spread the load uniformly over a length of 0.75 b.

$$\text{Load} = 0.5 \times q \times b \times \beta_1 a.$$

$$v_x = qb\{0.6667\beta_1 \tfrac{a}{b}\}$$

$$v_x = 0.354qb, \beta_{vy} = 0.354, b/a < 1.0$$

$$v_x = 0.3333\, qb, \beta_{vy} = 0.3333, b/a = 1.0$$

(b) Load on the longer beam: Spread the load on the beam uniformly over a length of 0.75 a and use $\beta_2 = 0.604$

Continuous end

$$v_y = 0.5 \times q \times \beta_2 b \times (2 - 2\beta_1)\, a$$

$$v_y = qb\{0.805(1 - \beta_1)\}$$

$$v_y = qb\{0.805(1 - 0.531\tfrac{b}{a})\}, \beta_{vy} = \{0.805(1 - 0.531\tfrac{b}{a}), b/a < 1.0$$

$$v_y = qb\{0.805(1 - 0.5)\}, \beta_{vy} = 0.403, b/a = 1.0$$

Simply supported end

$$v_y = 0.5 \times q \times (1 - \beta_2)\, b \times (2 - 2\beta_1)\, a$$

$$v_y = qb\{0.528(1 - \beta_1)\}$$

$$v_y = qb\{0.528(1 - 0.531\tfrac{b}{a})\}, \beta_{vy} = \{0.528(1 - 0.531\tfrac{b}{a}), b/a < 1.0$$

$$v_y = qb\{0.528(1 - 0.5)\}, \beta_{vy} = 0.264, b/a = 1.0$$

8.9.16.6 Slab with one short edge discontinuous (case 2 in nBS Table 3.14)

Fig.8.55 shows the collapse mode which is governed by two parameters β_1 and β_2. It can be shown that the basic equation for solving the problem is

$$[\frac{b}{a}]^2\{\frac{(\mu m + \gamma m')}{\beta_2} + \frac{\mu m}{\beta_1}\} + 4(m + m') = q\frac{b^2}{6}(3 - \beta_1 - \beta_2)$$

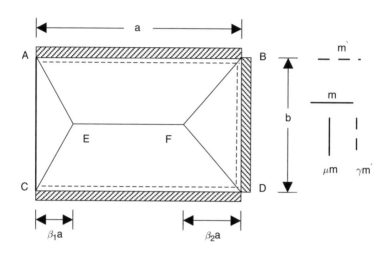

Fig.8.55 Collapse mode for a slab with one short edge discontinuous

(i) Square slab

$$a/b = 1, \mu m = m, \gamma m' = m' = 1.33 m.$$

$$m\{9.333 + \frac{1}{\beta_1} + \frac{2.333}{\beta_2}\} = q\frac{b^2}{6}(3 - \beta_1 - \beta_2)$$

Simplifying

$$m = \frac{qb^2}{6}\beta_1\beta_2\frac{(3 - \beta_1 - \beta_2)}{(9.333\beta_1\beta_2 + 2.333\beta_1 + \beta_2)}$$

For maximum m, $dm/d\beta_1 = 0$ and $dm/d\beta_2 = 0$.
$dm/d\beta_1 = 0$ leads to

$$A_1\beta_1^2 + B_1\beta_1 + C_1 = 0,$$
$$A_1 = 1 + 8\beta_2,$$
$$B_1 = -4\beta_2^2 + 0.85714\beta_2,$$
$$C_1 = 0.4286\beta_2^2 - 1.2857\beta_2$$

$dm/d\beta_2 = 0$ leads to

$$A_2\,\beta_2^2 + B_2\,\beta_2 + C_2 = 0,$$
$$A_2 = 1.0 + 9.333\beta_1,$$
$$B_2 = 4.6667\beta_1,$$
$$C_2 = 2.33\beta_1^2 - 7\beta_1$$

The values of β_1 and β_2 can be calculated as follows:
- Assume a value for β_2
- Calculate A_1, B_1 and C_1
- Solve the quadratic in β_1
- Using the calculated value of β_1, calculate A_2, B_2 and C_2
- Solve the quadratic in β_2
- Compare the assumed and calculated values of β_2
- Repeat calculations until the assumed and calculated values of β_2 differ by a very small value.
- Calculate the value of m/ (qb²)

Using this procedure, in this case
$\beta_1 = 0.4013$ and $\beta_2 = 0.5455$, m = 0.0213 qb², $m' = 0.0284$ qb².
Multiplying these values by 4/3,
$m = \mu m = 0.028$ qb², $\beta_{sx} = 0.028$ and $m' = \gamma m' = 0.038$ qb², $\beta_{sx} = 0.038$.

(ii) Rectangular slab: a/b > 1.0
Keep $m' = 1.33$ m, $\mu m' = 0.021$ qb² and $\gamma m' = 0.028$ qb² as constant for all values of a/b. Substituting these values in the equation for m and simplifying,

$$m = qb^2\{0.0536 - 0.01786\beta_1 - \frac{0.00228}{\beta_1}(\frac{b}{a})^2 - 0.01786\beta_2 - \frac{0.00533}{\beta_2}(\frac{b}{a})^2\}$$

$dm/d\beta_1 = 0$ leads to

$$-0.01786 + \frac{0.00228}{\beta_1^2}(\frac{b}{a})^2 = 0,\ \ \beta_1 = 0.3575\frac{b}{a}$$

$dm/d\beta_2 = 0$ leads to

$$-0.01786 + \frac{0.00533}{\beta_2^2}(\frac{b}{a})^2 = 0,\ \ \beta_2 = 0.5460\frac{b}{a}$$

Substituting these values of β_1 and β_2 in the equation for m and simplifying,

$$m = q\,b^2\{0.0536 - 0.0323(\frac{b}{a})\}$$

Multiplying the above value by 4/3, $m = q\,b^2\{0.0715 - 0.0431(\frac{b}{a})\}$

(i) Short span
(a) Positive moment

$$m = q\,b^2\{0.0715 - 0.0431(\frac{b}{a})\},\ \ \beta_{sx} = \{0.0715 - 0.0431(\frac{b}{a})\}, b/a < 1$$

$$m = q\,b^2 0.028\,, \quad \beta_{sx} = 0.028, b/a = 1$$

(b) Negative moment at short edge

$$m' = 1.33\,m = q\,b^2\{0.0953 - 0.0575(\frac{b}{a})\}\,, \quad \beta_{sx} = \{0.0953 - 0.0575(\frac{b}{a})\}, b/a < 1$$

$$m' = q\,b^2 0.038\,, \quad \beta_{sx} = 0.038, b/a = 1$$

(ii) Long span

$$\mu m = 0.028\ qb^2 \text{ and } \gamma m' = 0.038\ qb^2$$

For positive and negative moments, the moment coefficients are respectively.

$$\beta_{sy} = 0.028 \text{ and } \beta_{sy} = 0.038$$

(iii) Shear coefficients: Spread the load uniformly over a length of 0.75 b.

$$a/b = 1: \beta_1 = 0.4013 \text{ and } \beta_2 = 0.5455$$
$$a/b > 1.0: \beta_1 = 0.3575\ b/a \text{ and } \beta_2 = 0.546\ b/a.$$

Load on the shorter beam
(a) Continuous end

$$\text{Load} = 0.5 \times q \times b \times \beta_2\ a.$$

$$v_x = qb\{0.6667\beta_2 \frac{a}{b}\} = qb\{0.364\}, \beta_{vx} = 0.364$$

(b) Simply supported end

$$\text{Load} = 0.5 \times q \times b \times \beta_1\ a.$$

$$v_x = qb\{0.6667\beta_1 \frac{a}{b}\} = qb\{0.1589\}, \beta_{vx} = 0.1589, b/a < 1$$

$$v_x = qb\{0.6667\beta_1 \frac{a}{b}\} = qb\{0.2675\}, \beta_{vx} = 0.2675, b/a = 1$$

Load on the longer beam: Spread the load uniformly over a length of 0.75 a.

$$\text{Load} = 0.5 \times q \times 0.5b \times (2 - \beta_1 - \beta_2)\ a$$
$$v_y = qb\{0.3333(2 - \beta_1 - \beta_2)\}$$

$$v_y = qb\{0.333 \times (2 - 0.9035\frac{b}{a})\}, \quad \beta_{vy} = \{(0.667 - 0.301\frac{b}{a}), b/a < 1.0$$
$$v_y = qb\{0.333(2 - 0.3575 - 0.546)\}, \quad \beta_{vy} = 0.3655, b/a = 1.0$$

8.9.16.7 Slab with two adjacent edges discontinuous (case 4 in BS Table 3.14)

Fig.8.56 shows the collapse mode which is governed by three parameters β_1, β_2 and β_3. It can be shown that the basic equation for solving the problem is

$$[\frac{b}{a}]^2\{\frac{(\mu m + \gamma m')}{\beta_2} + \frac{\mu m}{\beta_1}\} + (m + m')\frac{1}{\beta_3} + m\frac{1}{(1 - \beta_3)} = q\frac{b^2}{6}(3 - \beta_1 - \beta_2)$$

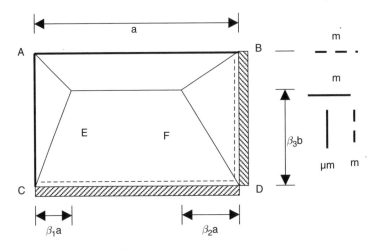

Fig.8.56 Collapse mode for a slab with two adjacent edges discontinuous.

(i) Square slab

$$a/b = 1, \mu m = m, \gamma m' = m' = 1.33 \text{ m.}$$

$$m\{\frac{1}{\beta_1} + \frac{2.333}{\beta_2} + \frac{2.333}{\beta_3} + \frac{1}{(1-\beta_3)}\} = q\frac{b^2}{6}(3 - \beta_1 - \beta_2)$$

For maximum m, $dm/d\beta_1 = 0$, $dm/d\beta_2 = 0$ and $dm/d\beta_3 = 0$.

$dm/d\beta_3 = 0$ leads to

$$-\frac{2.333}{\beta_3^2} + \frac{1}{(1-\beta_3)^2} = 0, \quad \beta_3^2 - 3.5\beta_3 + 1.75 = 0, \beta_3 = 0.604$$

Substituting this value of β_3 in the expression for m and simplifying,

$$m = \frac{qb^2}{6}\beta_1\beta_2 \frac{(3 - \beta_1 - \beta_2)}{(6.3884\beta_1\beta_2 + 2.333\beta_1 + \beta_2)}$$

For maximum m, $dm/d\beta_1 = 0$ and $dm/d\beta_2 = 0$.

$dm/d\beta_1 = 0$ leads to

$$A_1 \beta_1^2 + B_1 \beta_1 + C_1 = 0,$$
$$A_1 = 6.3384 \beta_2 + 2.3333,$$
$$B_1 = 2\beta_2,$$
$$C_1 = \beta_2^2 - 3\beta_2$$

$dm/d\beta_2 = 0$ leads to

$$A_2 \beta_2^2 + B_2 \beta_2 + C_2 = 0,$$
$$A_2 = 6.3384 \beta_1 + 1,$$
$$B_2 = 4.6667\beta_1,$$
$$C_2 = 2.33\beta_1^2 - 7\beta_1$$

The values of β_1 and β_2 can be calculated following the same procedure as in section 8.9.16.7.

In this case $\beta_1 = 0.39565$ and $\beta_2 = 0.64356$, m = 0.0261 qb^2, m' = 0.0348 qb^2.
Multiplying these values by 4/3,
$\quad$ m = μm = 0.035 qb^2, $\quad \beta_{sx} = 0.035$ and m' = γm' = 0.046 qb^2, $\quad \beta_{sx} = 0.046$.

(ii) Rectangular slab: a/b > 1.0
Keep μm = 0.0261 qb^2 and γm' = 0.0348 qb^2 as constant for all values of a/b.
Substituting these values in the equation for m and simplifying,

$$m = qb^2\{0.0783 - 0.0261\beta_1 - \frac{0.00407}{\beta_1}(\frac{b}{a})^2 - 0.0261\beta_2 - \frac{0.00955}{\beta_2}(\frac{b}{a})^2\}$$

dm/dβ_1 = 0 leads to

$$-0.0261 + \frac{0.00407}{\beta_1^2}(\frac{b}{a})^2 = 0, \ \beta_1 = 0.3949\frac{b}{a}$$

dm/dβ_1 = 0 leads to

$$-0.0261 + \frac{0.00955}{\beta_2^2}(\frac{b}{a})^2 = 0, \ \beta_2 = 0.6049\frac{b}{a}$$

Substituting these values of β_1 and β_2 in the equation for m and simplifying,

$$m = q\,b^2\{0.0783 - 0.0522(\frac{b}{a})\}$$

Multiplying the above value by 4/3,

$$m = q\,b^2\{0.1044 - 0.0696(\frac{b}{a})\}$$

Short span
(a) Positive moment

$$m = q\,b^2\{0.1044 - 0.0696(\frac{b}{a})\}, \ \beta_{sx} = \{0.1044 - 0.0696(\frac{b}{a})\}, b/a < 1$$

$$m = q\,b^2 0.035, \ \beta_{sx} = 0.035, b/a = 1$$

(b) Negative moment at short edge

$$m' = 1.33\,m = q\,b^2\{0.1392 - 0.0928(\frac{b}{a})\}, \ \beta_{sx} = \{0.1392 - 0.0928(\frac{b}{a})\}, b/a < 1$$

$$m' = 1.333 \times q\,b^2 0.035, \ \beta_{sx} = 0.047, b/a = 1$$

Long span
μm = 0.035 qb^2 and γm' = 0.046 qb^2
For positive and negative moments, the moment coefficients β_{sy} are 0.035 and 0.046 respectively.

(iii) Shear coefficients
$$a/b = 1, \beta_1 = 0.3957, \beta_2 = 0.6436$$
$$a/b > 1.0, \beta_1 = 0.395 \ b/a, \beta_2 = 0.605 \ b/a.$$

Shorter beam: Spread the load uniformly over a length of 0.75 b.

Continuous end

$$\text{Load} = 0.5 \times q \times b \times \beta_2\, a.$$

$$v_x = qb\{0.6667\beta_2\, \frac{a}{b}\}$$

$$v_x = qb\{0.6667\beta_2\, \frac{a}{b}\} = qb(0.403), \beta_{vx} = 0.403, b/a < 1.0$$

$$v_x = qb\{0.6667\beta_2\, \frac{a}{b}\} = qb(0.4291), \beta_{vx} = 0.4291, b/a = 1.0$$

Simply supported end

$$\text{Load} = 0.5 \times q \times b \times \beta_1\, a.$$

$$v_x = qb\{0.6667\beta_1\, \frac{a}{b}\} = qb\{0.264\}, \beta_{vx} = 0.264$$

Longer beam: Spread the load uniformly over a length of 0.75 a, and $\beta_3 = 0.604$.

Continuous end

$$\text{Load} = 0.5 \times q \times \beta_3\, b \times (2 - \beta_1 - \beta_2)\, a$$
$$v_y = qb\{0.4027(2 - \beta_1 - \beta_2)\}$$

$$v_y = qb\{0.4027(2 - 1.0007\frac{b}{a}\}, \beta_{vy} = 0.4027(2 - 1.0007\frac{b}{a}), b/a < 1.0$$

$$v_y = qb\{0.4027(2 - 1.0393\}, \beta_{vy} = 0.3869, b/a = 1.0$$

Simply supported end

$$\text{Load} = 0.5 \times q \times (1 - \beta_3)\, b \times (2 - \beta_1 - \beta_2)\, a$$
$$v_y = qb\{0.2640(2 - \beta_1 - \beta_2)\}$$

$$v_y = qb\{0.2640(2 - 1.0007\frac{b}{a}\}, \beta_{vy} = 0.2640(2 - 1.0007\frac{b}{a}), b/a < 1.0$$

$$v_y = qb\{0.4027(2 - 1.0393\}, \beta_{vy} = 0.3869, b/a = 1.0$$

8.9.16.8 Slab with only a short edge continuous (case 8 in BS Table 3.14)

Fig.8.57 shows the collapse mode which is governed by two parameters β_1 and β_2. It can be shown that the basic equation for solving the problem is

$$[\frac{b}{a}]^2 \{\frac{(\mu m + \gamma m')}{\beta_2} + \frac{\mu m}{\beta_1}\} + 4m = q\frac{b^2}{6}(3 - \beta_1 - \beta_2)$$

(i) Square slab

$$a/b = 1,\ \mu m = m,\ \gamma m' = 1.33\, m.$$

$$m\{\frac{1}{\beta_1} + \frac{2.333}{\beta_2} + 4\} = q\frac{b^2}{6}(3 - \beta_1 - \beta_2)$$

Simplifying,

$$m = \frac{qb^2}{6} \beta_1 \beta_2 \frac{(3 - \beta_1 - \beta_2)}{(4\beta_1\beta_2 + 2.333\beta_1 + \beta_2)}$$

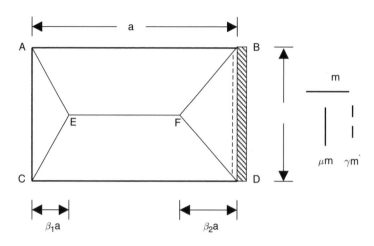

Fig.8.57 Collapse mode for a slab with one short edge continuous.

For maximum m, $dm/d\beta_1 = 0$ and $dm/d\beta_2 = 0$.
$dm/d\beta_1 = 0$ leads to

$$A_1 \beta_1^2 + B_1 \beta_1 + C_1 = 0,$$
$$A_1 = 4\beta_2 + 2.3333,$$
$$B_1 = 6\beta_2,$$
$$C_1 = \beta_2^2 - 3\beta_2$$

$dm/d\beta_2 = 0$ leads to

$$A_2 \beta_2^2 + B_2 \beta_2 + C_2 = 0,$$
$$A_2 = 4\beta_1 + 1,$$
$$B_2 = 4.6667\beta_1,$$
$$C_2 = 2.33\beta_1^2 - 7\beta_1$$

The values of β_1 and β_2 can be calculated following the same procedure as in section 8.9.16.6. In this case $\beta_1 = 0.2868$ and $\beta_2 = 0.65934$, m = 0.0311 qb^2, m` = 0.0414 qb^2. Multiplying these values by 4/3,
m = μm = 0.042 qb^2, $\beta_{sx} = 0.042$ and γm` = 0.055 qb^2, $\beta_{sx} = 0.055$.

(ii) Rectangular slab: a/b > 1.0. Keep μm = 0.031 qb^2 and γm` = 0.041 qb^2 as constant for all values of a/b. Substituting these values in the equation for m and simplifying,

$$m = qb^2\{0.125 - 0.0417\beta_1 - \frac{0.00776}{\beta_1}(\frac{b}{a})^2 - 0.0417\beta_2 - \frac{0.01811}{\beta_2}(\frac{b}{a})^2\}$$

$dm/d\beta_1 = 0$ leads to

$$-0.0417 + \frac{0.00776}{\beta_1^2}(\frac{b}{a})^2 = 0, \ \beta_1 = 0.4314\frac{b}{a}$$

$dm/d\beta_2 = 0$ leads to

$$-0.0417 + \frac{0.01811}{\beta_2^2}(\frac{b}{a})^2 = 0, \ \beta_2 = 0.659\frac{b}{a}$$

Substituting these values of β_1 and β_2 in the equation for m and simplifying,

$$m = qb^2\{0.125 - 0.0909(\frac{b}{a})\}$$

Multiplying the above value by 4/3,

$$m = qb^2\{0.1667 - 0.1212(\frac{b}{a})\}$$

Short span
Positive moment

$$m = qb^2\{0.1667 - 0.1212(\frac{b}{a})\}, \beta_{sx} = \{0.1667 - 0.1212(\frac{b}{a})\}, b/a<1,$$

$$m = qb^2 0.041, \beta_{sx} = 0.041, a/b=1$$

Long span
$\mu m = 0.041 \ qb^2$ and $\gamma m' = 0.055 \ qb^2$
For positive and negative moments, the moment coefficients β_{sy} are 0.041 and 0.055 respectively.

(iii) Shear coefficients
$a/b = 1$: $\beta_1 = 0.2868$ and $\beta_2 = 0.6593$
$a/b > 1.0$: $\beta_1 = 0.4314 \ b/a$ and $\beta_2 = 0.6590 \ b/a$.

Short beam: Spread the load uniformly over a length of 0.75 b.

Continuous end

$$\text{Load} = 0.5 \times q \times b \times \beta_2 \ a$$

$$v_x = qb\{0.6667\beta_2\frac{a}{b}\}, \beta_{vx} = 0.44$$

Simply supported end

$$\text{Load} = 0.5 \times q \times b \times \beta_1 \ a$$

$$v_x = qb\{0.6667\beta_1\frac{a}{b}\}, \beta_{vx} = 0.2876, b/a<1$$

$$v_x = qb\{0.6667\beta_1 \frac{a}{b}\}, \beta_{vx} = 0.1912, b/a = 1$$

Longer beam: Spread the load uniformly over a length of 0.75 a.

Load: $= 0.5 \times q \times 0.5 b \times (2 - \beta_1 - \beta_2) a$

$$v_y = qb\{0.3333(2 - \beta_1 - \beta_2)\}$$

$$\beta_{vy} = (0.6667 - 0.3634 \frac{b}{a}), b/a > 1, \quad \beta_{vy} = 0.3513, b/a = 1$$

8.9.16.9 Slab with only a long edge continuous (case 7 in BS Table 3.14)

Fig.8.58 shows the collapse mode which is governed by two parameters β_1 and β_2. It can be shown that the basic equation for solving the problem is

$$[\frac{b}{a}]^2 \{2\frac{\mu m}{\beta_1}\} + \frac{(m + m')}{\beta_2} + \frac{m}{(1 - \beta_2)} = q\frac{b^2}{6}(3 - 2\beta_1)$$

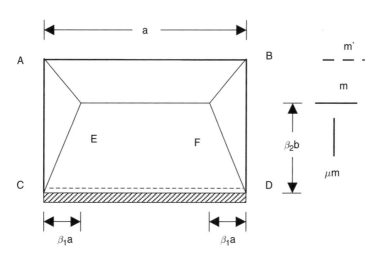

Fig.8.58 Collapse mode for a slab with one long edge continuous.

(i) Square slab

$$a/b = 1, \mu m = m, m' = 1.33 m.$$

$$m\{\frac{2}{\beta_1} + \frac{2.333}{\beta_2} + \frac{1.0}{(1 - \beta_2)}\} = q\frac{b^2}{6}(3 - 2\beta_1)$$

Simplifying:

$$m = \frac{qb^2}{6}\beta_1\beta_2(1 - \beta_2)\frac{(3 - 2\beta_1)}{\{2\beta_2(1 - \beta_2) - 1.333\beta_1\beta_2 + 2.333\beta_1\}}$$

$dm/d\beta_2 = 0$ leads to

$$\beta_2^2 - 3.5\beta_2 + 1.75 = 0, \quad \beta_2 = 0.604$$

Using this value of β_2, the expression for m is

$$m = \frac{qb^2}{6}\beta_1\frac{(3-2\beta_1)}{\{6.3885\beta_1+2.0\}}$$

dm/dβ_1 = 0 leads to

$$\beta_1^2 + 0.6261\beta_1 - 0.4696 = 0, \quad \beta_1 = 0.4404$$

Using β_1 = 0.4404 and β_2 = 0.604 in the expression for m, m = 0.032 qb^2, m´ = 0.043 qb^2.

Multiplying these values by 4/3,

$$m = \mu m = 0.043 \text{ qb}^2, \quad \beta_{sx} = 0.043 \text{ and m´} = 0.058 \text{ qb}^2, \quad \beta_{sx} = 0.058.$$

(ii) Rectangular slab: a/b > 1.0. Keep μm = 0.032 qb^2, m´ = 1.333m and β_2 = 0.604 as constant for all values of a/b. Substituting these values in the equation for m and simplifying,

$$m = qb^2\{0.0783 - 0.0522\beta_1 - \frac{0.010}{\beta_1}(\frac{b}{a})^2\}$$

dm/dβ_1 = 0 leads to

$$-0.0522 + \frac{0.010}{\beta_1^2}(\frac{b}{a})^2 = 0, \quad \beta_1 = 0.4377\frac{b}{a}$$

Substituting for β_1 in the equation for m and simplifying,

$$m = qb^2\{0.0783 - 0.0457(\frac{b}{a})\}$$

Dividing the above value by 0.75,

$$m = qb^2\{0.1044 - 0.0609(\frac{b}{a})\}$$

Short span

Positive moment

$$m = qb^2\{0.1064 - 0.0609(\frac{b}{a})\}, \beta_{sx} = \{0.1064 - 0.0609(\frac{b}{a})\}, b/a<1,$$

$$m = qb^20.043, \beta_{sx} = 0.041, a/b=1$$

Negative moment

$$m´ = 1.333m = qb^2\{0.1419 - 0.0812(\frac{b}{a})\}, \beta_{sx} = \{0.1419 - 0.0812(\frac{b}{a})\}, b/a<1$$

$$m´ = m = qb^20.0573, \beta_{sx} = 0.0573, a/b=1$$

Long span

$$\mu m = 0.043 \text{ qb}^2, \beta_{sy} = 0.058$$

(iii) Shear coefficients: Spread the load uniformly over a length of 0.75 b.
β_2 = 0.604 and β_1 = 0.4404, a/b = 1
β_2 = 0.604 and β_1 = 0.4382 b/a , a/b > 1.0

Shorter beam

$$\text{Load} = 0.5 \times q \times b \times \beta_1 \, a.$$

$$v_x = qb\{0.6667\beta_1 \frac{a}{b}\}$$

$$v_x = qb\{0.6667\beta_1 \frac{a}{b}\}, \beta_{vx} = 0.2921$$

Longer beam: Spread the load uniformly over a length of 0.75 a.

Continuous end

$$\text{Load} = 0.5 \times q \times \beta_2 b \times (2 - 2\beta_1)a$$
$$v_y = qb\{0.8053(1 - \beta_1)\}$$

$$v_y = qb\{0.8053 \times (1 - 0.4382\frac{b}{a}\}, \beta_{vy} = (0.8053 - 0.3529\frac{b}{a})$$

Simply supported end

$$\text{Load} = 0.5 \times q \times (1 - \beta_2) \, b \times (2 - 2\beta_1)a$$
$$v_y = qb\{0.5280(1 - \beta_1)\}$$

$$v_y = qb\{0.5280 \times (1 - 0.4382\frac{b}{a}\}, \beta_{vy} = (0.5280 - 0.2314\frac{b}{a})$$

8.10 HILLERBORG'S STRIP METHOD

This method of designing slabs is based on the Lower Bound Theorem of Plasticity. The basic idea of the method is to find a distribution of moments, which fulfils the equilibrium equations and designing the slab for these moments. Normally in a slab not only moments about two axes but also torsional moments exist. Analysis is complicated because of the presence of these torsional moments. Strip method simplifies analysis by in general completely ignoring torsional moments and assuming that the load is carried by a set of strips in bending only.

8.10.1 Simply Supported Rectangular Slab

As an example, consider the rectangular slab simply supported on four sides and subjected to a uniformly distributed load q as shown in Fig.8.59. It is reasonable to assume that the load in the triangular areas AEC and BFD are supported by reactions on the short sides AC and BD respectively. Similarly loads in the trapeziums AEFB and CEFD are supported by reactions on the sides closest to them, AB and CD respectively.

Once this assumption is made the slab is divided into a set of strips in the horizontal and vertical directions and the bending in these strips can be calculated and the strips reinforced.

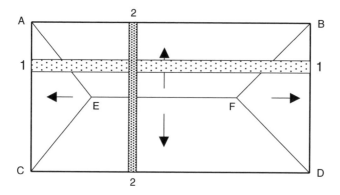

Fig.8.59 Load distribution in a simply supported slab.

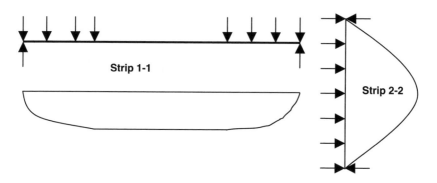

Fig.8.60 Bending moments in horizontal (1-1) and vertical (2-2) strips.

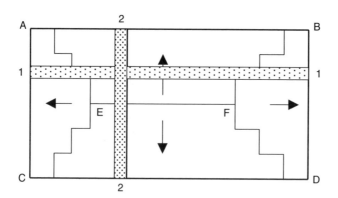

Fig 8.61 Step–wise load distribution to the supports.

As shown in Fig.8.60, for a typical strip 1-1 in the horizontal direction, the loading on the strip consists of uniformly distributed loading on the end portions only and for typical strip 2-2 in the vertical direction, with the loading on the strip consisting of uniformly distributed loading covering the entire span. The bending moments in these separate simply supported strips can be easily calculated and the slab may be reinforced accordingly.

The main difficulty in assuming the load distribution as shown in Fig.8.60 is that the loading on the strip across its width is not uniform. This difficulty can be avoided by assuming load distribution to the supports in a step–wise fashion as shown in Fig.8.61.

8.10.2 Clamped Rectangular Slab with a Free Edge

Fig.8.62 shows a slab clamped on three sides and free on one side. The load distribution to the supports is as indicated. If desired the step–wise load distribution can also be adopted. The strip 1-1 is a beam clamped at both ends while the strip 2-2 is a cantilever.

8.10.3 A Slab Clamped on Two Opposite Sides, One Side Simply Supported and One Edge Free

Fig.8.63 shows a slab clamped on two opposite sides and one side is simply supported while the opposite edge is free. The load distribution to the supports is as indicated.

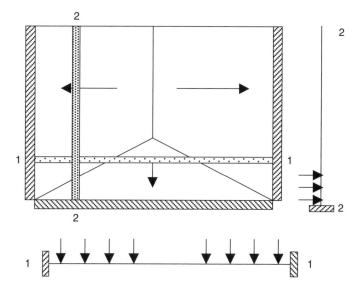

Fig.8.62 Load distribution in a slab one free edge.

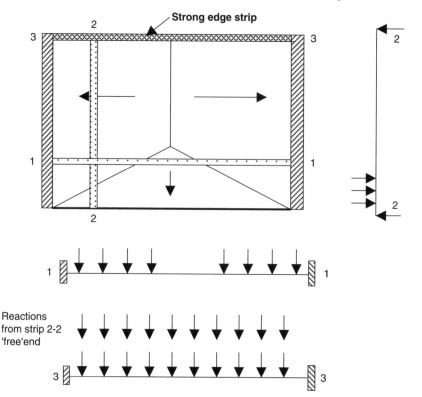

Fig.8.63 Slab with opposite edges simply supported and free.

The strip 1-1 is a beam clamped at both ends. In Fig.8.62, strip 2-2 was clamped at one end and could therefore act as a cantilever. In Fig.8.63, for the strip 2-2 to transmit any load to the simply supported end, it is necessary that there is a support at the 'free' end. Edge strip 3-3 provides this support. Therefore while designing strip 3-3, it is necessary to include not only the load applied directly onto the strip but also the reactions from strip 2-2 and similaras shown in Fig.8.63. Strip 3-3 acts like an edge beam by being more heavily reinforced than the rest of strips 1-1. Strip 3-3 could be thickened in order to allow sufficient depth of lever arm to the reinforcement.

8.10.4 Strong Bands

Fig.8.64 shows a rectangular slab simply supported on all sides and carrying a concentrated load W. The load is transmitted to the supports mainly through heavily reinforced strips in two directions. These strips are known as 'Strong Bands'. The strong bands act as beams and are more heavily reinforced compared to the rest of the slab. It is often convenient to increase the thickness in order to accommodate steel reinforcement and also to increase its lever arm. Distributed load on the rest of the slab can be distributed between the edge supports and strong bands. The load carried by the strong bands will be approximately in inverse proportion to the fourth power of the spans. Thus if the spans are a and b with $a \geq b$, then the load W_a and W_b carried by the strips in the a and b direction are

$$W_b = \frac{\alpha}{(1+\alpha)} W, \quad W_a = \frac{1}{(1+\alpha)} W, \quad \alpha = [\frac{a}{b}]^4$$

The concept of the strong band is also useful when designing slabs with holes or slabs with re–entrant corners.

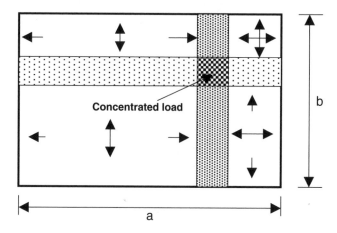

Fig.8.64 Strong band reinforcement.

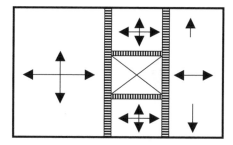

Fig.8.65 Slab with a hole.

Fig.8.65 shows a slab with a rectangular hole. By providing strong bands around the hole, edge beams are created and the loads can be distributed between the edge supports and strong bands. The two strong bands running between the supports also provide support for the short edge beams around the hole.

Fig.8.66 shows a slab with a re–entrant corner. By providing a strong band, the slab is conveniently divided into two rectangular slabs which can be effectively designed separately. The strong band acts as an additional support to the two slabs and allows the above simplification compared with the relatively complex distribution of moments obtained from an elastic analysis.

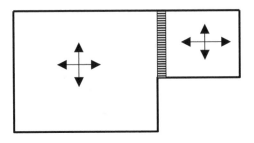

Fig.8.66 Slab with re–entrant corner.

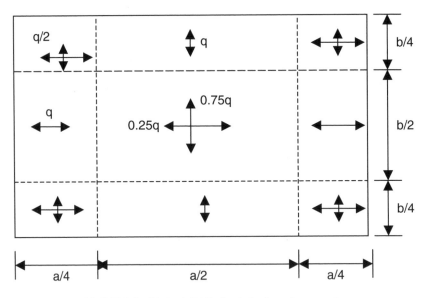

Fig.8.67 A feasible load distribution to the four edge supports.

8.10.5 Comments on The Strip Method

One of the main attractions of the Strip Method as compared with the Yield Line method is that apart from the fact that it is a Lower Bound method and therefore there is no need to investigate alternative mechanisms, the method not only gives the bending moments and shear forces at every point in the structure but also gives information on the loads and their distribution acting on the supporting beams. This is of great attraction to designers.

It is important to remember that the method ensures safety against bending failure only. It does not take account of the possibility of shear failure. Because of the fact that the emphasis is on safety at ultimate limit state, additional considerations are necessary to ensure that the design meets serviceability limit state conditions as well. For any given structure, it is possible to choose an infinite number of possible distributions of loads to the supports and the corresponding moments. As an example consider the load distribution on the rectangular slab simply supported on all edges shown in Fig.8.65. The proportion of the uniformly distributed load q against the arrows indicates the value of the load carried to the support in the direction indicated. This load distribution is different from the one shown in Fig.8.59. However from a serviceability limit state point of view, it is important to ensure that the chosen distribution of moments does not depart too far from the elastic distribution of moments.

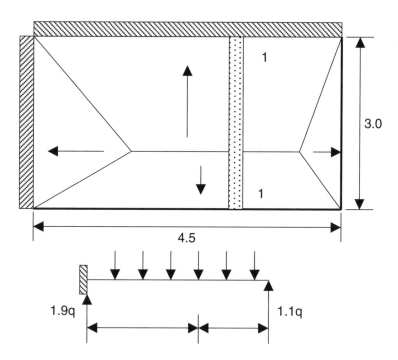

Fig.8.68 Zero shear lines to control load distribution to supports.

The following points should be borne in mind when deciding on the load distribution to the supports.

With fixed edges, the ratio between the proportion of load carried to the fixed edge and that carried to the simply supported edge should be increased by a factor of 1.6 to 1.8 compared to the case where both edges have same support conditions. If, for example, of the two opposite edges, one is fixed and the other is simply supported, then an appropriate load distribution should be as shown in Fig.8.68. The dividing lines between the regions can be treated as zero shear lines. For example, for the strip 1-1 shown in Fig.8.68, if the line of zero shear is at 1.1 from the simply supported end, then the reaction at the simply supported end is 1.1q and at the fixed end is 1.9q. The maximum bending moment in the span is at the point of zero shear and is equal to 0.605q and at the fixed end is 1.2q . Thus the ratio of reactions is 1.9/1.1 = 1.72 and the ratio of moments is 1.2/0.605 = 2.0. Thus by choosing the position of lines of zero shear, it is possible to control the moment distribution to correspond to the elastic values.

- Although the Strip Method assumes that torsional moments are zero, however where two simply supported edges meet, torsional moments do exist. In the absence of proper reinforcement, this will lead to cracking which is best limited by providing torsional reinforcement as suggested in codes of practice (Clause 3.5.3.5).
- The ratio between the support and span design moments in a strip fixed at both ends and subjected to uniformly distributed loading should be about 2.

8.11 DESIGN OF REINFORCEMENT FOR SLABS IN ACCORDANCE WITH A PREDETERMINED FIELD OF MOMENTS

With the wide spread availability of Finite Element programs to carry out elastic analysis of plates, it is necessary to have rules for designing reinforcement for a given set of bending and twisting moments in slabs. Fig.8.69 shows the bending moments M_x and M_y and twisting moment M_{xy} acting on an element of slab. The convention used in representing a moment by a double–headed arrow is that if the right hand thumb is pointed in the direction of the arrow head, then the direction of the moment is given by the direction the fingers of the right hand bend. Bending moments as shown in Fig.8.69, cause tension on the bottom face.

As shown in Fig.8.70, on a section inclined at an angle α to the y–axis, normal bending moment M_n and twisting moment M_{nt} act. It can be shown that
$$M_n = M_x \cos^2 \alpha + M_y \sin^2 \alpha + 2 M_{xy} \sin \alpha \cos \alpha$$
If the ultimate sagging moment of resistance provided by steel in x and y-directions are M^b_{xu} and M^b_{yu} respectively, then from Johansen's yield criterion (section 8.9.2), the normal moment of resistance on a section inclined at an angle α to the y–axis is given by
$$M^b_{nu} = M^b_{xu} \cos^2 \alpha + M^b_{yu} \sin^2 \alpha$$
Since it is desirable that the applied M_n must not be greater than the resistance M_{nu},
$$[\{M^b_{xu} \cos^2 \alpha + M^b_{yu} \sin^2 \alpha \} - \{M_x \cos^2 \alpha + M_y \sin^2 \alpha + 2M_{xy} \sin \alpha \cos \alpha\}] \geq 0$$
Dividing throughout by $\cos^2 \alpha$ and setting $t = \tan \alpha$, the above equation simplifies to

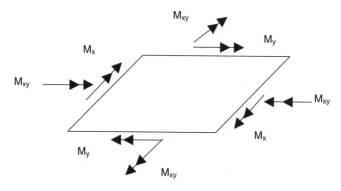

Fig 8.69 Bending and twisting moments on an element of slab.

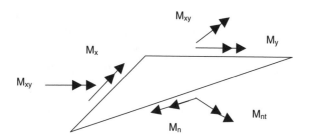

Fig 8.70 Normal bending moment and twisting moments on an element of slab.

$$\{(M^b_{xu} - M_x) + (M^b_{yu} - M_y)\, t^2 - 2M_{xy}\, t\} \geq 0$$

Yielding will take place when the difference between M_{nu} and M_n is a minimum. Differentiating with respect to t,

$$(M^b_{yu} - M_y)\, t - M_{xy} = 0$$

For the difference to be a minimum, the second derivative with respect to t must be positive. Therefore

$$(M^b_{yu} - M_y) \geq 0$$

Substituting the value of t and simplifying

$$(M^b_{xu} - M_x)(M^b_{yu} - M_y) - M_{xy}{}^2 = 0$$

This equation shows for what combination of bending and twisting moments a slab with a known moment of resistance in x and y-directions yields. This equation is known as the yield criterion for a slab. Note that the twisting moment term appears as a square indicating that the sign of M_{xy} is irrelevant.

From the yield criterion, the following special cases can be noted.

Case 1

$$\text{If } M^b_{xu} = 0, \text{ then } M^b_{yu} = M_y - M^2_{xy}/M_x$$

Case 2

$$\text{If } M^b_{yu} = 0, \text{ then } M^b_{xu} = M_x - M^2_{xy}/M_y$$

Case 3: If $M^b_{xu} \neq 0$ and $M^b_{yu} \neq 0$, then for economy$(M^b_{xu} + M^b_{yu})$ must be made a minimum. From the yield criterion

$$M^b_{xu} = M_x + M_{xy}^2 / (M^b_{yu} - M_y)$$
$$(M^b_{xu} + M^b_{yu}) = M_x + M_{xy}^2 / (M^b_{yu} - M_y) + M^b_{yu}$$

Minimizing the above expression with respect to M^b_{yu},

$$-M_{xy}^2 / (M^b_{yu} - M_y)^2 + 1 = 0$$
$$(M^b_{yu} - M_y) = \pm M_{xy}$$

Since $(M^b_{yu} - M_y) \geq 0$, choosing the positive sign,

$$M^b_{xu} = M_x + \left| M_{xy} \right|, \quad M^b_{yu} = M_y + \left| M_{xy} \right|$$

Note that only the numerical value of M_{xy} is used.

8.11.1 Rules for Designing Bottom Steel

In the following, positive bending moments are sagging moments which cause tension on the bottom face . The rules for calculating the moment of resistance required for flexural steel at bottom are as follows.

(a) If $\dfrac{M_x}{\left| M_{xy} \right|} \geq -1.0$ and $\dfrac{M_y}{\left| M_{xy} \right|} \geq -1.0$, then $M^b_{xu} = M_x + \left| M_{xy} \right|, \quad M^b_{yu} = M_y + \left| M_{xy} \right|$

(b) If $\dfrac{M_x}{\left| M_{xy} \right|} < -1.0$ and $M_y - \dfrac{M_{xy}^2}{M_x} > 1.0$ then $M^b_{xu} = 0, \quad M^b_{yu} = M_y - \dfrac{M_{xy}^2}{M_x}$

(c) If $\dfrac{M_y}{\left| M_{xy} \right|} < -1.0$ and $M_x - \dfrac{M_{xy}^2}{M_y} > 1.0$ then $M^b_{yu} = 0, \quad M^b_{xu} = M_x - \dfrac{M_{xy}^2}{M_y}$

(d) If none of the above conditions are valid, then $M^b_{yu} = M^b_{xu} = 0$

8.11.1.1 Examples of Design of Bottom Steel

The following examples illustrate the use of equations derived in section 8.11.1. The four criteria are checked to see which is the valid one for a specific combination of M_x, M_y and M_{xy}.

Example 1: $M_x = 30$ kNm/m, $M_y = 15$ kNm/m, $M_{xy} = 20$ kNm/m

Check criterion (a): $\dfrac{M_x}{\left| M_{xy} \right|} = 1.5 > -1.0$, $\dfrac{M_y}{\left| M_{xy} \right|} = 0.75 > -1.0$, Therefore criterion

(a) applies.

$$M^b_{xu} = 30 + 20 = 50 \text{ kNm/m}, \quad M^b_{yu} = 15 + 20 = 35 \text{ kNm/m}$$

Example 2: $M_x = -35$ kNm/m, $M_y = 15$ kNm/m, $M_{xy} = 20$ kNm/m

(a) $\dfrac{M_x}{\left| M_{xy} \right|} = -1.75 < -1.0$, $\dfrac{M_y}{\left| M_{xy} \right|} = 0.75 > -1.0$.

(b) $\dfrac{M_x}{|M_{xy}|} = -1.75 < -1.0$, $M_y - \dfrac{M_{xy}^2}{M_x} = 26.43 > 1.0$, Therefore criterion (b) applies.

$$M_{xu}^b = 0, \quad M_{yu}^b = M_y - \frac{M_{xy}^2}{M_x} = 26.43 \text{ kNm/m}$$

Example 3: $M_x = -15$ kNm/m, $M_y = -25$ kNm/m, $M_{xy} = 20$ kNm/m

(a) $\dfrac{M_x}{|M_{xy}|} = -0.75 > -1.0$, $\dfrac{M_y}{|M_{xy}|} = -1.25 < -1.0$

(b) $\dfrac{M_x}{|M_{xy}|} = -0.75 > -1.0$, $M_y - \dfrac{M_{xy}^2}{M_x} = 1.67 > 1.0$

(c) $\dfrac{M_y}{|M_{xy}|} = -1.25 < -1.0$, $M_x - \dfrac{M_{xy}^2}{M_y} = 1.0 > 0$, Therefore criterion (c) applies.

$$M_{yu}^b = 0, \quad M_{xu}^b = M_x - \frac{M_{xy}^2}{M_y} = 1.0 \text{ kNm/m}$$

Example 4: $M_x = -30$ kNm/m, $M_y = -40$ kNm/m, $M_{xy} = 20$ kNm/m

(a) $\dfrac{M_x}{|M_{xy}|} = -1.5 < -1.0$, $\dfrac{M_y}{|M_{xy}|} = -2.0 < -1.0$

(b) $\dfrac{M_x}{|M_{xy}|} = -1.5 < -1.0$, $M_y - \dfrac{M_{xy}^2}{M_x} = -26.67 < 0$

(c) $\dfrac{M_y}{|M_{xy}|} = -2.0 < -1.0$, $M_x - \dfrac{M_{xy}^2}{M_y} = -20.0 < 0$

Since none of the criteria (a) to (c) apply, therefore $M_{yu}^b = 0$, $M_{xu}^b = 0$. No steel is required at the bottom of the slab.

8.11.2 Rules for Designing Top Steel

In a manner similar to the determination of sagging moment of resistance, if the ultimate *hogging* moment of resistance provided by steel in x and y-directions is M_{xu}^t and M_{yu}^t respectively, then the rules for calculating the moment of resistance required for flexural steel at top are as follows. Note that the value of M_{xu}^t and M_{yu}^t are both negative indicating that they correspond to hogging bending moment requiring steel at the top of the slab.

(a) If $\dfrac{M_x}{|M_{xy}|} \le 1.0$ and $\dfrac{M_y}{|M_{xy}|} \le 1.0$, then $M_{xu}^t = M_x - |M_{xy}|$, $M_{yu}^t = M_y - |M_{xy}|$

(b) If $\dfrac{M_y}{|M_{xy}|} > 1.0$ and $M_x - \dfrac{M_{xy}^2}{M_y} < 0$ then $M_{yu}^t = 0$, $M_{xu}^t = M_x - \dfrac{M_{xy}^2}{M_y}$

(c) If $\dfrac{M_x}{|M_{xy}|} > 1.0$ and $M_y - \dfrac{M_{xy}^2}{M_x} < 0$ then $M_{xu}^t = 0,\ M_{yu}^t = M_y - \dfrac{M_{xy}^2}{M_x}$

(d) If none of the above conditions are true, then $M_{yu}^t = M_{xu}^t = 0$

8.11.2.1 Examples of design of top steel

Example 1: $M_x = -30,\ M_y = -40,\ M_{xy} = 20$ kNm/m

(a) $\dfrac{M_x}{|M_{xy}|} = -1.5 < 1.0$ and $\dfrac{M_y}{|M_{xy}|} = -2.0 < 1.0$. Criterion (a) applies.

$$M_{xu}^t = M_x - |M_{xy}| = -50 \text{ kNm/m},\ M_{yu}^t = M_y - |M_{xy}| = -60 \text{ kNm/m}$$

Example 2: $M_x = -30,\ M_y = 35,\ M_{xy} = 20$ kNm/m

(a) $\dfrac{M_x}{|M_{xy}|} = -1.5 < 1.0$ and $\dfrac{M_y}{|M_{xy}|} = 1.75 > 1.0$

(b) $\dfrac{M_y}{|M_{xy}|} = 1.75 > 1.0$ and $M_x - \dfrac{M_{xy}^2}{M_y} = -41.43 < 0$. Criterion (b) applies.

$$M_{yu}^t = 0,\ M_{xu}^t = M_x - \dfrac{M_{xy}^2}{M_y} = -41.43$$

Example 3: $M_x = 25,\ M_y = -45,\ M_{xy} = 20$ kNm/m

(a) $\dfrac{M_x}{|M_{xy}|} = 1.25 > 1.0$ and $\dfrac{M_y}{|M_{xy}|} = -2.25 < 1.0$

(b) $\dfrac{M_y}{|M_{xy}|} = -2.25 < 1.0$ and $M_x - \dfrac{M_{xy}^2}{M_y} = 33.89 > 0$.

(c) $\dfrac{M_x}{|M_{xy}|} = 1.25 > 1.0,\ M_y - \dfrac{M_{xy}^2}{M_x} = -61.0 < 0$. Criterion (c) applies.

$$M_{xu}^t = 0,\ M_{yu}^t = M_y - \dfrac{M_{xy}^2}{M_x} = -61.0 \text{ kNm/m}$$

8.11.3 Examples of Design of Top and Bottom Steel

In sections 8.11.1.1 and 8.11.2.1 examples were concerned with determining the required moment of resistance either at the top or the bottom face of the slab. However cases do arise where for a given combination of bending and twisting moments there is need to provide steel at both the faces. This case generally arises when twisting moments larger than bending moments are present.

Example 1: $M_x = 15$, $M_y = -18$, $M_{xy} = 20$ kNm/m

Bottom steel: $\dfrac{M_x}{|M_{xy}|} = 0.75 > -1.0$ and $\dfrac{M_y}{|M_{xy}|} = -0.9\} > -1.0$

$$M_{xu}^b = M_x + |M_{xy}| = 35 \text{ kNm/m}, \quad M_{yu}^b = M_y + |M_{xy}| = 2 \text{ kNm/m}$$

Top steel: $\dfrac{M_x}{|M_{xy}|} = 0.75 < 1.0$ and $\dfrac{M_y}{|M_{xy}|} = -0.9 < 1.0$

$$M_{xu}^t = M_x - |M_{xy}| = -5 \text{ kNm/m}, \quad M_{yu}^t = M_y - |M_{xy}| = -38 \text{ kNm/m}$$

This example shows that steel is required in both directions top and bottom.

Example 2: $M_x = 20$, $M_y = -20$, $M_{xy} = 20$ kNm/m

Bottom steel: $\dfrac{M_x}{|M_{xy}|} = 1.0 > -1$ and $\dfrac{M_y}{|M_{xy}|} = -1.0$

$$M_{xu}^b = M_x + |M_{xy}| = 40 \text{ kNm/m}, \quad M_{yu}^b = M_y + |M_{xy}| = 0$$

Top steel: $\dfrac{M_x}{|M_{xy}|} = 1.0$ and $\dfrac{M_y}{|M_{xy}|} = -1.0$

$$M_{xu}^t = M_x - |M_{xy}| = 0, \quad M_{yu}^t = M_y - |M_{xy}| = -40 \text{ kNm/m}$$

This example shows that steel is required in only y-direction at top and in only x-direction at bottom.

8.11.4 Comments on the design method using elastic analysis

The code BS 8110 in clause 3.5.2.1 permits the design of slabs using Johansen's Yield Line method which is an upper bound method or Hillerborg's Strip Method which is a lower bound method provided the ratio of moments between support and span moments are similar to those obtained by the use of elastic theory. If this cautionary note is not observed, slabs using these methods might prove unsuitable from a serviceability point of view because of early cracking. Using bending and twisting moments from elastic analysis to design slabs using the rules developed in sections 8.11.1 to 8.11.3 avoids this problem and leads to a very economical design. The main disadvantage is that the designed reinforcement will vary from point to point and some form of averaging is needed to convert the variable reinforcement into bands with constant reinforcement. The method is also highly amenable to computer aided design of general slab structures.

8.12 STAIR SLABS

8.12.1 Building Regulations

Statutory requirements are laid down in *Building Regulations and Associated Approved Documents,* Part H [Reference 9], where private and common stairways are defined. The private stairway is for use with one dwelling and the common stairway is used for more than one dwelling. Requirements from the Building Regulations are shown in Fig.8.71.

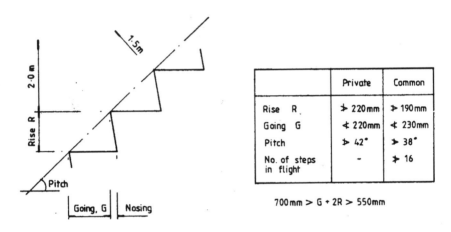

	Private	Common
Rise R	≯ 220mm	≯ 190mm
Going G	≮ 220mm	≮ 230mm
Pitch	≯ 42°	≯ 38°
No. of steps in flight	-	≯ 16

700mm > G + 2R > 550mm

Fig.8.71 Building regulation for dimensions of stairs.

8.12.2 Types of Stair Slab

Stairways are sloping one-way spanning slabs. Two methods of construction are used.

(a) Transverse spanning stair slabs
Transverse spanning stair slabs span between walls, a wall and stringer (an edge beam), or between two stringers. The stair slab may also be cantilevered from a wall. A stair slab spanning between a wall and a stringer is shown in Fig.8.72(a). The stair slab is designed as a series of beams consisting of one step with assumed breadth and effective depth shown in Fig.8.72(c). The moment reinforcement is generally one bar per step. Secondary reinforcement is placed longitudinally along the flight.

(b) Longitudinal spanning stair slab
The stair slab spans between supports at the top and bottom of the flight. The supports may be beams, walls or landing slabs. A common type of staircase is shown in Fig.8.73.

The effective span *l* lies between the top landing beam and the centre of support in the wall. If the total design load on the stair is *W* the positive design moment at mid-span and the negative moment over top beam B are both taken as *Wl/10*. The arrangement of moment reinforcement is shown in Fig.8.73. Secondary reinforcement runs transversely across the stair.

A staircase around a lift well is shown in Fig.8.74. The effective span *l* of the stair is defined in the code BS 8110 in section 3.10. This and other code requirements are discussed in section 8.12.3 below. The maximum moment near mid-span and over supports is taken as *Wl/10*, where *W* is the total design load on the span.

8.12.3 Code Design Requirements

(a) Imposed loading
The imposed loading on stairs is given in BS6399: Part 1:1996, Table 1. From this table the distributed loading is as follows:
1. dwelling not over three storeys, 1.5 kN/m^2
2. all other buildings, the same as the floors, between 3 kN/m^2 and 4 kN/m^2

(b) Design provisions
Provisions for design of staircases are set out in BS 8110: Part 1, section 3.10 and are summarized below.
1. The code states that the staircase may be taken to include a section of the landing spanning in the same direction and continuous with the stair flight;
2. The design ultimate load is to be taken as uniform over the plan area. When two spans intersect at right angles as shown in Fig.8.74 the load on the common area can be divided equally between the two spans;
3. When a staircase or landing spans in the direction of the flight and is built into the wall at least 110 mm along part or all of the length, a strip 150mm wide may be deducted from the loaded area (Fig.8.75);
4. When the staircase is built monolithically at its ends into structural members spanning at right angles to its span, the effective span is given by
$$l_a + 0.5(l_{b1} + l_{b2})$$
where l_a is the clear horizontal distance between supporting members. l_{b1} is the breadth of a supporting member at one end or 1.8 m whichever is the smaller and l_{b2} is the breadth of a supporting member at the other end or 1.8 m whichever is the smaller (Fig.8.74);
5. The effective span of simply supported staircases without stringer beams should be taken as the horizontal distance between centrelines of supports or the clear distance between faces of supports plus the effective depth whichever is less;
6. The depth of the section is to be taken as the minimum thickness perpendicular to the soffit of the stair slab;
7. The design procedure is the same as for beams and slabs (see provision 8 below);

8. For staircases without stringer beams when the stair flight occupies at least 60% of the span the permissible span-to-effective depth ratio may be increased by 15%.

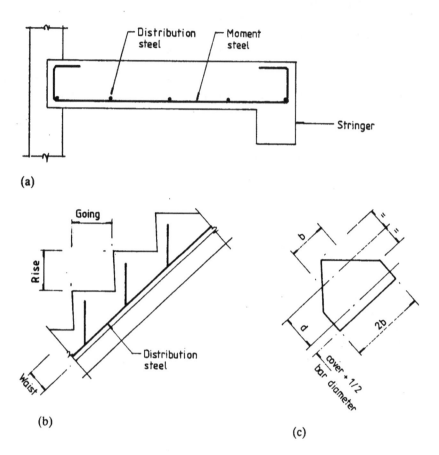

Fig.8.72 (a) Transverse section; (b) longitudinal section; (c) assumptions for design.

8.12.4 Example of Design of Stair Slab

(a) Specification
Design the side flight of a staircase surrounding an open stair well. A section through the stairs is shown in Fig.8.75(a).

The stair slab is supported on a beam at the top and on the landing of the flight at right angles at the bottom. The imposed loading is 5 kN/m². The stair is built 110mm into the sidewall of the stair well. The clear width of the stairs is 1.25m and the flight consists of eight risers at 180mm and seven goings of 220mm with 20mm nosing. The stair treads and landings have 15mm granolithic finish and the underside of the stair and landing slab has 15 mm of plaster finish. The materials are grade C30 concrete and grade 460 reinforcement.

(b) Loading and moment

Assume the waist thickness of structural concrete is 100 mm, the cover is 25 mm and the bar diameter is 10 mm. The loaded width and effective breadth of the stair slab are shown in section AA in Fig.8.75(a). The effective span of the stair slab is the clear horizontal distance (1540 mm) plus the distance of the stair to the centre of the top beam (235 mm) plus one-half of the breadth of the landing (625 mm), i.e. 2400 mm. The design ultimate loading on the stairs is calculated first.

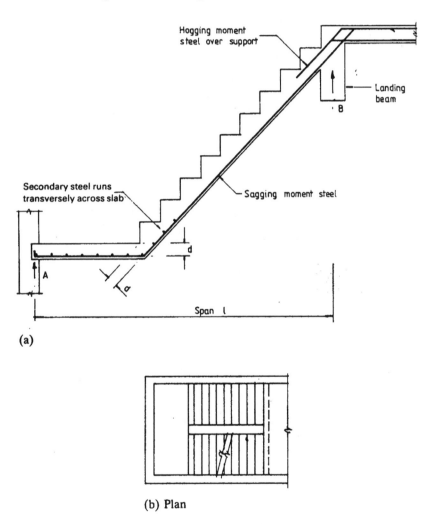

(a)

(b) Plan

Fig.8.73 A common type of stair case.

(i) Landing slab
The overall thickness including the top and underside finish is 130mm.

$$\text{dead load} = 0.13 \times 24 \times 1.4 = 4.4 \text{ kN/m}^2$$
$$\text{imposed load} = 5 \times 1.6 = 8.0 \text{ kN/m}^2$$
$$\text{total design ultimate load} = 12.4 \text{ kN/m}^2$$
$$50\% \text{ of landing load} = 0.5 \times 12.4 \times 0.625 \times 1.1 = 4.26 \text{ kN}$$

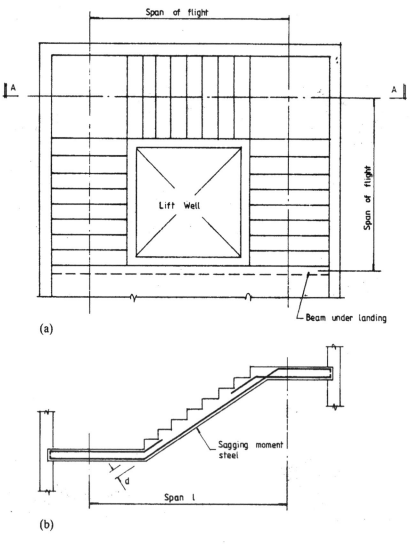

(a)

(b)

Fig.8.74 (a) Plan; (b) section AA.

One-half of the load on the landing slab is included for the stair slab under consideration. The loaded width is 1.1m.

(ii) Stair slab

The slope length is 2.29 m and the steps project 152 mm perpendicularly to the top surface of the waist. The average thickness including finishes is

$$100 + 152/2 + 30 = 206 \text{mm}$$

$$\text{Dead load} = 0.206 \times 24 \times 2.29 \times 1.1 \times 1.4 = 17.44 \text{ kN}$$
$$\text{Imposed load} = 5 \times 1.78 \times 1.1 \times 1.6 = 15.66 \text{ kN}$$
$$\text{Total load} = 33.1 \text{ kN}$$

The dead load is calculated using the slope length while the imposed load acts on the plan length. The loaded width is 1.1 m.
The total load on the span is

$$4.26 + 33.1 = 37.36 \text{ kN}$$

The maximum shear at the top support is 21.44 kN. The design moment for sagging moment near mid-span and the hogging moment over the supports are both $Wl/10$.

$$37.36 \times 2.4/10 = 8.97 \text{ kNm}$$

(c) Moment reinforcement

The effective depth

$$d = 100 - 25 - 5 = 70 \text{ mm}$$

The effective width b will be taken as the width of the stair slab, 1250 mm.

$$k = M/(bd^2 f_{cu}) = 8.97 \times 10^6/(1250 \times 70^2 \times 30) = 0.049 < 0.156$$
$$z/d = 0.5 + \sqrt{(0.25 - 0.049/0.9)} = 0.94 < 0.95$$
$$z = 0.94 \times 70 = 66 \text{ mm}$$
$$A_s = 8.97 \times 10^6/(0.95 \times 460 \times 66) = 311 \text{ mm}^2 \text{ for the full 1250 mm width.}$$

Provide eight 8 mm diameter bars to give an area of 402 mm². Space the bars at 180 mm centres. The same steel is provided in the top of the slab over both supports.
The minimum area of reinforcement is

$$(0.13/100) \times 100 \times 1000 = 130 \text{ mm}^2$$

Provide 8mm diameter bars at 300mm centres to give 167 mm²/m transversely as distribution steel.

(d) Shear resistance

$$\text{Shear} = 21.4 \text{ kN}$$
$$v = 21.4 \times 10^3/(1250 \times 70) = 0.25 \text{ N/mm}^2$$
$$100 A_s/(bd) = 100 \times 402/(1250 \times 70) = 0.46 < 3.0$$
$$400/d = 400/125 = 3.2 > 1.0$$
$$v_c = 0.79 \times (0.46)^{1/3}(3.2)^{1/4}(30/25)^{1/3}/1.25 = 0.69 \text{ N/mm}^2$$

The slab is satisfactory with respect to shear. Note that a minimum value of *d* of 125 mm is used in the formula.

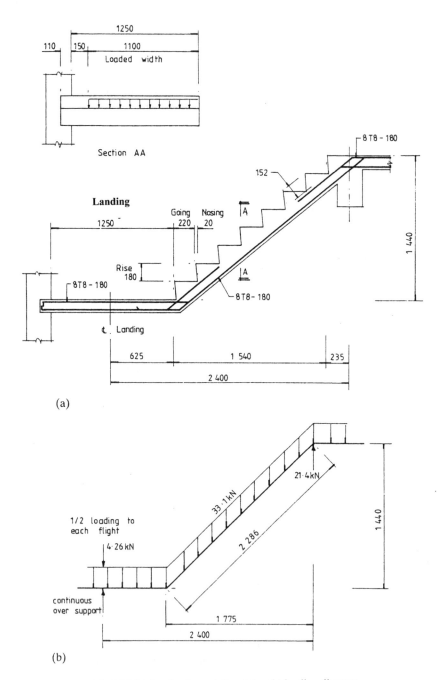

Fig.8.75 (a) Section through the stairs; (b) loading diagram.

(e) Deflection

The slab is checked for deflection.

The basic span/d ratio is 26 as the slb is effectively continuous at both ends of the 2.4m span considered.

The modification factor for tension reinforcement:

$$M/ (bd^2) = 8.97 \times 10^6/ (1250 \times 70^2) = 1.47$$
$$f_s = (2/3) \times 460 \times (311/402) = 237 \text{ N/mm}^2$$
$$0.55 + (477 - 237)/ (120 \times (0.9 + 1.47)) = 1.39 < 2.0$$
$$\text{allowable (span/d)} = 26 \times 1.39 \times 1.15 = 41.68$$
$$\text{actual (span/d)} = 2400/70 = 34.3$$

Note that the stair flight with a plan length of 1540 mm occupies 64% of the span and the allowable span/d ratio can be increased by 15% (BS 8110: Part 1, clause 3.10.2.2).

(f) Cracking

For crack control the clear distance between bars is not to exceed $3d = 210$ mm. The reinforcement spacing of 180 mm is satisfactory.

(g) Reinforcement

The reinforcement is shown in Fig.8.73(a).

8.13 REFERENCES

Jones, L.L. and Wood, R.H. 1967, *Yield Line analysis of slabs*, (Thames and Hudson).

Park, R. and Gamble W.L. 1980, *Reinforced Concrete Slabs*, (Wiley).

Cope, R.J. and Clark, L.A.1984, *Concrete Slabs: Analysis and Design*, (Elsevier).

Hillerborg, A. 1996, *Strip Method Design Handbook*, (E & FN Spon).

Hillerborg, A. 1975, *Strip Method of Design*, (A Viewpoint Publication).

Goodchild, C.H. 1997, *Economic concrete frame elements*, (British Cement Association).

Timoshenko, S. and Woinowsky–Krieger, S. 1959, *Theory of plates and shells*, (McGraw-Hill).

Building regulations and associated approved documents, (HMSO, 1985)

CHAPTER 9

COLUMNS

9. 1 TYPES, LOADS, CLASSIFICATION AND DESIGN CONSIDERATIONS

9.1.1 Types and Loads

Columns are structural members in buildings carrying roof and floor loads to the foundations. A column stack in a multi-storey building is shown in Fig.9.1 (a). Columns primarily carry axial loads, but most columns are subjected to moment as well as axial load. Referring to the part floor plan in the figure, the internal column A is designed for predominantly axial load while edge columns B and corner column C are designed for axial load and appreciable moment.

Design of axially loaded columns is treated first. Then methods are given for design of sections subjected to axial load and moment. Most columns are termed short columns and fail when the material reaches its ultimate capacity under the applied loads and moments. Slender columns buckle and the additional moments caused by deflection must be taken into account in design.

The column section is generally square or rectangular, but circular and polygonal columns are used in special cases. When the section carries mainly axial load it is symmetrically reinforced with four, six, eight or more bars held in a cage by links. Typical column reinforcement is shown in Fig.9.1 (b).

9.1.2 General Code Provisions

General requirements for design of columns are treated in BS 8110: Part 1, section 3.8.1. The provisions apply to columns where the greater cross-sectional dimension does not exceed four times the smaller dimension.

The minimum size of a column must meet the fire resistance requirements given in Fig. 3.2 of the code. For example, for a fire resistance period of 1.5 h, a fully exposed column must have a minimum dimension of 250 mm. The covers required to meet durability and fire resistance requirements are given in Tables 3.3 and 3.4 respectively of the code.

The code classifies columns first as

1. short columns when the ratios l_{ex}/h and l_{ey}/b are both less than 15 for braced columns and less than 10 for un-braced columns and
2. slender columns when the ratios are larger than the values given above

Here b is the width of the column cross-section, h is the depth of the column cross-section measured in the plane under consideration, l_{ex} is the effective height in respect of the major axis and l_{ey} is the effective height in respect of the minor axis.

In the second classification the code defines columns as braced or un-braced. The code states that a column may be considered to be braced in a given plane if lateral stability to the structure as a whole is provided by walls or bracing designed to resist all lateral forces in that plane. Otherwise the column should be considered as un-braced.

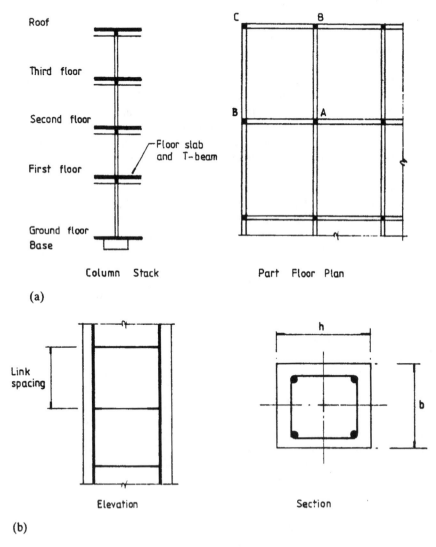

Fig.9.1 (a) Building column; (b) column construction.

Lateral stability in braced reinforced concrete structures is provided by shear walls, lift shafts and stairwells. In un-braced structures, resistance to lateral forces is provided by bending in the columns and beams in that plane. Braced and un-braced frames are shown in Figs 9.2(a) and 9.2(b) respectively.

Clause 3.8 1.4 of the code states that if a column has a sufficiently large section to resist the ultimate loads without reinforcement, it may be designed similarly to a plane concrete wall (section 10.4).

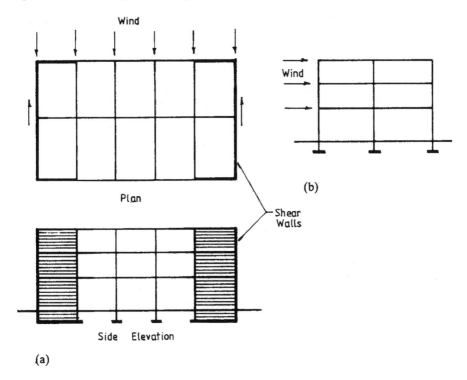

Fig.9.2 (a) Braced frame; (b) un-braced frame.

9.1.3 Practical Design Provisions

The following practical design considerations with regard to design of columns are extracted from BS 8110: Part 1, section 3.12. The minimum number of longitudinal bars in a column section is four. The main points from the code are as follows.

(a) Minimum percentage of reinforcement
The minimum percentage of reinforcement is given in Table 3.25 of the code for both grade 250 and grade 460 reinforcement as

$$100A_{sc}/A_{cc} = 0.4$$

where A_{sc} is the area of steel in compression and A_{cc} is the area of concrete in compression.

(b) Maximum area of reinforcement

Clause 3.12.6.2 states that the maximum area of reinforcement should not exceed 6% of the gross cross-sectional area of a vertically cast column except at laps where 10% is permitted. Maximum area of reinforcement should not exceed 8% of the gross cross-sectional area of a horizontally cast column.

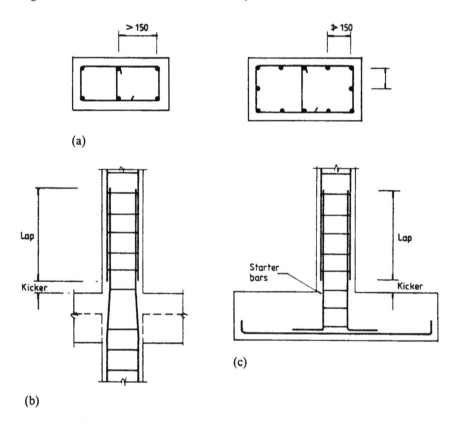

Fig.9.3 (a) Arrangement of links; (b) column lap; (c) column base.

(c) Requirements for links

Clause 3.12.7 covers containment of compression reinforcement:

1. The diameter of links should not be less than 6 mm or one-quarter of the diameter of the largest longitudinal bar;
2. The maximum spacing is to be 12 times the diameter of the smallest longitudinal bar;

3. The links should be arranged so that every corner bar and each alternate bar in an outer layer is supported by a link passing round the bar and having an included angle of not more than 135°. No bar is to be further than 150 mm from a restrained bar. These requirements are shown in Fig.9.3 (a).

(d) Compression laps and butt joints

Clause 3.12.8.15 of the code states that the length of compression laps should be 25% greater than the compression anchorage length. Compression lap lengths are given in Table 3.27 of the code (section 5.2.1 here). Laps in columns are located above the base and floor levels as shown in Fig.9.3 (b). Clause 3.12.8.16.1 of the code also states that the load in compression bars may be transferred by end bearing of square sawn cut ends held by couplers. Welded butt joints can also be made (clause 3.12.8.17).

9.2 SHORT BRACED AXIALLY LOADED COLUMNS

9.2.1 Code Design Expressions

Both longitudinal steel and all the concrete assist in carrying the load. The links prevent the longitudinal bars from buckling outwards. BS 8110: Part 1, clause 3.8.4.3 gives equation 38 for the ultimate load N that a short braced axially loaded column can support.

$$N = 0.4 f_{cu} A_c + 0.80 A_{sc} f_y \quad \text{(Code equation 38)}$$

where A_c is the net cross-sectional area of concrete in the column and A_{sc} is the area of vertical reinforcement. The expression allows for eccentricity due to construction tolerances but applies only to a column that cannot be subjected to significant moments. An example is column A in Fig.9.1 (a), which supports a symmetrical arrangement of floor beams. Note that for pure axial load the ultimate capacity N_{uz} of a column given in clause 3.8.3.1 of the code is

$$N_{uz} = 0.45 f_{cu} A_c + 0.95 A_{sc} f_y$$

Thus in the design equation for short columns the effect of the eccentricity of the load is taken into account by reducing the capacity for axial load by about 10%.

Clause 3.8.4.4 gives equation 39 for short braced columns supporting an approximately symmetrical arrangement of beams. These beams must be designed for uniformly distributed imposed loads and the spans must not differ by more than 15% of the longer span. The ultimate load is given by

$$N = 0.35 f_{cu} A_c + 0.70 A_{sc} f_y \quad \text{(Code equation 39)}$$

9.2.1.1 Examples of axially loaded short column

Example 1: A short braced axially loaded column 300 mm square in section is reinforced with four 25 mm diameter bars. Find the ultimate axial load that the column can carry and the pitch and diameter of the links required. The materials are grade C30 concrete and grade 460 reinforcement.

$$\text{Steel area } A_{sc} = 4T25 = 1963 \text{ mm}^2$$
$$\text{Concrete area } A_c = 300^2 - 1963 = 88037 \text{ mm}^2$$
$$N = (0.4 \times 30 \times 88037 + 0.80 \times 1963 \times 460) \times 10^{-3}$$
$$= 1056 + 722 = 1778 \text{ kN}$$

The links are not to be less than 6 mm in diameter or one-quarter of the diameter of the longitudinal bars. The spacing is not to be greater than 12 times the diameter of the longitudinal bars. Provide 8 mm diameter links at 300 mm centres. The column section is shown in Fig.9.4. From Table 3.4 of the code the cover for mild exposure is 25 mm.

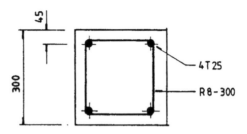

Fig.9.4 Designed column cross section.

Example 2: A short braced column has to carry an ultimate axial load of 1366 kN. The column size is 250 mm × 250 mm. Find the steel area required for the longitudinal reinforcement and select suitable bars. The materials are grade C30 concrete and grade 460 reinforcement.

Substitute in the expression for the ultimate load

$$1366 \times 10^3 = 0.4 \times 30(250^2 - A_{sc}) + 0.80 \times 460 A_{sc}$$
$$A_{sc} = 1730 \text{ mm}^2$$

Provide four 25 mm diameter bars to give a steel area of 1963 mm^2.

$$100 A_{sc}/(bh) = 100 \times 1963/(250 \times 250) = 3.14 < 6.0$$

This is satisfactory.

9.3 SHORT COLUMNS SUBJECTED TO AXIAL LOAD AND BENDING ABOUT ONE AXIS-SYMMETRICAL REINFORCEMENT

9.3.1 Code Provisions

The design of short columns resisting moment and axial load is covered in various clauses in BS 8110: Part 1, section 3.8. The main provisions are as follows

1. Clause 3.8.2.3 states that in column and beam construction in monolithic braced frames, the axial force in the column can be calculated assuming the beams are simply supported. If the arrangement of beams is symmetrical, the column can be designed for axial load only as set out in section 9.2 above. The column may also be designed for axial load and a moment due to the nominal eccentricity given in provision 2 below;

2. Clause 3.8.2.4 states that in no section in a column should the design moment be taken as less than the ultimate load acting at a minimum eccentricity e_{min} equal to 0.05 times the overall dimension of the column in the plane of bending, but not more than 20 mm;

3. Clause 3.8.4.1 states that in the analysis of cross-sections to determine the ultimte resistance to moment and axial force the same assumptions should be made as when analysing a beam. These assumptions are given in Clause 3.4.4.1 of the code;

4. Clause 3.8.4.2 states that design charts for symmetrically reinforced columns are given in BS 8110: Part 3 which uses γ_m for steel = 1.10 not 1.05;

5. Clause 3.8.4.3 states that it is usually only necessary to design short columns for the maximum design moment about one critical axis.
The application of the assumptions to analyse the section and construction of a design chart is given below.

9.3.2 Section Analysis

A reinforced column section subjected to the ultimate axial load N and ultimate moment M is shown in Fig.9.5. In most cases, columns are symmetrically reinforced because of the fact that the direction of the moment in most cases is reversible. An additional reason is with unsymmetrical reinforcement there is always the danger of the smaller amount of steel being wrongly placed on the face requiring the larger reinforcement. The moment M is equivalent to the axial load acting at an eccentricity e = M / N. Depending on the relative values of M and N, the following two main cases occur for analysis:

1. compression over the whole section where the neutral axis lies at the edge or outside the section as shown in Fig.9.6(a) with both rows of steel bars in compression.

2. compression on one side in the concrete and reinforcement and tension in the reinforcement on the other side with the neutral axis lying between the rows of reinforcement as shown in Fig.9.6(b)
For a given location of the neutral axis, the strains and stresses in both the concrete and the steel can be determined and from these the values of the internal

forces can be found. The resultant internal axial force and resistance moment can then be evaluated. For Fig.9.6 (b) the compression force C_c in the concrete is calculated as follows.

9.3.2.1 Parabolic–rectangular stress block

Let ε_0 be the strain at the end of parabolic variation of stress, where

$$\varepsilon_0 = 2.4\times10^{-4} \times \sqrt{\frac{f_{cu}}{(\gamma_m = 1.5)}}$$

let $\alpha = \varepsilon_0/0.0035$

In calculating the contribution to axial force and moment by the compressive stress in concrete, three cases have to be considered.

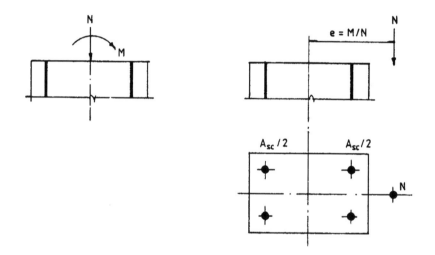

Fig.9.5 Column subjected to axial force and moment.

Case 1: $x \leq h$
In this case part of the concrete and compression steel are in compression and tension steel is in tension. As shown in Fig 9.6 (b), both the parabolic and the constant part of the stress lie inside the column cross-section. The compression force C_{c1} in the parabolic portion of depth αx is

$$C_{c1} = 0.45 f_{cu} \frac{2}{3} \{\alpha \times b\}$$

The factor (2/3) comes from the property of the parabola that the area of a parabola is equal to (2/3) the area of the enclosing rectangle.

From the property of the parabola, the centroid of the force C_{c1} is at a distance of (5/8) αx from the neutral axis.

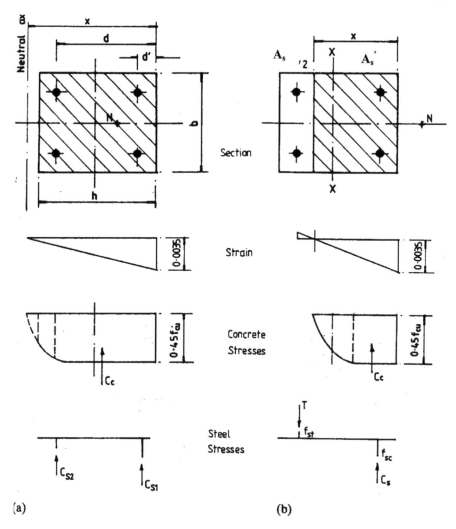

Fig.9.6 (a) Compression over whole section; (b) compression over part of section, tension in some steel.

The centroid of the force C_{c2} is at a distance of $0.5(1 - \alpha) \times$ from the compression face of the column.

The total compressive force C_c is

$$C_c = C_{c1} + C_{c2} = 0.45 f_{cu} (1 - \frac{1}{3}\alpha)bx = 0.45 f_{cu}bh (1 - \frac{1}{3}\alpha)\frac{x}{h}$$

Taking moments about the compression face, the centroid $\bar{a}$ of the compression force C_c from the compression face is

$$C_c \times \bar{a} = C_{c2} \, 0.5(1-\alpha)x + C_{c1}\{(1-\alpha)x + (\alpha x - \frac{5}{8}\alpha x)\}$$

Substituting for C_c, C_{c1} and C_{c2},

$$C_c \times \bar{a} = 0.45 f_{cu} \frac{2}{3}\alpha x \, b\{(1-\alpha)x + \frac{3}{8}\alpha x\} + 0.45 f_{cu}(1-\alpha)x b\{0.5(1-\alpha)x\}$$

$$C_c \times \bar{a} = 0.45 f_{cu} b x^2 [\frac{2}{3}\alpha(1-\frac{5}{8}\alpha) + \frac{1}{2}(1-\alpha)^2] = 0.45 f_{cu} b h^2 [\frac{6-4\alpha+\alpha^2}{12}] (\frac{x}{h})^2$$

$$\text{giving} \quad \bar{a} = h\frac{\{6-4\alpha+\alpha^2\}}{4(3-\alpha)} \frac{x}{h}$$

Case 2: x ≥ h/(1 – α)

In this case the entire cross-section is under a constant stress of 0.45 f_{cu}. Let the strain at the less compressed face be ε.

$$\varepsilon = \frac{0.0035}{x}(x-h) = \varepsilon_0$$

$$(x-h) \geq \alpha x, \quad x \geq \frac{h}{(1-\alpha)}$$

where $\alpha = \dfrac{\varepsilon_0}{0.0035}$

Case 3: $\dfrac{h}{(1-\alpha)} > x > h$

In this case only a part of the parabolic part of the stress–strain curve and the constant part are inside the column. The distance from the neutral axis to the less compressed face is (x – h) and the distance from the neutral axis to the point where the strain is ε_0 is αx.

Therefore the constant part of the stress is over a depth of (1 – α) x and the parabolic part is over the depth $\{\alpha x - (x-h)\} = h - (1-\alpha)x$

The compression force C_{c2} in the constant portion which is of depth (1 – α) x is

$$C_{c2} = 0.45 f_{cu}\{(1-\alpha)x\, b\}$$

The compression force C_{c1} in the parabolic portion of the stress block of depth $\{h-(1-\alpha)x\}$ is calculated as follows.

As shown in Fig 9.6(a), in the parabolic portion of the stress–strain diagram which is of depth αx, the stress σ at any distance y from the neutral axis is given by

$$\sigma = 0.45 f_{cu} \frac{y}{\alpha x}\{2 - \frac{y}{\alpha x}\}$$

This equation satisfies the boundary conditions as follows.

(σ = 0, y = 0) and (σ = 0.45 f_{cu}, y = αx)

$$\frac{d\sigma}{dy} = 0.45 f_{cu} \frac{2}{\alpha x}\{1 - \frac{y}{\alpha x}\} = 0, \text{ y} = \alpha x$$

The parabolic portion of interest extends from y = (x – h) to αx. The compressive force C_{c1} in the parabolic portion is obtained by integration as follows.

$$C_{c1} = b \int_{(x-h)}^{\alpha x} \sigma \, dy = 0.45 \, f_{cu} b \int_{(x-h)}^{\alpha x} \{2\frac{y}{\alpha x} - (\frac{y}{\alpha x})^2\} dy$$

$$C_{c1} = 0.45 \, f_{cu} \, b \, [\frac{y^2}{\alpha x} - \frac{1}{3}\frac{y^3}{(\alpha x)^2}]_{x-h}^{\alpha x}$$

$$C_{c1} = 0.45 \, f_{cu} \, b \, [\alpha x - \frac{\alpha x}{3} - \frac{(x-h)^2}{\alpha x} + \frac{(x-h)^3}{3(\alpha x)^2}]$$

Simplifying,

$$C_{c1} = 0.45 \, f_{cu} \, bx \, \frac{1}{3\alpha^2} \{2\alpha^3 - 3\alpha(1-\frac{h}{x})^2 + (1-\frac{h}{x})^3\}$$

$$C_c = C_{c1} + C_{c2} = 0.45 \, f_{cu} \, bh[\frac{\{3\alpha^2 - \alpha^3 - 3\alpha(1-\frac{h}{x})^2 + (1-\frac{h}{x})^3\}}{3\alpha^2}]\frac{x}{h}$$

The moment of the forces C_{c1} and C_{c2} from the face with the maximum compressive strain is

$$C_c \bar{a} = C_{c2} \frac{(1-\alpha x)}{2} + b \int_{(x-h)}^{\alpha x} \sigma \, dy(x-y)$$

$$= 0.45 f_{cu} b \frac{\{(1-\alpha)x\}^2}{2} + 0.45 f_{cu} \, b \int_{(x-h)}^{\alpha x} \{2\frac{y}{\alpha x} - (\frac{y}{\alpha x})^2\}(x-y)dy$$

Carrying out the integration and substituting the limits, after simplification

$$C_c \bar{a} = 0.45 f_{cu} \, b \, h^2 \, [\frac{(1-\alpha)^2}{2} +$$

$$\{\frac{8\alpha^3 - 5\alpha^4 - 12\alpha(1-\frac{h}{x})^2 + (8\alpha + 4)(1-\frac{h}{x})^3 - 3(1-\frac{h}{x})^4}{12\alpha^2}\}](\frac{x}{h})^2$$

9.3.2.2 Rectangular stress block

In the case of Rectangular stress block assumption, the corresponding equations are much simpler than the rectangular–parabolic stress block assumption.

Case 1: $0.9x/h \leq 1$

In this case part of the column cross section is in compression and part in tension.

$$C_c = 0.45 \, f_{cu} \, b \, 0.9x = 0.45 \, f_{cu} bh \, 0.9\frac{x}{h}$$

$$C_c /(bh) = 0.45 \, f_{cu} \{0.9\frac{x}{h}\}$$

Taking moments about the compression face, the centroid $\bar{a}$ of the compression force C_c from the compression face is

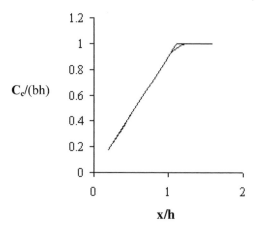

Fig.9.7 Comparison between compressive forces based on rigorous and simplified stress blocks.

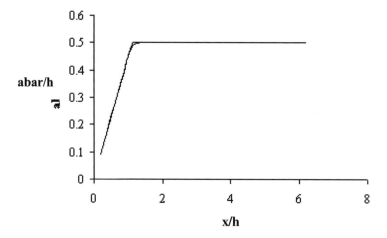

Fig.9.8 Comparison between resultant positions of compressive force based on rigorous and simplified stress blocks.

$$C_c \times \overline{a} = C_c \, 0.5 \times 0.9x = 0.45 \, f_{cu} \, bh^2 \, 0.405 \left(\frac{x}{h}\right)^2$$

$$C_c \times \overline{a} / (bh^2) = 0.45 \, f_{cu} \, 0.405 \left(\frac{x}{h}\right)^2$$

Case 2: 0.9x/h > 1

In this case the whole cross section is in compression.

$$C_c = 0.45 \, f_{cu} \, b h = 0.45 \, f_{cu} bh$$

Taking moments about the compression face, the centroid $\overline{a}$ of the compression force C_c from the compression face is

$$C_c \times \bar{a} = C_c \, 0.5h = 0.45 \, f_{cu} \, bh^2 \, 0.5$$

$$C_c \times \bar{a} / (bh^2) = 0.225 \, f_{cu}$$

Fig.9.7 and Fig.9.8 show respectively variation of $C_c/(bh)$ and $\bar{a}/h$ with x/h as calculated for the parabolic–rectangular stress block assumption and rectangular block assumption for $f_{cu} = 30$ N/mm². As can be seen, the differences are insignificant.

9.3.2.3 Stresses and strains in steel

For all positions of the neutral axis, the strains in the compression and tension steels are

$$\varepsilon_s' = 0.0035 \frac{(x - d')}{x} = 0.0035(1 - \frac{d'}{h} \frac{h}{x})$$

$$\varepsilon_s = 0.0035 \frac{(d - x)}{x} = 0.0035(\frac{d}{h} \frac{h}{x} - 1)$$

The stresses in the compression and tension steels are
$$f_{sc} = E \, \varepsilon_{sc} \le 0.95 \, f_y \text{ and } f_{st} = E \, \varepsilon_{st} \le 0.95 \, f_y$$
where E = Young's modulus for steel.

Note that when x > d, then ε_{st} becomes negative, indicating that the stress in the 'tension' steel is actually compressive. The force C_s in compression and force T in tension steel are

$$C_s = A_s` \, f_s', \ T = A_s \, f_s$$

9.2.3.4 Axial force N and moment M

The sum of the internal forces is
$$N = C_c + C_s - T$$
$$N = C_c + A_s' \, f_s' - A_s \, f_s$$

$$\frac{N}{bh} = \frac{C_c}{bh} + \frac{A_s'}{bh} f_s' - \frac{A_s}{bh} f_s$$

The sum of the moments of the internal forces about the centreline of the column is

$$M = C_c(0.5h - \bar{a}) + C_s(0.5h - d') + T(d - 0.5h)$$

$$M = 0.5h \, C_c - C_c \, \bar{a} + C_s(0.5h - d') + T(d - 0.5h)$$

$$\frac{M}{bh^2} = 0.5 \frac{C_c}{bh} - \frac{C_c \, \bar{a}}{bh^2} + \frac{A_s'}{bh} f_s'(0.5 - \frac{d'}{h}) + \frac{A_s}{bh} f_s(\frac{d}{h} - 0.5)$$

Using the above equations, it is not possible to directly design a section to carry a given load and moment. It is necessary to assume a trial section and the required amount of steel can be determined using design charts constructed using the above equations.

9.2.3.5 Example of a short column subjected to axial load and moment about one axis

Determine the ultimate axial load and moment about the centroidal axis that the column section with b = 300 mm, h = 400 mm, d` = 50 mm, d = 350 mm can carry when the depth to the neutral axis is 250 mm. Materials are grade C30 concrete and grade 460 reinforcement. Assume reinforcement consists of two 25 mm bars on each face.

(a) Using the Parabolic–Rectangular stress block
Calculate ε_0 for $f_{cu} = 30 \text{N/mm}^2$, where

$$\varepsilon_0 = 2.4 \times 10^{-4} \times \sqrt{\frac{f_{cu}}{(\gamma_m = 1.5)}}$$

$\varepsilon_0 = 1.0733 \times 10^{-3}$, $\alpha = \varepsilon_0 / 0.0035 = 0.3067$
h = 400 mm, b = 300 mm, d = 400 – 50 = 350 mm, d` = 50 mm.
x = 250 mm, x/h = 0.625 < 1.0, therefore using the formulae for Case 1 and substituting $f_{cu} = 30$, b = 300 mm, x = 250 mm,

$$C_c = 0.45 f_{cu} (1 - \frac{1}{3}\alpha)bx = 909 \text{ kN}$$

$$C_c \times \bar{a} = 0.45 f_{cu} b x^2 [\frac{2}{3}\alpha(1 - \frac{5}{8}\alpha) + \frac{1}{2}(1 - \alpha)^2] = 102.67 \text{ kNm}$$

Moment due to C_c about the centre of the column is = $C_c \times (0.5h) - C_c \times \bar{a}$.

The maximum strain in concrete at failure is 0.0035. The strains in the reinforcements are as follows:

Compression $\varepsilon_{sc} = 0.0035 \times 200/250 = 0.0028$

Tension $\varepsilon_{st} = 0.0035 \times 100/250 = 0.0014$

Taking E = 200 kN/mm² for steel,

$f_{sc} = \varepsilon_{sc} E = 0.0028 \times 200 \times 10^3 = 560 \text{ N/mm}^2 > (0.95 f_y = 437 \text{ N/mm}^2)$

$f_{sc} = 437 \text{ N/mm}^2$

$f_{st} = \varepsilon_{st} E = 0.0014 \times 200 \times 10^3 = 280 \text{ N/mm}^2 < (0.95 f_y = 437 \text{ N/mm}^2)$

$f_{sc} = 280 \text{ N/mm}^2$

Area of steel on each face = 2T25 = 981.5 mm². The steel forces are

Compression $C_s = 437 \times 981.5 \times 10^{-3} = 428.9 \text{ kN}$

Tension $T = 280 \times 981.5 \times 10^{-3} = 274.8 \text{ kN}$

The ultimate axial force is

$$N = 909.0 + 428.9 - 274.8 = 1063.1 \text{ kN}$$

The ultimate moment is found by taking moments of the internal forces about the centre of the column:

$$M = (909.3 \times 200 \times 10^{-3} - 102.7) + (428.9 + 274.8)\ 150 \times 10^{-3}$$
$$= 184.7 \text{ kNm}$$

(b) Using the rectangular stress block

Using the formulae for rectangular stress block,

$$C_c = 0.45\ f_{cu}\ b\ (0.9x) = 0.45 \times 30 \times 300 \times 0.9 \times 250 \times 10^{-3} = 911.25 \text{ kN}$$

$$C_c \times \bar{a} = 0.45 f_{cu} b\, 0.9x[0.45x] = 102.5 \text{ kNm}$$
$$N = 911.25 + 428.9 - 274.8 = 1065.4 \text{ kN}$$
$$M = (911.25 \times 200 \times 10^{-3} - 102.5) + (428.9 + 274.8)\ 150 \times 10^{-3} = 185.3 \text{ kNm}$$

Compared with the rectangular–parabolic stress block, errors in N and M are about 0.2% and 0.3% respectively. This is negligible.

9.3.3 Construction of Column Design Chart

A design curve can be drawn for a selected grade of concrete and reinforcing steel for a section with a given percentage of reinforcement, $100A_{sc}/(bh)$, symmetrically placed at a given location d/h. The curve is formed by plotting values of $N/(bh)$ against $M/(bh^2)$ for various positions of the neutral axis x. Other curves can be constructed for percentages of steel ranging from 0.4% to a maximum of 6% for vertically cast columns. The family of curves forms the design chart for that combination of materials and steel location. Separate charts are required for the same materials for different values of d/h which determines the location of the reinforcement in the section. Groups of charts are required for the various combinations of concrete and steel grades.

The process for construction of a design chart is demonstrated below.
1. Select materials: Concrete $f_{cu} = 30 \text{ N/mm}^2$, Reinforcement $f_y = 460 \text{ N mm}^2$
2. Select a value of $d/h = 0.85$, $d'/h = 0.15$.
3. Select a steel percentage $100A_{sc}/(bh) = 6$
Let the steel be symmetrically placed and $A`_s/\ bh = 0.03$ and $A_s/\ bh = 0.03$.
The design chart is constructed by selecting different values of x/h and calculating the corresponding $N/(bh)$ and $M/(bh^2)$.

$$\frac{N}{bh} = \frac{C_c}{bh} + \frac{A'_s}{bh} f'_s - \frac{A_s}{bh} f_s$$
$$N/(bh) = C_c\ /(bh) + 0.03\ (f_s' - f_s)$$

$$\frac{M}{bh^2} = 0.5\frac{C_c}{bh} - \frac{C_c\, \bar{a}}{bh^2} + \frac{A'_s}{bh} f'_s (0.5 - \frac{d'}{h}) + \frac{A_s}{bh} f_s(\frac{d}{h} - 0.5)$$

$$M\,/(bh^2) = 0.5 C_c\,/(bh) - C_c\, \bar{a}/(bh^2) + 0.03(0.35)\{f'_s + f_s\}$$

The equations for calculating $C_c/(bh)$ and $C_c \times \bar{a}/(bh^2)$ to be used depend up on the value of x/h and also on the type of stress block used. They can be summarised as follows.

(a) Parabolic–rectangular stress block assumption

$$C_c/(bh) = 0.45 f_{cu}\,(1 - \frac{1}{3}\alpha)\frac{x}{h}, \quad x/h \le 1.0$$

$$C_c/(bh) = 0.45 f_{cu}\,[\frac{\{3\alpha^2 - \alpha^3 - 3\alpha(1-\dfrac{h}{x})^2 + (1-\dfrac{h}{x})^3\}}{3\alpha^2}]\frac{x}{h}, \quad \frac{1}{(1-\alpha)} > x/h > 1.0$$

$$C_c/(bh) = 0.45 f_{cu}, \quad x/h \ge 1.0/(1-\alpha)$$

$$C_c \times \bar{a}/(bh^2) = 0.45 f_{cu}[\frac{6 - 4\alpha + \alpha^2}{12}](\frac{x}{h})^2, \quad x/h \le 1.0$$

$$C_c\,\bar{a}/(bh^2) = 0.45 f_{cu}\,[\frac{(1-\alpha)^2}{2} + \{8\alpha^3 - 5\alpha^4 - 12\alpha(1-\frac{h}{x})^2 + (8\alpha + 4)(1-\frac{h}{x})^3$$

$$- 3(1-\frac{h}{x})^4\} \times \{\frac{1}{12\alpha^2}\}](\frac{x}{h})^2, \quad \frac{1}{(1-\alpha)} > \frac{x}{h} > 1.0$$

$$C_c\,\bar{a}\,/(bh^2) = 0.225 f_{cu}, \quad x/h \ge 1.0/(1-\alpha)$$

(b) Rectangular stress block assumption

$$C_c/(bh) = 0.405 f_{cu}\frac{x}{h}, \quad x/h \le 1.1111$$

$$C_c/(bh) = 0.45 f_{cu}, \quad x/h > 1.1111$$

$$C_c \times \bar{a}/(bh^2) = 0.18225 f_{cu}\,(\frac{x}{h})^2, \quad x/h \le 1.1111$$

$$C_c \times \bar{a}/(bh^2) = 0.225 f_{cu}, \quad x/h > 1.1111$$

(c) For all positions of the neutral axis, the strains in the steel areas are

$$\varepsilon_s' = 0.0035\frac{(x-d')}{x} = 0.0035(1 - \frac{d'}{h}\frac{h}{x}) = 0.0035(1 - 0.15\frac{h}{x})$$

$$\varepsilon_s = 0.0035\frac{(d-x)}{x} = 0.0035(\frac{d}{h}\frac{h}{x} - 1) = 0.0035(0.85\frac{h}{x} - 1)$$

The stresses in the steel are

$$f_s' = E\,\varepsilon_s' \le 0.95 f_y \text{ and } f_s = E\,\varepsilon_s \le 0.95 f_y$$

where E = Young's modulus for steel.

9.3.3.1 Typical calculations for rectangular–parabolic stress block

Concrete: $f_{cu} = 30$ N/mm^2, $\varepsilon_0 = 1.0733 \times 10^{-3}$, $\alpha = \varepsilon_0 / 0.0035 = 0.3067$
Steel: $f_y = 460$ N/mm^2, $0.95\, f_y = 437$ N/mm^2, $E = 200$ kN/mm^2.
$$f_s' = E\,\varepsilon_s' \le (0.95\, f_y = 437 \text{ N/mm}^2)$$
$$f_s = E\,\varepsilon_s \le (0.95\, f_y = 437 \text{ N/mm}^2)$$

(a) x/h = 0.4

$$C_c /(bh) = 0.45\, f_{cu}\,(1 - \frac{1}{3}\alpha)\frac{x}{h} = 4.848$$

$$C_c \times \bar{a}/(bh^2) = 0.45\, f_{cu}[\frac{6 - 4\alpha + \alpha^2}{12}]\,(\frac{x}{h})^2 = 0.876$$

$$\varepsilon_s' = 0.0035(1 - \frac{d'}{h}\frac{h}{x}) = 0.002188$$
$$E\,\varepsilon_s' = 437.6 > 437.0,\ f_s' = 437 \text{ N/mm}^2$$
$$\varepsilon_s = 0.0035(\frac{d}{h}\frac{h}{x}) = 0.003938$$
$$E\,\varepsilon_s = 787.6 > 437.0,\ f_s = 437 \text{ N/mm}^2$$
$$N/(bh) = C_c /(bh) + 0.03\,(f_s' - f_s) = 4.848 + 0.0 = 4.848$$

$$M/(bh^2) = 0.5\,C_c /(bh^2) - C_c\,\bar{a}/(bh^2) + 0.03(0.35)\{f_s' + f_s\}$$
$$M/(bh^2) = 2.424 - 0.876 + 9.177 = \mathbf{10.725}$$

(b) x/h = 1.2

$$C_c /(bh) = 0.45\, f_{cu}\,[\frac{\{3\alpha^2 - \alpha^3 - 3\alpha(1 - \frac{h}{x})^2 + (1 - \frac{h}{x})^3\}}{3\alpha^2}]\frac{x}{h} = 13.3425$$

$$C_c\,\bar{a}/(bh^2) = 0.45\, f_{cu}[\frac{(1 - \alpha)^2}{2} + \{8\alpha^3 - 5\alpha^4 - 12\alpha(1 - \frac{h}{x})^2 +$$
$$(8\alpha + 4)(1 - \frac{h}{x})^3 - 3(1 - \frac{h}{x})^4\} \times \{\frac{1}{12\alpha^2}\}](\frac{x}{h})^2 = 6.5991$$

$$\varepsilon_s' = 0.0035(1 - \frac{d'}{h}\frac{h}{x}) = 0.003063$$
$$E\,\varepsilon_s' = 612.5 > 437.0,\ f_s' = 437 \text{ N/mm}^2$$
$$\varepsilon_s = 0.0035(\frac{d}{h}\frac{h}{x}) = -0.00102$$

$E\,\varepsilon_s = -204.2,\ f_s = -204$ N/mm^2 ('Tension' steel is in compression)
$$N/(bh) = C_c /(bh) + 0.03\,(f_{sc} - f_{st}) = 13.3425 + 19.236 = \mathbf{32.58}$$

$$M/(bh^2) = 0.5\,C_c /(bh^2) - C_c\,\bar{a}/(bh^2) + 0.03(0.35)\{f_s' + f_s\}$$
$$M/(bh^2) = 6.6713 - 6.5991 + 2.444 = \mathbf{2.5166}$$

(c) x/h = 2.0

$$C_c /(bh) = 0.45 \, f_{cu} = 13.5$$

$$C_c \bar{a} /(bh^2) = 0.225 \, f_{cu} = 6.75$$

$$\varepsilon_s' = 0.0035(1 - \frac{d'}{h}\frac{h}{x}) = 0.003238$$

$$\varepsilon_s = 0.0035(\frac{d}{h}\frac{h}{x}) = -0.00201$$

$$E \, \varepsilon_s' = 647.5 > 437, \quad f_s' = 437 \, \text{N/mm}^2$$
$$E \, \varepsilon_s = -402.5, \quad f_s = -403 \, \text{N/mm}^2$$
$$N/(bh) = C_c /(bh) + 0.03 \, (f_{sc} - f_{st}) = 13.50 + 25.185 = \mathbf{38.685}$$

$$M /(bh^2) = 0.5 \, C_c /(bh^2) - C_c \, \bar{a}/(bh^2) + 0.03(0.35)\{f_s' + f_s\}$$
$$M /(bh^2) = 6.75 - 6.75 + 0.362 = \mathbf{0.362}$$

9.3.3.2 Typical calculations for rectangular stress block

Note that only the expressions for concrete change. The expressions for strains and stresses in steel remain as for the case of rectangular–parabolic calculations.

(a) x/h = 0.4

$$C_c /(bh) = 0.405 \, f_{cu} \frac{x}{h} = 4.86$$

$$C_c \times \bar{a}/(bh^2) = 0.18225 \, f_{cu} \, (\frac{x}{h})^2 = 0.8748$$

$$f_s' = 437 \, \text{N/mm}^2, \quad f_s = 437 \, \text{N/mm}^2$$
$$N/(bh) = C_c /(bh) + 0.03 \, (f_{sc} - f_{st}) = 4.86 + 0.0 = \mathbf{4.86}$$

$$M /(bh^2) = 0.5 \, C_c /(bh^2) - C_c \, \bar{a}/(bh^2) + 0.03(0.35)\{f_s' + f_s\}$$
$$M /(bh^2) = 2.43 - 0.875 + 9.177 = \mathbf{10.732}$$

(b) x/h = 1.2

$$C_c /(bh) = 0.45 \, f_{cu} = 13.5$$

$$C_c \times \bar{a}/(bh^2) = 0.45 \, f_{cu} \, 0.5 = 6.75$$
$$f_s' = 437 \, \text{N/mm}^2, \quad f_s = -204 \, \text{N/mm}^2$$
$$N/(bh) = C_c /(bh) + 0.03 \, (f_{sc} - f_{st}) = 13.50 + 19.236 = \mathbf{32.74}$$

$$M /(bh^2) = 0.5 \, C_c /(bh^2) - C_c \, \bar{a}/(bh^2) + 0.03(0.35)\{f_s' + f_s\}$$
$$M /(bh^2) = 6.75 - 6.75 + 2.444 = \mathbf{2.44}$$

(c) x/h = 2.0

$$C_c /(bh) = 0.45 f_{cu} = 13.5$$

$$C_c \times a /(bh^2) = 0.45 f_{cu} \ 0.5 = 6.75$$

$$f_s^` = 437 \text{ N/mm}^2, \quad f_s = -403 \text{ N/mm}^2$$

$$N/(bh) = C_c /(bh) + 0.03 (f_{sc} - f_{st}) = 13.50 + 25.185 = \mathbf{38.685}$$

$$M /(bh^2) = 0.5 C_c /(bh^2) - C_c \ \bar{a}/(bh^2) + 0.03(0.35)\{f_s^` + f_s\}$$

$$M /(bh^2) = 6.75 - 6.75 + 0.362 = \mathbf{0.362}$$

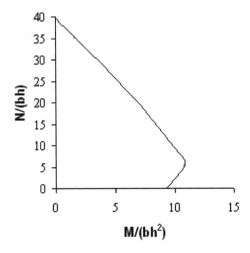

Fig.9.9 Column design chart for 6% steel.

In a similar manner calculations can be carried out for other values of x/h. Calculations are most conveniently done using a Spread Sheet. Using the results, a graph of N/(bh) versus M/(bh²) can be drawn as shown in Fig.9.9.

As is to be expected, when x > d, the 'tension' reinforcement goes into compression. This naturally increases the value of N/(bh) but drastically decreases the value of M/(bh²). When the entire column section is under a compressive stress of $0.45f_{cu}$ and the stress in both steels is $0.95f_y$ compression, then the maximum value of N/(bh) is attained and the corresponding value of M/(bh²) is equal to zero.

Curves for total steel percentages 0.4, 1,2,3,4, 5, 6, 7 and 8 can be plotted. The design chart is shown in Fig.9.10. Other charts are required for different values of the ratio d/h to give a series of charts for a given concrete and steel strength. A separate series of charts is required for each combination of materials used.

It has to be noted that any combination of {N/(bh), M(bh²)} which lies on or inside the curve corresponbding to a particular value of $A_{sc}/(bh)$ leads to a safe design.

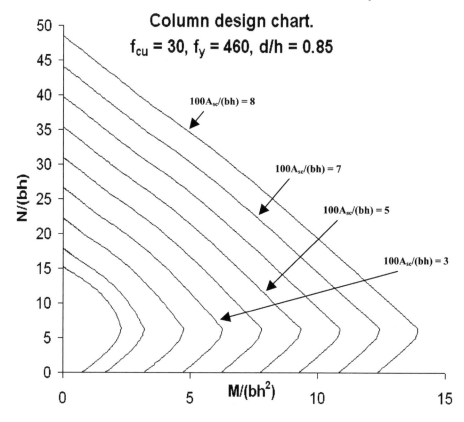

Fig.9.10 Column design chart

9.3.3.4 Column design using design charts

A short braced column is subjected to a design ultimate load of 1480 kN and an ultimate moment of 54 kNm. The column section is 300mm × 300mm. Determine the area of steel required. The materials are grade C30 concrete and grade 460 reinforcement.

Assume 25 mm diameter bars for the main reinforcement and 8 mm diameter links. The cover on the links is 25 mm.

$$b = h = 300 \text{ mm}$$
$$d = 300 - 25 - 8 - 12.5 = 254.5 \text{ mm}$$
$$d/h = 254.5/300 = 0.85$$

Use the chart shown in Fig.9.10 where d/h = 0.85.

$$N/(bh) = 1480 \times 10^3 / (300^2) = 16.4$$
$$M/(bh^2) = 54 \times 10^6 (300^3) = 2.0$$

For this combination of $\{N/(bh), M/(bh^2)\}$, the design chart gives $100A_{sc}/(bh) = 2$

$$A_{sc} = 2.0 \times 300^2/100 = 1800 \text{ mm}^2$$

Provide four 25 mm diameter bars to give an area of 1963 mm².
Calculations show that at $(x/h) = 0.94$, $f_s' = 437$ N/mm², $f_s = -67$ N/mm²
i. Rectangular–parabolic stress block: $N/(bh) = 16.43$, $M/(bh^2) = 2.15$
ii. Rectangular stress block: $N/(bh) = 16.46$, $M/(bh^2) = 2.17$

9.3.4 Further design chart

The design chart shown in Fig.9.10 strictly applies only to the case where the symmetrical reinforcement is placed on two opposite faces. Charts can be constructed for other arrangements of reinforcement. One such case is shown in Fig.9.11 where eight bars are spaced evenly around the perimeter of the column. The total steel A_{sc} is placed such that at the top and bottom rows steel is 0.375 A_{sc} (3 bars) and in the middle row it is 0.25 A_{sc}.(2 bars)

The contribution from concrete is calculated as in the previous section. There are however three strains to calculate.

$$\varepsilon_s' = 0.0035\frac{(x-d')}{x} = 0.0035(1 - \frac{d'}{h}\frac{h}{x}) \text{ (top layer)}$$

$$\varepsilon_{s1} = 0.0035\frac{(0.5h-x)}{x} = 0.0035(0.5\frac{h}{x} - 1) \text{ (Middle layer)}$$

$$\varepsilon_{s2} = 0.0035\frac{(d-x)}{x} = 0.0035(\frac{d}{h}\frac{h}{x} - 1) \text{ (bottom layer)}$$

The stresses in the compression and tension steels are calculated from strains as before.

$$C_s = A_s' f_s'$$
$$T_1 = A_{s1} f_{s1}, \quad T_2 = A_{s2} f_{s2}$$
$$N = C_c + C_s - T_1 - T_2$$
$$\frac{N}{bh} = \frac{C_c}{bh} + \frac{A_s'}{bh} f_s' - \frac{A_{s1}}{bh} f_{s1} - \frac{A_{s2}}{bh} f_{s2}$$

The sum of the moments of the internal forces about the centreline of the column is

$$M = C_c(0.5h - \bar{a}) + C_s(0.5h - d') + T_2(d - 0.5h)$$

$$= 0.5h\,C_c - C_c\,\bar{a} + C_s(0.5h - d') + T_2(d - 0.5h)$$

$$\frac{M}{bh^2} = 0.5\frac{C_c}{bh} - \frac{C_c\,\bar{a}}{bh^2} + \frac{A_s'}{bh} f_s'(0.5 - \frac{d'}{h}) + \frac{A_{s2}}{bh} f_{s2}(\frac{d}{h} - 0.5)$$

Note that the middle layer steel has zero lever arm about the centre line and hence does not contribute to moment of resistance.

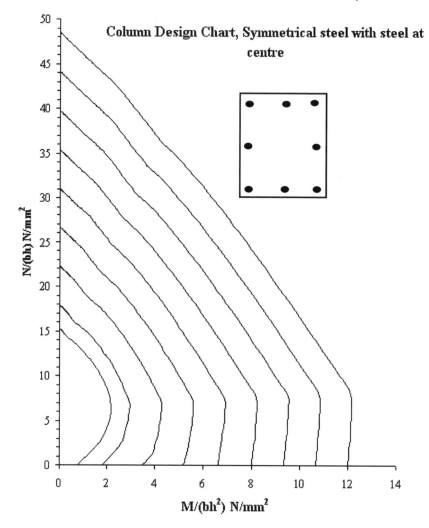

Fig.9.11 Column design chart for three layers of reinforcement. $f_{cu} = 30$, $f_y = 460$, $d/h = 0.9$.

9.4 SHORT COLUMNS SUBJECTED TO AXIAL LOAD AND BENDING ABOUT ONE AXIS: UNSYMMETRICAL REINFORCEMENT

An unsymmetrical arrangement of reinforcement provides the most economical solution for the design of a column subjected to a small axial load and a large moment about one axis. Such members occur in single storey reinforced concrete portals. Design charts for such cases can be constructed. If the total steel area is

6% say but is distributed such that the tension steel is 4% and compression steel is 2%, then the corresponding design chart is as shown in Fig.9.12.

When the ratio $(x/h) = 1.25$, the stress f_s in the 'tension' and compression steels are respectively -224 N/mm^2 and 437 N/mm^2. The compressive forces in the two steels are equal and the entire column is almost in a state of uniform compression and $N/(bh) = 22.27$. The moment contributed by the two steels is zero. At this stage the corresponding value of $M/(bh^2)$ is zero. Because of the smaller value of the compressive stress in the 'tension' steel, the value of $N/(bh)$ at this stage is smaller than in the case if the same amount of total steel is symmetrically distributed $(N/(bh) = 33.33)$.

When the ratio $(x/h) = 2.3$, the stresses in the two steels are equal to $-0.95f_y$ and the entire column is almost in a state of uniform compression and the maximum value of $N/(bh) = 26.61$ is reached. However because of the fact that the compressive forces in the two steel are not equal, the force in the 'tension' steel gives rise to a negative value of $M/(bh^2) = -1.53$. However if the reinforcement is symmetrically distributed, then $N/(bh) = 39.65$ and $M/(bh^2)$ will be zero.

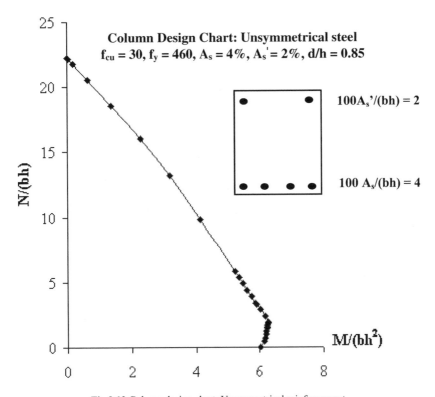

Fig.9.12 Column design chart: Unsymmetrical reinforcement.

9.4.1 Example of a Column Section Subjected to Axial Load and Moment: Unsymmetrical Reinforcement

Design a column subjected to an ultimate axial load of 230 kN and an ultimate moment of 244 kNm. Design the reinforcement required using an unsymmetrical arrangement. The concrete is grade C30 and the reinforcement is grade 460.

Assuming d'/h = 0.15 and d/h = 0.85 and because of the large moment, assume a rectangular section with b = 300 mm and h = 400 mm.

$$N/(bh) = 230 \times 10^3/(300 \times 400) = 1.92$$
$$M/(bh^2) = 244 \times 10^6/(300 \times 400^2) = 5.08$$

Assume $A_s'/(bh) = 1\%$ and $A_s/(bh) = 2\%$ and draw the design chart as shown in Fig.9.13.

Calculations show that at (x/h) = 0.52, $f_s' = 437$ N/mm^2 and $f_s = 437$ N/mm^2. Approximately only one half of the column cross-section is not in compression.

$$N/(bh) = 1.93, \quad M/(bh^2) = 6.26.$$

$A_s' = 0.01 \times 300 \times 400 = 1200$ mm^2, $A_s = 0.02 \times 300 \times 400 = 2400$ mm^2
Provide 3T25 on the compression face, $A_s' = 1473$ mm^2 and 5T25 on the tension face, $A_s = 2454$ mm^2.

If the column had been symmetrically reinforced, then for total of 3% steel, assuming $A_s'/(bh) = 1.5\%$ and $A_s/(bh) = 1.5\%$, calculations show that at (x/h) = 0.29, $f_s' = 338$N/mm^2, $f_s = 437$ N/mm^2. Approximately only one third of the column cross-section is in compression. For this steel arragement

$$N/(bh) = 2.03, M/(bh^2) = 5.37$$

Provide 4T25 on both faces. $A_s' = A_s = 1963$ mm^2. The total reinforcement is same as that for the unsymmetrical case. The unsymmetrical case provides a greater moment capacity

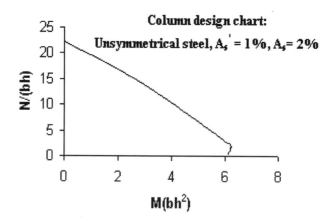

Fig.9.13 Column design chart: $A_s = 2\%$, $A_s' = 1\%$.

9.5 COLUMN SECTIONS SUBJECTED TO AXIAL LOAD
AND BIAXIAL BENDING

9.5.1 Outline of the Problem

When a column is subjected to an axial force and a bending moment about say x-axis, the neutral axis is parallel to the x-axis. However when a column is subjected to an axial force and moments about the two axes, the neutral axis is inclined to the x-axis as shown in Fig.9.14.

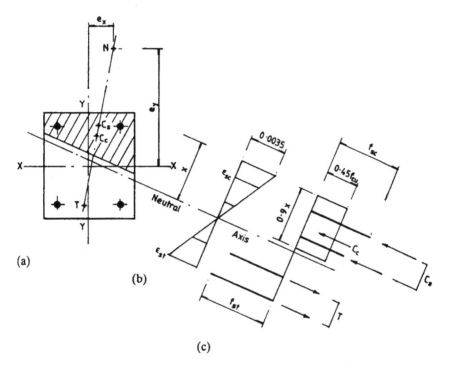

Fig.9.14 (a) Section; (b) strain diagram; (c) stresses and internal forces.

For a given location and direction of the neutral axis the strain diagram can be drawn with the maximum strain in the concrete of 0.0035. The strains in the compression and tension steel can be found and the corresponding stresses determined from the stress–strain diagram for the reinforcement. The resultant forces C_s and T in the compression and tension steel and the force C_c in the concrete can be calculated and their locations determined. The net axial force is
$$N = C_c + C_s - T$$
Moments of the forces C_c, C_s and T are taken about the XX and YY axes to give M_x and M_y.

Thus a given section can be analysed for a given location and direction of the neutral axis and the axial force and biaxial moments that it can support can be determined. As in the case of axial load with uni-axial bending moment, a failure surface can be constructed. It is generally found simple to use the rectangular stress block as opposed to parabolic–rectangular stress block. Calculations are naturally much more involved than in the case of axial load accompanied by uni-axial bending moment.

9.5.1.1 Expressions for contribution to moment and axial force by concrete

Fig.9.15 shows a rectangular column b × h and reinforced with four bars.

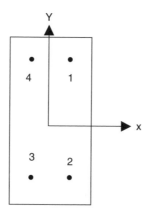

Fig.9.15 Column subjected to axial load and biaxial moments.

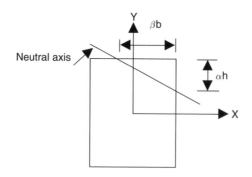

Fig.9.16 Column with the neutral axis inclined to x-axis.

Assuming the origin of coordinates at the centroid of the column cross section, the coordinates of the four bars can be calculated. The position of the neutral axis is governed by two parameters α and β as shown in Fig.9.16. Assuming that the maximum compressive strain is at the top right hand corner of the column, the normal strain in the cross-section is given by

$$\varepsilon = \varepsilon_u\,(C_1 + C_2\,(x/b) + C_3\,(y/h)), \quad \varepsilon_u = 0.0035$$

The constants can be calculated from the boundary conditions as follows:

$$\varepsilon = \varepsilon_u \text{ at } (x/b = 0.5,\ y/h = 0.5),$$
$$\varepsilon = 0 \text{ at } (x/b = (0.5 - \beta),\ y/h = 0.5),$$
$$\varepsilon = 0 \text{ at } (x/b = 0.5,\ y/h = (0.5 - \alpha))$$

Solving for the constants:

$$C_1 = 1 - 1/(2\beta) - 1/(2\alpha), \quad C_2 = 1/\beta, \quad C_3 = 1/\alpha h$$

$$\varepsilon = 0.0035\{1 + \frac{(\frac{x}{b} - 0.5)}{\beta} + \frac{(\frac{y}{h} - 0.5)}{\alpha}\}$$

The strain in the bars can be calculated by substituting the appropriate coordinates of the bars. The stress σ in the bars is equal to $\sigma = E\,\varepsilon$ but numerically not greater than $0.95\ f_y$.

Assuming a rectangular stress block with constant stress of $0.45f_{cu}$ and a depth equal to 0.9 times the depth of the neutral axis, the expressions for the compressive force and the corresponding moments about the x and y-axes due to the compressive stress in the column depends on the position of the neutral axis as follows.

Case 1: $0.9\,\beta \leq 1.0$ and $0.9\,\alpha \leq 1.0$
From the triangular shape of the stress block shown in Fig.9.17,

$$N_c = 0.45\ f_{cu}\ \{0.5 \times 0.9\ \alpha h \times 0.9\ \beta b\},$$
$$M_{xc} = N_c \times (0.5\ h - 0.9\ \alpha h/3),$$
$$M_{yc} = N_c \times (0.5\ b - 0.9\ \beta\ b/3)$$

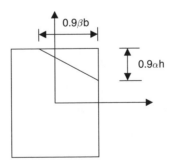

0.9βb

0.9αh

Fig.9.17 Neutral axis position for Case 1.

Case 2: $0.9\,\beta > 1.0$ and $0.9\,\alpha \leq 1.0$
From the trapezoidal stress block shown in Fig.9.18,

$$\alpha h_1 = \frac{\alpha h}{\beta b}(0.9\beta b - b) = \alpha h(0.9 - \frac{1}{\beta})$$
$$N_c = 0.45\ f_{cu}\ \{0.5\ (\alpha h_1 + 0.9\ \alpha h)\ b\},$$

$$M_{xc} = N_c \times (0.5 \, h - ybar),$$
$$M_{yc} = N_c \times (0.5 \, b - xbar)$$

Position of centroid from right face of the trapezium:
$$xbar = \frac{b}{3} \frac{(2h_1 + 0.9h)}{(h_1 + 0.9h)}$$

Position of centroid from top face of the trapezium:
$$ybar = \frac{\alpha}{3} \frac{(h_1^2 + 0.81\,h^2 + 0.9\,h_1\,h)}{(h_1 + 0.9\,h)}$$

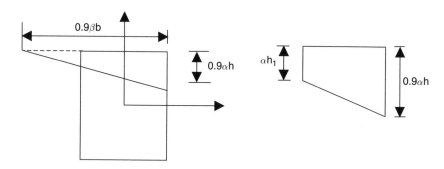

Fig.9.18 Neutral axis position for case 2.

Case 3: $0.9\,\beta \leq 1.0$ and $0.9\,\alpha > 1.0$

From the trapezoidal stress block shown in Fig.9.19

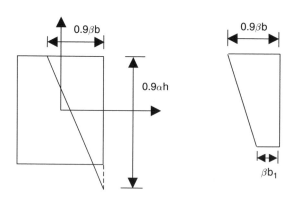

Fig.9.19 Neutral axis position for case 3.

$$\beta b_1 = \frac{\beta b}{\alpha h}(0.9\alpha h - h) = \beta b(0.9 - \frac{1}{\alpha})$$
$$N_c = 0.45 \, f_{cu} \{0.5 \, (\beta b_1 + 0.9 \, \beta b) \, h\},$$
$$M_{xc} = N_c \times (0.5 \, h - ybar),$$

$$M_{yc} = N_c \times (0.5\,b - xbar)$$

Position of centroid from right face of the trapezium:

$$xbar = \frac{\beta}{3}\frac{(b_1^2 + 0.81\,b^2 + 0.9\,b_1\,b)}{(b_1 + 0.9\,b)}$$

Position of centroid from top face of the trapezium:

$$ybar = \frac{h}{3}\frac{(2b_1 + 0.9b)}{(b_1 + 0.9b)}$$

Case 4: 0.9 β > 1.0 and 0.9 α > 1.0

The five sided stress block shown in Fig.9.20 can be considered as compression over the entire column cross section with tension in the triangular area in the left hand bottom corner. Compression over the entire column does not give rise to any moment. Moment is caused purely by the tension in the triangular area.

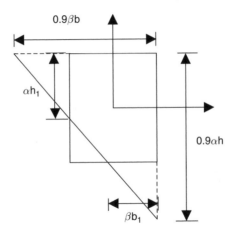

Fig.9.20 Neutral axis position for case 4.

$$\alpha h_1 = \{\frac{\alpha h}{\beta b}(0.9\beta b - b) = \alpha h(0.9 - \frac{1}{\beta})\} \le h$$

$$\beta b_1 = \{\frac{\beta b}{\alpha h}(0.9\alpha h - h) = \beta b(0.9 - \frac{1}{\alpha})\} \le b$$

$$N_c = 0.45\,f_{cu}\,\{bh - 0.5\,(h - \alpha h_1)\,(b - \beta b_1)\}$$

$$M_{xc} = 0.45 f_{cu}\frac{1}{2}(h - \alpha h_1)(b - \beta b_1)\{0.5h - \frac{1}{3}(h - \alpha h_1)\}$$

$$M_{yc} = 0.45 f_{cu}\frac{1}{2}(h - \alpha h_1)(b - \beta b_1)\{0.5b - \frac{1}{3}(b - \beta b_1)\}$$

9.5.1.2 Example of design chart for axial force and biaxial moments

Consider a rectangular column b × h and reinforced with four bars as shown in Fig.9.15. The total steel area A_{sc} = 2% of bh, f_{cu} = 30 N/mm^2, f_y = 460 N/mm^2, b/h = 0.5. The bars are located at 0.15h from top and bottom faces and at 0.3b from the sides. Calculate N/(bh), $M_x/(bh^2)$ and $M_y/(b^2h)$ for the following positions of the neutral axis.
The coordinates (x/b, y/h) of the four bars are:
1: (0.2, 0.35), 2: (0.2, –0.35), 3: (–0.2, –0.35), 4: (–0.2, 0.35)

(i) Assuming α =0.65, β = 0.95
Calculate the strains (positive is compressive) in steel from the equation

$$\varepsilon = 0.0035\{1 + \frac{(\frac{x}{b} - 0.5)}{\beta} + \frac{(\frac{y}{h} - 0.5)}{\alpha}\}$$

The strains in the four bars are respectively:
$$1.587 \times 10^{-3}, \quad -2.182 \times 10^{-3}, \quad -3.659 \times 10^{-3}, \quad 0.113 \times 10^{-3}$$
The corresponding stresses are:
$$317, \quad -436, \quad -437 \text{ and } 22.7 \text{ N/mm}^2$$
The contribution of the stresses in steel to:
$$N = (317 - 436 - 437 + 22.7) \times (0.02 \text{ bh}/4) = -2.667 \text{ bh (Tensile)}$$
$$M_x = (317 + 436 + 437 + 22.7) \times (0.02 \text{ bh}/4) \times 0.35h = 2.12 \text{ bh}^2$$
$$M_y = (317 - 436 + 437 - 22.7) \times (0.02 \text{ bh}/4) \times 0.2b = 0.295 \text{ b}^2 \text{ h}$$
The contribution of the compressive stress in concrete are, using:
$$0.9\alpha = 0.585, 0.9\beta = 0.855$$
$$N_c = 0.45 f_{cu} \times 0.5 \times (0.585 \text{ h} \times 0.855 \text{ b}) = 3.376 \text{ bh}$$
$$M_x = N_c \times (0.5 \text{ h} - 0.585h/3) = 1.030 \text{ bh}^2$$
$$M_y = N_c \times (0.5 \text{ b} - 0.855b/3) = 0.726 \text{ b}^2h$$
Adding the contribution of steel and concrete:
$$N/(bh) = 0.709, \quad M_x/(bh^2) = 3.15, \quad M_y/(b^2h) = 1.021$$

(ii) Assuming α =0.65, β = 1.35
The strains in the four bars are respectively:
$$1.915 \times 10^{-3}, \quad -1.85 \times 10^{-3}, \quad -2.89 \times 10^{-3}, \quad 0.877 \times 10^{-3}$$
The corresponding stresses are:
$$383, \quad -371, \quad -437 \text{ and } 175 \text{ N/mm}^2$$
The contribution of the stresses in steel to:
$$N = (383 - 371 - 437 + 175) \times (0.02 \text{ bh}/4) = -1.25 \text{ bh (Tensile)}$$
$$M_x = (383 + 371 + 437 + 175) \times (0.02 \text{ bh}/4) \times 0.35h = 2.39 \text{ bh}^2$$
$$M_y = (383 - 371 + 437 - 175) \times (0.02 \text{ bh}/4) \times 0.2b = 0.273 \text{ b}^2 \text{ h}$$
The contribution of the compressive stress in concrete are calculated using
$$0.9\alpha = 0.585, 0.9\beta = 1.215$$
From the trapezoidal stress block shown in Fig.9.18,

$$\alpha h_1 = \frac{\alpha h}{\beta b}(0.9\beta b - b) = \alpha h(0.9 - \frac{1}{\beta}) = 0.104 \text{ h}$$

$$N_c = 0.45\, f_{cu} \times \{0.5\,(\alpha h_1 + 0.9\,\alpha h)\,b\} = 4.65\,bh$$

$$ybar = \frac{\alpha}{3}\frac{(h_1^2 + 0.81 h^2 + 0.9\,h_1\,h)}{(h_1 + 0.9h)} = 0.2\,h$$

$$M_{xc} = N_c \times (0.5\,h - ybar) = 1.395\,bh^2$$

$$xbar = \frac{b}{3}\frac{(2h_1 + 0.9h)}{(h_1 + 0.9h)} = 0.3837\,b$$

$$M_{yc} = N_c \times (0.5\,b - xbar) = 0.541\,b^2 h$$

Adding the contribution of steel and concrete:

$$N/(bh) = 3.40, \quad M_x/(bh^2) = 3.78, \quad M_y/(b^2 h) = 0.814$$

(iii) Assuming $\alpha = 1.2$, $\beta = 0.65$

The strains in the four bars are respectively:

$$1.447 \times 10^{-3}, \ -0.59 \times 10^{-3}, \ -2.75 \times 10^{-3}, \ -0.71 \times 10^{-3}$$

The corresponding stresses are:

$$289, \ -119, \ -437 \text{ and } -141 \text{ N/mm}^2$$

The contribution of the stresses in steel to:

$$N = (289 - 119 - 437 - 141) \times (0.02\,bh/4) = -2.04\,bh \text{ (Tensile)}$$
$$M_x = (289 + 119 + 437 - 141) \times (0.02\,bh/4) \times 0.35h = 1.232\,bh^2$$
$$M_y = (289 - 119 + 437 + 141) \times (0.02\,bh/4) \times 0.2b = 0.748\,b^2\,h$$

The contribution of the compressive stress in concrete to the forces are, using

$$0.9\alpha = 1.08, \ 0.9\beta = 0.65$$

From the trapezoidal stress block shown in Fig.9.19,

$$\beta b_1 = \frac{\beta b}{\alpha h}(0.9\alpha h - h) = \beta b(0.9 - \frac{1}{\alpha}) = 0.0433\,b$$

$$N_c = 0.45\,f_{cu} \times \{0.5\,(\beta b_1 + 0.9\,\beta b)\,h\}, = 4.24\,bh$$

$$ybar = \frac{h}{3}\frac{(2b_1 + 0.9b)}{(b_1 + 0.9b)} = 0.3565\,h$$

$$M_{xc} = N_c \times (0.5\,h - ybar) = 0.608\,bh^2$$

$$xbar = \frac{\beta}{3}\frac{(b_1^2 + 0.81 b^2 + 0.9\,b_1\,b)}{(b_1 + 0.9b)} = 0.1962\,b$$

$$M_{yc} = N_c \times (0.5\,b - xbar) = 1.288\,b^2 h$$

Adding the contribution of steel and concrete:

$$N/(bh) = 2.20, \quad M_x/(bh^2) = 1.84, \quad M_y/(b^2 h) = 2.04$$

(iv) Assuming $\alpha = 1.3$, $\beta = 1.5$

Calculate the strains (positive is compressive) in steel.

The strains in the four bars are respectively:

$$2.396 \times 10^{-3}, \ 0.512 \times 10^{-3}, \ -0.42 \times 10^{-3}, \ 1.463 \times 10^{-3}$$

The corresponding stresses are:

$$437, \ 102, \ -84 \text{ and } 293 \text{ N/mm}^2$$

The contribution of the stresses in steel to:

$$N = (437 + 102 - 84 + 293) \times (0.02\,bh/4) = 3.74\,bh$$
$$M_x = (437 - 102 + 84 + 293) \times (0.02\,bh/4) \times 0.35h = 1.25\,bh^2$$

$$M_y = (437 + 102 + 84 - 293) \times (0.02 \text{ bh}/4) \times 0.2b = 0.331 \text{ b}^2 \text{ h}$$

The contribution of the compressive stress in concrete to the forces are, using

$$0.9\alpha = 1.17, 0.9\beta = 1.35$$

From the trapezoidal stress block shown in Fig.9.20,

$$\alpha h_1 = \{\frac{\alpha h}{\beta b}(0.9\beta b - b) = \alpha h(0.9 - \frac{1}{\beta})\} \le h \ , \alpha h_1 = 0.3033 \text{ h}$$

$$\beta b_1 = \{\frac{\beta b}{\alpha h}(0.9\alpha h - h) = \beta b(0.9 - \frac{1}{\alpha})\} \le b \ , \beta b_1 = 0.1962 \text{ b}$$

$$N_c = 0.45 \text{ f}_{cu} \{bh - 0.5 (h - \alpha h_1) (b - \beta b_1)\} = 9.72 \text{ bh}$$

$$M_{xc} = 0.45 f_{cu} \frac{1}{2}(h - \alpha h_1)(b - \beta b_1)\{0.5h - \frac{1}{3}(h - \alpha h_1)\} = 1.012 \text{ bh}^2$$

$$M_{yc} = 0.45 f_{cu} \frac{1}{2}(h - \alpha h_1)(b - \beta b_1)\{0.5b - \frac{1}{3}(b - \beta b_1)\} = 0.877 \text{ b}^2\text{h}$$

Adding the contribution of steel and concrete:
$N/(bh) = 13.46, \quad M_x/(bh^2) = 2.26, \quad M_y/(b^2h) = 1.208$

9.5.1.3 *Axial force biaxial moment interaction curve*

Calculations similar to that in the previous section can be done and the corresponding interaction curves as shown in Fig.9.21 can be constructed for $M_x/(bh^2) = 2.0$ and 3.0..

9.5.2 Approximate method given in BS 8110

In the absence of interaction diagram as described in section 9.5.1, an approximate design method given in BS8110: Part 1, clause 3.8.4.5 can be used. The method reduces the biaxial bending case to a uni-axial one by designing for a larger value of the moment than applied. The amount of increase depends on the ratio of the axial load to the capacity under axial load only. The applied moment and dimensions are shown in Fig.9.22, where

M_x design ultimate moment about the XX axis
M_x' effective uni-axial design moment about the XX axis
M_y design ultimate moment about the YY axis
M_y' effective uni-axial design moment about the YY axis
h overall depth perpendicular to the XX axis
h' effective depth perpendicular to the XX axis
b overall width perpendicular to the YY axis
b' effective width perpendicular to the YY axis

$$\text{If } M_x/h' \ge M_y/b, \quad M_x' = M_x + \beta \frac{h'}{b'} M_y \quad \text{(Code equation 40)}$$

If $M_x/h' < M_y/b$, $M_y' = M_y + \beta \dfrac{b'}{h'} M_x$ (Code equation 41)

The coefficient β is taken from Table 3.22 of the code. It depends on the value of $N/(bhf_{cu})$, e.g. for $N/(bhf_{cu}) = 0$, 0.3, and $\geq$ 0.6, $\beta = 1.0$, 0.65, 0.3 respectively.

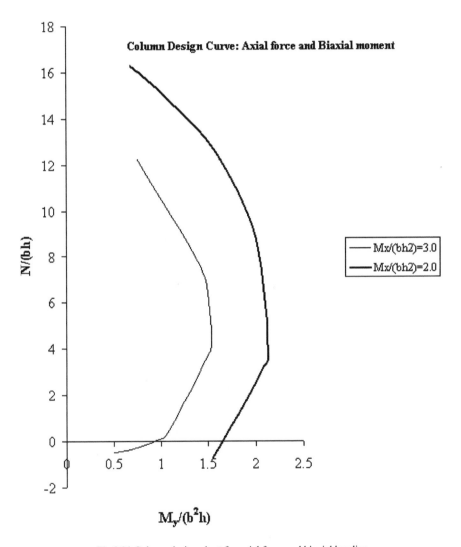

Fig.9.21 Column design chart for axial force and biaxial bending.

9.5.2.1 Example of design of column section subjected to axial load and biaxial bending: BS 8110 method

Design the reinforcement for the column section shown in Fig.9.23
It is subjected to the following actions at ULS:

$$N = 950 \text{ kN}$$
$$M_x \text{ about xx-axis} = 95 \text{ kN m}$$
$$M_y \text{ about yy-axis} = 65 \text{kNm}$$

The materials are grade C30 concrete and grade 460 reinforcement. Assume the cover is 25 mm, links are 8 mm in diameter and main bars are 25 mm in diameter.

$$h' = 400 - 25 - 8 - 12.5 = 355 \text{ mm, say 350mm}$$
$$b' = 300 - 25 - 8 - 12.5 = 255 \text{ mm, say 250 mm}$$
$$M_x/h' = 95/0.35 = 271.4$$
$$M_y/b' = 65/0.25 = 260$$
$$M_x/h' > M_y/b'$$
$$N/(bh\ f_{cu}) = 950 \times 10^3 /(400 \times 300 \times 30) = 0.264$$
$$\beta = 0.693 \text{ (Table 3.22 of the code)}$$
$$M_x{}' = 95 + 0.693 \times (350/250) \times 65 = 158.1 \text{ kNm}$$
$$N/(bh) = 950 \times 10^3 /(400 \times 300) = 7.92$$
$$M_x{}'/(bh^2) = 158.1 \times 10^6/(300 \times 400^2) = 3.29$$

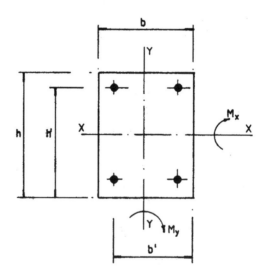

Fig.9.22 BS 8110 method for biaxial bending design.

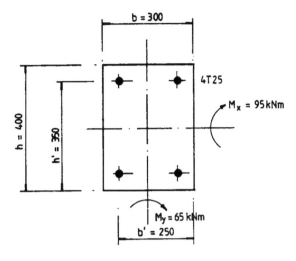

Fig.9.23 Column section under axial load and biaxial moments.

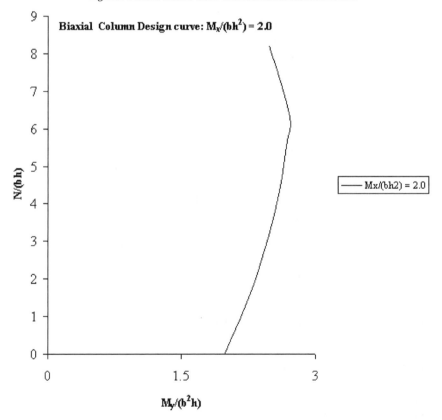

Fig.9.24 Biaxial design chart for column in Fig.9.24.

Using $A_s = A_s' = 2T25 = 982$ mm^2, $d/h = 0.85$, $1\ 00\ A_{sc}\ (bh) = 1.64$, calculations show that at $(x/h) = 0.58$, $f_s' = 437$ N/mm^2, $f_s = 326$ N/mm^2,

$$N/(bh) = 7.94,\ M/(bh^2) = 3.86$$

The reinforcement is shown in Fig.9.23.

If it is decided to use the exact column design for biaxial moment, then a corresponding column design chart as shown in Fig.9.24 need to be constructed. As a check, using $\alpha = 1.3$, $\beta = 1.02$ and the coordinates of the bars as ($\pm 0.375h$, $\pm 0.333b$)

Calculate the strains (positive is compressive) in steel.

The strains in the four bars are respectively:

$$2.592 \times 10^{-3},\ \ 0.573 \times 10^{-3},\ \ -1.71 \times 10^{-3},\ \ 0.306 \times 10^{-3}$$

The corresponding stresses are:

$$437,\ \ 115,\ \ -437 \text{ and } 61 \text{ N/mm}^2$$

The contribution of the stresses in steel to:

$$N = (437 + 115 - 343 + 61) \times (0.016\ bh/4) = 1.08\ bh$$
$$M_x = (437 - 115 + 343 + 61) \times (0.016\ bh/4) \times 0.375h = 1.09\ bh^2$$
$$M_y = (437 + 115 + 343 - 61) \times (0.016\ bh/4) \times 0.333b = 1.11\ b^2 h$$

The contribution of the compressive stress in concrete to the forces are calculated using $0.9\alpha = 1.17$ and $0.9\beta = 0.92$.

From the trapezoidal stress block shown in Fig.9.19,

$$\beta b_1 = \frac{\beta b}{\alpha h}(0.9\alpha h - h) = \beta b(0.9 - \frac{1}{\alpha}) = 0.1334\ b$$

$$N_c = 0.45\ f_{cu} \times \{0.5\ (\beta b_1 + 0.9\ \beta b)\ h\}, = 7.104\ bh$$

$$ybar = \frac{h}{3}\frac{(2b_1 + 0.9b)}{(b_1 + 0.9b)} = 0.3755\ h$$

$$M_{xc} = N_c \times (0.5\ h - ybar) = 0.885\ bh^2$$

$$xbar = \frac{\beta}{3}\frac{(b_1^2 + 0.81b^2 + 0.9b_1\ b)}{(b_1 + 0.9b)} = 0.3119\ b$$

$$M_{yc} = N_c \times (0.5\ b - xbar) = 1.336\ b^2 h$$

Adding the contribution of steel and concrete:

$$N/(bh) = 8.18,\ \ M_x/(bh^2) = 1.98,\ \ M_y/(b^2h) = 2.45$$

The required values are: $N = 950$ kN, $M_x = 95$ kNm, $M_y = 65$ kNm. If $b = 300$ mm and $h = 400$ mm, the section is safe because

$$N/(bh) = 7.92 < 8.18,\ \ M_x/(bh^2) = 1.98,\ \ M_y/(b^2h) = 1.81 < 2.45$$

9.6 EFFECTIVE HEIGHTS OF COLUMNS

9.6.1 Braced and Un-braced Columns

An essential step in the design of a column is to determine whether the proposed dimensions and framing arrangement will result in the column being 'short' or a 'slender'. If the column is slender, additional moments due to deflection must be

added to the moments from the primary analysis. In general columns in buildings are 'short',
Clause 3.8.1.3 of the code defines short and slender columns as follows:

1. For a braced structure, the column is considered as short if both the slenderness ratios l_{ex}/h and l_{ey}/b are less than 15. If either ratio is greater than 15, the column is considered as slender.

2. For an un-braced structure, the column is considered as short if both the slenderness ratios l_{ex}/h and l_{ey}/b are less than 10. If either ratio is greater than 10 the column is considered as slender.

Here h is the column depth perpendicular to the XX axis, b is the column width perpendicular to the YY axis, l_{ex} is the effective height in respect of the XX axis and l_{ey} is the effective height in respect of the YY axis.

The code states that the columns can be considered braced in a given plane if the structure as a whole is provided with stiff elements such as shear walls which are designed to resist all the lateral forces in that plane. The bracing system ensures that there is no significant lateral displacement between the ends of the columns. If the above conditions are not met the column should be considered as un-braced. Examples of braced and un-braced columns are shown in Fig.9.26.

9.6.2 Effective Height of a Column

The effective height of a column depends on

1. the actual height between floor beams, base and floor beams or lateral supports
2. the column section dimensions $h \times b$
3. the end conditions such as the stiffness of beams framing into the columns or whether the column to base connection is designed to resist moment
4. whether the column is braced or un-braced

The effective height of a pin-ended column is its actual height. The effective height of a general column is the height of an equivalent pin-ended column of the same buckling capacity as the actual member. Theoretically the effective height is the distance between the points of inflexion along the member length. These points may lie within the member as in a braced column or on an imaginary line outside the member as in an un-braced column. Some effective heights for columns are shown in Fig.9.25.

For a braced column the effective height will always be less than or equal to the actual height. In contrast, the effective height of an un-braced column will always be greater than the actual height except in the case where sway occurs without rotation at the ends (Fig.9.25). It is important to note that the effective heights of a column in two plan directions may well be different. Also, the column may be braced in one direction but un-braced in the other direction.

9.6.3 Effective Height Estimation From BS 8110

Two methods are given in the code to determine the effective height of a column:
1. simplified recommendations given in BS 8110: Part 1, clause 3.8.1.6, that can be used in normal cases
2. a more rigorous method given in BS8110: Part 2, section 2.5
Clause 3.8.1.6.1 states that the general equation for obtaining effective heights is:
$$l_e = \beta \, l_0$$
where l_0 is the clear height between end restraints and β is a coefficient from Tables 3.19 and 3.20 of the code for braced and un-braced columns; β is a function of the end condition. In Tables 3.19 and 3.20 the end conditions are defined in

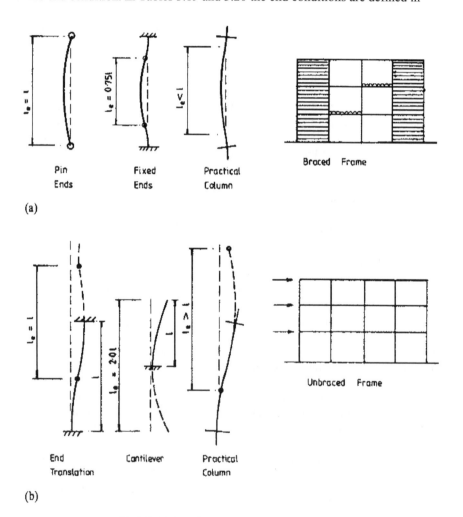

Fig.9.25 (a) Braced columns; (b) un-braced columns.

terms of a scale from 1 to 4. An *increase* in the scale corresponds to a *decrease* in end fixity. The four end conditions are as follows.

Condition 1. The end of the column is connected monolithically to beams on either side which are at least as deep as the overall dimension of the column. When the column is connected to a foundation structure this should be designed to carry moment.

Condition 2 The end of the column is connected monolithically to beams or slabs on either side which are shallower than the overall dimension of the columns.

Condition 3 The end of the column is connected to members that, while not designed specifically to provide restraint, do provide some nominal restraint.

Condition 4 The end of the column is unrestrained against both lateral movement and rotation, i.e. it is the free end of a cantilever.

Some values of β from Tables 3.19 and 3.20 of the code are as follows:

Braced column:
> Top end, condition 1 $\beta = 0.75$
> Bottom end, condition 1   Essentially fixed ends

Un-braced column:
> Bottom end, condition 4 $\beta = 2.2$
> Top end, condition 1   Essentially a cantilever

The more accurate assessment of effective heights from BS 8110: Part 2, section 2.5, is set out below. The derivation of the equations is based on a limited frame consisting of the columns concerned, column lengths above and below if they exist and the beams top and bottom on either side if they exist. The symbols used are defined as follows:

> I second moment of area of the section
> l_e effective height in the plane considered
> l_0 clear height between end restraints
> α_{c1} = ratio of the sum of the column stiffnesses to the sum
> of the beam stiffnesses at the lower end
> α_{c2} = ratio of the sum of the column stiffnesses to the sum of
> the beam stiffnesses at the upper end
> $\alpha_{c\,min}$ the lesser of α_{c1} and α_{c2}

Only members properly framed into the column are considered. The stiffness is I/l_0. In specific cases the following simplifying assumptions may be made:

1. In the flat slab construction the beam stiffness is based on the section forming the column strip;

2. For simply supported beams framing into a column, $\alpha_c = 10$;

3. For the connection between column and base designed to resist only nominal moment, $\alpha_c = 10$;

4. For the connection between column and base designed to resist column moment, $\alpha_c = 1.0$.

The effective heights for framed structures are as follows:

1. For braced columns the effective height is the lesser of
$$l_e = l_0 [0.7 + 0.05(\alpha_{c1} + \alpha_{c2})] < l_0$$
$$l_e = l_0 [0.85 + 0.05 \, \alpha_{cmin}] < l_0$$

2. For un-braced columns the effective height is the lesser of
$$l_e = l_0 [1.0 + 0.15(\alpha_{c1} + \alpha_{c2})] < l_0$$
$$l_e = l_0 [2.0 + 0.3 \, \alpha_{cmin}] < l_0$$

9.6.4 Slenderness Limits for Columns

The slenderness limits for columns are specified in clauses 3.8.1.7 and 3.8.1.8, as follows.

1. Generally the clear distance l_0 between end restraints is not to exceed 60 times the minimum thickness of the column;
2. For un-braced columns, if in any given plane one end is unrestrained, e.g. a cantilever, its clear height l_0 should not exceed
$$\ell_0 = \frac{100 \, b^2}{h} \leq 60 \, b$$
where h and b are the larger and smaller dimensions of the column.

9.6.4.1 Example of calculating the effective heights of column by simplified and rigorous methods

(a) Specification
The lengths and proposed section dimensions for the columns and beams in a multi-storey building are shown in Fig.9.27. Determine the effective lengths and slenderness ratios for the XX and YY axes for the lower column length AB, for the two cases where the structure is braced and un-braced. The connection to the base and the base itself are designed to resist the column moment. Use both the rigorous and the simplified methods.

(b) Simplified method

(i) *YY axis buckling*: End conditions:
Top: The top end is connected monolithically to beams with a depth (500 mm) greater than the column dimension (400 mm), i.e. condition 1
Bottom: The base is designed to resist moment i.e. condition 1.

1. Braced column:
slenderness: $\beta = 0.75$ (Table 3.21 of the code) and l_0, the clear height between end restraints, is 4750 mm;
$$l_{ey}/h = 0.75 \times 4750/400 = 8.9 < 15$$
The column is 'short'.

2. Un-braced column:
slenderness: $\beta = 1.2$ (Table 3.22);
$$l_{ey}/h = 1.2 \times 4750/400 = 14.25 > 10, \text{ i.e. the column is 'slender'.}$$

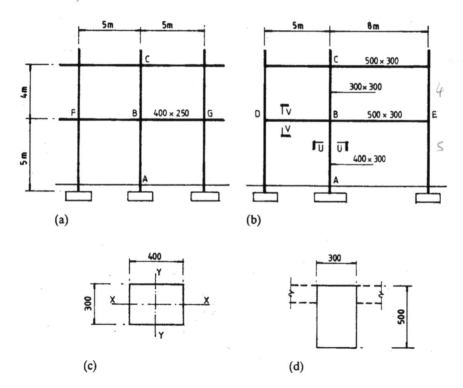

Fig.9.26 Multi storey building(a) Side elevation; (b) transverse frame; (c) column UU; (d) beam VV.

(ii) *XX axis buckling*: End conditions:

Top: condition 1;
Base: condition l.

1. Braced column
   slenderness: $l_{ex}/b = 0.75 \times 4750/300 = 11.9 < 15$; the column is 'short'.

2. Un-braced column
   slenderness: $l_{ex}/b = 1.2 \times 4750/300 = 19 > 10$; the column is 'slender'.

(c) Rigorous method
(i)YY *axis buckling*: The stiffness I/L of
   Column AB : $(300 \times 400^3/12)/5000 = 320 \times 10^3$
   Column BC : $(300 \times 300^3/12)/4000 = 169 \times 10^3$
It is conservative practice to base beam moments of inertia on the beam depth multiplied by the rib width.
   Beam BD: $(300 \times 500^3/12)/5000 = 625 \times 10^3$
   Beam BE: $(300 \times 500^3/12)/8000 = 391 \times 10^3$

   Joint A: $\alpha_{c1} = 1.0$ (fixed end)
   Joint B: $\alpha_{c2} = (320 + 169)/(625 + 391) = 0.48$

1. Braced column:
Slenderness: l_0, the clear height between end restraints, is 4750 mm and l_{ey} is the lesser of
   $4750[0.7 + 0.05(1.0 + 0.48)] = 3677$ mm
   $4750(0.85 + 0.05 \times 0.48) = 4152$ mm but must be less than l_0;
   $l_{ey}/h = 3677/400 = 9.19 < 15.0$
The column is 'short'.

2.Un-braced column
Slenderness: l_{ey} is the lesser of
   $4750[1.0 + 0.15(1.0 + 0.48)] = 5805$ mm
   $4570(2.0 + 0.3) = 9500$ mm
   $l_{ey}/h = 5805/400 = 14.5 > 10.0$
The column is 'slender'.

(ii) *XX axis buckling*: The stiffness I/L
   Column AB $= (400 \times 300^3/12)/5000 = 180 \times 10^3$
   Column BC $= (300 \times 300^3/12)/4000 = 169 \times 10^3$
It is conservative practice to base beam second moments of area on the beam depth multiplied by the rib width. For beam BF and BG
   $I = (250 \times 400^3/12) = 1.33 \times 10^9$ mm^4, $I/l = 1.33 \times 10^9/5000 = 267 \times 10^3$
   Joint A $= \alpha_{c1} = 1.0$ (fixed end)
   Joint B $= \alpha_{c2} = (180 + 169)/(267 + 267) = 0.65$

1. Braced column slenderness
$$4750[0.7 + 0.05(1.0 + 0.65)] = 3717 \text{ mm}$$
$$4750(0.85 + 0.05 \times 0.65) = 4192 \text{ mm but must be less than } l_0; \text{ thus}$$
$$l_{ey}/ b = 3717/300 = 12.4 < 15.0. \text{ The column is 'short'.}$$

2. Un-braced column slenderness: l_{ey} is the lesser of
$$4750 \times [1.0 + 0.15(1.0 + 0.65)] = 5926 \text{ mm}$$
$$4570 \times (2.0 + 0.3) = 9500 \text{ mm}$$
$$l_{ex}/b = 5926/300 = 19.8 > 10.0. \text{ The column is 'slender',}$$

(d) Comment and summary
The two methods give the same outcome. These may be summarized as Braced column is 'short' with respect to both axes and Un-braced column is 'slender' with respect to both axes. The maximum slenderness ratio is 19.8.

9.7 DESIGN OF SLENDER COLUMNS

9.7.1 Additional Moments Due to Deflection

In the primary analysis of the rigid frames the secondary moments due to deflection are ignored. This effect is small for short columns but with slender columns significant additional moments occur. A simplified discussion is given.

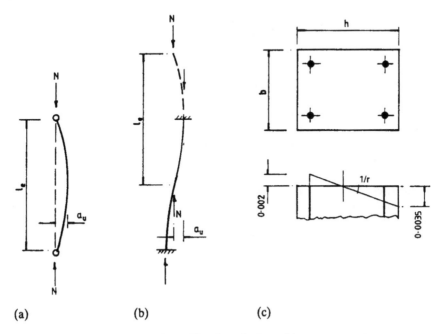

(a)   **(b)**   **(c)**

Fig.9.27 (a) Braced column; (b) un-braced column; (c) curvature at centre.

If the pin-ended column shown in Fig.9.27 (a) is bent such that the curvature $1/r$ is uniform, the deflection at the centre can be shown to be $a_u = l^2/8r$. If the curvature is taken as varying uniformly from zero at the ends to a maximum of $1/r$ at the centre, $a_u = l^2/12r$. For practical columns a_u is taken as the mean $l_e^2/10r$. The same value is used for the un-braced column at the centre of the buckled length, as shown in Fig.9.27(b).

The curvature at the centre of the buckled length of the column is assessed when the concrete in compression and steel in tension are at their maximum strains. The curvature for this case is shown in Fig.9.27(c). The concrete strain shown is increased to allow for creep and a further increase is made to take account of slenderness. The maximum deflection for the case set out above is given in the code by the expression

$$a_u = \beta_a \, K \, h$$

where

$$\beta_a = \frac{1}{2000}(\frac{\ell_e}{b'})^2$$

and b' is the smaller dimension of the column, equal to b if b is less than h. K is a reduction factor that corrects the curvature and so the resulting deflection for the cases where steel strain in tension is less than its maximum value of 0.002 or where compression exists over the whole section. The value of K is

$$K = \frac{N_{uz} - N}{N_{uz} - N_{bal}} \leq 1.0$$

where

$$N_{uz} = 0.45 \, f_{cu} \, A_c + 0.95 \, f_y \, A_{sc}$$

This is the capacity under pure axial load.

N_{bal} = Design ultimate load capacity of a balanced section. For this case, when the maximum strain in concrete is 0.0035, the strain in steel is also at yield. The strain in steel at yield is 0.95 f_y/E = 0.95 × 460/(200 × 10^3) = 0.0022

The depth x of neutral axis is therefore equal to

$$x = d/(1 + 0.0022/0.0035) = 0.614 \, d$$

The stress block depth is 0.9x

The compression steel normally yields. N_{bal} is given by

$$N_{bal} = 0.45 \, f_{cu} \, (0.9 \times 0.614 \, d) \, b + 0.95 \, f_y \, A_s' - 0.95 \, f_y \, A_s$$

For symmetrically reinforced rectangular sections, $A_s' = A_s$.

$$N_{bal} = 0.25 \, f_{cu} \, bd$$

The assessment is first made with $K = 1$ and then K is calculated from the above formula and a second iteration is made. The value of K converges quickly to its final result.

Referring to Fig.9.27, the deflection causes an additional moment in the column given by

$$M_{add} = N \, a_u$$

The additional moment is added to the initial moment M_i from the primary analysis to give the total design moment M_t:

$$M_t = M_i + M_{add}$$

In a braced column the maximum additional moment occurs in the centre of the column whereas in the un-braced column it occurs at the end of the column.

9.7.2 Design Moments in a Braced Column Bending About a Single Axis

The distribution of moments over the height of a typical braced column in a concrete frame from Fig. 3.20 in the code is shown in Fig.9.28. The maximum additional moment occurs at the centre of the column where the deflection due to buckling is greatest. The initial moment at the point of maximum additional moment is given in clause 3.8.3.2 of the code by

$$M_i = 0.4M_1 + 0.6M_2 \geq 0.4M_2$$

where M_1 is the smaller initial end moment and M_2 is the larger initial end moment.

The column will normally be bending in double curvature in a building frame and M_1 is to be taken as negative and M_2 as positive. The code states that the maximum design moment is the greatest of the following four values (Fig.9.29): M_2; $(M_i + M_{add})$, $(M_1 + M_{add}/2)$; $(e_{min} N)$, where e_{min} is 0.05h or 20mm maximum.

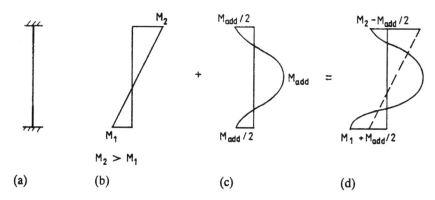

Fig.9.28 Slender braced column (a) End conditions; (b) initial moments; (c) additional moments; (d) final design moments.

9.7.3 Further Provisions for Slender Columns

Further important provisions regarding the design of slender columns set out in BS8110: Part 1, clauses 3.8.3.33.8.3.6, are as follows.

(a) Slender columns bent about a single axis (major or minor)
If the longer side h is less than three times the shorter side b for columns bent about the major axis and $l_e/h < 20$, the design moment is $M_i + M_{add}$ as set out above.

(b) Slender columns where $l_e/h > 20$ bent about the major axis
The section is to be designed for biaxial bending. The additional moment occurs about the minor axis.

(c) Slender columns bent about their major axis
If $h > 3b$ (see 9.7.3(a) above), the section is to be designed for biaxial bending as in 9.7.3(b) above.

(d) Slender columns bent about both axes
Additional moments are to be calculated for both directions of bending. The additional moments are added to the initial moments about each axis and the column is designed for biaxial bending.

9.7.4 Unbraced Structures

The distribution of moments in an un-braced column is shown in Fig. 3.21 of the code (see Fig.9.29 below). The additional moment is assumed to occur at the stiffer end of the column. The additional moment at the other end is reduced in proportion to the ratio of joint stiffnesses at the ends.

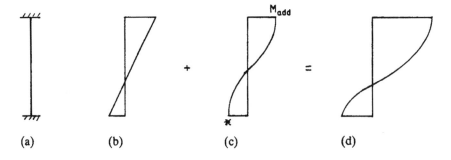

Fig.9.29 Slender unbraced column (a) End conditions; (b) initial moments; (c) additional moments; (d) design moments. The asterisk indicates that M_{add} is reduced in proportion to the ratio of end stiffnesses.

9.7.4.1 Example of design of a slender column

(a) Specification
Design the column length AB in the building frame shown in Fig.9.27 for the two cases where the frame is braced and un-braced. The bending moment diagrams for the column bent about the YY axis and the axial loads for unfactored dead, imposed and wind loads are shown in Fig.9.33. The materials are grade C30 concrete and grade 460 reinforcement.

(b) Column AB braced

In the case of a braced column, the wind load is resisted by shear walls. Referring to the example (9.6.4.1), the above the column is short with respect to both axes. The design loads and moments at the top of the column are

$$N = (1.4 \times 765) + (1.6 \times 305) = 1559 \text{ kN}$$
$$M = (1.4 \times 48) + (1.6 \times 28) = 112 \text{ kN m}$$

The cover is 25 mm, the links are 8 mm and the bars are 25 mm in diameter. The inset of the bars is approximately 50 mm.

$$d/h = 350/400 = 0.875$$
$$N/(bh) = 1559 \times 10^3 /(300 \times 400) = 13.0$$
$$M/(bh^2) = 112 \times 10^6 /(300 \times 400^2) = 2.33$$

Using column design chart, $100A_{sc}/bh = 1.6\%$.
Calculations show that for a column with symmetrical steel equal to

$$100A_{sc}/(bh) = 1.6, \text{ at } x/h = 0.82 \; f_s{}' = 437 \text{ N/mm}^2, \quad f_s = 26 \text{ N/mm}^2 \text{ and}$$
$$N/(bh) = 13.22, \quad M/(bh^2) = 2.58.$$

$A_{sc} = 1.6 \times 300 \times 400/100 = 1920 \text{ mm}^2$. Provide 4T25 bars of area 1963 mm^2.

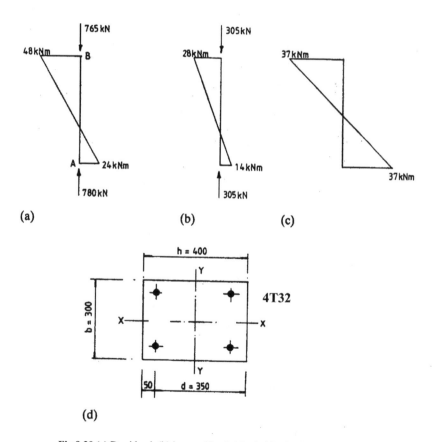

Fig.9.30 (a) Dead load; (b) imposed load; (c) wind load; (d) column section.

(c) Column AB un-braced:

Calculate the design loads and moments taking wind load into account.

$$N = 1.2 \times (765 + 305) = 1284 \text{ kN}$$
$$M = 1.2 \times (48 + 28 + 37) = 135.6 \text{ kNm}$$

The column must be checked for the following.

Case 1: dead + imposed load

$$N = 1559 \text{ kN}, M = 112 \text{ kNm}$$

Case 2: dead + imposed + wind load

$$N = 1284 \text{ kN}, M = 135.6 \text{ kNm}$$

The column is slender with respect to both axes. The maximum slenderness ratio from example in section 9.6.4.1 is

$$l_{ex}/b = 19.8$$

The column is bent about the major axis and l_e/h does not exceed 20.

1. Trial 1: Assume the factor $K = 1$ initially.

$$\beta_a = 19.8^2 / 2000 = 0.196$$

The deflection is

$$a_u = 0.196 \times 1 \times 400 = 78.41 \text{ mm}$$

(i) Case 1: dead + imposed load

Calculate additional moment:

$$M_{add} = 1559 \times 78.41 \times 10^{-3} = 122.2 \text{ kNm}$$

Calculate design moment as the sum of applied and additional moments:

$$M_t = 112 + 122.2 = 234.2 \text{ kNm}$$
$$N/(bh) = 1559 \times 10^3 /(300 \times 400) = 13.0$$
$$M/(bh^2) = 234.2 \times 10^6 /(300 \times 400^2) = 4.88$$

Calculate the required steel area using the charts. Calculations show that for a column with symmetrical steel equal to $100A_{sc}/(bh) = 3.4$, at $x/h = 0.70$, $f_s' = 437 \text{ N/mm}^2$, $f_s = 150 \text{ N/mm}^2$, $N/(bh) = 13.36$, $M/(bh^2) = 5.05$.

$$A_{sc} = 3.4 \times 300 \times 400/100 = 4080 \text{ mm}^2$$

2. Trial 2: Calculate the reduction factor K:

$$N_{uz} = 0.45 \times 30 \times (400 \times 300 - 4080) \times 10^{-3} + (0.95 \times 460 \times 4080) \times 10^{-3}$$
$$= 3347.9 \text{ kN}$$
$$N_{bal} = 0.25 \times 30 \times 300 \times 350 \times 10^{-3} = 787.5 \text{ kN}$$
$$K = (3347.9 - 1559)/(3347.9 - 787.5) = 0.699$$
$$M_t = 112 + (0.699 \times 122.2) = 197.4 \text{ kN m}$$
$$N/(bh) = 13.0, M_t/(bh^2) = 4.11$$

Calculate the required steel area using the charts. Calculations show that for a column with symmetrical steel equal to $100A_{sc}/(bh) = 2.68$, at $x/h = 0.73$, $f_s' = 437 \text{ N/mm}^2$, $f_s = 126 \text{ N/mm}^2$, $N/(bh) = 13.0$, $M/(bh^2) = 4.11$.

$$A_{sc} = 2.69 \times 300 \times 400/100 = 3228 \text{ mm}^2$$
$$N_{uz} = 0.45 \times 30 \times (400 \times 300 - 3228) \times 10^{-3} + (0.95 \times 460 \times 3228) \times 10^{-3}$$
$$= 2987 \text{ kN}$$
$$K = (2987.0 - 1559)/(2987.0 - 787.5) = 0.649$$

This value of K is almost same as K at the start of Trial 2. Therefore convergence has taken place. Calculate the total moment and finalize design.

$$M_t = 112 + (0.649 \times 122.2) = 191.3 \text{ kN m}$$
$$M_t/(bh^2) = 191.3 \times 10^6/(300 \times 400^2) = 3.99$$
$$N/(bh) = 13.0, M_t/(bh^2) = 3.99$$

Provide 4T32, to give an area of 3217 mm^2.
$$100A_{sc}/(bh) = 100 \times 3217/(400 \times 300) = 2.68$$

(ii) Case 2; dead + imposed + wind load

Calculate the additional moment:

$$M_{add} = 1284 \times 78.41 \times 10^{-3} = 100.7 \text{ kNm}$$
$$M_t = 135.6 + 100.7 = 236.3 \text{ kNm}$$
$$N/(bh) = 1284 \times 10^3/(300 \times 400) = 10.7$$
$$M/(bh^2) = 236.3 \times 10^6/(300 \times 400^2) = 4.92$$
$$100A_{sc}/bh = 3.0$$

Case 1 gives marginally the more severe design condition. The column reinforcement is shown in Fig.9.30.

CHAPTER 10

WALLS IN BUILDINGS

10.1 FUNCTIONS, TYPES AND LOADS ON WALLS

All buildings contain walls the function of which is to carry loads, enclose and divide space, exclude weather and retain heat. Walls maybe classified into the following types:

1. internal non-load-bearing walls of block-work or light movable partitions that divide space only
2. external curtain walls that carry self-weight and lateral wind loads
3. external and internal infill walls in framed structures that may be designed to provide stability to the building but do not carry vertical building loads; the external walls would also carry lateral wind loads
4. load-bearing walls designed to carry vertical building loads and horizontal lateral and in-plane wind loads and provide stability

Type 4 structural concrete walls are considered.
   The role of the wall is seen clearly through the type of building in which it is used. Building types and walls provided are as follows:

(a) framed buildings: wall types 1, 2 or 3
(b) load-bearing and shear wall building with no frame: wall types 1, 2 and 4
(c) combined frame and shear wall building: wall types 1, 2 and 4

Type (c) is the normal multi-storey building.
   A wall is defined in BS8110: Part 1, clause 1.2.4, as a vertical load-bearing member whose length exceeds four times its thickness. This definition distinguishes a wall from a column.
   Loads are applied to walls in the following ways:

1. vertical loads from roof and floor slabs or beams supported by the wall
2. lateral loads on the vertical wall slab from wind, water or earth pressure
3. horizontal in-plane loads from wind when the wall is used to provide lateral stability in a building as a shear wall

10.2 TYPES OF WALL AND DEFINITIONS

Structural concrete walls are classified into the following two types defined in clause 1.2.4 of the code:

1. A reinforced concrete wall is a wall containing at least the minimum quantity of reinforcement given in clause 3.12.5 (section 10.3 below). The reinforcement is taken into account in determining the strength of the wall.

2. A plain concrete wall is a wall containing either no reinforcement or insufficient reinforcement to comply with clause 3.12.5. Any reinforcement in the wall is ignored when considering strength. Reinforcement is provided in most plain walls to control cracking.

Also in accordance with clause 1.2.4 mentioned above, walls are further classified as follows:

1. **A braced wall** is a wall where reactions to lateral forces are provided by lateral supports such as floors or cross-walls;
2. **An un-braced wall** is a wall providing its own lateral stability such as a cantilever wall;
3. **A stocky wall** is a wall where the effective height divided by the thickness, l_e/h, does not exceed 15 for a braced wall or 10 for an unbraced wall;
4. **A slender wall** is a wall other than a stocky wall.

10.3 DESIGN OF REINFORCED CONCRETE WALLS

10.3.1 Wall Reinforcement

(a) Minimum area of vertical reinforcement
The minimum amount of reinforcement required for a reinforced concrete wall from Table 3.25 of the code expressed by the term $100A_{SC}/A_{CC}$ is 0.4 where A_{SC} is the area of steel in compression and A_{cc} is the area of concrete in compression.

(b) Area of horizontal reinforcement
The area of horizontal reinforcement in walls where the vertical reinforcement resists compression and does not exceed 2% is given in clause 3.12.7.4 as

$$f_y = 250\text{N/mm}^2 \quad 0.3\% \text{ of concrete area}$$
$$f_y = 460\text{N/mm}^2 \quad 0.25\% \text{ of concrete area}$$

(c) Provision of links
If the compression reinforcement in the wall exceeds 2% links must be provided through the wall thickness (clause 3.12.7.5).

10.3.2 General Code Provisions For Design

The design of reinforced concrete walls is discussed in section 3.9.3 of the code. The general provisions are as follows.

(a) Axial loads

The axial load in a wall may be calculated assuming the beams and slabs transmitting the loads to it are simply supported.

(b) Effective height

Where the wall is constructed monolithically with adjacent elements, the effective height l_e should be assessed as though the wall were a column subjected to bending at right angles to the plane of the wall.

If the construction transmitting the load is simply supported, the effective height should be assessed using the procedure for a plain wall (section 10.4(b))

(c) Transverse moments

For continuous construction transverse moments can be calculated using elastic analysis. If the construction is simply supported, the eccentricity and moment may be assessed using the procedure for a plain wall. The eccentricity is not to be less than $h/20$ or 20mm where h is the wall thickness (section 10.4.1(g) below).

(d) In-plane moments

Moments in the plane of a single shear wall can be calculated from statics. When several walls resist forces the proportion allocated to each wall should be in proportion to its stiffness.

Consider two shear walls connected by floor slabs and subjected to a uniform horizontal load, as shown in Fig.10.1. The walls deflect by the same amount

$$\delta = pH^3/8EI$$

Thus the load is divided between the walls in proportion to their moments of inertia:

$$\text{wall 1: } p_1 = p\frac{l_1^3}{l_1^3 + l_2^3} \text{, wall 2: } p_2 = p - p_1$$

A more accurate analysis for connected shear walls is given in Chapter 15.

(e) Reinforcement for walls in tension

If tension develops across the wall section the reinforcement is to be arranged in two layers and the spacing of bars in each layer should comply with the bar spacing rules in section 3.12.11 of the code.

10.3.3 Design of Stocky Reinforced Concrete Walls

The design of stocky reinforced concrete walls is covered in section 3.9.3.6 of the code. The provisions in the various clauses are as follows.

(a) Walls supporting mainly axial load

If the wall supports an approximately symmetrical arrangement of slabs, the design axial load capacity n_w per unit length of wall is given by

$$n_w = 0.35 f_{cu} A_c + 0.67 A_{sc} f_y$$

where A_c is the gross area of concrete per unit length of wall and A_{sc} is the area of compression reinforcement per unit length of wall. The expression applies when the slabs are designed for uniformly distributed imposed load and the spans on either side do not differ by more than 15%.

(b) Walls supporting transverse moment and uniform axial load
Where the wall supports a transverse moment and a uniform axial load, a unit length of wall can be designed as a column using column design charts discussed in Chapter 9

(c) Walls supporting in-plane moments and axial load
The design for this case is set out in section 10.3.4 below.

(d) Walls supporting axial load and transverse and in-plane moments
The code states that the effects are to be assessed in three stages.

(i) In-plane Axial force and in-plane moments are applied. The distribution of force along the wall is calculated using elastic analysis assuming no tension in the concrete.

(ii) Transverse The transverse moments are calculated using the procedure set out in section 10.3.2(c).

(iii) Combined The effects of all actions are combined at various sections along the wall. The sections are checked using the general assumptions for beam design.

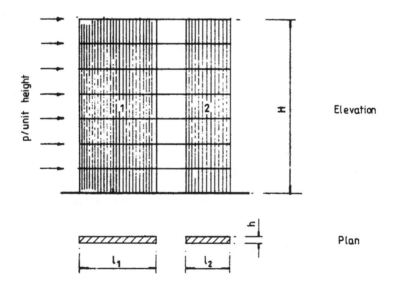

Fig.10.1 Shearwalls connected by floor slabs.

10.3.4 Walls Supporting In-plane Moments and Axial Loads

(a) Wall types and design methods

Some types of shear wall are shown in Fig.10.2. The simplest type is the straight wall with uniform reinforcement as shown in 10.2(a). In practice the shear wall includes columns at the ends as shown in 10.2(e). Channel-shaped walls are also common as shown in 10.2(d), and other arrangements are used.

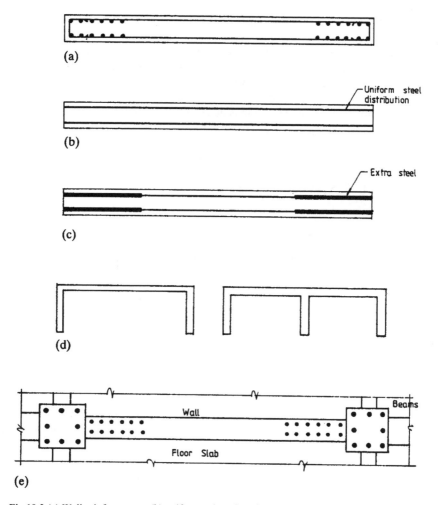

Fig.10.2 (a) Wall reinforcement: (b) uniform strips of steel; (c) extra reinforcement in end zones; (d) channel-shaped shear walls; (e) shear wall between columns.

Three design procedures are discussed.

1. using an interaction chart
2. assuming a uniform elastic stress distribution
3. assuming that end zones resist moment

The methods are discussed briefly below. Examples illustrating their use are given.

(b) Interaction chart

The chart construction is based on the assumptions for design of beams given in section 3.4.4.1 of the code. A straight wall with uniform reinforcement is considered. For the purpose of analysis the vertical bars are replaced by uniform strips of steel running the full length of the wall as shown in Fig.10.2(b). The chart is shown in Fig.10.4. The chart is constructed using the following equations:

Assuming $f_y = 460$ N/mm^2 and Young's modulus for steel is 200 kN/mm^2, then the strain ε_y when the stress is 0.95 f_y is given by

$$\varepsilon_y = 0.95 \, f_y/(200 \times 10^3) = 2.185 \times 10^{-3}$$

If the maximum compressive strain in concrete is 0.0035 and the neutral axis depth is x, the strain in steel is equal to ε_y at a depth c from the neutral axis, where

$$c = (\varepsilon_y /0.0035) \, x = 0.6243 \, x,$$
$$(x - c) = 0.3757x$$

Case 1: If $(x/h) \leq 0.6157$, as shown in Figure 10.3 (a),

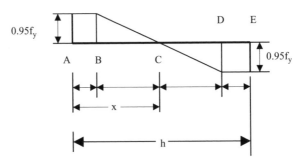

Fig.10.3(a) Neutral axis position for Case 1.

$$AB = 0.3757x, \quad BC = 0.6243x, \quad CD = 0.6243x, \quad DE = (h - 1.6243x)$$

Using the rectangular compressive stress block, if the thickness of the wall is b, the compressive force due to concrete = $0.45f_{cu} \, (0.9x) \, b$

If the steel in the wall is A_{sc} mm^2/m, then

$$N = [0.45 f_{cu} \, b \, 0.9x + (AB + 0.5 \, BC - 0.5 \, CD - DE) \times 10^{-3} \times 0.95 f_y \, A_{sc}] \times 10^{-3}$$

where AB etc. are in mm and N in kN.

Taking moments about the centre of the wall,

$$M =[0.45\,f_{cu}\,b\,0.9x(0.5h-0.45x)+\{AB(0.5h-0.5AB)+0.5\,BC(0.5h-AB-\frac{BC}{3})$$

$$-0.5\,CD(0.5h-AB-BC-\frac{2}{3}CD)$$

$$+DE((0.5h-AB-BC-CD-0.5\,DE))\}\times10^{-3}\times0.95f_{y}A_{sc}]\times10^{-6}$$

where M is in kNm

Case 2: If $1.0 \geq$ (x/h) > 0.6157, as shown in Figure 10.3 (b)

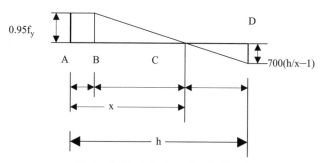

Fig.10.3(b) Neutral axis position for Case 2.

$$AB = 0.3757x, \quad BC = 0.6243x, \quad CD = h-x$$

$$N =[0.45f_{cu}\,b\,0.9x+\{(AB+0.5\,BC)0.95f_{y}-0.5\,CD\times700(\frac{h}{x}-1)\}\times10^{-3}\,A_{sc}]\times10^{-6}$$

$$M =[0.45\,f_{cu}\,b\,0.9x(0.5h-0.45x)+\{AB(0.5h-0.5AB)0.95f_{y}$$

$$+0.5\,BC(0.5h-AB-\frac{BC}{3})0.95f_{y}$$

$$-0.5\,CD\times700(\frac{h}{x}-1)(0.5h-AB-BC-\frac{2}{3}CD)\,\}\times10^{-3}\,A_{sc}]\times10^{-6}$$

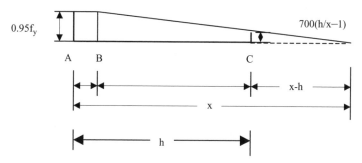

Fig.10.3(c) Neutral axis position for Case 3.

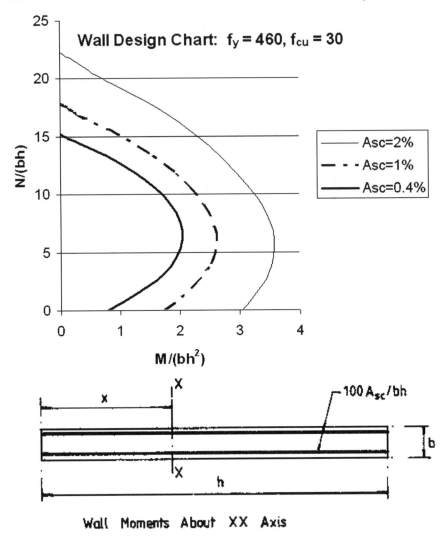

Fig.10.4 Wall design chart.

Case 3: If $2.6617 \geq (x/h) > 1.0$, then as shown in Figure 10.3 (c),
$$AB = 0.3757x, \quad BC = h - 0.3757\,x$$
$$N = [0.45 f_{cu}\, b(0.9x \leq h) + \{AB \times 0.95 f_y + 0.5\,BC \times (0.95 f_y + \sigma)\} \times 10^{-3} A_{sc}\,] \times 10^{-3}$$
$$M = [0.45 f_{cu}\, b(0.9x \leq h)\{0.5h - (0.45x \leq 0.5h)\} + \{AB(0.5h - 0.5AB)0.95 f_y$$
$$+\, 0.5\,BC(0.95 f_y + \sigma)(0.5h - AB - r)\} \times 10^{-3} A_{sc}\,] \times 10^{-6}$$

$$r = \frac{BC}{3}(1 + \frac{\sigma}{\sigma + 0.95 f_y}), \quad \sigma = 700(1 - \frac{h}{x})$$

A chart could also be constructed for the case where extra steel is placed in two zones at the ends of the walls as shown in Fig.10.2(c). Charts could also be constructed for channel-shaped walls.

In design the wall is assumed to carry the axial load applied to it and the overturning moment from wind. The end columns, if existing, are designed for the loads and moments they carry.

(c) Elastic stress distribution

A straight wall section, including columns if desired, or a channel-shaped wall is analyzed for axial load and moment using the properties of the gross concrete section in each case. The wall is divided into sections and each section is designed for the average direct load on it. Compressive forces are resisted by concrete and reinforcement. Tensile stresses are resisted by reinforcement only.

(d) Assuming that end zones resist moment

Reinforcement located in zones at each end of the wall is designed to resist the moment. The axial load is assumed to be distributed over the length of the wall.

10.3.4.1 *Example of design of a wall subjected to axial load and in-plane moments using design chart*

(a) Specification

The plan and elevation for a braced concrete structure are shown in Fig.10.5.

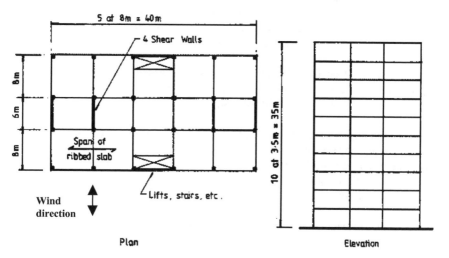

Fig.10.5 Framing arrangement.

The total dead load of the roof and floors is 6 kN/m^2. The imposed load on roof is 1.5 kN/m^2 and that for each floor is 3.0 kN/m^2. The wind speed is 20 m/s and the building is located in a city centre. Design the transverse shear walls as straight walls without taking account of the columns at the ends. The load bearing part of the wall is 160 mm thick with 20 mm thick decorative tiles on both faces. The materials are grade C30 concrete and grade 460 reinforcement.

Refer to BS 6399-1:1996: *Loading for buildings, Part 1: Code of practice for dead and imposed loads*, for more information

(b) Type of wall: slenderness

The wall is 160 mm thick structurally and is braced. The slenderness is calculated as for columns. Referring to Table 3.19 in the code the end conditions are as follows:

1. At the top the wall is connected to a ribbed slab 350 mm deep, i.e. condition 1;
2. At the bottom the connection to the base is designed to carry moment, i.e. condition 1.

From Table 3.21, $\beta = 0.75$. The clear height is 3150 mm, say. The slenderness is
$$0.75 \times 3150/160 = 14.8 < 15$$
The wall is 'stocky'.

(c) Dead and imposed loads on wall

The dead load on the wall, given that the wall is 200 mm thick including finishes, is as follows.

Note that there are 10 floors including the roof and the plan area of each floor is 8 × 6 m. Total height of the building = 35 m.
$$\text{Roof and floor slabs: } 10 \times (6 \times 8) \times 6 \text{ kN/m}^2 = 2880 \text{ kN}$$
$$\text{Wall, 200 mm thick: } (0.2 \times 6 \times 35) \times 24 \text{ kN/m}^3 = 1008 \text{ kN}$$
$$\text{Total dead load at base: } 2880 + 1008 = 3888 \text{ kN}$$

The wall carries load from 10 floors. Therefore imposed load can be reduced by 40% in accordance with BS6399: Part 1, Table 2. Imposed load is
$$(1 - 0.4) \times \{1.5 \text{ (Roof load)} + 3.0 \times 9 \text{ floors}\} (6 \times 8) = 820.8 \text{ kN}$$

(d) Dead and imposed loads at each end of the wall from one transverse beam

The slabs span between the beams which are supported on the walls. On any transverse beam, all the load acting on area 8 x 8 acts. The beam reaction acts on the column.
$$\text{Roof and floor slab: } 10 \times \{(1/2) \times (8 \times 8)\} \times 6 = 1920 \text{ kN}$$
$$\text{Column (500 mm × 500 mm) at wall ends: } 35 \times 0.5 \times 0.5 \times 24 \text{ kN/m}^3 = 210.0 \text{ kN}$$
$$\text{Imposed load: } (1 - 0.4) \times \{(1/2) \times (8 \times 8)\} \times (1.5 + 3.0 \times 9) = 547.2 \text{ kN}$$

(e) Wind load

Wind loads are specified in BS 6399-2:1997 *Loading for buildings-Part 2: Code of practice for wind loads*. For normal calculations the so called Standard Method is used.

The case for which wind load is calculated is wind acting normal to the 40 m width. The maximum height H = 35 m above the ground. In the following wind load is calculated using the code. For explanation of the symbols used see *BS 6399-2:1997* for more details.

Reference height H_r = 35 m

Effective height H_e = H_r = 35 m

Building type factor K_b = 1.0 (Table 1of Code)

Dynamic augmentation factor, $C_r \approx 0.04$ (Fig. 3 of Code)

The code rules apply.

Basic wind speed, V_b = 20 m/s (Assumed)

Altitude factor, $S_a \approx 1.0$

Direction factor, S_d = 1.0

Seasonal factor, S_s = 1.0

Probability factor, S_p = 1.0

Site wind speed, $V_s = V_b \times S_a \times S_d \times S_s \times S_p$ = 20 m/s

Terrain and building factor, S_b: Site in town with the closest distance to sea upwind greater than 100 km. Using Table 4 of Code and interpolating between 1.85 for 30 m and 1.95 for 50 m, S_b = 1.88

Effective wind speed, $V_e = V_s \times S_b$ = 20 × 1.88 = 37.6 m/s

Dynamic pressure, $q_s = 0.613\ V_e^2 = 0.613 \times 37.6^2 = 866.64$ N/m^2 = 0.87 kN/m^2

External surface pressure coefficient, C_{pe} :

Smaller dimension of the building, D = 22 m

H = 35 m

D/H = 0.63 < 1.0

C_{pe} {windward (front) face} = 0.85,

C_{pe} {Leeward (rear) face} = –0.5 (Table 5 of Code)

Size factor, C_a: Site in town with the closest distance to sea upwind greater than 100 kM. Category B,

$C_a \approx 0.85$ (Fig. 4 of Code)

External surface pressure, (windward face):

$p_e = q_s \times C_{pe} \times C_a = 0.87 \times 0.85 \times 0.85 = 0.63$ kN/m^2

External surface pressure, (leeward face):

$p_e = q_s \times C_{pe} \times C_a = 0.87 \times (-0.5) \times 0.85 = -0.37$ kN/m^2

Total pressure on the building = 0.63 – (–0.37) = 1.0 kN/m^2

The wind loads are *assumed* to be resisted equally by four shear walls. The horizontal wind loads and corresponding moments at the base are as follows:

Total horizontal load per wall = {0.85 × 1.0 × (1 + 0.04)} × (40 × 35) /4 = 309.4 kN

Moment = 309.4 × 35/2 = {0.80 × 40 × 35^2/2} /4 = 5414.5 kNm

(f) Load combination

(i) Case 1   1.2(Dead + Imposed + Wind)

$$N = 1.2 \times [3888 + 820.8 + 2(1920 + 210.0 + 547.2)] = 12075.8 \text{ kN}$$
$$M = 1.2 \times 5414.5 = 6497.4 \text{ kNm}$$

(ii) Case 2   1.4(Dead + Wind)
$$N = 1.4 \times [3888 + 2(1920 + 210.0)] = 11407.2 \text{ kN}$$
$$M = 1.4 \times 5414.5 = 7580.3 \text{ kNm}$$

(iii) Case 3   1.0 × Dead + 1.4 × Wind
$$N = 3888 + 2(1920 + 210.0) = 8148.0 \text{ kN}$$
$$M = 1.4 \times 5414.5 = 7580.3 \text{ kNm}$$

(g) Wall design for load combinations in (f)

The wall is 160 mm thick by 6000 mm long. The design is made using the chart in Fig.10.4. The steel percentages for the three load cases are given in Table 10.1, from which

$$A_{sc} = \{\frac{0.4}{100}(160 \times 6000)\}\frac{1}{6} = 640 \, \text{mm}^2/\text{m}$$

Table 10.1 Load combinations, wall design

	Case 1	Case 2	Case 3
N/ (bh)	12.58	11.88	8.49
M/ (bh^2)	1.13	1.32	1.32
100 A$_{sc}$/ (bh)	0.4	0.4	<0.4

Table 10.2 Load combinations, actual designed wall capacity

	Case 1	Case 2	Case 3
N/ (bh)	12.58	11.88	8.49
M/ (bh^2)	3.29	3.42	3.87
x/h	0.695	0.670	0.568

Provide two rows one on each face of 10 mm diameter bars at 200 mm centres to give a total steel area of 784 mm^2 /m. Table 10.2 shows that the capacity of the wall for the steel area provided is adequate.

10.3.4.2 Example of design of a wall subjected to axial load and in-plane moments with concentrated steel in end zones/columns

The plan of the structure is shown in Fig.10.6. In the absence of a design chart to cover this case, the following approximate design procedure can be used.

Consider the load case 1.4(dead + wind). Assume that the end columns resist the moment due to wind. The lever arm is 6.0 m. The equivalent axial force due to moment caused by wind is

$$\pm 1.4 \times 5414.5/6.0 = \pm 1263.4 \text{ kN}$$

The self weight of the column is (500 mm × 500 mm) at the wall ends:

Self weight of columns = $35 \times 0.5 \times 0.5 \times 24$ kN/m^3 = 210.0 kN

The total column load due to dead load and the additional equivalent force due to wind is:

$$1.4 \times (1920 + 210.0) \pm 1263.4 = 4245.4 \text{ or } 1718.6 \text{ kN}$$

The axial load capacity of the column:

$$4245.4 = 0.45 \times f_{cu} \times 500^2 \times 10^{-3} + 0.95 \, A_{sc} \, f_y \times 10^{-3}$$

Solving, $A_{sc} = 1717$ mm^2. Provide 4T25 = 1964 mm^2.

$$100 \times 1964/500^2 = 0.79 > 0.4 \text{ (minimum)}$$

The axial force due to dead load on the wall is

$$= 1.4 \times 3888.0 = 5443.2 \text{ kN}$$

The axial force capacity of the wall is

$$5443.2 = 0.45 \times f_{cu} \times 5500 \times 160 \times 10^{-3} + 0.95 \, A_{sc} \, f_y \times 10^{-3}$$
$$5443.2 = 11880 \text{ kN} + 0.95 \, A_{sc} \, f_y \times 10^{-3}$$

Only minimum steel of 0.4% is required.

$$A_{sc} = (0.4/100) \times (1000 \times 160) = 640 \text{ mm}^2\text{/m}$$

Provide T10 at 200 mm centres on both faces to give an area of 784 mm^2/m.

Fig 10.7 shows the reinforcement details in the wall and the end columns

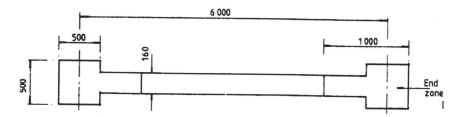

Fig.10.6 Wall with end columns.

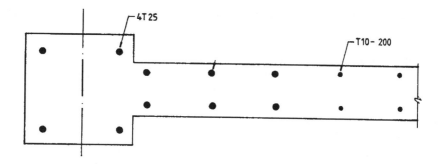

Fig.10.7 Reinforcement in the wall.

Figure 10.8 shows the design chart for the designed wall with end columns assuming

$f_{cu} = 30$ N/mm^2, $f_y = 460$ mm^2 and the bar centre in the columns inset from the edges by 60 mm. Table 10.3 shows the required capacity in brackets and also the designed capacity showing that all three load combinations considered are in the safe region.

Table 10.3 Forces and designed capacities.

	Case 1	Case 2	Case 3
N	12076 (11894)	11407 (11196)	8148 (7997)
M	6497.4 (5880.0)	7580.3 (6860)	7580.3 (6860)

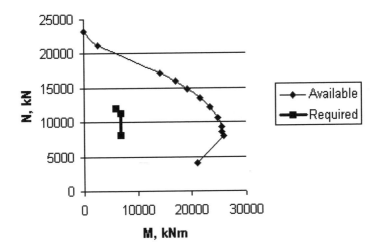

Fig.10.8 Design chart for a wall with end columns.

The procedure used for the construction of the chart is the same as explained in 10.3.4 (b). As an example choosing x = 5000 mm, the stress distribution in the steel is as shown in Fig.10.9.

(a) Left hand column: Area of steel = (4T25)= 1963.5 mm^2
        Compression in concrete = $0.45 \times 30 \times 500^2 \times 10^{-3} = 3375$ kN
        Compression in steel = (4T25) $\times 0.95 \times 460 \times 10^{-3} = 858.05$ kN

(b) Wall: (Area of steel = 784 mm^2/m)
    Length of the wall in compression = 0.9 x $- 500 = 0.9 \times 5000 - 500 = 4000$ mm
        Compression in concrete: = $0.45 \times 30 \times 4000 \times 160 \times 10^{-3} = 8640$ kN
    Length of wall in which the steel stress varies linearly = $0.6243 \times 5000 = 3122$ mm
        Compressive force = $3122 \times 10^{-3} \times 784 \times (0.5 \times 0.95 \times 460) \times 10^{-3} = 534.8$ kN
    Length of wall in which the steel stress is constant = $5000 - 500 - 3122 = 1378$ mm
        Compressive force = $1378 \times 10^{-3} \times 784 \times (0.95 \times 460) \times 10^{-3} = 472.1$ kN

Portion of the wall in tension = $6500 - 5000 - 500 = 1000$ mm

Strain in the steel at 1000 mm from neutral axis = $0.0035 \times 1000/x = 7.0 \times 10^{-4}$

Stress in the steel = $7.0 \times 10^{-4} \times (E = 200 \times 10^3) = 140$ N/mm^2

Tensile force due to steel = $1000 \times 10^{-3} \times 784 \times (0.5 \times 140) \times 10^{-3} = 54.9$ kN

Average stress in the steel in column = $140 \times (1250/1000) = 175$ N/mm^2

Tensile force due to steel in column = $1963.5 \times 175 \times 10^{-3} = 343.6$ kN

Total axial force is

$N = 3375.0 + 858.05 + 8640.0 + 534.8 + 472.1 - 54.9 - 343.6 = 13481.5$ kN

Taking moments about the centre of the wall, the resistant moment is

$M = \{(3375.0 + 858.05) \times (3250 - 250) + 8640 \times (3250 - 4000/2 - 500)$
$+ 534.8 \times (3250 - 500 - 1378 - 3122/3) + 472.1 \times (3250 - 500 - 1378/2)$
$+ 54.9 \times (3250 - 1000/3 - 500) + 343.6 \times (3250 - 500/2)\} \ 10^{-3} = 21489$ kNm

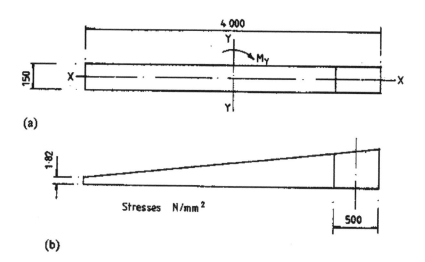

6.0 m

1.379   3.12   1.0

Fig.10.9 (a) Plan; (b) stress distribution in steel.

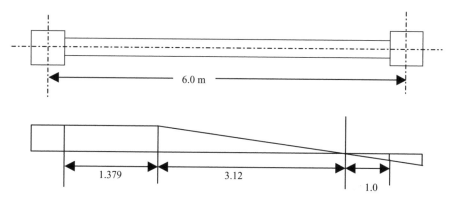

4 000

Y

M$_Y$

150

X — X

Y

(a)

1·82

Stresses   N/mm^2

500

(b)

Fig.10.10 (a) Wall section; (b) longitudinal stress distribution..

10.3.4.3 Example of design of a wall subjected to axial load, transverse and in-plane moments

(a) Specification
The section of a stocky reinforced concrete wall shown in Fig.10.10(e) is subject to the following actions:

$$N = 4300 \text{ kN}$$
$$M_y = 2100 \text{ kNm}$$
$$M_x = 244 \text{ kNm}$$

Design the reinforcement for the heaviest loaded end zone 500 mm long. The materials are grade C30 concrete and grade 460 reinforcement.

(b) Stresses

From an elastic analysis the stresses in the section due to moment M_y are calculated as follows.

$$A = 150 \times 4000 = 6 \times 10^5 \text{ mm}^2$$
$$I = 150 \times 4000^3/12 = 8 \times 10^{11} \text{ mm}^4$$
$$\text{fibre stress} = \frac{4300 \times 10^3}{6 \times 10^5} + \frac{2100 \times 10^6 \times 2000}{8 \times 10^{11}}$$
$$= 7.17 + 5.25 = 12.42 \text{N/mm}^2$$

The stress at 250 mm from the end is

$$7.17 + 5.25 \times 1750/2000 = 11.76 \text{ N/mm}^2$$

The load on the end zone is

$$11.76 \times 150 \times 500 \times 10^{-3} = 882 \text{ kN}$$

Design the end zone for an axial load of

$$N = 882 \text{ kN}$$
$$M = 224\,(500/4000) = 28 \text{ kN m:}$$
$$N/\,(bh) = 11.76$$
$$M/\,(bh^2) = 28 \times 10^6/\,(500 \times 150^2) = 2.5$$
$$100\,A_{sc}/bh = 1.5$$
$$A_{sc} = 1125 \text{ mm}^2$$

Provide 4T20 to give an area of 1263 mm^2

10.3.5 Slender Reinforced Walls

The following provisions are summarized from section 3.9.3.7 of the code.

1. The design procedure is the same as in section 10.3.3(d) above.
2. The slenderness limits are as follows:
   braced wall, steel area $A_s < 1\%$, $l_e/h \leq 40$
   braced wall, steel area $A_s > 1\%$, $l_e/h \leq 45$
   unbraced wall, steel area $A_s > 0.4\%$, $l_e/h \leq 30$

10.3.6 Deflection of Reinforced Walls

The code states that the deflection should be within acceptable limits if the above recommendations are followed. The code also states that the deflection of reinforced shear walls should be within acceptable limits if the total height does not exceed 12 times the length.

10.4 DESIGN OF PLAIN CONCRETE WALLS

10.4.1 Code Design Provisions

A plain wall contains either no reinforcement or less than 0.4% reinforcement. The reinforcement is not considered in strength calculations. The design procedure is summarized from section 3.9.4 of the code.

(a) Axial loads
The axial loads can be calculated assuming that the beams and slabs supported by the wall are simply supported.

(b) Effective height

(i) Unbraced plain concrete wall: The effective height l_e for a wall supporting a roof slab spanning at right angles is $1.5 \, l_0$, where l_0 is the clear height between the lateral supports. The effective height l_e for other walls, e.g. a cantilever wall, is $2 \, l_0$.

(ii) Braced plain concrete wall: The effective height l_e when the lateral support resists both rotation and lateral movement is 0.75 times the clear distance between the lateral supports or twice the distance between a support and a free edge. l_e is measured vertically where the lateral restraints are horizontal floor slabs. It is measured horizontally if the lateral supports are vertical walls.

The effective height l_e when the lateral supports resist only lateral movement is the distance between the centres of the supports or 2.5 times the distance between a support and a free edge. The effective heights defined above are shown in Fig.10.11.

The lateral support must be capable of resisting the applied loads plus 2.5% of the vertical load that the wall is designed to carry at the point of lateral support.

The resistance of a lateral support to rotation only exists where both the lateral support and the braced wall are detailed to resist rotation and for precast or in situ floors where the bearing width is at least two-thirds of the thickness of the wall.

(c) Slenderness limits
The slenderness ratio l_e/h should not exceed 30 whether the wall is braced or unbraced.

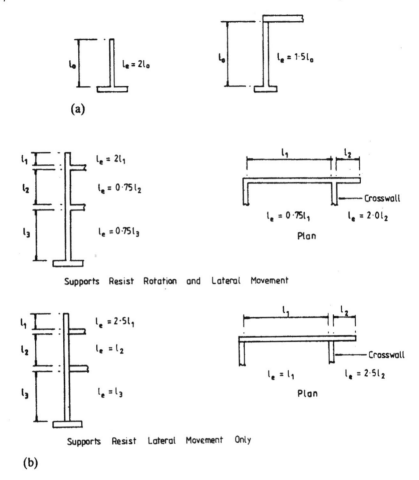

Fig.10.11 (a) Unbraced walls; (b) braced walls.

(d) Minimum transverse eccentricity
The minimum transverse eccentricity should not be less than $h/20$ or 20mm. Further eccentricity due to deflection occurs in slender walls.

(e) In-plane eccentricity
The in-plane eccentricity can be calculated by statics when the horizontal force is resisted by several walls. It is shared between walls in proportion to their stiffnesses provided that the eccentricity in any wall is not greater than one-third of its length. If the eccentricity is greater than one-third of the length the stiffness of that wall is taken as zero.

(f) Eccentricity of loads from a concrete floor or roof

The design loads act at one-third of the depth of the bearing width from the loaded face. Where there is an in situ floor on either side of the wall the common bearing area is shared equally (Fig.10.12). Loads may be applied through hangers at greater eccentricities (Fig.10.12).

(g) Transverse eccentricity of resultant forces

The eccentricity of forces from above the lateral support is taken as zero. From Fig.10.12, where the force R from the floor is at an eccentricity of h/6 and the force P from above is taken as axial, the resultant eccentricity is

$$e_R = \frac{Rh}{6(P+R)}$$

(h) Concentrated loads

Concentrated loads from beam bearings or column bases may be assumed to be immediately dispersed if the local stress under the load does not exceed 0.6 f_{cu} for concrete grade 25 or above.

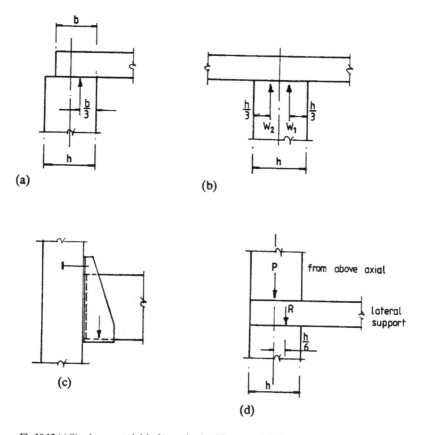

Fig.10.12 (a) Simply supported slab; (b) cast-in-situ slab over wall; (c) hanger; (d) resultant eccentricity.

(i) Design load per unit length

The design load per unit length should be assessed on the basis of a linear distribution of load with no allowance for tensile strength.

(j) Maximum unit axial load for a stocky braced plain wall

The maximum ultimate load per unit length is given by

$$n_w = 0.3(h - 2e_x) f_{cu}$$

where e_x is the resultant eccentricity at right angles to the plane of the wall (minimum value h/20). In this equation the load is considered to be carried on part of the wall with the section in tension neglected. The stress block is rectangular with a stress value of $0.3f_{cu}$ (Fig.10.13).

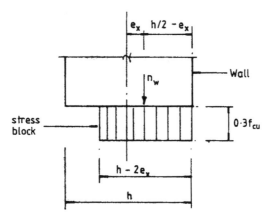

Fig.10.13 Stress block for a stocky braced wall.

(k) Maximum design axial load for a slender braced plain wall

The ultimate load per unit length is given by

$$n_w \leq 0.3(h - 1.2e_x - 2e_a) f_{cu}$$

where $e_a = l_e^2 / (2500\ h)$ is the additional eccentricity due to the wall's out of plane deflection and l_e is the effective height of the wall.

(l) Maximum design axial load for unbraced plain walls

The ultimate load per unit length should satisfy the following:

$$n_w \leq 0.3(h - 2e_{x1}) f_{cu}$$
$$n_w \leq 0.3(h - 2e_{x2} - 2e_a) f_{cu}$$

where e_{x1} is the resultant eccentricity at the top of the wall and e_{x2} is the resultant eccentricity at the bottom of the wall.

(m) Shear strength

The shear strength need not be checked if one of the following conditions is satisfied:

1. The horizontal design shear force is less than one-quarter of the design vertical load;

2. The shear stress does not exceed 0.45 N/mm² over the whole wall cross-section.

(n) Cracking

Reinforcement may be necessary to control cracking due to flexure or thermal and hydration shrinkage. The quantity in each direction should be at least 0.25% of the concrete area for grade 460 steel and 0.3% of the concrete area for grade 250 steel. Other provisions regarding 'anti-crack' reinforcement are given in the code.

(o) Deflection of plain concrete walls

The deflection should be within acceptable limit, if the preceding recommendations are followed.

The deflection of plain concrete shear walls should be within acceptable limits if the total height does not exceed ten times its length.

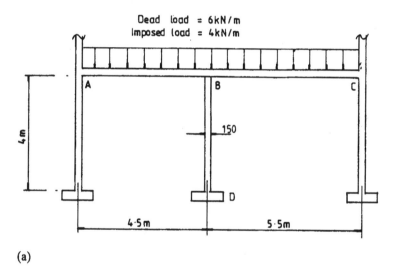

(a)

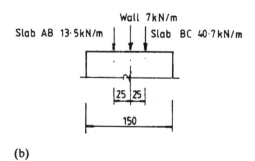

(b)

Fig.10.14 (a) Section through building: (b) section at mid-height.

45°, as shown in Fig.11.1 (a). Otherwise a reinforced concrete pad is required (Fig.11.1 (b)).

Assumptions to be used in the design of pad footings are set out in clause 3.11.2 of the code:

1. When the base is axially loaded the load may be assumed to be uniformly distributed. The actual pressure distribution depends on the soil type; refer to soil mechanics textbooks;

2. When the base is eccentrically loaded, the reactions may be assumed to vary linearly across the base.

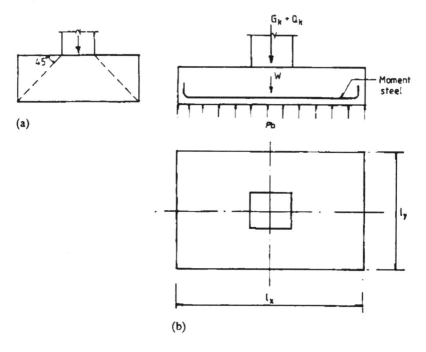

Fig.11.1 (a) Mass concrete foundation; (b) reinforced concrete pad foundation.

11.2.2 Axially Loaded Pad Bases

Refer to the axially loaded pad footing shown in Fig.11.1 (b) where the following symbols are used:

$$G_k = \text{characteristic dead load from the column (kN)}$$
$$Q_k = \text{characteristic imposed load from the column (kN)}$$
$$W = \text{weight of the base (kN)}$$
$$\ell_x, \ell_{y=} \text{ base length and breadth (m)}$$
$$P_b = \text{safe bearing pressure (kN/m}^2)$$

The required area is found from the characteristic loads including the weight of the base:

$$\text{Base area} = (G_k + Q_k + W)/P_b = \ell_x \; \ell_y \; m^2$$

The design of the base is made for the ultimate load delivered to the base by the column shaft, i.e. the design load is $(1.4G_k + 1.6Q_k)$.

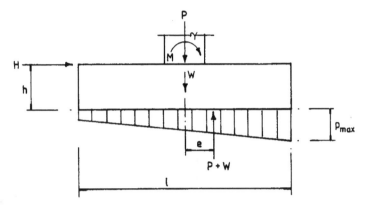

Fig.11.2 Eccentrically loaded base.

The critical sections in design are set out in clauses 3.11.2.2 and 3.11.3 of the code and are as follows.

(a) Bending
The critical section is at the face of the column on a pad footing or the wall in a strip footing. The moment is taken on a section passing completely across a pad footing and is due to the ultimate loads on one side of the section. No redistribution of moments should be made. The critical sections are XX and YY in Fig.11.3 (a).

(b) Distribution of reinforcement
Refer to Fig.11.3 (b). The code states arbitrarily that where l_c is half the spacing between column centres (if more than one) or the distance to the edge of the pad, whichever is the greater, exceeds $(3/4)(c + 3d)$, two-thirds of the required reinforcement for the given direction should be concentrated within a zone from the centreline of the column to a distance $1.5d$ from the face of the column. Here c is the column width and d is the effective depth of the base slab. Otherwise the reinforcement may be distributed uniformly over l_c. The reason for this is that although base pressure is assumed to be uniform, the bending moment tends to be somewhat higher towards the column than away from it. The concentration of reinforcement below the column area allows for this higher bending moment.

The arrangement of reinforcement is shown in Fig.11.3 (b)

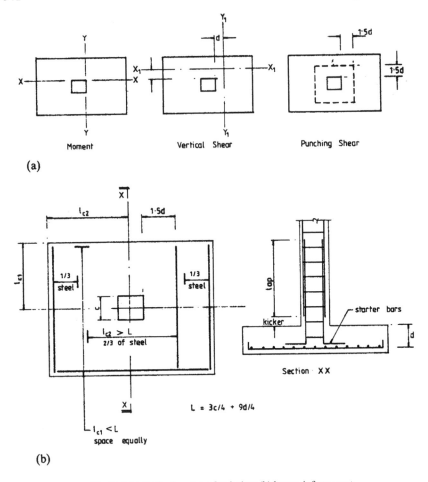

Fig.11.3 (a) Critical sections for design; (b) base reinforcement.

(c) Shear on vertical section across full width of base

Refer to Fig.11.3 (a). The vertical shear force is the sum of the loads acting outside the section considered. The shear stress is

$$v = V/ (l\,d)$$

where ℓ is the length or width of the base.

Refer to clause 3.4.5.10 (Enhanced shear strength near supports, simplified approach). If the shear stress is checked at d from the support and v is less than the value of v_c from Table 3.8 of the code, no shear reinforcement is required and no further checks are needed. If shear reinforcement is required, refer to Table 3.16 of the code. It is normal practice to make the base sufficiently deep so that shear reinforcement is not required. The depth of the base is often controlled by the design for shear.

(d) Punching shear around the loaded area

The punching shear force is the sum of the loads outside the periphery of the critical section. Refer to clause 3.7.7.6 of the code and Chapter 5, section 5.1.8 dealing with the design of flat slabs for shear. The shear stress is checked on the perimeter at 1.5d from the face of the column. If the shear stress v is less than the value of v_c in Table 3.8 no shear reinforcement is needed and no further checks are required. If shear reinforcement is required refer to clause 3.7.7.5 of the code. The critical perimeter for punching shear is shown in Fig.11.3 (a). The maximum shear at the column face must not exceed $0.8\sqrt{f_{cu}}$ or 5 N/mm^2.

(e) Anchorage of column starter bars

Refer to Fig.11.3 (b). The code states in clause 3.12.8.8 that the compression bond stresses that develop on starter bars within bases do not need to be checked provided that

1. the starter bars extend down to the level of the bottom reinforcement
2. the base has been designed for the moments and shears set out above.

(f) Cracking

See the rules for slabs in clause 3.12.11.2.7 of the code. The bar spacing is not to exceed 3d or 750 mm, but much lesser spacing is possible depending on the amount of flexural steel supplied.

(g) Minimum grade of concrete

The minimum grade of concrete to be used in foundations is grade C35.

(h) Nominal cover

Clause 3.3.1.4 of the code states that the minimum cover should be 75mm if the concrete is cast directly against the earth, or 40mm if cast against adequate blinding. Table 3.2 of the code classes non-aggressive soil as a moderate exposure condition.

11.2.2.1 Example of design of an axially loaded base

(a) Specification

A column 400 mm × 400 mm carries a dead load of 800 kN and an imposed load of 300 kN. The safe bearing pressure is 200 kN/m^2. Design a square base to resist the loads. The concrete is grade C35 and the reinforcement grade 460.

The condition of exposure is moderate from Table 3.2 for non-aggressive soil. The nominal cover is 40 mm for concrete cast against blinding.

(b) Size of base

Assume the weight is 80 kN.

$$\text{service load} = 800 + 300 + 80 = 1180 \text{ kN}$$
$$\text{area of base} = 1180/200 = 5.9 \text{ m}^2.$$
$$\text{Make the base 2.5 m} \times 2.5 \text{ m.}$$

(c) Moment steel

ultimate load = $(1.4 \times 800) + (1.6 \times 300) = 1600$ kN

ultimate base pressure = $1600/6.25 = 256$ kN/m^2

Note: The self weight of the footing is not included because the self weight and the corresponding base pressure will cancel themselves out.

The critical section YY at the column face is shown in Fig.11.4 (a).

$M_{yy} = 256 \times 1.05 \times 2.5 \times 1.05/2 = 352.8$ kNm

Try an overall depth of 500 mm with 16 mm bars both ways.

The weight of the footing = $2.5 \times 2.5 \times 0.5 \times 24 = 75$ kN

75 kN < 80 kN assumed in design.

The effective depth of the top layer of steel is

$d = 500 - 40 - 16 - 16/2 = 436$ mm

$k = M/ (bd^2 f_{cu}) = 352.8 \times 10^6/ (2500 \times 436^2 \times 35) = 0.021 < 0.156$

$z/d = 0.5 + \sqrt{(0.25 - 0.021/0.9)} = 0.98 > 0.95$

$A_s = 352.8 \times 10^6/ (0.95 \times 436 \times 0.95 \times 460) = 1949$ mm^2

Number of 16 mm bars = $1949/201.1 = 9.7$

Provide 10T16 bars, $A_s = 2010$ mm^2.

The distribution of the reinforcement is determined to satisfy the rule.

$3/4(c+ 3d) = 0.75 (400 + 3 \times 436) = 1281$ mm

$\ell_c = 2500/2 = 1250$ mm < 1281 mm

The bars can be spaced equally at 270 mm centres.

The full anchorage length required past the face of the column is $38 \times 16 = 608$mm. Adequate anchorage is available.

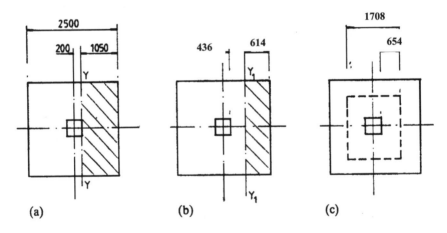

Fig.11.4 (a) Moment; (b) vertical shear; (c) punching shear.

(d) Vertical shear

The critical section Y_1Y_1 at $d = 436$ mm from the face of the column is shown in Fig.11.4 (b).

$$V = 256 \times 2.5 \times (1050 - 436) \times 10^{-3} = 392.96 \text{ kN}$$
$$v = 392.96 \times 10^3 / (2500 \times 436) = 0.36 \text{ N/mm}^2$$

The bars extend 565 mm, i.e. more than d, beyond the critical section and so the steel is effective in increasing the shear stress.

$$100 \, A_s / (bd) = 100 \times 2010 / (2500 \times 436) = 0.18,$$
$$400/d = 400/436 < 1.0, \text{ take as } 1.0$$
$$f_{cu} = 35 < 40$$
$$v_c = 0.79 \times (0.18)^{1/3} (1.0)^{1/4} (35/25)^{1/3}/1.25 = 0.40 \text{ N/mm}^2$$
$$(v = 0.36) < (v_c = 0.40)$$

and so the shear stress is satisfactory and no shear reinforcement is required.

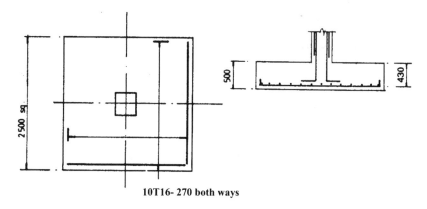

10T16- 270 both ways

Fig.11.5 Reinforcement arrangement.

(e) Punching shear

Punching shear is checked on a perimeter $1.5d = 654$ mm from the column face. The critical perimeter is shown in Fig.11.4(c).

$$c + 3d = 400 + 3 \times 436 = 1708$$
$$\text{perimeter, } u = 1708 \times 4 = 6832 \text{ mm}$$
$$\text{shear force } V = 256(2.5^2 - 1.708^2) = 853.2 \text{ kN}$$
$$v = 853.2 \times 10^3 / (6832 \times 436) = 0.29 \text{ N/mm}^2$$

The reinforcing bars extend 396 mm beyond the critical section. Even if the steel is discounted, $v_c = 0.34 \text{N/mm}^2$ from Table 3.8 of the code for $100A_s / (bd) \leq 0.15$ for grade C25 concrete. The base is clearly satisfactory and no shear reinforcement is required. The bars will be anchored by providing a standard 90° bend at the ends.

Check the maximum shear stress at the face of the column:

$$V = 256 \times (2.5^2 - 0.4^2) = 1559 \text{ kN}$$
$$u_0 = 4 \times 400 = 1600 \text{ mm}$$
$$v_{max} = 1559 \times 10^3 / (1600 \times 436) = 2.23 \text{ N/mm}^2 < (0.8\sqrt{35} = 4.73 \text{ N/mm}^2)$$

This is satisfactory.

(f) Cracking
The bar spacing does not exceed 750 mm and the flexural reinforcement supplied
is 0.18% which is less than 0.3%. No further checks are required.

(g) Reinforcement
The arrangement of reinforcement is shown in Fig.11.5. Note that in accordance
with clause 3.12.8.8 of the code the compression bond on the starter bars need not
be checked.

11.3 ECCENTRICALLY LOADED PAD BASES

11.3.1 Vertical Pressure

Clause 3.11.2 of the code states that the base pressure for eccentrically loaded pad
bases may be assumed to vary linearly across the base for design purposes.

The characteristic loads on the base are the axial load P, moment M and
horizontal load H as shown in Fig.11.2. The base dimensions are length ℓ, width b
and depth h.

$$\text{Base area } A = b\ell$$
$$\text{section modulus } Z = b\ell^2/6$$

The total load is P + W and the moment at the underside of the base is $M + Hh$.
The maximum earth pressure is

$$p_{max} = \frac{(P+W)}{A} + \frac{(M+Hh)}{Z}$$

This should not exceed the safe bearing pressure. The eccentricity e of the
resultant reaction is

$$e = \frac{(M+Hh)}{(P+W)}$$

If $e < \ell/6$ there is pressure over the whole of the base, as shown in Fig.11.2. If $e >$
$\ell/6$ part of the base does not bear on the ground, as shown in Fig.11.6 (a). In this
case

$$c = \ell/2 - e$$

and the length in bearing is $3c$. The maximum pressure is

$$p_{max} = \frac{2(P+W)}{3bc}$$

Although the code does not prohibit it, this situation is not recommended and must
be generally avoided.

Sometimes a base can be set eccentric to the column by, say, e_l to offset the
moments due to permanent loads and give uniform pressure, as shown in
Fig.11.6(b):

$$\text{Eccentricity } e_l = (M+Hh)/P$$

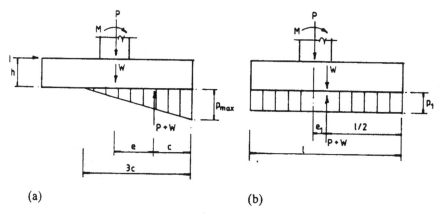

Fig.11.6 Eccentrically loaded pads: (a) Bearing on part of base; (b) base set eccentric to column.

11.3.2 Resistance to Horizontal Loads

Horizontal loads applied to bases are resisted by passive earth pressure against the end of the base, friction between the base and ground for cohesion-less soils such as sand, or adhesion for cohesive soils such as clay. In general, the load will be resisted by a combination of all actions. The ground floor slab can also be used to resist horizontal load. The forces are shown in Fig.11.7 (a).

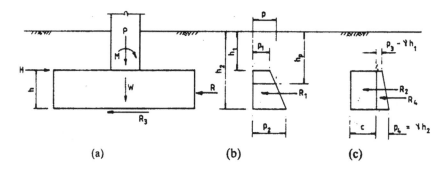

Fig.11.7 (a) Base; (b) cohesion less soil; (c) cohesive soil.

Formulae from soil mechanics for calculating the resistance forces are given for the two cases of cohesionless and cohesive soils.

(a) Cohesionless soils
Refer to Fig.11.7 (b). Denote the angle of internal soil friction ϕ and the soil density γ. The passive earth pressure p at depth h is given by

$$p = \gamma\, h\, k_p$$
$$k_p = (1 + \sin \phi)/(1 - \sin \phi)$$

If p_1 and p_2 are passive earth pressures at the top and bottom of the base, then the passive resistance

$$R_1 = 0.5\, b\, h\, (p_1 + p_2)$$

where

$$h \times b = \text{base depth} \times \text{base breadth}$$

If μ is the coefficient of friction between the base and the ground, generally taken as tan ϕ, the frictional resistance is

$$R_3 = \mu\, (P + W)$$

(b) Cohesive soils

Refer to Fig.11.7(c). For cohesive soils $\phi = 0$. Denote the cohesion at zero normal pressure c and the adhesion between the base and the load β. The resistance of the base to horizontal load is

$$R = R_2 + R_4 + R_3$$
$$R = 2cbh + 0.5b\, h\, (p_3 + p_4) + \beta\, \ell\, b$$

where the passive pressure p_3 at the top is equal to γh_1, the passive pressure p_4 at the bottom is equal to γh_2 and ℓ is the length of the base. The resistance forces to horizontal loads derived above should exceed the factored horizontal loads applied to the foundation.

In the case of portal frames it is often helpful to introduce a tie beam between bases to take up that part of the horizontal force due to portal action from dead and imposed loads as in the pinned base portal shown in Fig.11.8b. Wind load has to be resisted by passive earth pressure, friction or adhesion.

Pinned bases should be used where ground conditions are poor and it would be difficult to ensure fixity without piling. It is important to ensure that design assumptions are realized in practice.

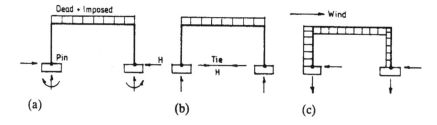

Fig.11.8 (a) Portal base reactions; (b) force *H* taken by tie; (c) wind load and base reactions.

11.3.3 Structural Design

The structural design of a base subjected to ultimate loads is carried out for the ultimate loads and moments delivered to the base by the column shaft. Pinned and fixed bases are shown in Fig.11.9.

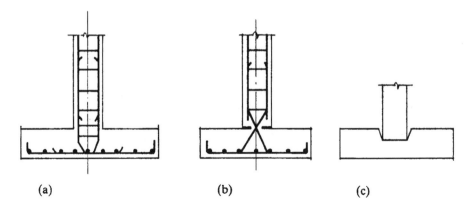

Fig.11.9 (a) Fixed base; (b) pinned base; (c) pocket base.

11.3.3.1 Example of design of an eccentrically loaded base

(a) Specification

The characteristic loads for an internal column footing in a building are given in Table 11.1. The proposed dimensions for the 450 mm square column and base (3600 × 2800 mm) are shown in Fig.11.10. The base supports a ground floor slab 200 mm thick. The soil is firm well drained clay with the following properties:

Unit weight = 18 kN/m³,
Safe bearing pressure = 150 kN/m²,
Cohesion = 60 kN/m²

The materials to be used in the foundation are grade C35 concrete and grade 460 reinforcement.

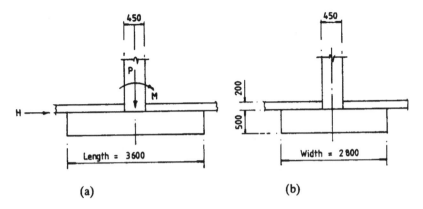

Fig.11.10 (a) Side elevation; (b) end elevation.

Table 11.1 Applied column loads and moments

	Vertical load, kN	Horizontal load, kN	Moment, kNm
Dead	770	35	78
Imposed	330	15	34

(b) Maximum base pressure on soil
The maximum base pressure is checked for the service loads.
$$\text{weight of base + slab} = 0.7 \times 3.6 \times 2.8 \times 24 = 169.3 \text{ kN}$$
$$\text{total axial load} = 770 + 330 + 169.3 = 1269.3 \text{ kN}$$
$$\text{total moment} = 78 + 34 + 0.5(35 + 15) = 137 \text{ kN m}$$
$$\text{base area } A = 2.8 \times 3.6 = 10.08 \text{ m}^2$$
$$\text{section modulus} = Z = 2.8 \times 3.6^2/6 = 6.05 \text{ m}^3$$
$$\text{maximum base pressure} = 1269.3/10.08 + 137/6.05 = 125.7 + 22.6 = 148.6 \text{ kN/m}^2$$
$$\text{Maximum base pressure} < (\text{safe bearing pressure} = 150 \text{ kN/m}^2)$$

(c) Resistance to horizontal load
Check the passive earth resistance assuming no ground slab.
No adhesion, $\beta = 0$, ($h_1 = 0$, $p_3 = 0$), ($h_2 = 0.5$, $p_4 = 18 \times 0.5 = 9$)
The passive resistance is
$$= 2c \, b \, h + 0.5 \, b \, h \, (p_3 + p_4) + \beta \, \ell \, b$$
$$= \{2 \times 60 \times 2.8 \times 0.5\} + \{0.5 \times 2.8 \times 0.5 \times (0 + 9.0)\} + 0$$
$$= 168 + 6.3 = 174.3 \text{ kN}$$
$$\text{factored horizontal load} = (1.4 \times 35) + (1.6 \times 15) = 73 \text{ kN}$$
$$\text{Passive resistance} > 73 \text{ kN.}$$
The resistance to horizontal load is satisfactory.
    The reduction in moment on the underside of the base due to the horizontal reaction from the passive earth pressure has been neglected.

(d) Design of the moment reinforcement
The design is carried out for the ultimate loads from the column.
(i) *Long span moment steel*
$$\text{axial load } N = (1.4 \times 770) + (1.6 \times 330) = 1606 \text{ kN}$$
$$\text{horizontal load } H = (1.4 \times 35) + (1.6 \times 15) = 73 \text{ kN}$$
$$\text{moment } M = (1.4 \times 78) + (1.6 \times 34) + (0.5 \times 73) = 223.5 \text{ kNm}$$
$$\text{maximum pressure} = 1606/10.08 + 223.5/6.05 = 196.3 \text{ kN/m}^2$$
$$\text{minimum pressure} = 1606/10.08 - 223.5/6.05 = 122.4 \text{ kN/m}^2$$
The pressure distribution is shown in Fig.11.11 (a).
At the face of the column pressure is
$$\text{Pressure} = 122.4 + (196.3 - 122.4) \times (3.6 - 1.575)/3.6 = 164.0 \text{ kN/m}^2$$
Moment at the face of the column is
$$M_y = 164 \times 2.8 \times 1.575^2/2 + 0.5(196.3 - 164.0) \times 2.8 \times 1.575 \times (2/3) \times 1.575$$
$$= 644.3 \text{ kNm}$$
If the cover is 40 mm and 20 mm diameter bars are used, the effective depth for the bottom layer is
$$d = 500 - 40 - 10 = 450 \text{ mm}$$

$$k = M/ (bd^2 f_{cu}) = 644.3 \times 10^6/ (2800 \times 450^2 \times 35) = 0.033 < 0.156$$
$$z/d = 0.5 + \sqrt{(0.25 - 0.033/0.9)} = 0.968 > 0.95$$
$$A_s = 644.3 \times 10^6/ (0.95 \times 450 \times 0.95 \times 460) = 3449 \text{ mm}^2$$

Provide 18T16. $A_s = 3619 \text{ mm}^2$.
$$0.75 \ (c + 3d) = 0.75 \ (450 + 3 \times 450) = 1350 \text{ mm},$$
$$\ell_c = 2800/2 = 1400 \text{ mm}$$
$$0.75 \ (c + 3d) < \ell_c$$

The difference between 1350mm and 1400mm is small enough to be ignored and steel can be distributed uniformly. Provide 18 bars at 160 mm centres to give a total steel area of 3619 mm².

Check minimum steel: $100 \ A_s/(bd) = 100 \times 3619/(2800 \times 450) = 0.29 > 0.13$

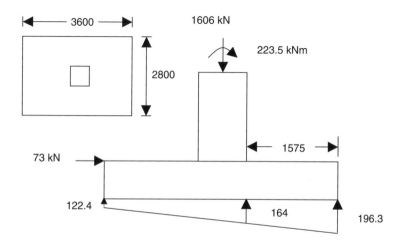

Fig.11.11 (a) Base pressures; (b) plan.

(ii) Short span moment steel

$$\text{average pressure} = 0.5 \times (196.3 + 122.4) = 159.4 \text{ kN/m}^2$$
$$\text{moment } M_x = 159.4 \times 3.6 \times 1.175^2/2 = 396.1 \text{ kNm}$$
$$\text{effective depth } d = 500 - 40 - 20 - 10 = 430 \text{ mm}$$
$$k = M/ (bd^2 f_{cu}) = 396.1 \times 10^6/ (3600 \times 430^2 \times 35) = 0.017 < 0.156$$
$$z/d = 0.5 + \sqrt{(0.25 - 0.017/0.9)} = 0.98 > 0.95$$
$$A_s = 396.1 \times 10^6/ (0.95 \times 430 \times 0.95 \times 460) = 2219 \text{ mm}^2$$

The minimum area of steel from Table 3.25 of the code is 0.13%.
$$A_s = (0.13/100) \times 3600 \times 500 = 2340 \text{ mm}^2$$

Provide 21T12. A_s provided = 2375 mm².
$$0.75(c + 3d) = 1350 < \ell_c = 1800 \text{ mm}$$

Place two-thirds of the bars (14 bars) in the central zone 1350 mm wide. Provide 14T12-100mm. In the outer strips 1125 mm wide provide 4T12 at 350 mm centres. The arrangement of bars is shown in Fig.11.12.

(e) Vertical shear

Long span: The vertical shear stress is checked at $d = 450$ mm from the face of the column.

Pressure $= 122.4 + (196.3 - 122.4) \times (3.6 - 1.575 + 0.450)/3.6 = 173.2$ kN/m^2

Shear at a distance d from the face of the column is

$$V_y = 0.5(173.2 + 196.3) \times 2.8 \times (1.575 - 0.450) = 582.0 \text{ kN}$$
$$v = 582.0 \times 10^3/(2800 \times 450) = 0.46 \text{ N/mm}^2$$
$$100A_s/(bd) = 100 \times 3619/(2800 \times 450) = 0.29 < 3.0$$
$$400/d = 400/450 < 1.0, \text{ take as } 1.0$$
$$v_c = 0.79 \times (0.29)^{1/3} (1.0)^{1/4} (35/25)^{1/3}/1.25 = 0.47 \text{ N/mm}^2$$
$$v < v_c$$

No shear reinforcement is required.

Short span:

$$\text{average pressure} = 0.5(196.3 + 122.4) = 159.4 \text{ kN/m}^2$$

The average pressure acts over an area of dimensions

$$\{(2800 - 450)/2 - 450 = 725 \text{ mm}\} \times 3600 \text{ mm}$$

Shear at a distance d from the face of the column is

$$V_y = 159.4 \times 3.6 \times 0.725 = 416.0 \text{ kN}$$
$$v = 416.0 \times 10^3/(3600 \times 430) = 0.27 \text{ N/mm}^2$$
$$100A_s/(bd) = 100 \times 2375/(3600 \times 430) = 0.15 < 3.0$$
$$400/d = 400/430 < 1.0$$
$$v_c = 0.79 \times (0.15)^{1/3} (1.0)^{1/4} (35/25)^{1/3}/1.25 = 0.38 \text{ N/mm}^2$$
$$v < v_c$$

The shear stress is satisfactory and no shear reinforcement is required.

(f) Punching shear and maximum shear

The punching shear is checked on a perimeter 1.5d from the face of the column. Using the data from (e) above,

$$\text{Average } d = 0.5 (450 + 430) = 440 \text{ mm}$$
$$\text{Average } 100A_s/(bd) = 0.5(0.29 + 0.15) = 0.22$$
$$400d = 400/440 < 1.0, \text{ take as } 1.0$$
$$v_c = 0.79 \times (0.22)^{1/3} (1.0)^{1/4} (35/25)^{1/3}/1.25 = 0.43 \text{ N/mm}^2$$
$$c + 3d = 450 + 3 \times 440 = 1770 \text{ mm}$$
$$\text{perimeter } u = 1770 \times 4 = 7080 \text{ mm}$$
$$\text{Average pressure} = 0.5(196.3 + 122.4) = 159.4 \text{ kN/m}^2$$

Check the maximum shear stress at the face of the column:

$$\text{shear } V = 159.4 \times (3.6 \times 2.8 - 0.45^2) = 1574.0 \text{ kN}$$
$$u_0 = 4 \times 440 = 1760 \text{ mm}$$
$$v_{max} = 1574 \times 10^3/(1760 \times 440) = 2.03 \text{ N/mm}^2 < (0.8\sqrt{35} = 4.73 \text{ N/mm}^2)$$

$$\text{shear } V = 159.4 (3.6 \times 2.8 - 1.77^2) = 1107.0 \text{ kN}$$
$$v = 1107.0 \times 10^3/(7080 \times 440) = 0.36 \text{ N/mm}^2 < (v_c = 0.43 \text{ N/mm}^2)$$

The base has adequate shear strength to resist punching shear failure.

(g) Sketch of reinforcement
The reinforcement is shown in Fig.11.12.

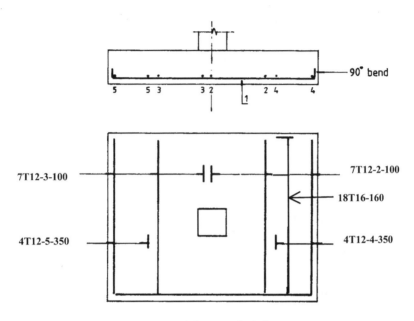

Fig.11.12 Reinforcement in the base.

11.3.3.2 Example of design of a footing for pinned base steel portal

(a) Specification
The column base reactions for a pinned base rigid steel portal for various load cases are shown in Fig.11.13. Determine the size of foundation for the two cases of independent bases and tied bases. The soil is firm clay with the following properties:

$$\text{Unit weight} = 18 \text{ kN/m}^3,$$
$$\text{Safe bearing pressure} = 150 \text{ kN/m}^2,$$
$$\text{Cohesion and adhesion} = 50 \text{ kN/m}^2$$

(b) Independent base
The base is first designed for (dead + imposed load). The proposed arrangement of the base is shown in Fig.11.14 (a). The base is 2 m long by 1.2 m wide by 0.5 m deep. The finished thickness of the floor slab is 180 mm. The unfactored loads on the soil are:

$$\text{vertical load} = 103 + 84 + \{(0.5 + 0.18) \times 2 \times 1.2 \times 24\} = 226.2 \text{ kN}$$
$$\text{horizontal load} = 32.4 + 40.3 = 72.7 \text{ kN}$$
$$\text{moment} = 72.7 \times 0.5 = 36.4 \text{ kN m}$$

$$area = 2 \times 1.2 = 2.4m^2$$
$$section\ modulus = 1.2 \times 2^2/6 = 0.8\ m^3$$

The maximum vertical pressure is
$$226.2/2.4 + 36.4/0.8 = 139.8\ kN/m^2$$

The resistance to horizontal load is
$$= 2\ c\ b\ h + 0.5\ b\ h\ (p_3 + p_4) + \beta\ \ell\ b$$
$$= (2 \times 50 \times 1.2 \times 0.5) + 0.5 \times 1.2 \times 0.5 \times (0 + 18 \times 0.5) + 50 \times 2 \times 1.2$$
$$= 60 + 2.7 + 120 = 182.7\ kN$$

The maximum factored horizontal load is
$$(1.4 \times 32.4) + (1.6 \times 40.3) = 109.9\ kN < 182.7\ kN$$

The base is satisfactory with respect to resistance to sliding.

Check the (dead + imposed + wind load internal suction) on the right hand side base.

$$vertical\ load = 103 + 84 - 29.4 = 157.6\ kN$$
$$horizontal\ load = 32.4 + 40.3 + 2.4 = 75.1\ kN$$
$$moment = 75.1 \times 0.5 = 37.6\ kNm$$
$$maximum\ pressure = 157.6/2.4 + 37.6/0.8 = 112.7\ kN/m^2$$

The reinforcement for the base can be designed and the shear stress checked as in the previous example.

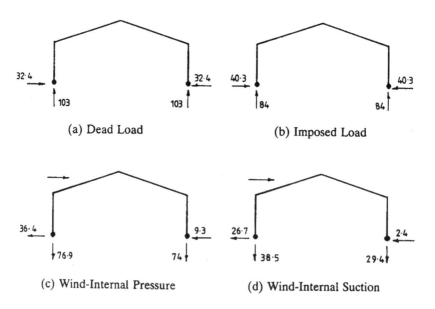

(a) Dead Load

(b) Imposed Load

(c) Wind-Internal Pressure

(d) Wind-Internal Suction

Fig.11.13 Pinned portal frame reactions (characteristic reactions)(a) Dead load; (b) imposed load; (c) wind, internal pressure; (d) wind, internal suction.

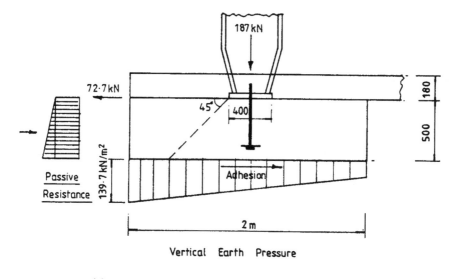

(a)

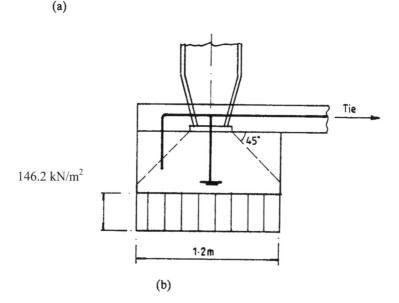

(b)

Fig 11.14 Unfactored loads and base pressures (a) independent base; (b) tied base.

(c) Tied base

The proposed base is shown in Fig.11.14 (b). The trial size for the base is 1.2 m × 1.2 m × 0.5 m deep and tie rods are provided in the ground slab. The horizontal tie resists the reaction from the dead and imposed loads. For this case

$$\text{vertical load} = 103 + 84 + \{24 \times 1.2^2 \times (0.5 + 0.18)\} = 210.5 \text{ kN}$$
$$\text{maximum pressure} = 210.5/1.2^2 = 146.2 \text{ kN/m}^2$$

The main action of the wind load is to cause uplift and the slab has to resist a small compression from the net horizontal load when the dead load and wind load internal pressure are applied at left hand base.

(d) Design of tie
To find the steel area for the tie using grade 460 reinforcement,

$$\text{ultimate load} = (1.4 \times 32.4) + (1.6 \times 40.3) = 109.84 \text{ kN}$$
$$A_s = 109.84 \times 10^3 / (0.95 \times 460) = 251.4 \text{ mm}^2$$

Provide two 16 mm diameter bars to give an area of 402 mm^2.

If the steel column base bearing plate is 400 mm × 400 mm, the underside of the base lies within the 45° load dispersal lines. Theoretically no reinforcement is required but 12 mm bars at 160 mm centres each way would provide minimum reinforcement.

11.4 WALL, STRIP AND COMBINED FOUNDATIONS

11.4.1 Wall Footings

Typical wall footings are shown in Figs 11.15(a) and 11.15(b). In Fig.11.15 (a) the wall is cast integral with the footing. The critical section for moment is at Y_1Y_1, the face of the wall, and the critical section for shear is at Y_2Y_2, d from the face of the wall. A 1m length of wall is considered and the design is made on similar lines to that for a pad footing.

If the wall is separate from the footing, e.g. a brick wall, the base is designed for the maximum moment at the centre and maximum shear at the edge, as shown in Fig.11.15 (b). The wall distributes the load W/t per unit length to the base and the base distributes the load W/b per unit length to the ground, where W is the load per unit length of wall, t is the wall thickness and b is the base width. The maximum shear at the edge of the wall is

$$W (b - t)/ (2b)$$

The maximum moment at the centre of the wall is

$$\frac{W}{b}\frac{b}{2}\frac{b}{4} - \frac{W}{t}\frac{t}{2}\frac{t}{4} = \frac{W}{8}(b-t)$$

11.4.2 Shear Wall Footing

If the wall and footing resist an in-plane horizontal load, e.g. when the wall is used as a shear wall to stabilize a building, the maximum pressure at one end of the wall is found assuming a linear distribution of earth pressure (Fig.11.16). The footing is designed for the average earth pressure on, say, 0.5 m length at the end subjected to maximum earth pressure. Define the following variables:

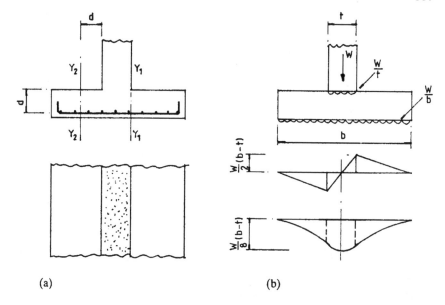

Fig.11.15 (a) Wall and footing integral; (b) wall and footing separate.

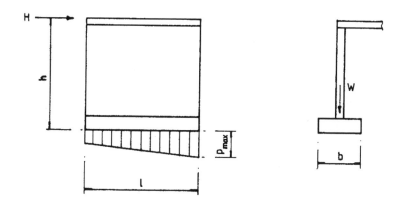

Fig 11.16 Shear wall footing.

W = total load on the base
H = horizontal load at the top of the wall
h = height of the wall
b = width of the base
ℓ = length of the wall and base
base area $A = b\ell$
section modulus $Z = b\ell^2/6$
maximum pressure $= W/A + H\,h/Z$

If the footing is on firm ground and is sufficiently deep so that the underside of the base lies within 45° dispersal lines from the face of the wall, reinforcement need not be provided. However, it would be very advisable to provide at least minimum reinforcement at the top and bottom of the footing to control cracking in case some settlement should occur.

11.4.3 Strip Footing

A continuous strip footing is used under closely spaced rows of columns as shown in Fig.11.17 where individual footings would be close together or overlap.

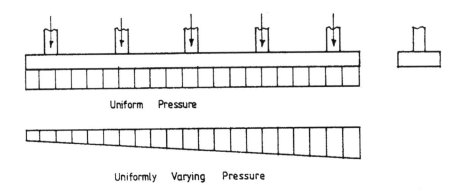

Fig.11.17 Continuous strip footing.

If the footing is concentrically loaded, the pressure is uniform. If the column loads are not equal or not uniformly spaced and the base is assumed to be rigid, moments of the loads can be taken about the centre of the base and the pressure distribution can be determined assuming that the pressure varies uniformly. These cases are shown in Fig.11.17.

In the longitudinal direction, the footing may be analysed for moments and shears by the following methods.

1. Assume a rigid foundation. Then the shear at any section is the algebraic sum of the column forces acting down and the base pressure acting up on one side of the section, and the moment at the section is the corresponding sum of the moments of the forces on one side of the section;
2. A more accurate analysis may be made if the flexibility of the footing and the assumed elastic response of the soil are taken into account. The footing is analysed as a so called beam on an elastic foundation.

In the transverse direction the base may be designed along lines similar to that for a pad footing.

11.4.4 Combined Bases

Where two columns are close together and separate footings would overlap, a combined base can be used as shown in Fig.11.18 (a). Again, if one column is close to an existing building or sewer it may not be possible to design a single pad footing, but if it is combined with that of an adjacent footing a satisfactory base can engineered. This is shown in Fig.11.18 (b).

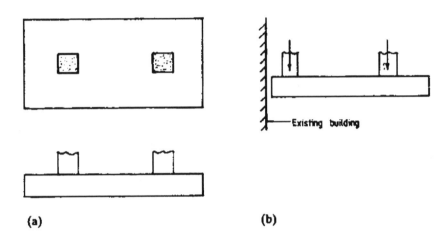

(a)   **(b)**

Fig.11.18 (a) Combined base; (b) column close to existing building.

If possible, the base is arranged so that its centreline coincides with the centre of gravity of the loads because this will give a uniform pressure on the soil. In a general case with an eccentric arrangement of loads, moments of forces are taken about the centre of the base and the maximum soil pressure is determined from the total vertical load and moment at the underside of the base. The pressure is assumed to vary uniformly along the length of the base.

In the longitudinal direction the actions for design may be found from statics. At any section the shear is the algebraic sum of the forces and the moment the sum of the moments of all the forces on one side of the section. In the transverse direction, the critical moment and shear are determined in the same way as for a pad footing. Punching shears at the column face and at 1.5 times the effective depth from the column face must also be checked.

11.4.4.1 Example of design of a combined base

(a) Specification
Design a rectangular base to support two columns carrying the following loads:
Column 1: Dead load = 310 kN, imposed load = 160 kN
Column 2: Dead load = 430 kN, imposed load = 220 kN

The columns are each 350 mm square and are spaced at 2.5 m centres. The width of the base is not to exceed 2.0 m. The safe bearing pressure on the ground is 160 kN/m². The materials are grade C35 concrete and grade 460 reinforcement.

(b) Base arrangement and soil pressure
Assume the weight of the base is 130 kN. Various load conditions are examined.

(i) Case 1: Dead + imposed load on both columns
$$\text{total vertical load} = (310 + 160) + (430 + 220) + 130 = 1250 \text{ kN}$$
$$\text{area of base} = 1250/160 = 7.81 \text{ m}^2$$
$$\text{length of base} = 7.81/2.0 = 3.91 \text{ m}$$
Choose 4.5 m × 2.0 m × 0.6 m deep base.
$$\text{The weight of the base is } (4.5 \times 2.0 \times 0.6 \times 24) = 129.6 \text{ kN.}$$
$$\text{area} = 4.5 \times 2.0 = 9.0 \text{ m}^2$$
$$\text{section modulus} = 2.0 \times 4.5^2/6 = 6.75 \text{ m}^3$$
The base is arranged so that the centre of gravity of the loads coincides with the centreline of the base, in which case the base pressure will be uniform. This arrangement will be made for the maximum ultimate loads.
The ultimate loads are
$$\text{column 1:load} = 1.4 \times 310 + 1.6 \times 160 = 690 \text{ kN}$$
$$\text{column 2: load} = 1.4 \times 430 + 1.6 \times 220 = 954 \text{ kN}$$
The distance of the centre of gravity from column 1 is
$$x = (954 \times 2.5)/ (690 + 954) = 1.45 \text{ m}$$
The base arrangement is shown in Fig.11.19.

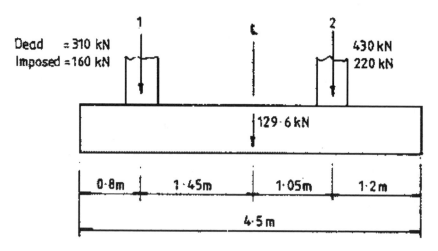

Fig.11.19 Combined base dimensions and column loads.

The soil pressure is checked for service loads for case 1:
$$\text{direct vertical load} = 310 + 160 + 430 + 220 + 129.6 = 1249.6 \text{ kN}$$

Since the centroid of the loads does not exactly coincide with the centroid of the base, check for maximum pressure which is non-uniform. The moment about the centreline of the base is

$$M = (430 + 220) \times 1.05 - (310 + 160) \times 1.45 = 1.0 \text{ kNm}$$
$$\text{maximum pressure} = 1249.6/9.0 + 1.0/6.75 = 138.9 \text{ kN/m}^2 < 160.0$$

Maximum pressure towards the column 2 side.

(ii) Case 2: Column 1, dead + imposed load; column 2, dead load

$$\text{direct load} = (310+160) + (430 + 0) + 129.6 = 1029.6 \text{ kN}$$
$$\text{moment} = M = (430+0) \times 1.05 - (310 + 160) \times 1.45 = -230 \text{ kNm}$$
$$\text{maximum pressure} = 1029.6/9.0 + 230.0/6.75 = 148.5 \text{ kN/m}^2 < 160.0 \text{ kN/m}^2$$

Maximum pressure towards the column 1 side.

(iii) Case 3: Column 1: dead load; column 2: dead + imposed load

$$\text{direct load} = (310+0) + (430 + 220) + 129.6 = 1089.6 \text{ kN}$$
$$\text{moment} = M = (430+220) \times 1.05 - (310 + 0) \times 1.45 = 233 \text{ kN m}$$
$$\text{maximum pressure} = 1089.6/9.0 + 233.0/6.75 = 155.6 \text{ kN/m}^2 < 160.0 \text{ kN/m}^2$$

Maximum pressure towards the column 2 side.
The base is satisfactory with respect to soil pressure.

(c) Analysis for actions in longitudinal direction

The cover is 40 mm, and the bars, say, 20 mm in diameter, giving an effective depth d of 550 mm. Using the Macaulay bracket notation, the shear force V and moment M in the longitudinal direction due to ultimate loads are calculated by statics.

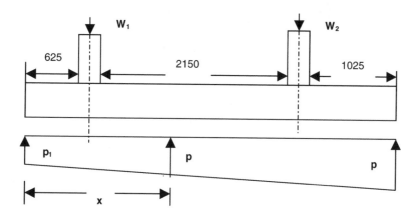

Fig.11.20 Notation Macaulay bracket formulae.

The notation used is shown in Fig.11.20 where
p_1 and p_2 are the base pressure at left and right hand ends respectively
p base pressure at a distance $\times$ from left hand end
W_1 and W_2 are the column loads.

$$p = p_1 + (p_2 - p_1)\frac{x}{4.5}$$

$$V = 2\{\frac{(p_1 + p)}{2}x\} - W_1\langle x - 0.8\rangle^0 - W_2\langle x - 3.3\rangle^0$$

$$M = 2\{p_1\frac{x^2}{2} + \frac{(p - p_1)}{2}\frac{x^2}{3}\} - W_1\langle x - 0.8\rangle - W_2\langle x - 3.3\rangle$$

The maximum design moments are at the column face and between the columns, and maximum shears are at d from the column face. Calculations are best done using spreadsheets. The load cases are as follows. The weight of the base is ignored as the corresponding base pressure will cancel each other out.

Case 1: Maximum load on both columns

W_1: 1.4 dead + 1.6 imposed = 690 kN
W_2: 1.4 dead + 1.6 imposed = 954 kN
$W_1 + W_2 = 690+954 = 1644.0$ kN,
moment $M = 954 \times 1.05 - 690 \times 1.45 = 1.20$ kN m
$p_1 = 1644.0/9.0 - 1.2/6.75 = 182.48$ kN/m^2
$p_2 = 1644.0/9.0 + 1.2/6.75 = 182.84$ kN/m^2

The results are shown in Table 11.1. Fig. 11.21 and Fig.11.22 show respectively the shear force and bending moment diagrams.

Table 11.1 Pressure, shear and moment calculation for case 1.

x	p	V	M	Remarks
0.075	182.49	27.37	1.03	d from left face of column 1
0.625	182.53	228.13	71.29	Left face of column 1
0.975	182.56	−334.09	52.75	Right face of column 1
1.525	182.60	−133.25	−75.78	d from right face of column 1
1.89	182.63	0	**−100.08**	Maximum negative moment
2.575	182.69	**250.30**	−14.34	d from left face of column 2
3.125	182.73	451.28	178.59	Left face of column 2
3.475	182.76	−374.80	**191.98**	Right face of column 2
4.025	182.80	−173.74	41.13	d from right face of column 2

Design values: Shear force = 250.30 kN, Moment = 191.98 kNm and −101.08 kNm

Case 2: Maximum load on column 1 and minimum load on column 2

W_1: 1.4 dead + 1.6 imposed = 690 kN
W_2: 1.0 dead = 430 kN
$W_1 + W_2 = 690+430 = 1120.0$ kN,
moment $M = 340 \times 1.05 - 690 \times 1.45 = -549.0$ kN m
$p_1 = 1120.0/9.0 + 549/6.75 = 205.78$ kN/m^2
$p_2 = 1120.0/9.0 - 549.0 /6.75 = 43.11$ kN/m^2

The results are shown in Table 11.2. Fig. 11.23 and Fig.11.24 show respectively the shear force and bending moment diagrams.

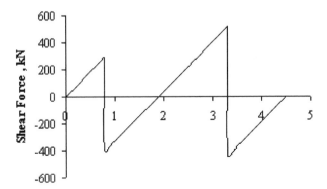

Fig.11.21 Shear force diagram for case 1.

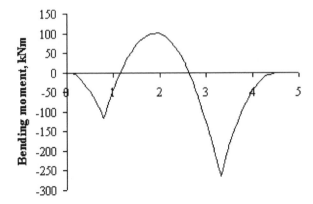

Fig.11.22 Bending moment diagram for case 1.

Case 3: Minimum load on column 1 and maximum load on column 2

W_1: 1.0 dead = 310 kN

W_2: 1.4 dead + 1.6 imposed = 954 kN

$W_1 + W_2 = 310+954 = 1264.0$ kN

moment $M = 954 \times 1.05 - 310 \times 1.45 = 552.2$ kNm

$p_1 = 1264.0/9.0 - 552.2/6.75 = 58.66$ kN/m^2

$p_2 = 1264.0/9.0 + 552.2/6.75 = 222.25$ kN/m^2

The results are shown in Table 11.3. Fig.11.25 and Fig.11.26 show respectively the shear force and bending moment diagrams.

Table 11.2 Pressure, shear and moment calculation for case 2.

x	p	V	M	Remarks
0.075	203.69	30.66	1.15	d from left face of column 1
0.625	183.19	243.10	*77.44*	Left face of column 1
0.975	170.54	−323.09	63.70	Right face of column 1
1.525	150.66	*−146.44*	−64.42	d from right face of column 1
2.04	131.92	0	*−101.52*	Maximum negative moment
2.575	112.70	130.08	−66.03	d from left face of column 2
3.125	92.82	243.11	37.60	Left face of column 2
3.475	80.16	−126.35	58.29	Right face of column 2
4.025	60.28	−49.11	11.04	d from right face of column 2

Design values: Shear force = 146.44 kN, Moment = 77.44 kNm and −101.52 kNm

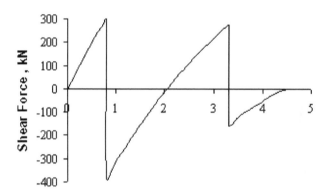

Fig.11.23 Shear force diagram for case 2.

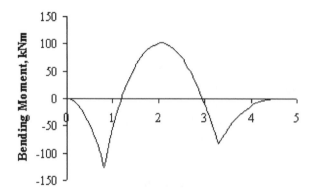

Fig.11.24 Bending moment diagram for case 2.

Table 11.3 Pressure, shear and moment calculation for case 3.

x	p	V	M	Remarks
0.075	61.39	9.00	0.34	d from left face of column 1
0.625	81.38	87.53	25.87	Left face of column 1
0.975	94.11	−161.06	12.75	Right face of column 1
1.525	114.10	−46.54	−45.35	d from right face of column 1
1.73	121.29	0	*−50.00*	Maximum negative moment
2.575	152.27	*233.14*	45.60	d from left face of column 2
3.125	172.26	411.64	*221.91*	Left face of column 2
3.475	184.99	−417.32	220.65	Right face of column 2
4.025	204.98	−202.84	49.10	d from right face of column 2

Design values: Shear force = 233.14 kN, Moment = 221.91 kNm and −50.0 kNm.

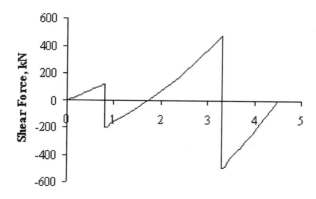

Fig.11.25 Shear force diagram for case 3.

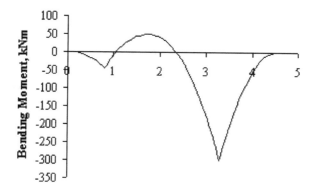

Fig 11.26 Bending moment diagram for case 3.

(d) Design of longitudinal reinforcement

(i) Bottom Steel
The maximum moment from case 3 is (Fig.11.26)
$$M = 221.91 \text{ kNm}$$
$$k = 221.91 \times 10^6/ (2000 \times 550^2 \times 35) = 0.011 < 0.156$$
$$z/d = 0.5 + \sqrt{(0.25 - 0.011/0.9)} = 0.99 > 0.95$$
$$A_s = 221.91 \times 10^6/ (0.95 \times 550 \times 0.95 \times 460) = 971.9 \text{ mm}^2$$
The minimum area of reinforcement is
$$(0.13/100) \times 2000 \times 600 = 1560 \text{ mm}^2 > \text{required steel}.$$
Provide minimum reinforcement.
Provide 16T12-125 mm centres to give a total area of 1808 mm².
$$(\ell_c = 1000 \text{ mm}) < \{0.75(c + 3d) = 0.75(350+ 3 \times 550) = 1500 \text{ mm}\}$$
Reinforcement should be spread uniformly across the width.

(ii) Top steel
The maximum moment from case 2 is (Fig.11.24),
$$M = 101.53 \text{ kNm}$$
Provide minimum reinforcement as above in each direction.

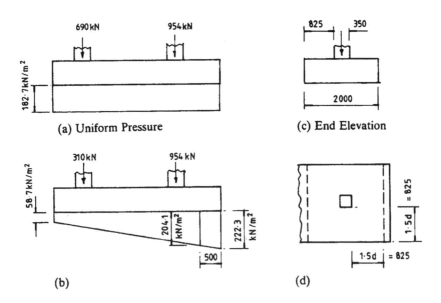

(a) Uniform Pressure   (c) End Elevation

(b)   (d)

Fig.11.27 (a) Uniform pressure; (b) case 3, varying pressure; (c) end elevation; (d) punching shear.

(e) Transverse reinforcement
As shown in Fig.11.27 (b), at ULS the maximum pressures under the base for load case 3 is 222.3 kN/m².

The pressure at 0.5 m from the end is
$$= 58.66 + (222.25 - 58.66) \times 4.0/4.5 = 204.1 \text{ kN/m}^2$$
The average pressure on a 0.5 m length at the heavier end is
$$= (222.25 + 204.1)/2 = 213.2 \text{ kN/m}^2$$
The moment at the face of the columns on a 0.5 m length at the heaviest loaded end is
$$M = \{213.2 \times (0.5 \times 0.825) \times 0.825/2 = 36.3 \text{ kNm}$$
$$k = 36.3 \times 10^6 / (500 \times 550^2 \times 35) = 0.007 < 0.156$$
$$z/d = 0.5 + \sqrt{(0.25 - 0.007/0.9)} = 0.99 > 0.95$$
$$A_s = 36.3 \times 10^6 / (0.95 \times 550 \times 0.95 \times 460) = 159.0 \text{ mm}^2$$
Provide minimum reinforcement in the transverse direction over the length of the base.

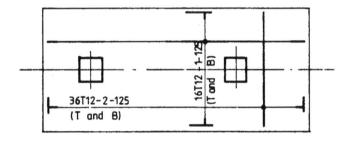

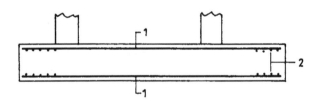

Fig.11.28 Reinforcement in the base slab.

(f) Vertical shear
The maximum vertical shear from case 1 is
$$V = 250.3 \text{ kN}$$
$$v = 250.3 \times 10^3 / (2000 \times 550) = 0.23 \text{ N/mm}^2$$
$$100A_s/(bd) = 100 \times 1808/(2000 \times 550) = 0.164 < 3.0$$
$$400/d = 400/450 < 1.0. \text{ Take as } 1.0.$$
$$v_c = 0.79 \times (0.164)^{1/3} (1.0)^{1/4} (35/25)^{1/3}/1.25 = 0.39 \text{ N/mm}^2$$
$$v < v_c$$
No shear reinforcement is required.

(g) Punching shear
Punching shear is checked in a perimeter at 1.5d from the face of a column. The perimeter for punching shear which just touches the sides of the base is shown in Fig.11.27 (d). The punching shear is less critical than the vertical shear in this case.

(h) Sketch of reinforcement
The reinforcement is shown in Fig.11.28. A complete mat has been provided at the top and bottom. Some U-spacers are required to fix the top reinforcement in position.

11.5 PILED FOUNDATIONS

11.5.1 General Considerations

When a solid bearing stratum such as rock is deeper than about 3 m below the base level of the structure a foundation supported on end-bearing piles will provide an economical solution. Foundations can also be carried on friction piles by skin friction between the pile sides and the soil where the bedrock is too deep to obtain end bearing.
The main types of piles are as follows (Fig.11.29 (a)):

1. Precast reinforced or prestressed concrete piles driven into the required position;
2. Cast-in-situ reinforced concrete piles placed in holes formed either by
        (a) driving a steel tube with a plug of dry concrete or packed aggregate at the end into the soil or;
        (b) boring a hole and lowering a steel tube to follow the boring tool as a temporary liner. (Other methods are also used. See Foundation Analysis and Design by Joseph E Bowles, 5th Edition, 1995, McGraw-Hill.)

A reinforcement cage is inserted and the tube is withdrawn after the concrete is placed.
Short bored plain concrete piles are used for light loads such as carrying ground beams to support walls. Deep cylinder piles are used to carry large loads and can be provided under basement and raft foundations. A small number of cylinder piles can give a more economical solution than a large number of ordinary piles.
The safe load that a pile can carry can be determined by

1. test loading a pile;
2. using a pile formula that gives the resistance calculated from the energy of the driving force and the final set or penetration of the pile per blow

In both cases the ultimate load is divided by a factor of safety of from 2 to 3 to give the safe load. Safe loads depend on the size and depth and whether the pile is of

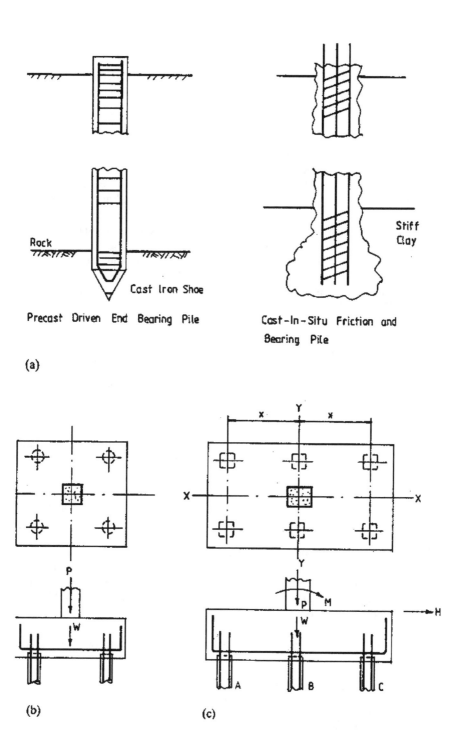

Rock

Cast Iron Shoe

Precast Driven End Bearing Pile

Stiff Clay

Cast-In-Situ Friction and Bearing Pile

(a)

P

W

(b)

X

Y

X

Y

P

M

H

W

A

B

C

(c)

Fig.11.29 (a) Pile types; (b) small pile cap, vertical load; (c) pile group resisting axial load and moment.

the end-bearing or friction type. The pile can be designed as a short column if lateral support from the ground is adequate. However, if ground conditions are unsatisfactory it is better to use test load results. The group action of piles should be taken into account because the group capacity can be considerably less than the summed capacities of the individual piles.

The manufacture and driving of piles are carried out by specialist firms who guarantee to provide piles with a given bearing capacity on the site. Safe loads for precast and *cast-in-situ* piles vary from 100 kN to 1500 kN. Piles are also used to resist tension forces and the safe load in withdrawal is often taken as one-third of the safe load in bearing. Piles in groups are generally spaced at 0.8–1.5 m apart. Sometimes piles are driven at an inclination to resist horizontal loads in poor ground conditions. Rakes of 1 in 5 to 1 in 10 are commonly used in building foundations.

In an isolated foundation, the pile cap transfers the load from the column shaft to the piles in the group. The cap is cast around the tops of the piles and the piles are anchored into it by projecting bars. Some arrangements of pile caps are shown in Figs 11.29(b) and 11.29(c).

11.5.2 Loads in Pile Groups

In general pile groups are subjected to axial load, moment and horizontal loads. The pile loads are as follows.

(a) Axial load
When the load is applied at the centroid of the group it is assumed to be distributed uniformly to all piles by the pile cap, which is taken to be rigid. This gives the load per pile

$$F_a = (P + W)/N$$

where P is the axial load from the column, W is the weight of the pile cap and N is the number of piles (Fig.11.29(b)).

(b) Moment on a group of vertical piles
The pile cap is assumed to rotate about the centroid of the pile group and the pile loads resisting moment vary uniformly from zero at the centroidal axis to a maximum or minimum for the piles farthest away. Referring to Fig.11.29(c), the second moment of area about the YY axis is

$$I_y = 2(x^2 + x^2) = 4x^2$$

where x is the pile spacing. The maximum load due to moment on piles A in tension and C in compression is

$$F_m = \pm \frac{M\,x}{I_y} = \pm \frac{M}{4x}$$

For a symmetrical group of piles spaced at $\pm x_1, \pm x_2 \ldots \pm x_n$ perpendicular to the centroidal axis YY, the second moment of area of the piles about the YY axis is

$$I_y = 2(x_1^2 + x_2^2 + ... + x_n^2)$$

The maximum pile load is

$$F_m = \pm \frac{M x_n}{I_y}$$

If the pile group is subjected to bending about both the XX and YY axes moments of inertia are calculated for each axis. The pile loads from bending are calculated for each axis as above and summed algebraically to give the resultant pile loads. The loads due to moment are combined with those due to vertical load.

(c) Horizontal load
In building foundations where the piles and pile cap are buried in the soil, horizontal loads can be resisted by friction, adhesion and passive resistance of the soil. Ground slabs that tie foundations together can be used to resist horizontal reactions due to rigid frame action and wind loads by friction and adhesion with the soil and so can relieve the pile group of horizontal load. However, in the case of isolated foundations in poor soil conditions where the soil may shrink away from the cap in dry weather or in wharves and jetties where the piles stand freely between the deck and the sea bed, the piles must be designed to resist horizontal load.

Pile groups resist horizontal loads by

1. bending in the piles
2. using the horizontal component of the axial force in inclined piles

These cases are discussed below.

(d) Pile in bending
A group of vertical piles subjected to a horizontal force *H* applied at the top of the piles is shown in Fig.11.30. The piles are assumed to be fixed at the top and bottom. The deflection of the pile cap is shown in 11.30(b).

Shear per pile $V = H/N$
Moment M_1 in each pile $= Hh_1/(2N)$

where N is the number of piles and h_1 is the length of pile between fixed ends.

The horizontal force is applied at the top of the pile cap of depth h_2 and this causes a moment Hh_2 at the pile tops. When vertical load and moment are also applied the resultant pile loads are a combination of those caused by the three actions. The total vertical load $P + W$ is distributed equally to the piles. The total moment $M + Hh_2$ is resisted by vertical loads in the piles and the analysis is carried out as set out in 11.5.2(b) above. The pile is designed as a reinforced concrete column subjected to axial load and moment. If the pile is clear between the cap and ground, additional moment due to slenderness may have to be taken into account. If the pile is in soil, complete or partial lateral support may be assumed.

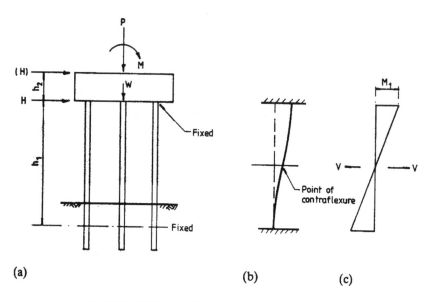

Fig.11.30 (a) Pile group; (b) deflection; (c) moment diagram.

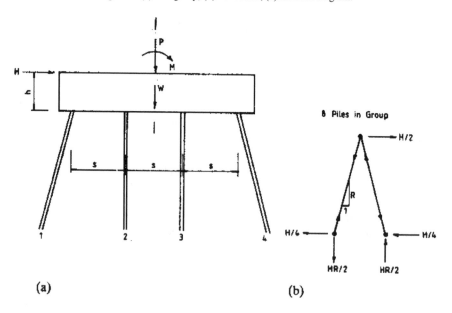

Fig.11.31 (a) Pile group; (b) resistance to horizontal load.

(e) Resistance to horizontal load by inclined piles

An approximate method used to determine the loads in piles in a group subjected to axial load, moment and horizontal load where the horizontal load is resisted by inclined piles is set out. In Fig.11.31 the foundation carries a vertical load P, moment M and horizontal load H. The weight of the pile cap is W.

The loads F in the piles are calculated as follows.

(i) Vertical loads, pile loads F_v

The sum of vertical loads is P+W.

$$F_{v2} = F_{v3} = \frac{(P+W)}{8}, \quad F_{v1} = F_{v4} = \frac{(P+W)}{8} \frac{\sqrt{(R^2+1)}}{R}$$

(ii) Horizontal loads, pile loads F_H

The horizontal load is assumed to be resisted by pairs of inclined piles as shown in Fig.11.31 (b). The sum of the horizontal loads is H.

$$F_{H2} = F_{H3} = 0, \quad -F_{H1} = F_{H4} = \frac{H}{4}\sqrt{(R^2+1)}$$

(iii) Moments, pile loads F_M

The second moment of area is
$$I_y = 2[(0.5S)^2 + (1.5S)^2]$$
The sum of the moments is
$$M' = M + Hh$$

$$-F_{M2} = F_{M3} = \frac{0.55S\,M'}{I_y}, \quad -F_{M1} = F_{M4} = \frac{1.55S\,M'}{I_y}\frac{\sqrt{(R^2+1)}}{R}$$

The maximum pile load is $F_{v4} + F_{H4} + F_{M4}$

11.5.2.1 Example of loads in pile group

The analysis using the approximate method set out above is given for a pile group to carry the loads and moment from a 6m long shear wall similar to the one designed in Chapter 10, section 10.3.4.1.

The design actions for service loads are assumed to be as follows:

Axial load = 9592 kN,

Moment = 5657 kNm,

Horizontal load = 281 kN

The proposed pile group consisting of 18 piles inclined at 1 in 6 as shown in Fig.11.32.

The weight of the base is 610 kN. For the vertical loads F_{V1} to F_{V6}

$$\frac{(9592+610)}{16}\frac{\sqrt{(6^2+1)}}{6} = 574.6\,kN$$

For the horizontal loads

$$-F_{HI} = -F_{H2} = -F_{H3} = F_{H4} = F_{H5} = F_{H6} = 281 \times \sqrt{(6^2 + 1)}/18 = 94.9 \text{ kN}$$

The second moment of area is

$$I_y = 3 \times 2(0.6^2 + 1.8^2 + 3.0^2) = 76.6$$

The moment at the pile top is

$$5657 + (281 \times 1.2) = 5994 \text{ kNm}$$

$$-F_{M1} = F_{M6} = \frac{5994 \times 3}{76.6} \frac{\sqrt{(6^2 + 1)}}{6} = 238 \text{ kN}$$

$$-F_{M2} = F_{M5} = \frac{5994 \times 1.8}{76.6} \frac{\sqrt{(6^2 + 1)}}{6} = 142.8 \text{ kN}$$

$$-F_{M3} = F_{M4} = \frac{5994 \times 0.6}{76.6} \frac{\sqrt{(6^2 + 1)}}{6} = 47.6 \text{ kN}$$

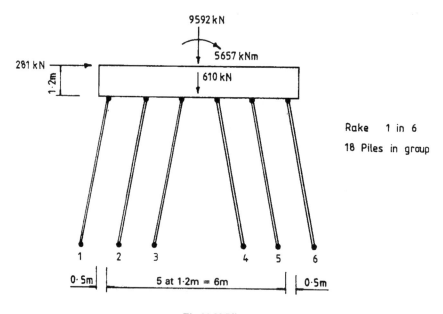

Fig.11.32 Pile group

The maximum pile load is

$$F_6 = 574.6 + 94.9 + 238 = 907.5 \text{ kN}$$

The pile group and pile cap shown in Fig.11.32 can be analysed using a plane frame computer program. The large size of the cap in comparison with the piles ensures that it acts as a rigid member. The pile may be assumed to be pinned or fixed at the ends.

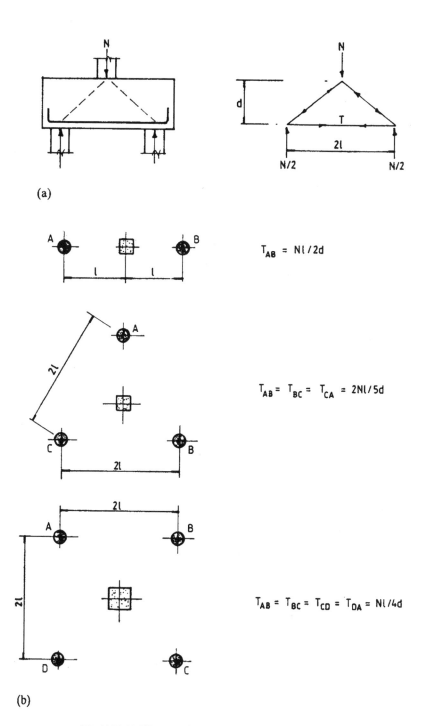

$T_{AB} = Nl/2d$

$T_{AB} = T_{BC} = T_{CA} = 2Nl/5d$

$T_{AB} = T_{BC} = T_{CD} = T_{DA} = Nl/4d$

Fig 11.33 (a) Pile cap and truss; (b) tensile forces in pile caps

11.5.3 Design of Pile Caps

The design of pile caps is covered in section 3.11.4 of the code. The design provisions are as follows.

(a) General
Pile caps are designed either using bending theory or using the truss analogy also called the Strut–Tie method. When the Strut–Tie method is used, the truss should be of triangulated form with a node at the centre of the loaded area. The lower nodes are to lie at the intersection of the centrelines of the piles with the tensile reinforcement. Tensile forces in pile caps for some common cases are shown in Fig.11.33.

(b) Strut–Tie method with widely spaced piles
Where the spacing exceeds 3 times the pile diameter only reinforcement within 1.5 times the diameter from the centre of the pile should be considered to form a tension member of a truss.

(c) Shear forces
The shear strength of a pile cap is normally governed by the shear on a vertical section through a full width of the cap. The critical section is taken at 20% of the pile diameter inside the face of the pile as shown in Fig.11.34. The whole of the force from the piles with centres lying outside this line should be considered.

(d) Design shear resistance
The shear check may be made in accordance with provisions for the shear resistance of solid slabs given in clauses 3.5.5 and 3.5.6 of the code. The following limitations apply with regard to pile caps.

1. The distance a_v from the face of the column to the critical shear plane is as defined in 11.5.3(c) above. The enhanced shear stress is $2dv_c/a_v$
where v_c is taken from Table 3.8 of the code. The maximum shear stress at the face of the column must not exceed $0.8\sqrt{f_{cu}}$ or 5 N/mm^2.
2. Where the pile spacing is less than or equal to three times the pile diameter ϕ, the enhancement can be applied over the whole of the critical section. Where the spacing is greater, the enhancement can only be applied to strips of width equal to 3ϕ centred on each pile. Minimum stirrups are not required in pile caps where $v < v_c$ (enhanced if appropriate).
3. The tension reinforcement should be provided with a full anchorage in accordance with section 3.12.8 of the code. The tension bars must be anchored by bending them up the sides of the pile cap.

(e) Punching shear
The following two checks are required:

1. The design shear stress on the perimeter of the column is not to exceed $0.8\sqrt{f_{cu}}$ or 5 N/mm^2;

2. If the spacing of the piles is greater than 3 times the pile diameter, punching shear should be checked on the perimeter shown in Fig.11.34.

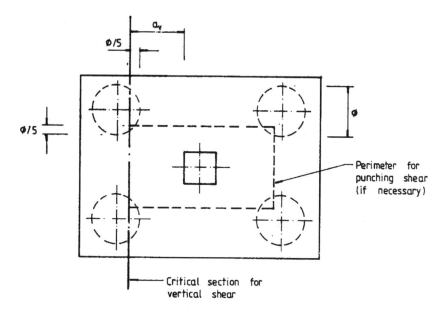

Fig.11.34 Critical shear section and perimeter.

11.5.3.1 Example of design of pile cap

Design a four–pile cap to support a factored axial column load of 3600 kN. The column is 400 mm × 400 mm and the piles are 400 mm in diameter spaced at 1200 mm centres. The materials are grade C35 concrete and grade 460 reinforcement. The pile cap is shown in Fig.11.35. The overall depth is taken as 900 mm and the effective depth as 800 mm.

(a) Moment reinforcement
Refer to Fig.11.33 for a four–pile cap
$$\text{Tension } T = (3600 \times 600)/(4 \times 800) = 675 \text{ kN}$$
$$A_s = 675 \times 10^3/(0.95 \times 460) = 1545 \text{ mm}^2 \text{ per tie.}$$
Provide 12T20 bars, area = 3770 mm^2, for reinforcement across the full width for two ties.
The minimum reinforcement from Table 3.26 of the code is
$$(0.13/100) \times 1900 \times 900 = 2223 \text{ mm}^2 < 3770 \text{ mm}^2 \text{ provided.}$$

The pile spacing is not greater than three times the pile diameter, so the reinforcement can be spaced evenly across the pile cap. The spacing is 160 mm.

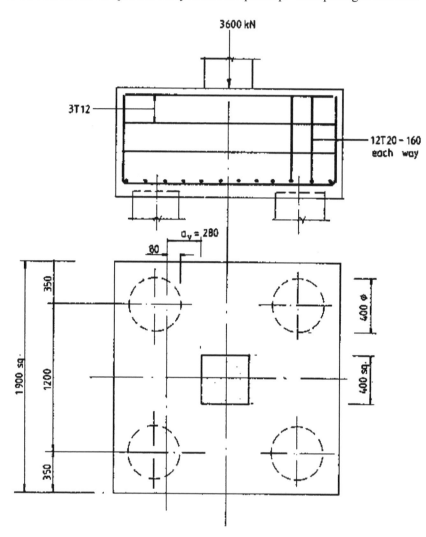

Fig.11.35 Reinforcement in the pile cap.

(b) Shear

The critical section for shear lies $\phi/5$ inside the pile as shown in Fig.11.35, giving $a_v = 280$ mm. The shear V is

$$V = 1800 \text{ kN}.$$
$$v = 1800 \times 10^3/ (1900 \times 800) = 1.18 \text{ N/mm}^2$$
$$100A_s/ (bd) = 100 \times 3770/ (1900 \times 800) = 0.25 < 3.0$$
$$400/d = 400/800 < 1.0, \text{ take as } 1.0.$$

$$v_c = 0.79 \times (2.48)^{1/3} (1.0)^{1/4} (35/25)^{1/3} / 1.25 = 0.44 \text{ N/mm}^2$$

The enhanced design shear stress is
$$= v_c (2 \, d/a_v)$$
$$= 0.44 \times 2 \times 800/280 = 2.53 \text{ N/mm}^2$$
The shear stress at the critical section is satisfactory for vertical shear and no shear reinforcement is required.

Check the shear stress on the perimeter of the column:
$$v = 3600 \times 10^3 / (1600 \times 800) = 2.81 \text{ N/mm}^2 < (0.8\sqrt{35} = 4.73 \text{ N/mm}^2)$$
This is satisfactory. The pile spacing is not greater than 3 times the pile diameter and so no check for punching shear is required.

(c) Arrangement of reinforcement
The anchorage of the main bars is 38 diameters (760 mm), from Table 3.27 of the code. Since the spacing of 160 mm does not exceed 750 mm it is satisfactory. Secondary steel, 12 mm diameter bars, is required on the sides of the pile cap. The reinforcement is shown in Fig.11.35.

11.7 REFERENCES

Bowles, Joseph. E. (1995), *Foundation analysis and design.*, 5th Edition, (McGraw-Hill).

CHAPTER 12

RETAINING WALLS

12.1 WALL TYPES AND EARTH PRESSURE

12.1.1 Types of Retaining Wall

Retaining walls are structures used to retain mainly earth but also other materials which would not be able to stand vertically unsupported. The wall is subjected to overturning due to pressure of the retained material.

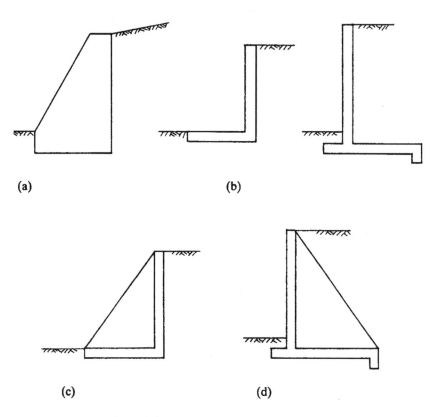

(a)   (b)

(c)   (d)

Fig.12.1 (a) Gravity wall; (b) cantilever walls; (c) buttress wall; (d) counterfort wall.

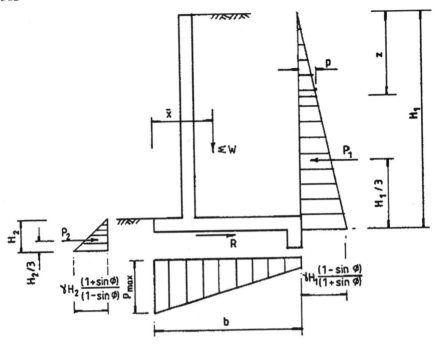

Fig 12.2 (a) Earth pressure: cohesion less soil (c = 0).

The types of retaining wall are as follows:

1. In a **gravity wall** stability is provided by the weight of concrete in the wall;
2. In a **cantilever wall** the wall slab acts as a vertical cantilever. Stability is provided by the weight of structure and earth on an inner base or the weight of the structure only when the base is constructed externally;
3. In **counterfort and buttress walls** the vertical slab is supported on three sides by the base and counterforts or buttresses. Stability is provided by the weight of the structure in the case of the buttress wall and by the weight of the structure and earth on the base in the counterfort wall.

Examples of retaining walls are shown in Fig.12.1. Detailed designs for cantilever and counterfort retaining walls are given.

12.1.2 Earth Pressure on Retaining Walls

(a) Active soil pressure
The relevant code for calculating earth pressure on retaining walls is
BS 8002: 1994: *Code of Practice for earth retaining structures*

Active soil pressures are given for the two extreme cases of soil such as a cohesionless soil like sand and a cohesive soil like clay (Fig.12.2). General formulae are available for intermediate cases. The formulae given apply to drained soils and reference should be made to textbooks on soil mechanics for pressure where the water table rises behind the wall. The soil pressures given are those due to a level backfill. If there is a surcharge of q kN/m² on the soil behind the wall, this is equivalent to an additional soil depth of $z = q/\gamma$ where γ is the unit weight in kN/m³. The textbooks give solutions for cases where there is sloping backfill.

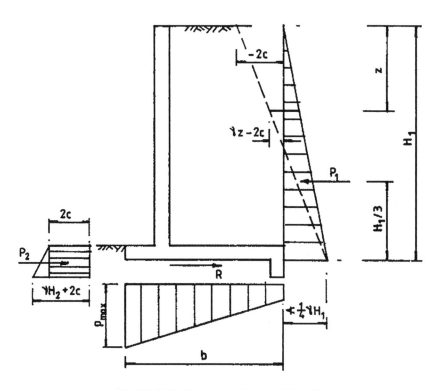

Fig 12.2 (b) Earth pressure: cohesive soil ($\phi = 0$).

(i) Cohesionless soil, c = 0 (Fig.12.2 (a)): The horizontal pressure at any depth z is given by

$$p = K_a\,(\gamma z + q)$$
$$K_a = \frac{(1 - \sin\phi)}{(1 + \sin\phi)}$$

where γ is the unit weight of soil in kN/m³, q = uniformly distributed surcharge in kN/m², ϕ is the angle of internal friction and K_a is the coefficient of active earth pressure.

The horizontal force P_1 on the wall of height H_1 is

$$P_1 = \frac{1}{2}K_a\gamma H_1^2 + K_a q H_1$$

(ii) Cohesive soil, $\phi = 0$ (Fig.12.2 (b)): The pressure at any depth z is given theoretically by

$$p = \gamma z + q - 2c$$

where c is the cohesion at zero normal pressure. This expression gives negative values near the top of the wall. In practice there are cracks at the top of the soil normally filled with water.

(b) Wall stability against overturning

Referring to Fig.12.2 the vertical loads are made up of the weight of the wall stem and base and the weight of backfill on the base. Front fill on the outer base has been neglected. Surcharge would need to be included if present. If the centre of gravity of these loads is x from the *toe* of the wall, the stabilizing moment with respect to overturning about the toe is ΣWx with a beneficial partial safety factory $\gamma_f = 1.0$. The overturning moment due to the active earth pressure is $\gamma_f P_1 H_1/3$ with an adverse partial safety factor $\gamma_f = 1.4$. The stabilizing moment from passive earth pressure has been neglected. For the wall to satisfy the requirement of stability

$$\Sigma Wx \geq \gamma_f P_1 H_1/3$$

(c) Vertical pressure under the base

The vertical pressure under the base is calculated for service loads. For a 1 m length of cantilever wall with base width b transmitting forces to the foundation

$$\text{area } A = b \text{ m}^2 \text{ , section modulus } Z = b^2/6 \text{ m}^3.$$

If ΣM is the sum of the moments of all vertical forces ΣW about the centre of the base and of the active pressure on the wall, then

$$\Sigma M = \Sigma W(x - b/2) - P_1 H_1/3$$

where x = the centre of gravity of vertical loads from the *toe* of the wall.

The passive pressure in front of the base has again been neglected.   The maximum pressure is

$$P_{max} = \frac{\Sigma W}{A} + \frac{\Sigma M}{Z}$$

This should not exceed the safe bearing pressure on the soil.

(d) Resistance to sliding (Fig.12.2)

The resistance of the wall to sliding is as follows.

(i) Cohesionless soil: The friction R between the base and the soil is $\mu \Sigma W$, where μ is the coefficient of friction between the base and the soil ($\mu = \tan \phi$). The passive earth pressure force P_2 against the front of the wall from a depth H_2 of soil is

$$P_2 = \frac{1}{2}\gamma H_2^2 \frac{(1+\sin\phi)}{(1-\sin\phi)}$$

(ii) Cohesive soils: The adhesion R between the base and the soil is βb where β is the adhesion in kN/m^2. The passive earth pressure P_2 is

$$P_2 = 0.5\gamma H_2^2 + 2cH_2$$

A downstand nib can be added, as shown in Fig.12.2 to increase the resistance to sliding through passive earth pressure.

For the wall to be safe against sliding

$$1.4\,P_1 < P_2 + R$$

where P_1 is the horizontal active earth pressure on the wall.

12.2 DESIGN OF CANTILEVER WALLS

12.2.1 Initial Sizing of the Wall

The initial dimensions of the wall can be determined from the following equations.

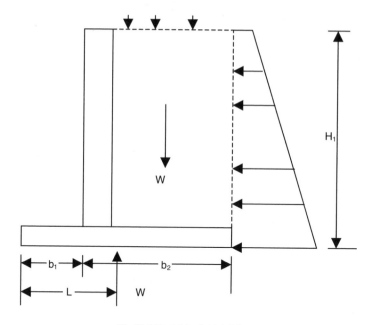

Fig.12.3 Model for initial sizing.

Ignoring the difference in unit weight between soil and concrete and the weight of the toe slab of width b_1, for a unit length of wall the total gravity load W is approximately given by

$$W = \gamma\, b_2\, H_1 + q\, b_2$$

The total horizontal force P_1 is given by

$$P_1 = 0.5\, K_a\gamma\, H_1^2 + K_a\, q\, H_1$$

where q = surcharge in kN/m^2 and K_a = coefficient of active earth pressure

(i) Resistance to sliding:

Ignoring any contribution from passive earth pressure and using a load factor γ_f equal to 1.4 on P_1 as it is an adverse load and 1.0 on W as it is a beneficial load, for resistance against sliding,

$$\mu\, W \geq 1.4\, P_1$$

Substituting for W and P_1,

$$\{\mu\, \gamma\, b_2\, H_1 + q\, b_2\} \geq 1.4\, \{0.5\, K_a\gamma\, H_1^2 + K_a q \times H_1\}$$

Simplifying,

$$\frac{b_2}{H_1}\{\mu + \frac{q}{\gamma H_1}\} \geq 1.4\, K_a\{0.5 + \frac{q}{\gamma H_1}\}$$

(ii) Zero tension in the base pressure

Taking moments about the toe of the wall,

$$W\,(b_1 + b_2/2) - 0.5\, K_a\gamma\, H_1^2\,(H_1/3) - K_a q\, H_1\,(H_1/2) = W\,L$$
$$L = b_1 + 0.5\, b_2 - b_2\, K_a\,(H_1/b_2)^2\{1/6 + q/\,(2\,\gamma H_1)\}/\,[1 + q/\,(\gamma H_1)]$$

Eccentricity e of W with respect to the centre of the base is

$$e = 0.5(b_1 + b_2) - L$$
$$e = (1/6)\, b_2\, K_a\,(H_1/b_2)^2\{1 + 3q/\,(\gamma H_1)\}/\,[1 + q/\,(\gamma H_1)] - 0.5\, b_1$$

For no tension to develop at the heel, W must lie in the middle third of the base. Therefore

$$e \leq (b_1 + b_2)/6$$

$$\frac{b_1}{b_2} \geq 0.25\, K_a\,(\frac{H_1}{b_2})^2[\frac{1 + 3\dfrac{q}{\gamma H_1}}{1 + \dfrac{q}{\gamma H_1}}] - 0.25$$

12.2.2 Design Procedure for a Cantilever Retaining Wall

For a given height of earth to be retained, the steps in the design of a cantilever retaining wall are as follows.

1. Assume a breadth for the base. This can be calculated from the equations developed in section 12.2.1. A nib is often required to increase resistance to sliding.
2. Calculate the horizontal earth pressure on the wall. Considering all forces, check stability against overturning and the vertical pressure under the base of the wall. Calculate the resistance to sliding and check that this is satisfactory. A partial safety factor γ_f of 1.4 is applied to the horizontal loads for the overturning and sliding check. The maximum vertical pressure is calculated using service loads and this should not exceed the safe bearing pressure.
3. Reinforced concrete design for the wall is made for ultimate loads using appropriate load factors. Surcharge if present may be classed as either dead or imposed load depending on its nature.

Referring to Fig.12.4 the structural design consists of the following.

(a) Cantilever wall: calculate shear forces and moments caused by the horizontal earth pressure. Design the vertical moment steel for the inner (earth side) face and check the shear stresses. Minimum secondary steel is provided in the horizontal direction for the inner face and both vertically and horizontally for the outer face.

(b) Inner footing (heel slab): The net moment due to vertical loads on the top and earth pressure on the bottom face causes tension in the top and reinforcement is designed for this position.

(c) Outer footing (toe slab): The moment due to earth pressure at the bottom face causes tension in the bottom face.

The moment reinforcement for the three parts is shown in Fig. 12.4.

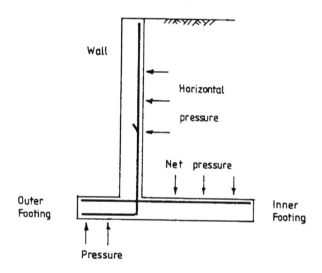

Fig.12.4 Three parts of the cantilever retaining wall.

12.2.3 Example of Design of a Cantilever Retaining Wall

(a) Specification

Design a cantilever retaining wall to support a bank of earth 3.5 m high. The top surface is horizontal behind the wall but it is subjected to a dead load surcharge of 15 kN/m². The soil behind the wall is well-drained sand with the following properties:

$$\text{Unit weight } \gamma = 17.6 \text{ kN/m}^3$$
$$\text{angle of internal friction } \phi = 30°$$

The material under the wall has a safe bearing pressure of 100 kN/m². The coefficient of friction μ between the base and the soil is 0.5. Design the wall using grade 30 concrete and grade 460 reinforcement.

Active earth pressure coefficient:
$$K_a = (1 - \sin \phi)/ (1 + \sin \phi) = (1 - 0.5)/ (1 + 0.5) = 0.3333$$

(b) Check preliminary sizing

(i) Check minimum stem thickness
For 1 m length of the wall, bending moment M at the base of the cantilever is
$$M = 0.5\, K_a \gamma\, H^2\, (H/3) + K_a\, q\, H\, (H/2)$$
Substituting $K_a = 0.333$, $\gamma = 17.6$ kN/m³, H = 3.5 m, q = 15 kN/m²,
$$M = 0.5 \times 0.3333 \times 17.6 \times 3.5^2 \times 3.5/3 + 0.3333 \times 15 \times 3.5 \times 3.5/2$$
$$M = 41.92 + 30.62 = 72.54 \text{ kN m/m}$$
In order for there to be no need for compression steel,
$$M < 0.156\, bd^2\, f_{cu}$$
Taking b = 1000 mm, $f_{cu} = 30$ N/mm²,

$$d > \left\{ \sqrt{\frac{M}{0.156 \times b \times f_{cu}}} = \sqrt{\frac{72.54 \times 10^6}{0.156 \times 1000 \times 30}} = 125 \right\}$$

Take a value of d much larger than this to reduce the amount of steel required. Assume total stem thickness of 250 mm. Same thickness is assumed for the base slab as well.

(ii) Check resistance to sliding
$$H_1 = 3.5 + 0.25 = 3.75 \text{ m},$$
$$q/ (\gamma\, H_1) = 15.0/ (17.6 \times 3.75) = 0.227$$
$$K_a = 0.333$$
$$\mu = 0.5$$
Use equation from section 12.2.1.1, to calculate width b_2.
$$\frac{b_2}{H_1}\{\mu + \frac{q}{\gamma H_1}\} \geq 1.4\, K_a\{0.5 + \frac{q}{\gamma H_1}\}$$
$$(b_2/H_1)\{0.5 + 0.227\} \geq 1.4 \times 0.333 \times (0.5 + 0.227)$$
$$b_2/H_1 \geq 0.467$$
$$b_2 \geq 1.75 \text{ m},$$
$$\text{Take } b_2 = 2.05 \text{ m},$$
$$b_2 / H_1 = 0.55$$

(iii) Check eccentricity
$$b_2 = 2.05,$$
$$H_1/b_2 = 1.829$$
$$q/ (\gamma\, H_1) = 0.227$$

$$K_a = 0.333$$

Use equation from section 12.2.1.1, to calculate width b_1.

$$\frac{b_1}{b_2} \geq 0.25 \, K_a \, (\frac{H_1}{b_2})^2 [\frac{1 + 3\,\dfrac{q}{\gamma H_1}}{1 + \dfrac{q}{\gamma H_1}}] - 0.25$$

$$(b_1/b_2) \geq 0.25 \times 0.333 \times (1.829)^2 \, [\{1 + 3 \times 0.227\} / \{1 + 0.227\}] - 0.25$$

$$(b_1/b_2) \geq 0.132,$$

$$b_1 \geq 0.27 \text{ m, Take } b_1 = 0.8 \text{ m.}$$

The proposed arrangement of the wall is shown in Fig.12.5. Wall and base thicknesses are assumed to be 250 mm. A nib has been added under the wall to assist in the prevention of sliding.

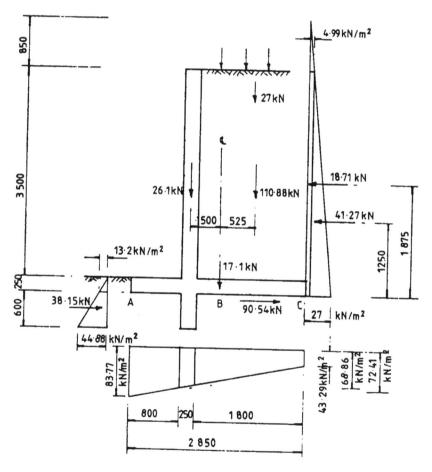

Fig. 2.5 Forces acting on the retaining wall.

(c) Wall stability

Consider 1 m length of wall. The horizontal pressure at depth z from the top is

$$p = K_a(\gamma z + q) \quad = 0.333(17.6\,z + 15)$$

The horizontal pressure at the base ($z = 3.75$ m) $= 27$ kN/m^2

The horizontal pressure at the top ($z = 0$) $= 5$ kN/m^2.

The weight of wall, base and earth and the corresponding moments about the toe of the wall for stability calculations are given in Table 12.1. Clockwise moments are taken as positive.

Table 12.1 Stability calculations (Cantilever wall)

Type of Load	Load (kN)	Distance to centroid from A, m	Moment about A (kNm)
HORIZONTAL (Active earth pressure)			
Surcharge	$5 \times 3.75 = 18.75$	$3.75/2 = 1.875$	-35.16
Triangular	$0.5 \times 3.75 \times$ $(27 - 5) = 41.25$	$3.75/3 = 1.25$	-51.56
Σ	$18.75 + 41.25 = 60.0$		$-35.16 - 51.56 =$ -86.72
VERTICAL (Gravity)			
Wall + Nib	$(3.75 + 0.6) \times 0.25 \times$ $24 = 26.1$	$0.8 + 0.25/2 = 0.925$	24.14
Base	$2.85 \times 0.25 \times 24 =$ 17.1	$2.85/2 = 1.425$	24.37
Back fill	$1.8 \times 3.5 \times 17.6 =$ 110.88	$0.8 + 0.25 + 1.8/2 =$ 1.95	216.22
Surcharge	$15 \times 1.8 = 27$	$0.8 + 0.25 + 1.8/2 =$ 1.95	52.65
Σ	181.08		317.38

(i) Maximum soil pressure

Width of base b = 2.85 m

For 1 m length of wall, area

$$A = 2.85 \text{ m}^2$$
$$\text{section modulus Z} = 2.85^2/6 = 1.35 \text{ m}^3$$

Taking moments of all forces about the toe A, the centroid of the base pressure from A is at a distance L.

$$L \times 181.08 = 317.38 - 86.72 = 230.66$$
$$L = 230.66/181.08 = 1.273 \text{ m},$$
$$\text{eccentricity, e} = B/2 - L = 2.85/2 - 1.273 = 0.15 < 2.85/6.$$

Hence no tension is developed at C.

The base is acted on by

$$\text{vertical load} = 181.08 \text{ kN}$$
$$\text{moment M} = 181.08 \times e = 27.16 \text{ kNm}.$$

The maximum soil pressure at A calculated for *service load* is
$$181.08/ (A = 2.85) + 27.16/ (Z = 1.35) = 83.7 \text{ kN/m}^2 < 100 \text{ kN/m}^2$$
This is satisfactory, as the maximum pressure is less than the safe bearing capacity of soil.

(ii) Stability against overturning
The stabilizing (beneficial) moment due to gravity loads about the toe A of the wall has a partial safety factor $\gamma_f = 1.0$ and the disturbing (adverse) moment due to horizontal loads has a partial safety factor $\gamma_f = 1.4$. The net stabilizing moment is
$$(317.38 \times 1.0 - 86.72 \times 1.4) = 195.97 > 0$$
The wall is considered as safe against overturning.

(iii) Resistance to sliding
The forces resisting sliding are the friction under the base and the passive resistance for a depth of earth of 850 mm to the top of the base. The gravity loads are beneficial loads but the horizontal load is an adverse load. Ignoring the passive pressure, for the wall to be safe against sliding
$$(\mu = 0.5) \{181.08 \times (\gamma_f = 1.0)\} > (\gamma_f = 1.4) \times 60.0,$$
$$\text{ie. } 90.54 > 84.0$$
The resistance to sliding is satisfactory. There was no need for the nib but is included for additional protection.

(iv) Overall comment
The wall section is satisfactory. The maximum soil pressure under the base controls the design.

(d) Structural design of wall, heel and toe slabs

(1) Cantilever wall slab

(i) Bending design
At serviceability limit state, the horizontal pressure at the base ($z = 3.5$ m) is 25.53 kN/m^2 and at the top ($z = 0$) is 5 kN/m^2.
$$\text{Average pressure} = 0.5 \times (25.33 + 5.0) = 15.17 \text{ kN/m}^2$$
At ultimate limit state using $\gamma_f = 1.4$, at the base of the cantilever, shear force V and moment M are
$$V = 15.17 \times 3.5 \times (\gamma_f = 1.4) = 74.8 \text{ kN}$$
$$M = \{(25.53 - 5.0) \times 0.5 \times 3.5 \times 3.5/3 + 5.0 \times 3.5 \times 3.5/2\} \times (\gamma_f = 1.4)$$
$$M = 101.56 \text{ kNm}$$
Assume that the cover is 40 mm and the diameter of the bars is 16 mm. Effective depth d is
$$d = 250 - 40 - 8 = 202 \text{ mm}$$
$$k = M/ (bd^2 f_{cu}) = 101.56 \times 10^6/ (1000 \times 202^2 \times 30) = 0.083 < 0.156$$
$$z/d = 0.5 + \sqrt{(0.25 - 0.083/0.9)} = 0.897 < 0.95$$
$$A_s = 101.56 \times 10^6/ (0.897 \times 202 \times 0.95 \times 460) = 1282 \text{ mm}^2/\text{m}$$
Provide 16 mm diameter bars at 150 mm centres to give a steel area of 1340 mm^2/m.

Check minimum percentage of steel.

$$100A_s/ (bh) = 100 \times 1340/ (1000 \times 250) = 0.54 > 0.13$$

Provided steel is greater than the minimum percentage of steel.

Check maximum spacing of steel permitted.

$$100A_s/ (bd) = 100 \times 1340/ (1000 \times 202) = 0.66$$

Maximum spacing allowed from Table 3.28 and clause 3.12.11.2.7 (b) is

$$155/ (0.66) = 235 \text{ mm}.$$

(ii) Curtailment of flexural steel Determine the depth z from the top where the 16 mm diameter bars can be reduced to a diameter of 12 mm.
12 mm bars at 150 mm centres provides an area of steel equal to

$$A_s = 753 \text{ mm}^2/\text{m}$$

The corresponding moment of resistance is approximately

$$M = 101.56 \times (753/1282) = 59.65 \text{ kNm}$$

This moment occurs at a depth z from top given by

$$59.65 = 1.4 \times K_a \, (\gamma \, z^3/6 + 15x \, z^2/2)$$
$$59.65 = 1.4 \times 0.333 \times (17.6 \times z^3/6 + 15 \times z^2/2)$$

Solving by trial and error, z = 2.84 m,

$$d = 250 - 40 - 6 = 204 \text{ mm}$$
$$M = 59.65$$
$$k = 59.65 \times 10^6/ (1000 \times 204^2 \times 30) = 0.048 < 0.156$$
$$z/d = 0.5 + \sqrt{(0.25 - 0.048/0.9)} = 0.944 < 0.95$$
$$A_s = 59.65 \times 10^6/ (0.944 \times 204 \times 0.95 \times 460) = 709 \text{ mm}^2/\text{m} < 753 \text{ mm}^2/\text{m}$$

An alternative possibility is to use 16 mm bars at 300 mm c/c. The main advantage of this system is that by curtailing alternate bars provided at the base at 150 mm, a spacing of 300 mm is obtained without any need to lap bars of a different diameter. This is a more efficient arrangement from a construction point of view.

T16 at 300 mm centres gives $A_s = 670 \text{ mm}^2/\text{m}$. The moment of resistance provided is approximately

$$M = 101.56 \times 670/1282 = 50.08 \text{ kNm}$$

This moment occurs at a depth z from top given by

$$50.08 = 1.4 \times 0.333 \times (17.6 \times z^3/6 + 15 \times z^2/2)$$

Solving by trial and error, z = 2.65 m,

$$k = 50.08 \times 10^6/ (1000 \times 204^2 \times 30) = 0.04 < 0.156$$
$$z/d = 0.5 + \sqrt{(0.25 - 0.04/0.9)} = 0.953 > 0.95$$
$$A_s = 50.08 \times 10^6/ (0.95 \times 204 \times 0.95 \times 460) = 591 \text{ mm}^2/\text{m} < 670 \text{ mm}^2/\text{m}$$

Check minimum percentage of steel.

$$100A_s/ (bh) = 100 \times 670/ (1000 \times 250) = 0.27 > 0.13$$

Check maximum spacing between bars.

$$100A_s/ (bd) = 100 \times 670/ (1000 \times 202) = 0.33$$

Maximum spacing allowed from Table 3.28 and clause 3.12.11.2.7 (b) is
$$155/ (0.33) = 470 \text{ mm} > 300 \text{ mm proposed.}$$
Referring to the anchorage requirements in BS8110: Part 1, clause 3.12.9.1, bars are to extend an anchorage length beyond the theoretical cut off point. The anchorage length from Table 3.27 of the code for grade C30 concrete is (section 5.2.1)
$$\text{anchorage length} = 40 \times 16 = 640 \text{ mm}$$
Alternate bars need to continue up to a distance from top of
$$= 2650 - 640 = 2010 \text{ mm}$$
Stop bars off bars at a distance fro base equal to
$$= 3500 - 2010 = 1490 \text{ mm, say } 1500 \text{ mm.}$$

(iii) Shear check

The shear force V at the base of the wall is
$$V = 15.17 \times 3.5 \times (\gamma_f = 1.4) = 74.8 \text{ kN}$$
$$v = 74.8 \times 10^3 / (1000 \times 202) = 0.37 \text{ N/mm}^2$$
$$100 \, A_s / (bd) = 100 \times 1340 / (1000 \times 202) = 0.66 < 3.0$$
$$400/d = 400/202 = 1.98 > 1.0, \, f_{cu} = 30 < 40.$$

$$v_c = \frac{0.79}{(\gamma_m = 1.25)} (0.66)^{\frac{1}{3}} (1.98)^{\frac{1}{4}} (\frac{30}{25})^{\frac{1}{3}} = 0.70 \text{ N/mm}^2$$

$$v < v_c$$

The shear stress is satisfactory.

(iv) Deflection and cracking

The deflection need not be checked. For control of cracking the bar spacing must not exceed the limitations in clause 3.12.11.2.7. This is satisfied.

(v) Distribution steel

For distribution steel provide the minimum area of 0.13% from Table 3.25 of the code.
$$A_s = (0.13/100) \times 1000 \times 250 = 325 \text{ mm}^2/\text{m}$$
Provide 10 mm diameter bars at 240 mm centres horizontally on the inner face. For crack control on the outer face provide 10 mm diameter bars at 240 mm centres each way.

(2) Inner footing (Heel slab)

In order to determine the appropriate load factors to be used, it is necessary to consider the effect of gravity loads and earth pressure loads on the bending moment caused in the heel slab.

From Table 12.1, gravity loads provide:
$$\text{Vertical load} = 181.08 \text{ kN,}$$
$$\text{Moment about the Toe A} = 317.38 \text{ kNm (Clockwise)}$$
The centroid of the base pressure due to *gravity loads only* from A is at a distance L.
$$L \times 181.08 = 317.38, \text{ giving } L = 1.75 \text{ m,}$$
$$\text{eccentricity, e} = 2.85/2 - L = 0.325$$

The base is acted on by a
$$\text{vertical load} = 181.08 \text{ kN}$$
$$\text{moment M} = 181.08 \times e = 58.85 \text{ kNm (clockwise)}$$
$$181.08/A = 63.54 \text{ kN/m}^2$$
$$M/Z = 43.59 \text{ kN/m}^2$$

On the top of the heel slab there is surcharge of 15 kN/m^2 and a height of soil equal to 3.5 m and self weight of slab of 250 mm. The total downward load is
$$\{15 + 3.5 \times (\gamma = 17.6) + 0.25 \times 24\} = 82.6 \text{ kN/m}^2$$

The bending moment on the base due to horizontal earth pressure is
$$M = 86.72 \text{ kNm/m (anticlockwise)}.$$
$$M/Z = 64.06 \text{ kN/m}^2$$
The pressures due to gravity loads and horizontal earth pressure are shown in Fig.12.6.

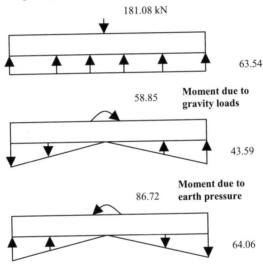

Fig.12.6 Forces on the base slab due to gravity and earth pressure forces.

The net effect of gravity loads is to produce tension on the bottom of the slab while the base pressure due to horizontal loads produces tension on the top of the slab. Therefore gravity loads are beneficial loads with a load factor of 1.0 while earth pressure loads are adverse with a load factor of 1.4 to be applied. Using these load factors, the base pressure at right and left ends of the base slab are

$$\text{Left end} = 63.54 - 43.59 + 64.06 \times 1.4 = 109.63 \text{ kN/m}^2$$
$$\text{Right end} = 63.54 + 43.59 - 64.06 \times 1.4 = 17.45 \text{ kN/m}^2$$

The base pressure at the junction of the heel slab and cantilever is
$$= 17.45 + (109.63 - 17.45) \times (1.8/2.85) = 75.67 \text{ kN/m}^2$$
Fig.12.7 shows the forces acting on the heel slab.

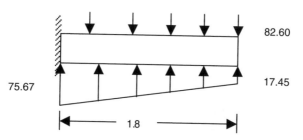

Fig.12.7 Pressures in kN/m² acting on heel slab.

(i) Bending design

Referring to Fig.12.7, moment M at the face of the wall is

$$M = 0.5 \times (82.6 - 17.45) \times 1.8^2 - 0.5 \times (75.67 - 17.45) \times 1.8 \times 1.8/3$$
$$= 74.10 \text{ kN m/m}$$
$$k = M/(bd^2 f_{cu}) = 74.10 \times 10^6/(1000 \times 202^2 \times 30) = 0.061 < 0.156$$
$$z/d = 0.5 + \sqrt{(0.25 - 0.061/0.9)} = 0.927 < 0.95$$
$$A_s = 74.10 \times 10^6/(0.927 \times 202 \times 0.95 \times 460) = 905 \text{ mm}^2/\text{m}$$

Provide 12 mm diameter bars at 120 mm centres to give 942 mm²/m

Check minimum steel percentage:

$$100A_s/(bh) = 100 \times 942/(1000 \times 250) = 0.38 > 0.13$$

Check maximum permitted spacing:
$$100A_s/(bd) = 100 \times 942/(1000 \times 202) = 0.47$$
Maximum spacing allowed $= 155/(0.47) = 330$ mm > 120 mm.
Spacing is satisfactory.

(ii) Shear Check

Referring to Fig.12.7, the shear force V at the face of the wall is

$$V = (82.6 - 17.45) \times 1.8 - 0.5 \times (75.67 - 17.45) \times 1.8 = 64.87 \text{ kN}$$
$$v = 64.87 \times 10^3/(1000 \times 202) = 0.32 \text{ N/mm}^2$$
$$100 A_s/(bd) = 100 \times 942/(1000 \times 202) = 0.47 < 3.0$$
$$400/d = 400/202 = 1.98 > 1.0,$$
$$f_{cu} = 30 < 40.$$

$$v_c = \frac{0.79}{(\gamma_m = 1.25)}(0.47)^{\frac{1}{3}}(1.98)^{\frac{1}{4}}(\frac{30}{25})^{\frac{1}{3}} = 0.62 \text{ N/mm}^2$$

The shear stress is satisfactory.

(iii) Deflection and cracking The deflection need not be checked. For control of cracking the bar spacing must not exceed the limitations in clause 3.12.11.2.7 and this is satisfied.

(iii) Distribution steel For distribution steel provide the minimum area of 0.13% from Table 3.25 of the code.

$$A_s = (0.13/100) \times 1000 \times 250 = 325 \text{ mm}^2/\text{m}$$

Provide 10 mm diameter bars at 240 mm centres horizontally on the inner face. For crack control on the outer face provide 10 mm diameter bars at 240 mm centres each way.

(3) Outer Footing (Toe slab)

As shown in Fig.12.6, both gravity and horizontal loads acting on the base slab produce tension on the bottom of the slab. Therefore both loads are adverse and take a load factor of 1.4. The only beneficial load is due to self weight. Using these load factors, the base pressure at right and left ends of the base slab are

Left end = $(63.54 - 43.59 + 64.06) \times 1.4 = 117.61 \text{ kN/m}^2$
Right end = $(63.54 + 43.59 - 64.06) \times 1.4 = 60.30 \text{ kN/m}^2$

The base pressure at the junction of the toe slab and cantilever

$= 60.30 + (117.61 - 60.30) \times (1.8 + 0.25)/2.85 = 101.52 \text{ kN/m}^2$

The self weight load = $0.25 \times 24 = 6 \text{ kN/mm}^2$

Fig.12.8 shows the forces acting on the toe slab

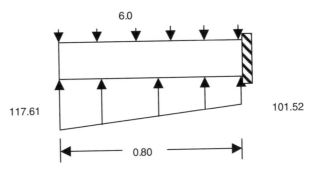

Fig.12.8 Pressures in kN/m² acting on toe slab.

The shear and moment at the face of the wall are as follows:

shear = $(101.52 - 6) \times 0.8 + (117.61 - 101.52) \times 0.5 \times 0.8 = 82.86 \text{ kN}$
moment = $0.5 \times (101.52 - 6) \times 0.8^2 + 0.5 \times (117.61 - 101.52) \times 0.8 \times (2/3) \times 0.8$
$= 34.0 \text{ kN m/m}$

Reinforcement from the wall which is designed for a moment of 101.56 kNm/m will be anchored in the toe slab and will provide the moment steel here. The anchorage length required is 640 mm and this will be provided by the bend and a straight length of bar along the outer footing. The radius of the bend is determined to limit the bearing stress to a safe value. The permissible bearing stress inside the bend is

$$\text{bearing stress} = \frac{F_{bt}}{r\phi} \le \frac{2f_{cu}}{1+2\dfrac{\phi}{a_b}} = \frac{2\times30}{1+2\dfrac{16}{150}} = 49.5\,\text{N/mm}^2$$

where a_b is the bar spacing, 150 mm.
The force in the bar
$$F_{bt} = 0.95 \times 460 \times \text{Area of 16 mm bar} = 87.86\,\text{kN}$$
The minimum internal radius r of the bend is
$$r = 87.86 \times 10^3 / (49.5 \times 16) = 110.9\,\text{mm}$$
Make the radius of the bend 150 mm

Shear stress: The flexural steel and the dimensions of the toe slab are same as for the stem which is safe for a shear force of 125.7 kN. This is satisfactory. The distribution steel is 10 mm diameter bars at 240 mm centres.

(4) Nib
The passive earth pressure coefficient $K_p = 1/K_a = 3.0$.
The earth pressure at the top and bottom of the nib are
$$\text{Top: } K_p\,\gamma\,z = 3 \times 17.6 \times 0.25 = 13.2\,\text{kN/m}^2$$
$$\text{Bottom: } K_p\,\gamma\,z = 3 \times 17.6 \times 0.85 = 44.88\,\text{kN/m}^2$$
Referring to Fig.12.5 the shear and moment in the nib using a load factor of 1.4 are as follows:
$$\text{shear} = 1.4 \times (13.2 + 44.88) \times 0.6/2 = 24.39\,\text{kN}$$
$$\text{moment} = 1.4 \times \{13.2 \times 0.6^2/2 + (44.88 - 13.2) \times 0.5 \times 0.6 \times (2/3) \times 0.6\}$$
$$= 8.65\,\text{kNm}$$
The forces are quite small. The minimum reinforcement is 0.13% or 325 mm^2/m. Provide 10 mm diameter bars at 150 mm centres ($A_s = 524$ mm^2/m) to lap onto the main wall steel. The distribution steel is 10 mm diameter bars at 240 mm centres.

(e) Sketch of the wall reinforcement: A sketch of the wall with the reinforcement designed above is shown in Fig.12.9. Note that the reinforcement is organized to produce a 3–D cage which can be easily fabricated.

12.3 COUNTERFORT RETAINING WALLS

12.3.1 Stability Check and Design Procedure

A counterfort retaining wall is shown in Fig.12.10. The spacing of the counterforts is usually made equal to the height of the wall. The following comments are made regarding the design.

(a) Stability
Consider as one unit a centre-to-centre length of panels taking into account the weight of the counterfort. The horizontal earth acting on this unit together with the

gravity loads must provide satisfactory resistance to overturning and sliding. The calculations are made in a similar way to those for a cantilever wall.

(b) Wall slab

The slab is thinner than that required for a cantilever wall. It is built in on three edges and free at the top. It is subjected to a triangular load due to the active earth pressure. The lower part of the wall cantilevers vertically from the base and the upper part spans horizontally between the counterforts. A load distribution

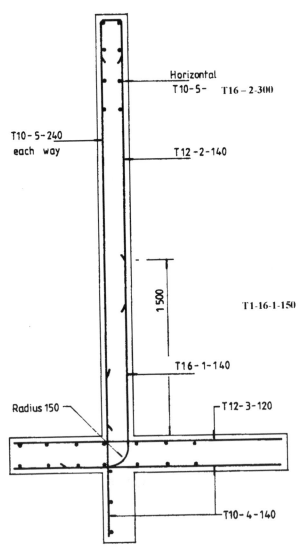

Fig.12.9 Reinforcement detail in the cantilever retaining wall.

commonly adopted between vertically and horizontally spanning elements is shown in Fig.12.10. The Finite Element Method could also be used to analyse the wall to determine the moments for design. Yield line analysis and Hillerborg's Strip methods are used in the example that follows.

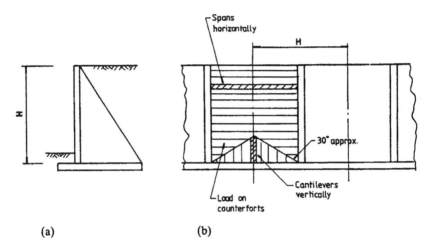

Fig.12.10 (a) Section: (b) back of wall.

(c) Base slab

Like the wall slab, the base slab behind the vertical wall is built-in on three sides and free on the fourth. The loading is trapezoidal in distribution across the base due to the net effect of the weight of earth down and earth pressure under the base acting upwards. As in the case of the wall slab, near the junction with the wall, the forces are resisted by cantilever action while away from this junction, the load is resisted by beam action with the strips spanning between the counterforts. Like the wall slab, the moments in the base slab can be determined using yield line analysis or Hillerborg's Strip methods.

(b) Outer footing (Toe slab)

If provided, it is designed as a cantilever in a manner similar to cantilever retaining wall.

(d) Counterforts

Counterforts support the wall and base slabs and are designed as vertical cantilevers of varying T-beam section. The load on the counterforts is from the wall slab spanning between the counterforts. A design is made for the section at the base and one or more sections up the height of the counterfort. Link reinforcement must be provided between the wall slab and inner base slab and the counterfort to transfer the loading. Reinforcement for the counterfort is shown in Fig.12.11(c).

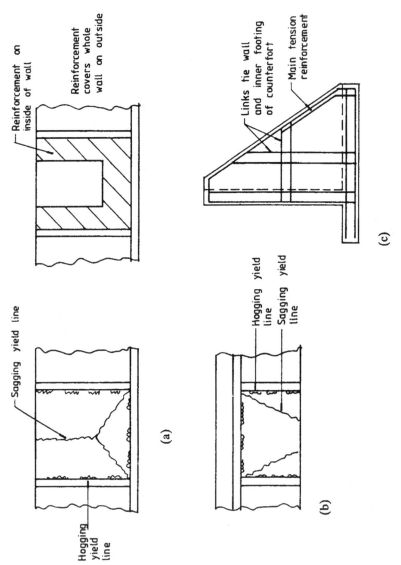

Fig.12.11 Counterfort wall (a) Yield line pattern and reinforcement in wall; (b) yield line pattern in base slab; (c) reinforcement in counterfort.

12.3.2 Example of Design of a Counterfort Retaining Wall

(a) Specification
A counterfort retaining wall has a height from the top to the underside of the base of 5 m and a spacing of counterforts of 5 m. The backfill is level with the top of the wall. The earth in the backfill is granular with the following properties:

Unit weight $\gamma = 15.7$ kN/m^3,
Angle of internal friction $\phi = 30°$,
Coefficient of active earth pressure $K_a = 0.333$,
Coefficient of friction between the soil and concrete $\mu = 0.5$,
Safe bearing pressure of the soil under the base = 150 kN/m^2

The construction materials are grade C35 concrete and grade 460 reinforcement.

(b) Trial section
The proposed section for the counterfort retaining wall is shown in Fig.12.12. Wall slab is made 180 mm thick and the counterfort and base slab are both 250 mm thick.

(c) Stability
Consider a 5 m length of wall centre to centre of counterforts. The horizontal earth pressure at depth z is

$$K_a \, \gamma \, z = 0.333 \times 15.7 \times z = 5.23 \, z \text{ kN/m}^2$$
$$\text{The pressure at } z = 5 \text{ m is } 26.15 \text{ kN/m}^2.$$

The loads are shown in Fig.12.12. The stability calculations are given in Table 12.2. Clockwise moments are considered as positive.

Table 12.2 Stability calculations (Counterfort wall). All loads unfactored.

Type of Load	Load, (kN)	Distance to centroid from A, (m)	Moment about A, (kNm)
HORIZONTAL (Active earth pressure)			
Triangular	$0.5 \times 5 \times 5 \times 26.15 =$ 326.88	$5/3 = 1.67$	-544.79
VERTICAL (Gravity)			
Wall	$5 \times 0.18 \times 4.75 \times 24$ $=102.6.$	$0.18/2 = 0.09$	9.23
Base	$5 \times 0.25 \times 3.5 \times 24 =$ 105.0	$3.5/2 = 1.75$	183.75
Back fill	$4.75 \times 3.32 \times 4.75 \times 15.7$ $= 1176.05$	$0.18 + 3.32/2 = 1.84$	2163.93
Counterfort	$0.5 \times 3.32 \times 4.75 \times 0.25$ $\times 24 = 47.31$	$0.18 + 3.32/3 = 1.29$	60.87
Σ	1430.96		2417.78

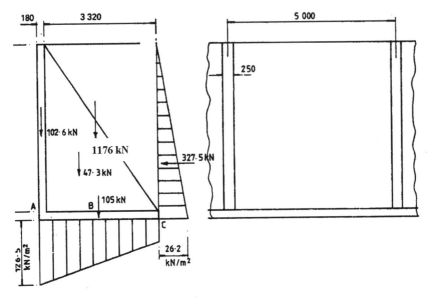

Fig.12.12 Forces acting on the structure.

(i) Maximum soil pressure
The properties of the base are as follows:
$$\text{area } A = 3.5 \times 5 = 17.5 \text{ m}^2$$
$$\text{section modulus } Z = 5 \times 3.5^2/6 = 10.21 \text{ m}^3$$
Taking moments of all forces about the toe A, the centroid of the base pressure from A is at a distance L.
$$L \times 1430.96 = 2417.78 - 544.79, L = 1.31 \text{ m}$$
$$\text{eccentricity, } e = 3.5/2 - L = 0.44 < 3.5/6$$
No tension is developed at C.
The base is acted on by
$$\text{vertical load} = 1430.96 \text{ kN, moment} = 1430.96 \times e = 629.62 \text{ kNm.}$$
The maximum soil pressure at *A* calculated for *service load* is
$$1430.96/ (A = 17.5) + 629.62/ (Z = 10.21) = 143.44 \text{ kN/m}^2 < 150 \text{ kN/m}^2$$
Width of base is sufficient to prevent bearing capacity failure.

(ii) Stability against overturning
The stabilizing moment due to gravity loads about the toe A of the wall has a partial safety factor $\gamma_f = 1.0$ and the disturbing moment due to horizontal loads has a partial safety factor $\gamma_f = 1.4$.
$$(2417.78 \times 1.0 - 544.79 \times 1.4) = 1655.1 > 0$$
The wall is very safe against overturning.

(iii) Resistance to sliding

The forces resisting sliding are the friction under the base. For the wall to be safe against sliding

$$(\mu = 0.5) \{1430.96 \times (\gamma_f = 1.0)\} > (\gamma_f = 1.4) \times 326.88$$
$$715.48 > 457.63$$

The resistance to sliding is satisfactory.

(iv) Overall comment

The wall section is satisfactory. The maximum soil pressure under the base controls the design.

12.3.3 Design of Wall Slab using Yield Line Method

The yield line solution is given for a square wall with a triangular load with the yield line pattern shown in Fig.12.13 (a). Parameter α, locating point F, controls the collapse pattern. Deflection at F is Δ. It is assumed that the slab will be isotropically reinforced with moment of resistance for both positive (tension on the outer face) and negative (tension on the earth face) being equal to m.

(i) Energy dissipated in the yield lines

1. Rigid region AEFD and BEFC: Both rotate about y-axis only.

    (i). Negative yield lines: $\ell_y = a$, $m_y = m$, $\theta_y = \Delta/(0.5a)$
    (ii). Positive yield lines: $\ell_y = a$, $m_y = m$, $\theta_y = \Delta/(0.5a)$

Energy dissipated

$$E_1 = 2\{m \times a \times \Delta/(0.5a) + m \times a \times \Delta/(0.5a)\} = 8\, m\, \Delta$$

2. Rigid region DFC: Rotates about x-axis only.

    (i). Negative yield lines: $\ell_x = a$, $m_x = m$, $\theta_x = \Delta/(\alpha a)$
    (ii). Positive yield lines: $\ell_x = a$, $m_x = m$, $\theta_x = \Delta/(\alpha a)$

Energy dissipated

$$E_2 = \{m \times a \times \Delta/(\alpha a) + m \times a \times \Delta/(\alpha a)\} = (2/\alpha)\, m\, \Delta$$

3. Total energy dissipated

$$E = E_1 + E_2 = (8 + 2/\alpha)\, m\, \Delta$$

(ii) External work done

Set up the coordinate axes (x, y) with origin at A.

1. Rigid region AEGF:

    Pressure at any point $= \gamma\, y$
    Rotation θ_y about the y-axis $= \Delta/(0.5a)$
    Deflection at any point $= \theta_y\, x$

Work done is

$$W_1 = \gamma\theta_y \left[\int_0^{0.5a} x \left\{ \int_0^{(1-\alpha)a} y\, dy \right\} dx \right] = \gamma\theta_y \left[\frac{x^2}{2} \right]_0^{0.5a} \left\{ \frac{y^2}{2} \right\}_0^{(1-\alpha)a} = \frac{1}{8}\gamma a^3 \Delta (1-\alpha)^2$$

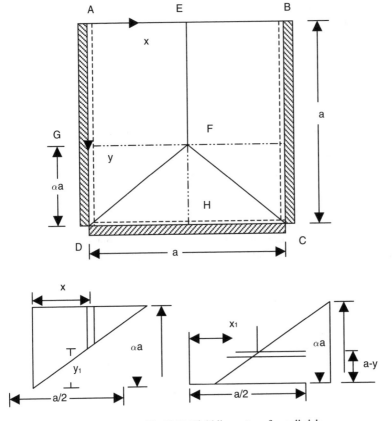

Fig.12.13 Yield line pattern for wall slab.

2. Rigid region GFD:

Along a vertical strip at a distance x,

$$y_1 = \alpha a \, x / (0.5a) = 2\alpha x$$

Pressure at any point $= \gamma \, y$

Rotation θ_y about the y-axis $= \Delta / (0.5a)$

Deflection at any point $= \theta_y \, x$

The limits for y are $(a - \alpha a)$ and $\{(a - y_1) = a - 2 \, \alpha x\}$

The limits for x are 0 and a/2.

$$W_2 = \gamma \theta_y \int_0^{0.5a} x\{ \int_{a(1-\alpha)}^{(a-2\alpha x)} y \, dy\} \, dx = \gamma \theta_y \int_0^{0.5a} x[\{\frac{y^2}{2}\}_{a(1-\alpha)}^{a-2\alpha x}] dx$$

$$= \frac{\gamma \theta_y}{2} \int_0^{0.5a} x[\{(a - 2\alpha x)^2 - a^2(1-\alpha)^2\}] dx$$

$$=\frac{\gamma\theta_y}{2}\int_0^{0.5a}x[\,\{4\alpha^2x^2-4a\alpha x-a^2\alpha(\alpha-2)]\,dx$$

$$=\frac{\gamma\theta_y}{2}\int_0^{0.5a}[\,\{4\alpha^2x^3-4a\alpha x^2-a^2\alpha(\alpha-2)x]\,dx$$

$$=\frac{\gamma\theta_y}{2}\{4\alpha^2\frac{x^4}{4}-4a\alpha\frac{x^3}{3}-a^2\alpha(\alpha-2)\frac{x^2}{2}\}_0^{0.5a}$$

$$=\frac{\gamma\theta_y}{2}a^4\frac{(4\alpha-3\alpha^2)}{48}$$

$$W_2=\gamma\Delta\,a^3\frac{(4\alpha-3\alpha^2)}{48}$$

3. Rigid region FDH:

Along a horizontal strip at a distance y,

$$x_1/\,(0.5a)=(a-y)/(\alpha\,a)$$
$$x_1=(a-y)/(2\alpha)$$

Pressure at any point $=\gamma\,y$

Rotation θ_x about the x-axis $=\Delta/\,(\alpha a)$

Deflection at any point $=\theta_x\,(a-y)$

The limits for x are $\{x_1=(a-y)/\,(2\alpha)\}$ and a/2

The limits for y are $(a-\alpha a)$ and a

$$W_3=\gamma\theta_x\int_{a(1-\alpha)}^a y(a-y)\{\int_{\frac{a-y}{2\alpha}}^{0.5a}dx\}\,dy=\gamma\theta_x\int_{a(1-\alpha)}^a y(a-y)[\,\{x\}_{\frac{a-y}{2\alpha}}^{0.5a}]\,dy$$

$$=\gamma\theta_x\int_{a(1-\alpha)}^a y(a-y)[\,0.5a-\frac{0.5}{\alpha}(a-y)]\,dx$$

$$=\frac{\gamma\theta_x}{2\alpha}\int_{a(1-\alpha)}^a y(a-y)[\,\alpha a-(a-y)]\,dy$$

$$=\frac{\gamma\theta_x}{2\alpha}\int_{a(1-\alpha)}^a[\,\{-(1-\alpha)a^2y+(2-\alpha)a\,y^2-y^3]\,dy$$

$$=\frac{\gamma\theta_x}{2\alpha}[-(1-\alpha)a^2\frac{y^2}{2}+(2-\alpha)a\frac{y^3}{3}-\frac{y^4}{4}]_{a(1-\alpha)}^a$$

$$=\frac{\gamma\theta_x}{2\alpha}a^4[-(1-\alpha)\frac{1}{2}+(2-\alpha)\frac{1}{3}-\frac{1}{4}+(1-\alpha)\frac{(1-\alpha)^2}{2}-(2-\alpha)\frac{(1-\alpha)^3}{3}+\frac{(1-\alpha)^4}{4}]$$

$$=\frac{\gamma\theta_x}{24\alpha}a^4[(2\alpha-1)+(1-\alpha)^3(1+\alpha)]$$

$$=\frac{\gamma\theta_x}{24\alpha}a^4(2\alpha^3-\alpha^4)$$

$$W_3=\frac{\gamma\Delta}{24}a^3(2\alpha-\alpha^2)$$

3. Total work done W

$$W = 2(W_1 + W_2 + W_3)$$

$$W = \frac{\gamma\Delta}{24}a^3(6 - 4\alpha + \alpha^2)$$

4. Moment m:

Equating the work done by external loads to the energy dissipated at the yield lines,

$$m = \frac{\gamma a^3}{48}\frac{(6\alpha - 4\alpha^2 + \alpha^3)}{(4\alpha + 1)}$$

For a maximum m, $dm/d\alpha = 0$

$$(4\alpha + 1)(6 - 8\alpha + 3\alpha^2) - 4(6\alpha - 4\alpha^2 + \alpha^3) = 0$$

Simplifying,

$$6 - 8\alpha - 13\alpha^2 + 8\alpha^3 = 0, \ \alpha = 0.483312$$
$$m = 0.014762 \ \gamma a^3,$$

Substituting a = 5.0 m, γ = 15.7,
$$m = 28.97 \text{ kNm/m}$$

Using a load factor on the earth pressure of 1.4 and also increasing the calculated moment by 10% to account for the formation of corner levers,
$$m = (28.97 \times 1.4) \times 1.1 = 44.62 \text{ kNm/m}$$

5. Reinforcement

Use 16 mm diameter bars and 40 mm cover.
$$d = 180 - 40 - 16/2 = 132 \text{ mm}$$
$$k = M/(bd^2 f_{cu}) = 44.62 \times 10^6/(1000 \times 132^2 \times 30) = 0.085 < 0.156$$
No compression steel is required
$$z/d = 0.5 + \sqrt{(0.25 - 0.085/0.9)} = 0.894 < 0.95$$
$$A_s = M/(z \times 0.95 f_y)$$
$$= 44.62 \times 10^6/(0.89 \times 132 \times 0.95 \times 460) = 869 \text{ mm}^2/\text{m}$$

Provide 16 mm diameter bars at 200 mm centres to give a steel area of 1150 mm²/m. The same steel is provided in each direction on the outside and inside of the wall.

Minimum steel percentage:

$$100 \times 1150/(1000 \times 180) = 064 > 0.13$$

The steel on the outside of the wall covers the whole area. On the inside of the wall the steel can be cut off at 0.3 times the span from the bottom and from each counterfort support in accordance with the simplified rules for curtailment of bars in slabs.

Alternatively, the points of cut-off of the bars on the inside of the wall may be determined by finding the size of a slab simply supported on three sides and one edge free that has the same ultimate moment of resistance m = 44.62 kNm/m as the whole wall. This slab has the same yield line pattern as the wall slab.

Maximum Spacing: The clear spacing of the bars does not exceed 3d and the slab depth is not greater than 200 mm and so the slab is satisfactory with respect to cracking.

In the above only one mode of collapse has been investigated. As Yield line method is an upper bound method other possible yield line patterns need to be investigated before finalizing the reinforcement.

12.3.4 Design of Base Slab using Yield Line Method

(i) Base pressure calculation at the ultimate
The properties of the base are:
$$\text{area } A = 17.5 \text{ m}^2$$
$$\text{section modulus } Z = 10.21 \text{ m}^3$$
The forces at SLS are shown in Table 12.2. Taking moments of all forces about the toe A, the centroid of the base pressure from A is at a distance L.

Case (a): Load factor is 1.4 for earth pressure and 1.0 for gravity load,
$$L \times 1430.96 = 2417.78 - 544.79 \times 1.4, L = 1.16 \text{ m}$$
$$\text{eccentricity, } e = 3.5/2 - L = 0.59 > 3.5/6.$$
Tension is developed at C.
The base is acted on by a
$$\text{vertical load} = 1430.96 \text{ kN}$$
$$\text{moment} = 1430.96 \times e = 849.11 \text{ kNm}$$
The maximum soil pressure at A and minimum soil pressure at C are
$$1430.96 / 17.5 \pm 849.11/ 10.21 = 164.94 \text{ and } -1.40 \text{ kN/m}^2$$
The negative pressure is very small and can be neglected.

Case (b): Load factor is 1.0 for earth pressure and 1.4 for gravity load,
$$L \times 1430.96 \times 1.4 = 2417.78 \times 1.4 - 544.79 \times 1.0, L = 1.42 \text{ m}$$
$$\text{eccentricity, } e = 3.5/2 - L = 0.33 < 3.5/6.$$
No tension develops at C.
The base is acted on by
$$\text{vertical load} = 1430.96 \ 1.4 = 2003.34 \text{ kN}$$
$$\text{moment} = 2003.34 \text{ x } e = 665.26 \text{ kNm}$$
The maximum soil pressure at A and minimum soil pressure at C are
$$2003.34 /17.5 \pm 665.26 / 10.21 = 179.63 \text{ and } 49.32 \text{ kN/m}^2$$

Case (c): Using load factor of 1.4 for earth pressure and 1.4 for gravity load,
$$L \times 1430.96 \times 1.4 = 2417.78 \times 1.4 - 544.79 \times 1.4, L = 1.31 \text{ m}$$
$$\text{eccentricity, } e = 3.5/2 - L = 0.44 < 3.5/6$$
No tension develops at C.
The base is acted on by
$$\text{vertical load} = 1430.96 \times 1.4 \text{ kN} = 2003.34$$
$$\text{moment} = 1430.96 \times 1.4 \times e = 883.67 \text{ kNm}$$
The maximum soil pressure at A and minimum soil pressure at C are

$$1430.96 \times 1.4 / 17.5 \pm 907.18 / 10.21 = 201.03 \text{ and } 27.92 \text{ kN/m}^2$$

Using Case (b), the yield line solution is given for a rectangular base slab with a trapezoidal load due to base pressure on the bottom face and a uniform load due to self weight of the slab and soil on the slab at the top face. The uniformly distributed load at the top is

$$\text{Base slab} = 0.25 \times 24 \times 1.4 = 8.4 \text{ kN/m}^2$$
$$\text{Weight of soil} = 4.75 \times 15.7 \times 1.4 = 104.41 \text{ kN/m}^2$$
$$\text{Total} = 112.81 \text{ kN/m}^2$$

Two modes will be investigated.

Mode 1: The yield line pattern shown in Fig.12.14.

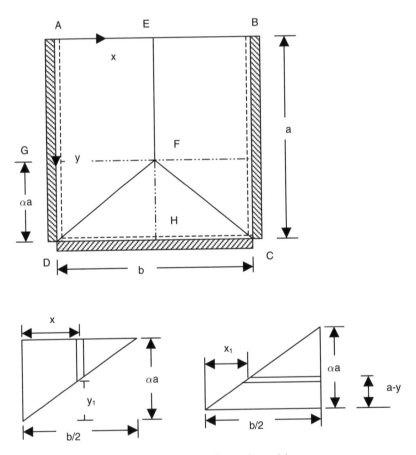

Fig.12.14 Mode 1 collapse of base slab.

One parameter α, locating the position of point F, controls the pattern.. Deflection at F is Δ. It is assumed that the slab will be isotropically reinforced with moment of resistance for both positive (tension at bottom) and negative (tension at top) being equal to m.

The pressure at a point distant y from the free edge is

$$p = 112.81 - \{49.32 + (179.63 - 49.32)y/3.5\} = 63.49 - 37.23\, y$$

(i) Energy dissipated in the yield lines

1. Rigid region AEFD and BEFC: Both rotate about y-axis only.

(i). Negative yield lines: $\ell_y = a$, $m_y = m$, $\theta_y = \Delta/(0.5b)$

(ii). Positive yield lines: $\ell_y = a$, $m_y = m$, $\theta_y = \Delta/(0.5b)$

Energy dissipated E_1 is

$$E_1 = 2\{m \times a \times \Delta/(0.5b) + m \times a \times \Delta/(0.5b)\} = 8\, m\,(a/b)\,\Delta$$

2. Rigid region DFC: Rotates about x-axis only.

(i) Negative yield lines: $\ell_x = b$, $m_x = m$, $\theta_x = \Delta/(\alpha a)$

(ii) Positive yield lines: $\ell_x = a$, $m_x = m$, $\theta_x = \Delta/(\alpha a)$

Energy dissipated E_2 is

$$E_2 = \{m \times b \times \Delta/(\alpha a) + m \times b \times \Delta/(\alpha a)\} = (2b/\alpha a)\, m\, \Delta$$

3. Total energy dissipated E is

$$E = E_1 + E_2 = (8a/b + 2b/\alpha a)\, m\, \Delta$$

Substituting a = 3.5, b = 5.0, total energy dissipated $E = (5.6 + 2.8571/\alpha)\, m\, \Delta$

(ii) External work done

Set up the coordinate axes (x, y) with origin at A.

(1) Rigid region AEGF:

Pressure at any point = $63.49 - 37.23\, y$

Rotation θ_y about the y-axis = $\Delta/(0.5b)$

Deflection at any point = $\theta_y\, x$

Work done is W_1:

$$W_1 = \theta_y[\ \int_0^{0.5b} x\{\ \int_0^{(1-\alpha)a} (63.49 - 37.23 y)\, dy\}\ dx]$$

$$W_1 = \frac{\Delta}{0.5b}[\frac{x^2}{2}]_0^{0.5b}\{63.49 y - 37.23\frac{y^2}{2}\}_0^{(1-\alpha)a}$$

Substituting a = 3.5, b = 5.0

$$W_1 = 277.77 \times (1-\alpha) - 285.04 \times (1-\alpha)^2$$

$$W_1 = 292.31\alpha - 285.04\alpha^2 - 7.27$$

(2) Rigid region GFD:

Along a vertical strip at a distance x,

$$y_1 = \alpha a\, x/(0.5b) = (2\alpha a/b)\, x$$

Pressure at any point = $63.49 - 37.23\, y$

Rotation θ_y about the y-axis = $\Delta/(0.5b)$

$$\text{Deflection at any point} = \theta_y \, x$$

The limits for y are:

$$(a - \alpha a) \text{ and } \{(a - y_1) = a - (2\alpha a/b)\, x\}$$

The limits for x are 0 and b/2.

$$W_2 = \theta_y \int_0^{0.5b} x \{ \int_{a(1-\alpha)}^{(a-2\alpha\frac{a}{b}x)} (63.49 - 37.23y)\, dy \}\, dx$$

$$W_2 = \theta_y \int_0^{0.5b} x [\{63.49y - 37.23 \frac{y^2}{2}\}_{a(1-\alpha)}^{a-2\alpha\frac{a}{b}x}]\, dx$$

$$W_2 = \theta_y \int_0^{0.5b} x [63.49(a\alpha - 2\alpha\frac{a}{b}x) - 18.615(4\alpha^2\frac{a^2}{b^2}x^2 - 4\alpha\frac{a^2}{b}x - \alpha^2 a^2 + 2a^2 \alpha]\, dx$$

$$= \theta_y \int_0^{0.5b} [63.49(a\alpha x - 2\alpha\frac{a}{b}x^2) - 18.615(4\alpha^2\frac{a^2}{b^2}x^3 - 4\alpha\frac{a^2}{b}x^2 - \alpha^2 a^2 x + 2a^2 \alpha x]\, dx$$

$$= \theta_y [63.49(a\alpha\frac{x^2}{2} - 2\alpha\frac{a}{b}\frac{x^3}{3}) - 18.615(\alpha^2\frac{a^2}{b^2}x^4 - 4\alpha\frac{a^2}{b}\frac{x^3}{3} - \alpha^2 a^2\frac{x^2}{2} + a^2 \alpha x^2]_0^{0.5b}$$

$$= \Delta[63.49(a\alpha b\frac{1}{12}) - 18.615 a^2 b(\frac{\alpha}{6} - \frac{\alpha^2}{8})]$$

$$= \Delta[92.59\alpha - 47.51(4\alpha - 3\alpha^2)]$$

$$W_2 = \Delta[142.53\alpha^2 - 97.45\alpha]$$

(3) Rigid region FDH:

Along a horizontal strip at a distance y,

$$x_1/(0.5b) = (a - y)/(\alpha a)$$

$$x_1 = (0.5b/\alpha a)(a - y)$$

Pressure at any point $= 63.49 - 37.23\, y$

Rotation θ_x about the x-axis $= \Delta/(\alpha a)$

Deflection at any point $= \theta_x (a - y)$

The limits for x are $\{x_1 = (0.5b/\alpha a)(a - y)\}$ and b/2

The limits for y are $(a - \alpha a)$ and a

$$W_3 = \theta_x \int_{a(1-\alpha)}^{a} (63.49 - 37.23y)(a - y)\{ \int_{\frac{0.5b}{\alpha a}(a-y)}^{0.5b} dx \}\, dy$$

$$= \theta_x \int_{a(1-\alpha)}^{a} (63.49 - 37.23y)(a - y)[\{x\}_{\frac{0.5b}{\alpha a}(a-y)}^{0.5b}]\, dy$$

$$= \theta_x \int_{a(1-\alpha)}^{a} (63.49 - 37.23y)(a - y)[0.5b - \frac{0.5b}{\alpha a}(a - y)]\, dy$$

$$= \theta_x \frac{0.5b}{\alpha a} \int_{a(1-\alpha)}^{a} (63.49 - 37.23 y)(a - y)[\alpha a - a + y)] \, dy$$

$$= \Delta \frac{0.5b}{(\alpha a)^2} \int_{a(1-\alpha)}^{a} (63.49 - 37.23 y)[\alpha a^2 - \alpha a y - a^2 + 2ay - y^2)] \, dy$$

$$= \Delta b \, a[5.29\alpha + 1.55 a \alpha (\alpha - 2)]$$

$$= \Delta b a[5.425\alpha^2 - 5.56\alpha]$$

$$W_3 = \Delta[94.94\alpha^2 - 97.3\alpha]$$

(iii) Total work done

$$W = 2(W_1 + W_2 + W_3)$$

$$W = \Delta(-14.54 + 195.06\alpha - 95.14\alpha^2)$$

(iv) Moment m

Equating the work done by external loads to the energy dissipated at the yield lines,

$$m = \frac{(-14.54\alpha + 195.06\alpha^2 - 95.14\alpha^3)}{(5.6\alpha + 2.8571)}$$

For a maximum m, $\alpha = 1.0$, m = 10.10 kNm/m

Mode 2: The yield line pattern shown in Fig.12.15. Making the same assumptions as for Mode 1, the pressure p at any point y from the free edge is

$$p = 63.49 - 37.23 \, y$$

(i) Energy dissipated in the yield lines

1. Rigid region AFD and BEC: Both rotate about y-axis only.
(i). Negative yield lines: $\ell_y = a$, $m_y = m$, $\theta_y = \Delta / (\alpha b)$
(ii). Positive yield lines: $\ell_y = a$, $m_y = m$, $\theta_y = \Delta / (\alpha b)$
Energy dissipated E_1 is
$$E_1 = 2\{m \times a \times \Delta / (\alpha b) + m \times a \times \Delta / (\alpha b)\} = 4m \, a/ (\alpha b) \, \Delta$$

2. Rigid region DFEC: Rotates about x-axis only.
(i). Negative yield lines: $\ell_x = b$, $m_x = m$, $\theta_x = \Delta/a$
(ii) Positive yield lines: $\ell_x = 2\alpha b$, $m_x = m$, $\theta_x = \Delta/a$
Energy dissipated E_2 is
$$E_2 = \{m \times b \times \Delta/a + m \times 2 \, \alpha b \times \Delta/a\} = (2\alpha + 1) \, (b/a) \, m \, \Delta$$

3. Total energy dissipated
$$E = E_1 + E_2 = \{(4a/ \alpha b + (2 \, \alpha + 1) \, (b/a)\} \, m \, \Delta$$
Substituting a = 3.5, b = 5.0, Energy dissipated is
$$E = (2.8/ \alpha + 1.4286 + 2.8571 \, \alpha) \, m \, \Delta$$

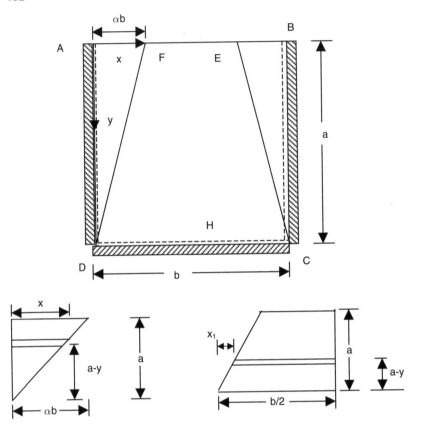

Fig.12.15 Mode 2 collapse of base slab.

(ii) External work done
Set up the coordinate axes (x, y) with origin at A.

1. Rigid region ADF:

Along a horizontal strip at a distance y, $x_1 = (\alpha b/a)(a-y)$

Pressure at any point $= 63.49 - 37.23\,y$

Rotation θ_y about the y-axis $= \Delta/(\alpha b)$

Deflection at any point $= \theta_y\,x$

The limits for x are 0 and x_1

The limits for y are 0 and a.

$$W_1 = \theta_y \int_0^a (63.49 - 37.23y)\,dy\{\int_0^{\frac{b}{a}x(a-y)} x\ dx\} = \theta_y \int_0^{0.5b} [(63.49 - 37.23y)\{\frac{x^2}{2}\}_0^{\alpha\frac{b}{a}(a-y)}]\,dy$$

$$= \theta_y \alpha^2 \frac{1}{2} \frac{b^2}{a^2} \int_0^a (63.49 - 37.23\, y)\{a^2 - 2ay + y^2\}\, dy$$

$$= \theta_y \alpha^2 \frac{1}{2} \frac{b^2}{a^2} \int_0^a [63.49\{a^2 - 2ay + y^2\} - 37.23\{a^2 y - 2ay^2 + y^3\}]\, dy$$

$$W_1 = \Delta ba[63.49\frac{\alpha}{6} - 37.23\, a\frac{\alpha}{24}]$$

2. Rigid region FDEC:

Considering only a symmetrical half on the left,
Along a horizontal strip at a distance y,

$$x_1/(\alpha b) = (a - y)/a,$$
$$x_1 = (\alpha b/a)\,(a - y)$$

Pressure at any point $= 63.49 - 37.23\, y$
Rotation θ_x about the x-axis $= \Delta/a$
Deflection at any point $= \theta_x\,(a - y)$
The limits for x are x_1 and b/2
The limits for y are a and 0

$$W_2 = \theta_x \int_0^a (63.49 - 37.23 y)(a - y)\{ \int_{\frac{\alpha b}{a}(a-y)}^{0.5b} dx \}\, dy$$

$$= \theta_x \int_0^a (63.49 - 37.23 y)(a - y)[\, \{x\}_{\frac{\alpha b}{a}(a-y)}^{0.5b}\,]\, dy$$

$$= \theta_x \int_0^a (63.49 - 37.23 y)(a - y)[\, 0.5b - \frac{\alpha b}{a}(a - y)]\, dy$$

$$= \theta_x \frac{b}{a} \int_0^a (63.49 - 37.23 y)(a - y)[\, 0.5a - \alpha a + \alpha y)]\, dy$$

$$= \Delta \frac{b}{a^2} \int_0^a (63.49 - 37.23 y)[\, 0.5a^2 - \alpha a^2 - 0.5ay + 2ay - \alpha y^2)]\, dy$$

$$W_2 = \Delta b\, a[63.49\frac{(3 - 4\alpha)}{12} - 37.23 a\frac{(1 - \alpha)}{12}]$$

Total work done $W = 2(W_1 + W_2)$

$$W = \Delta b\, a[63.49\frac{(3 - 2\alpha)}{6} - 37.23 a\frac{(2 - \alpha)}{12}]$$

Substituting a = 3.5, b = 5.0,

$$W = \Delta(175.48 - 180.33\alpha)$$

3. Moment m

Equating the work done by external loads to the energy dissipated at the yield lines,

$$m = \frac{(175.48\alpha - 180.33\alpha^2)}{(2.8 + 1.4286\alpha + 2.8571\alpha^2)}$$

For a maximum m, $\alpha = 0.3788$, m = 10.82 kNm/m

Mode 2 gives marginally higher value of m = 10.82 kNm/m. Increasing the calculated moment by 10% to account for the formation of corner levers, m = 10.82 x 1.1 = 11.90 kNm/m.

4. Reinforcement
Use 10 mm diameter bars and 40 mm cover.

$$d = 250 - 40 - 10/2 = 205 \text{ mm}$$
$$k = M/(bd^2 f_{cu}) = 11.90 \times 10^6/(1000 \times 205^2 \times 30) = 0.094 < 0.156$$

No compression steel is required

$$z/d = 0.5 + \sqrt{(0.25 - 0.095/0.9)} = 0.99 > 0.95$$
$$A_s = M/(z \times 0.95 \ f_y) = 11.90 \times 10^6/(0.95 \times 205 \times 0.95 \times 460) = 140 \text{ mm}^2/\text{m}$$

Minimum steel:
$$A_s = (0.13/100) \ 250 \times 1000 = 325 \text{ mm}^2/\text{m}$$
Minimum steel governs. Provide 10 mm diameter bars at 200 mm centres to give a steel area of 392 mm²/m. The same steel is provided in each direction on the upper and lower faces of the slab.

Maximum Spacing
Steel percentage is less than 0.3%. Maximum spacing limited to 3d. Spacing does not exceed this limit so the slab is satisfactory with respect to cracking.

12.3.5 Base Slab Design using Hillerborg's Strip Method

Although Yield line Method was used in design in the previous sections, it is not the ideal method. Hillerborg's Strip Method offers a better alternative. At the junction of the base slab with the wall slab, load is resisted by cantilever action but at a distance away from the base, load is resisted by clamped beam action with the slab spanning between the counterforts. As shown in Fig.12.16, the 3.5 m × 5.0 m base slab is divided into a set of 14 strips each 125 mm wide. The pressure at any level y from the free edge is given by the equation,

$$p = 63.49 - 37.23 \ y$$

It is assumed that load lying in a triangle with the sides at an inclination of approximately 30° to the horizontal (see Fig.12.16) is resisted by 'vertical' cantilever action. The rest of the load is resisted by 'horizontal' beam action.

12.3.5.1 Horizontal strips in base slab

The 'horizontal' strips span between the counterforts. The strips towards the base are loaded as shown in Fig.12.17. The bending moment at the support and midspan of each strip is calculated using the equation

$$\text{Moment at support} = \frac{qL^2}{12}C_1 \, , \, C_1 = \alpha^2(6-4\alpha),$$

$$\text{Moment at midspan} = \frac{qL^2}{24}C_2 \, , \, C_2 = 8\alpha^3$$

where q = uniformly distributed load in kN/m, L = span = 5 m, α = a/L, a = loaded length.

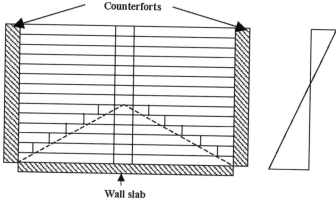

Fig.12.16 Division of base slab into 'horizontal' strips.

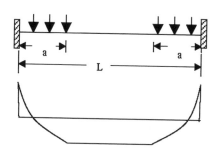

Fig.12.17 Load on 'horizontal' strips.

Detailed calculations are shown in Table 12.3. Fig.12.18 shows the bending moment distribution in the 'horizontal' strips.

At the *support* the maximum bending moment causing tension on the *bottom* of the slab is 122.58 kNm/m and the maximum bending moment at *midspan* causing tension on the *top* of the slab is 61.29 kNm/m in strip no.1. Similarly, at the support the maximum bending moment causing tension on the *top* of the slab is

26.93 kNm/m in strip no.10 and the maximum bending moment causing tension on the *bottom* of the slab is 9.42 kNm/m in strip no.9.

$$d = 250 - 40 - 10/2 = 205 \text{ mm}$$
$$\text{Width of strip} = 250 \text{ mm,}$$
$$M = 122.58 \times 0.25 = 30.65 \text{ kNm}$$
$$k = M/(bd^2 f_{cu}) = 30.65 \times 10^6/(250 \times 205^2 \times 30) = 0.097 < 0.156$$

No compression steel is required
$$z/d = 0.5 + \sqrt{(0.25 - 0.097/0.9)} = 0.88 < 0.95$$
$$A_s = M/(z \times 0.95 \, f_y) = 30.65 \times 10^6/(0.88 \times 205 \times 0.95 \times 460) = 389 \text{ mm}^2$$

Provide 2T16. $A_s = 402 \text{ mm}^2$. This steel is required over a width of 250 mm.
Minimum steel $A_s = (0.13/100) \, 250 \times 250 = 81.3 \text{ mm}^2$
Similar calculations can be done for the required steel in other strips

Table 12.3: Bending moments (kNm/m) in horizontal strips in base slab

Strip No.	y	p	a	α	C_1	C_2	M Supp	M Midspan
1	0.125	58.84	2.5	0.5	1	0.5	122.58	61.29
2	0.375	49.53	2.5	0.5	1	0.5	103.18	51.59
3	0.625	40.22	2.5	0.5	1	0.5	83.79	41.90
4	0.875	30.91	2.5	0.5	1	0.5	64.40	32.20
5	1.125	21.61	2.5	0.5	1	0.5	45.01	22.51
6	1.375	12.30	2.5	0.5	1	0.5	25.62	12.81
7	1.625	2.99	2.5	0.5	1	0.5	6.23	3.12
8	1.875	−6.32	2.5	0.5	1	0.5	−13.16	−6.58
9	2.125	−15.62	2.08	0.42	0.75	0.29	−24.49	−9.42
10	2.375	−24.93	1.67	0.33	0.52	0.15	−26.93	−7.69
11	2.625	−34.24	1.25	0.25	0.31	0.06	−22.29	−4.46
12	2.875	−43.55	0.83	0.17	0.15	0.02	−13.44	−1.68
13	3.125	−52.85	0.42	0.08	0.04	0.00	−4.33	−0.25
14	3.375	−62.16	0.00	0.00	0.00	0.00	0	0

12.3.5.2 Cantilever moment in base slab

The cantilever moment is determined by taking a series of vertical strips. The strips cantilever from the wall slab. The cantilever moment is greatest in the middle vertical strip. Pressures occur only in strips 9-14. The bending moment M at the base of the cantilever is given by the product of the pressures on the 250 mm wide strips and the distance from the base to the centre of the strips. Pressures at the centre of strips are given in Table 12.3.

$$M = 0.250 \times \{15.62 \times 1.375 + 24.93 \times 1.125 + 34.24 \times 0.875$$
$$+ 43.55 \times 0.625 + 52.85 \times 0.375 + 62.16 \times 0.125\} = 33.58 \text{ kNm/m}$$
$$\text{Horizontal width of strip} = 830 \text{ mm}$$

M on the strip = 33.58 × 0.83 = 27.87 kNm

$$k = M/ (bd^2 f_{cu}) = 27.87 × 10^6/ (830 × 205^2 × 30) = 0.027 < 0.156$$

No compression steel is required

$$z/d = 0.5 + \sqrt{(0.25 - 0.027/0.9)} = 0.97 > 0.95$$

$$A_s = M/ (z × 0.95 f_y) = 27.87 × 10^6/ (0.95 × 205 × 0.95 × 460) = 328 \text{ mm}^2$$

Provide 2-T16. A_s = 402 mm². This steel is required over a width of 830 mm. Similar calculations can be done for the required steel in other strips.

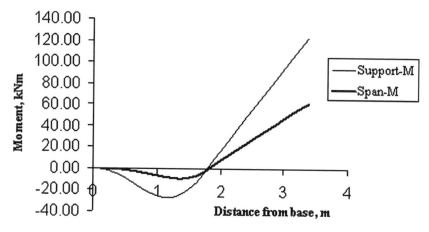

Fig.12.18 Bending moment (kNm/m) in 'horizontal' strips in base slab.

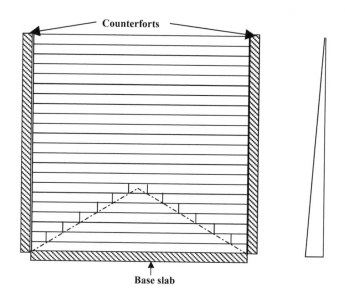

Fig.12.19 Horizontal strips in vertical wall slab.

Table 12.4 Bending moment (kNm/m) in horizontal strips of the wall.

Strip No.	y	p	a	α	M-Sup.	M-Midspan
1	0.125	0.92	2.5	0.5	1.91	0.95
2	0.375	2.75	2.5	0.5	5.72	2.86
3	0.625	4.58	2.5	0.5	9.54	4.77
4	0.875	6.41	2.5	0.5	13.36	6.68
5	1.125	8.24	2.5	0.5	17.17	8.59
6	1.375	10.07	2.5	0.5	20.99	10.49
7	1.625	11.91	2.5	0.5	24.80	12.40
8	1.875	13.74	2.5	0.5	28.62	14.31
9	2.125	15.57	2.5	0.5	32.44	16.22
10	2.375	17.40	2.5	0.5	36.25	18.13
11	2.625	19.23	2.5	0.5	40.07	20.03
12	2.875	21.06	2.5	0.5	43.88	21.94
13	3.125	22.90	2.5	0.5	47.70	23.85
14	3.375	24.73	2.5	0.5	51.52	25.76
15	3.625	26.56	2.08	0.42	41.63	16.01
16	3.875	28.39	1.67	0.33	30.67	8.76
17	4.125	30.22	1.25	0.25	19.68	3.94
18	4.375	32.05	0.83	0.17	9.89	1.24
19	4.625	33.89	0.42	0.08	2.78	0.16
20	4.875	35.72	0	0	0.00	0.00

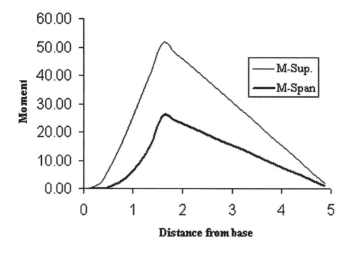

Fig.12.20 Bending moment (kNm/m) in the horizontal strips in the wall slab.

12.3.6 Wall Design using Hillerborg's Strip Method

Wall design is done similar to the base design. As the height of the wall is 5.0 m, it is divided into 20 strips each 250 mm wide as shown in Fig.12.19.

The pressure at any level y from the top is given by the equation, $1.4\,\gamma K_a y$, where $\gamma = 15.7$ kN/m^3, K_a = coefficient of earth pressure = 0.33, 1.4 is the load factor for earth pressure. Therefore P = 7.33 y. It is assumed that load lying in a triangle with the sides at an inclination of approximately 30° to the horizontal is resisted by vertical cantilever action. Calculations are shown in Table 12.4 and Fig.12.20 shows the bending moment distribution in the horizontal strips. The calculation of steel in the strips is done as for the base.

12.3.6.1 Cantilever moment in wall slab

The cantilever moment is greatest in the central vertical strip. Pressures occur only in strips 15-20. The bending moment M at the base of the cantilever is given by the product of the pressures at the 250 mm wide strips and the distance from the base to the centre of the strips.

$$M = 0.250 \times (26.56 \times 1.375 + 28.39 \times 1.125 + 30.22 \times 0.875 + 32.05 \times 0.625$$
$$+ 33.89 \times 0.375 + 35.72 \times 0.125) = 33.03 \text{ kNm/m}$$
$$\text{Width of vertical strip} = 830 \text{ mm,}$$
$$\text{M on the strip} = 33.03 \times 0.83 = 27.41 \text{ kNm}$$

Steel required can be calculated in the usual way.

12.3.7 Counterfort Design using Hillerborg's Strip Method

The reactions from the horizontal strips of the wall slab act as horizontal forces on the counterfort. At any level, the force R on the counterfort from the 250 mm wide strips on either side of the counterfort is (Fig.12.17)

$$R = 2 \times p \times a \times 0.250$$

This is calculated at the centre of each strip. From the calculated value of R, shear force and bending moment at different levels in the counterfort can be calculated. The detailed calculations are shown in Table 12.5. The distribution of shear force and bending moment are shown in Fig.12.21 and Fig.12.22.

The depth of the counterfort is 180 mm at top and increasing to 3500 mm at the bottom. Assuming 40 mm cover and 16 mm bars, the effective depth at different levels is calculated. The width of the counterfort is 250 mm. The back of the counterfort is inclined at an angle θ to the horizontal where from Fig.12.12,

$$\theta = \tan^{-1}(4750/33200) = 55°.$$

At any level, the area of steel required is given by

$$A_s = M/ (z \times 0.95 \times f_y \times \sin 55)$$

Note that because of the fact that the tension steel is placed parallel to the back of the counterfort as shown in Figure 12.11, only the vertical component the force in the steel is taken into account. The required area of steel is very small because of the very large effective depth of the counterfort.

Table 12.5 Shear force and bending moment in counterfort

SF (kN)	M (kNm)	h (mm)	D (mm)	k	z/d	A_s (mm²)
0	0	267	219	0	0.95	0
1.14	0.29	442	394	0.000	0.95	0.2
4.58	1.43	617	569	0.001	0.95	0.7
10.30	4.01	792	744	0.001	0.95	1.6
18.32	8.59	966	918	0.001	0.95	2.7
28.62	15.74	1141	1093	0.002	0.95	4.2
41.21	26.04	1316	1268	0.002	0.95	6.0
56.09	40.07	1491	1443	0.003	0.95	8.1
73.27	58.38	1665	1617	0.003	0.95	10.6
92.73	81.57	1840	1792	0.003	0.95	13.4
114.48	110.19	2015	1967	0.004	0.95	16.4
138.52	144.82	2189	2141	0.004	0.95	19.8
164.85	186.03	2364	2316	0.005	0.95	23.6
193.47	234.40	2539	2491	0.005	0.95	27.6
224.38	290.49	2714	2666	0.005	0.95	32.0
252.04	353.50	2888	2840	0.006	0.95	36.5
275.70	422.43	3063	3015	0.006	0.95	41.1
294.59	496.08	3238	3190	0.007	0.95	45.6
307.95	573.06	3413	3365	0.007	0.95	50.0
315.01	651.82	3587	3539	0.007	0.95	54.0

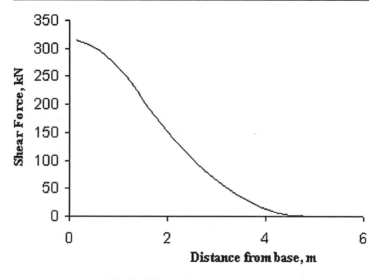

Fig.12.21 Shear force in counterfort.

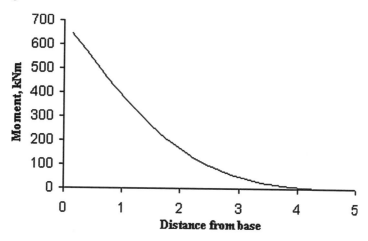

Fig.12.22 Bending moment in counterfort.

As shown in Fig.12.11, it is essential to tie the counterfort and the wall slab together by horizontal links to resist the force R in tension. Similarly, the counterforts must be anchored to the base slab by vertical links as\shown in Fig.12.11.

CHAPTER 13

DESIGN OF STATICALLY INDETERMINATE STRUCTURES

13.1 INTRODUCTION

Design of structures in structural concrete involves satisfying
- The serviceability limit state (SLS) criteria
- The ultimate limit state (ULS) criteria

Design for ULS is concerned with safety and this means ensuring that the ultimate load of the structure is at least equal to the design ultimate load. The theoretical principles used in design at ULS are based on Classical Theory of Plasticity which was developed for the design of steel structures with unlimited ductility. Fig.13.1 shows the moment-curvature relationship for a steel section. As can be seen, once the ultimate or plastic moment capacity is reached, for further changes in curvature and hence increasing deformation, the moment capacity is maintained provided that the compression flanges do not buckle.

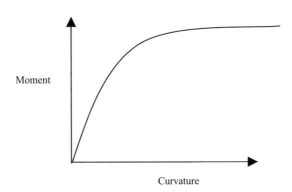

Moment

Curvature

Fig.13.1 Moment-curvature relationship for a steel section.

Assuming that unlimited ductility can be relied upon, then according to the Theory of Plasticity, at ultimate load the state of stress has to satisfy the following three conditions.
- Equilibrium condition: The state of stress must be in equilibrium with the ultimate load. One convenient way of obtaining a set of stresses in equilibrium with external loads is to do an elastic analysis of a structure under a load equal to the ultimate load. It does not in any way imply that the designed structure behaves elastically under the applied ultimate load. Theoretically it is permissible to use elastic analysis or any variation of it

as long as the stresses are in equilibrium with the external loads. The implication of this statement for the design of reinforced concrete structures will be discussed later.

- Yield Condition: The state of stress must not violate the yield condition for the material. This means for example, that for any combinations of bending moment and axial force, the capacity of the column should not exceeded the limits as defined by column design chart (section 9.3, Chapter 9). In members in framed structures primarily subjected to bending moment and shear forces, adequate reinforcement is provided such that the moment and shear capacity of the section is at least equal to the design forces at that section.

- Mechanism Condition: Sufficient yielded zones must be present to convert the structure in to a mechanism, indicating that there is no reserve load capacity left. In the case of framed structures this means that there must be sufficient plastic hinges and in the case of plate structures sufficient 'Yield lines' (Chapter 8) to convert the structure in to a mechanism.

When using the methods based on the classical theory of plasticity to design structures in structural concrete, it is important to recognize the fact that unlike steel, reinforced concrete is a material of very limited ductility. Fig.13.2 shows by the discontinuous line the moment-curvature relationship for a reinforced concrete section. After the maximum moment capacity is reached, the capacity is maintained for a limited increase in curvature beyond the curvature at maximum capacity. For curvature beyond this value, the moment capacity *decreases*. It is therefore necessary to ensure at no section is the curvature so large that the moment capacity decreases significantly before the structure collapses.

The need to pay attention to ductility and its effect on ultimate strength as well as serviceability behaviour is explained by two examples.

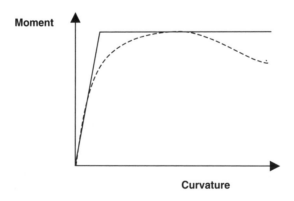

Fig.13.2 Idealized and actual moment-curvature relationship.

13.2 DESIGN OF A PROPPED CANTILEVER

Consider the design of a propped cantilever of 6 m span as shown in Fig.13.3. It is required to support at mid-span an ultimate load W equal to 100 kN. The design can be carried out in several ways as follows.

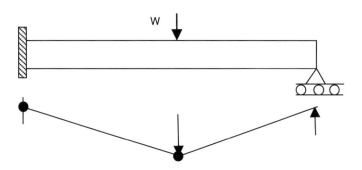

Fig.13.3 Propped cantilever.

(i) Design 1 based on elastic bending moment distribution

In a propped cantilever supporting a mid-span load W over a span L, from elastic analysis, the moments at support and mid-span are respectively 3WL/16 and 5WL/32. If W = 100 kN and L = 6 m, the corresponding moments are 112.5 kNm and 93.75 kNm respectively. If the beam is designed for these moments, then assuming for simplicity that moment-curvature is elastic-perfectly plastic as shown by full line in Fig.13.2, plastic hinges will form *simultaneously* at the support and mid-span sections and the beam will collapse. Up to the collapse load, there is no rotation at the built-in support and the beam behaves as an elastic structure right up to collapse.

Of course this is a very simplified picture as to what really happens when a beam is tested, because cracking and other non-linear behaviour start almost right from the beginning and the moment-curvature is more like that shown by dotted line in Fig.13.2. However the grossly simplified elastic-perfectly assumption for moment-curvature relationship is sufficient for the present discussion.

(ii) Design 2 based on modified elastic bending moment distribution

Instead of designing the beam using the elastic moment distribution, let the beam be designed for a support moment equal to 80% of elastic value of 112.5 kNm, i.e. 90 kNm. The moment at mid-span for equilibrium at the ultimate load is given by (WL/4 – Support moment /2) = 100 × 6/4 – 90/2 = 105 kNm which is 112% of the corresponding moment of resistance at mid-span in design 1.

In the elastic state, the maximum bending moment is at the support. Since the design moment at the support is 90 kNm, which is only 80% of the corresponding elastic moment at a load of 100 kN, the first plastic hinge will form at the support at a load of 80 kN. Up to the stage when the first plastic hinge forms, the beam behaves as an elastic propped cantilever and the rotation at the built in support is zero. The moment at the mid-span is $5/32 \times (80 \times 6) = 75$ kNm which is less than the moment capacity of the section which is 105 kNm.

For a load greater than 80 kN since a plastic hinge has formed at the support, the moment there cannot increase any further but moment at mid-span can increase until a second plastic hinge forms at mid-span and the beam collapses. Therefore for the load stage from 80 kN to 100 kN, the beam behaves as if loaded by a concentrated load at mid-span and a support moment equal to 90 kNm. The *additional behaviour* of the beam beyond a load of 80 kN can be computed by treating the beam as a simply supported beam. During this stage, the support section continues to rotate. The elastic rotation θ at the support in a simply supported beam of flexural rigidity EI and loaded at mid-span by a load P is given by

$$\theta = P \, L^2 / (16 \, EI)$$

Substituting

$$P = (100 - 80) = 20 \text{ kN and } L = 6, EI \, \theta = 45$$

At this stage the moment at mid-span is equal to the moment capacity of 105 kNm and the beam collapses by the formation of plastic hinges at the support and at mid-span.

Comparing the two designs, both beams collapse by the formation of plastic hinges at support and at mid-span. However in Design 1, the two plastic hinges form simultaneously and there was no rotation at the built in support right up to collapse. However in Design 2 with the support moment capacity of only 80% of the elastic value as used in Design 1, the support section has to rotate from the load equal to 80 kN at which the first plastic hinge forms right up to collapse load of 100 kN with the moment at the support remaining at the value of 90 kNm. The support section had to undergo substantial rotation while continuing to maintain a moment of 90 kNm. In other words, the section needs to have sufficient ductility between 80 kN to ultimate load of 100 kN to ensure that there is no decrease in moment capacity.

(iii) Design 3 based on greater modification to elastic bending moment distribution than Design 2

In this case the beam is designed for a support moment of 67.5 kNm (60% of elastic value of 112.5 kNm). The moment at mid-span for equilibrium at the ultimate load is equal to $(100 \times 6/4 - 67.5/2) = 116.25$ kNm. Carrying out the calculations as was done for Design 2, the first plastic hinge forms at the support at a load of 60 kN. Up to the stage when the first plastic hinge forms, the beam behaves as propped cantilever and the rotation at the support is zero. The moment at mid-span is $5 \times 60 \times 6/32 = 56.25$ kNm which is less than the capacity of the section which is 116.25 kNm. Since a plastic hinge has formed at the support at 60 kN, the moment there cannot increase any further. However, since the ultimate load to be supported is 100 kN, for the load stage from 60 kN to 100 kN, the beam behaves as if loaded by the concentrated load at mid-span and a support moment equal to 67.5 kNm. Substituting $P = (100 - 60) = 40$ kN and $L = 6$, EI $\theta = 90$. At this stage the moment at the mid-span also reaches a value equal to the moment capacity of 116.25 kNm and the beam collapses by the formation of plastic hinges at the support and mid-span.

Comparing the three designs, all three beams collapse by the formation of plastic hinges at support and at mid-span. However at the stage when the load on the beam is at its ultimate value, the rotation θ at the built in support for the three designs considered are EI θ = 0, 45 and 90 respectively. Thus the smaller the designed support moment capacity is compared with the elastic value, the larger is the rotation at the support. This is shown in Fig.13.4.

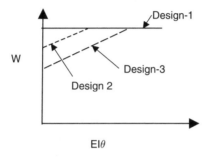

Fig.13.4 Load-support rotation relationship.

During the stage when the support is rotating from load at which first plastic hinge forms to ultimate load, the moment at the support has to remain constant at the designed value. The larger the load range, the larger the resulting rotation and greater is the demand placed on the ductility of the section. Sections that yield earlier in the loading history are the ones where there is the possibility of moment capacity reducing due to increasing curvature. The greater the difference between the load at which the first plastic hinge forms and the ultimate load, the greater will be the required plastic hinge rotation. It is important therefore that the difference between the ultimate load and the load at which the first section yields is made as small as possible.

What the above example has demonstrated is that it is perhaps possible to design a structure using a bending moment distribution different from the elastic moment distribution, provided sufficient ductility could be assured. Otherwise the assumption that the moment will remain constant during the rotation of the plastic hinge becomes invalid leading to unsafe design. *It is therefore desirable that while designing, that the `stress distribution` used in design departs from elastic `stress distribution` as little as possible.*

13.3 DESIGN OF A CLAMPED BEAM

The idea that although a design might satisfy the ULS criteria, it might be unacceptable from an SLS point of view is demonstrated by the following example. Consider the design of a beam spanning a distance L between two walls and subjected to a uniformly distributed load q. Fig.13.5 shows three bending moment diagrams, all of which are in equilibrium with a load of q. From an ultimate limit state (ULS) point of view, one can design the beam using any one of the three bending moment distributions.

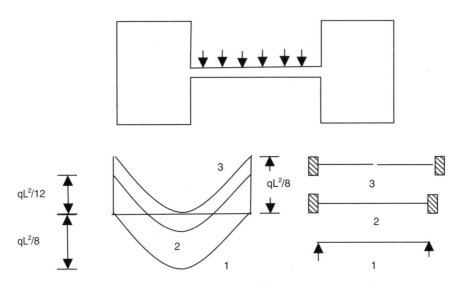

Fig.13.5 Alternative designs for a clamped beam.

Design 1: Assume that the beam behaves as a simply supported beam. The bending moment at mid-span is $qL^2/8$. In this case clearly only steel at the bottom face is required. The moment of resistance at the support is zero and the first plastic hinges at the supports form at essentially zero load while the plastic hinge at mid-span forms at the ultimate load. The support hinge starts rotating right from the start leading to large cracks there. While this design is satisfactory from a ULS point of view, it is clearly an unsatisfactory design from a serviceability limit state (SLS) point of view.

Design 2: Assume that the beam behaves as a clamped beam. From elastic analysis, bending moment at the junction with the wall is $qL^2/12$ and at mid-span is $qL^2/24$. The plastic hinges at support and at mid-span form simultaneously. This design is satisfactory from both the ULS and SLS points of view, because the design corresponds to the behaviour of the beam taking into account the proper boundary conditions.

Design 3: Assume that the beam behaves as a pair of cantilevers. Bending moment at the junction with the wall is $qL^2/8$. In this case clearly only steel at the top face is required. The moment of resistance at the mid-span is zero and the first plastic hinge at mid-span forms at essentially zero load while the plastic hinges at supports form at the ultimate load. The mid-span hinge starts rotating right from the start leading to large cracks there. While this design is satisfactory from a ULS point of view, it is clearly an unsatisfactory design from a serviceability limit state (SLS) point of view.

As shown in Fig.13.5, the bending moment distribution in Design 2 is the elastic distribution and requires both top and bottom reinforcement. The bending moment distributions used in Design 1 requires only bottom reinforcement and Design 3 requires only top reinforcement. They are extreme variations of the elastic moment distribution. This example demonstrates the need to pay particular attention to both ULS and SLS aspects, keeping in mind the rather limited ductility available in the case of reinforced concrete sections. Once again the example demonstrates that using elastic distribution of moments is likely to lead to satisfactory designs from both ULS and SLS points of view.

13.4 WHY USE ANYTHING OTHER THAN ELASTIC VALUES IN DESIGN?

One question that naturally arises is why not simply use the elastic values of moments and avoid all problems of ductility? The reason for using values of moments other than the elastic values is purely a matter of convenience. Generally at support sections in frame structures, flat slabs and such structures there is considerable congestion of steel due to the fact that flexural steel in two directions at top and bottom of the beam or slab are required. In addition, steel in the column and shear links need to be accommodated in the same congested area. Therefore any reduction of steel in this zone is an advantage from the point of view of detailing. Using moments at supports smaller than the elastic values helps in mitigating the problem. Elastic 'stresses fields' often contain zones of stress concentration and it is useful to modify these stress distributions in the interests of a 'smoothed out' stress distribution which leads to a more convenient detailing of reinforcement.

13.5 LIMITS ON DEPARTURE FROM ELASTIC MOMENT DISTRIBUTION IN BS 8110

Considerable experimental evidence shows that a satisfactory design can be obtained on the basis of reasonably small adjustments to the elastic bending moment distribution. In general in framed structures reductions *of moments up to 30% of the elastic moments can be tolerated* without making excessive demands on the ductility of the structure. It is worth pointing out that as demonstrated in section 13.2, *ductility demand is increased by the use of moment values smaller than the elastic values*. However as the ductility demand is unaffected by values of moment above the elastic values, *there is no limit to the use of moment values larger than the elastic values*. In the case of flexural members, one way of ensuring that sufficient ductility is available is to limit the maximum depth of neutral axis. *Larger reduction in moments from the elastic values will require smaller maximum depth of neutral axis so that steel yields well before concrete reaches maximum strain.*

To take account of these factors, the code BS 8110 sets out the procedure for adjusting the elastic moment distribution for design. This process is called moment redistribution and the constraints on redistribution are set out in section 3.2.2. This section states that a redistribution of moments obtained by a rigorous elastic analysis or by other simplified methods set out in the code may be carried out provided that the following hold:

1. Equilibrium between internal and external forces is maintained under all appropriate combinations of design ultimate load.
2. Where the design ultimate resistance moment at a section is reduced by redistribution from the largest moment within that region, the neutral axis depth x should satisfy the condition

$$(x/d) \le (\beta_b - 0.4)$$

where β_b = (Moment after redistribution/Moment before redistribution) ≤ 1.0

The moments before and after redistribution at a section are to be taken from the respective maximum moment diagrams. This provision ensures that there is adequate rotation capacity at the section for redistribution to take place.

13.5.1 Moment of Resistance

In the case of rectangular beams *without compression reinforcement*, the maximum depth of stress block is

$$0.9x = 0.9 (\beta_b - 0.4) d$$

The average compressive stress in the stress block is $0.447 f_{cu}$

Total compressive force = $0.447 f_{cu} (0.9x) b$, Lever arm = $d - 0.45 x$

The moment of resistance is

$$M = 0.447 f_{cu} (0.9x) b \{d - 0.45x\}$$

Substituting for x,

$$M = bd^2 f_{cu} \{0.402 (\beta_b - 0.4) - 0.18 (\beta_b - 0.4)^2\}$$
$$k = M/ (bd^2 f_{cu}) = 0.402 (\beta_b - 0.4) - 0.18 (\beta_b - 0.4)^2$$

Table 13.1 shows the variation of k with β_b. The larger the percentage of redistribution the smaller the value of moment that singly reinforced sections can resist before the need for compression steel arises.

Table 13.1 Variation k with β_b

β_b	x/d	k
0.70	0.30	0.1044
0.75	0.35	0.1187
0.80	0.40	0.1320
0.85	0.45	0.1445
0.90	0.50	0.1560

13.5.2 Serviceability Considerations

Elastic bending moment at ULS: Fig13.6 shows the elastic bending moment distribution in a uniformly loaded clamped beam. If the ultimate design load is q, then from elastic analysis the bending moments at the support and mid-span are respectively $qL^2/12$ and $qL^2/24$. The points of contra-flexure are at 0.211L from the fixed ends.

Elastic bending moment at SLS: At ULS the load factor for dead and live loads are respectively 1.4 and 1.6 or an average value of approximately 1.5. The load at SLS is q/1.5 = 0.7 q. Since at SLS, the beam is more likely to behave elastically, the bending moments at SLS is 0.7 times the bending moment values calculated by elastic analysis at ULS.

Redistributed bending moment at ULS: If the bending moments at the supports are redistributed by 30%, then at ULS the bending moments at the support and mid-span after redistribution are respectively $0.7qL^2/12$ and $1.6 \ qL^2/24$. For this distribution of bending moments the points of contra flexure are at 0.135L from the fixed end.

Fig.13.6 shows the SLS and redistributed bending moments at ULS. Because of the shift in the position of contra flexure points, at certain sections in the beam the bending moment at SLS is *larger* than the redistributed bending moments at ULS. During design, it is necessary to ensure that the moment of resistance is equal to larger of the SLS and redistributed ULS moment at the section. It is for this reason that the code in clause 3.2.2.1 (c) states that the resistance moment at any section should be at least 70% of the moment at that section obtained from an elastic maximum moments diagram covering all appropriate combinations design ultimate load. This condition also implies that the maximum redistribution permitted is 30%, i.e. the largest moment given in the elastic maximum moments diagram may be reduced by up to 30%.

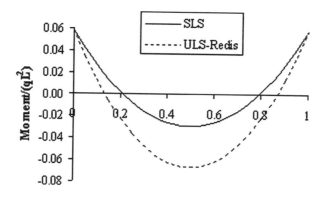

Fig.13.6 Bending moment distribution at ULS and SLS.

13.6 CONTINUOUS BEAMS

13.6.1 Continuous Beams in *in-situ* Concrete Floors

Continuous beams are a common element in *cast-in-situ* construction. A reinforced concrete floor in a multi-storey building is shown in Fig.13.7. The floor action to support the loads is as follows:

1. The one-way slab is supported on the edge frame, intermediate T-beams and centre frame. It spans transversely across the building.

2. Intermediate T-beams on line AA span between the transverse end and interior frames to support the floor slab.

3. Transverse end frames DD and interior frames EE span across the building and carry loads from intermediate T-beams and longitudinal frames.

4. Longitudinal edge frames CC and interior frame BB support the floor slab.

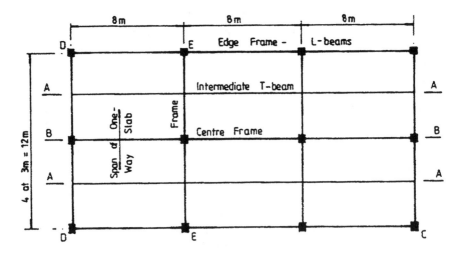

Fig.13.7 Floor in a multi-storey building.

The horizontal members of the rigid frames may be analysed as part of the rigid frame. This is discussed in of BS 8110: Part 1, section 3.2.1. The code gives a continuous beam simplification in clause 3.2.1.2.4 where moments and shears may be obtained by taking the members as continuous beams over supports with the columns providing no restraint to rotation (Chapter 3, section 3.4.2). The steps in design of continuous beams are the same as those set out in Chapter 4, section 4.4.3 for simple beams except for the limit on the depth of the neutral axis depending on the amount of redistribution done.

13.6.2 Loading on Continuous Beams

13.6.2.1 Arrangement of loads to give maximum moments

The loading is to be applied to the continuous beam to give the most adverse conditions at any section along the beam. If G_k is the characteristic dead load and Q_k is the characteristic imposed load. qualitative Influence lines obtained using Muller–Breslau's principle show that in any continuous beam, the following two basic loading patterns need to be investigated.

1. maximum moment in a span of a beam occurs when that span and every alternate span are loaded by $(1.4G_k + 1.6Q_k)$ and the rest of the spans by $1.0\ G_k$

2. maximum moment at a support in a beam occurs when spans on either side of the support and every alternate span are loaded by $(1.4G_k + 1.6Q_k)$ and the rest of the spans by $1.0\ G_k$

In order to reduce the number of load cases to be analysed, BS 8110: Part 1, clause 3.2.1.2.2 prescribes only the following load cases to be analysed:

1. All spans are loaded with the maximum design ultimate load $1.4G_k + 1.6Q_k$;
2. Alternate spans are loaded with the maximum design ultimate load $(1.4G_k + 1.6Q_k)$ and all other spans are loaded with the minimum design ultimate load $1.0G_k$.

13.6.2.2 Example of critical loading arrangements

The total dead load on the floor in Fig.13.7 including an allowance for the ribs of the T-beams, screed, finishes, partitions, ceiling and services is 6.6 kN/m² and the imposed load is 3 kN/m². Calculate the design load and set out the load arrangements to comply with BS 8110: Part 1, clause 3.2.1.2.2, for the continuous T-beam on lines AA and BB.

$$G_k = 3 \times 6.6 = 19.8 \text{ kN/m}$$
$$Q_k = 3 \times 3 = 9 \text{ kN/m}$$
$$1.4\ G_k + 1.6\ Q_k = (1.4 \times 19.8) + (1.6 \times 9) = 42.12 \text{ kN/m}$$

The loading arrangements are shown in Fig.13.8.

13.6.2.3 Loading from one-way slabs

Continuous beams supporting slabs designed as spanning one-way can be considered to be uniformly loaded. The slab is assumed to consist of a series of beams as shown in Fig.13.9. Note that some two-way action occurs at the corners of one-way slabs.

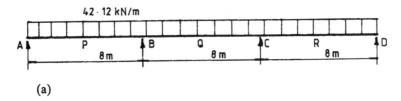

(a)

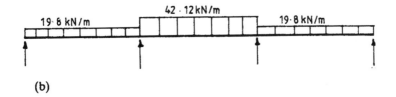

(b)

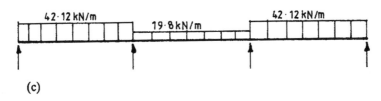

(c)

Fig.13.8 (a) Case 1, all spans loaded with $1.4G_k + 1.6Q_k$; (b) case 2, central span loaded with $(1.4G_k+1.6Q_k)$; (c) case 3, end spans loaded with $1.4G_k + 1.6Q_k$.

13.6.2.4 Loading from two-way slabs

If the beam is designed as spanning two-ways, the four edge beams assist in carrying the loading. The load distribution normally assumed for analyses of the edge beams is shown in Fig.13.10 where lines at 45° are drawn from the corners of the slab. This distribution gives triangular and trapezoidal loads on the edge beams as shown in the Fig.13.10.

The fixed end moments for the two load cases shown in Fig.13.10(b) and Fig.13.10(c) are as follows.

(i) *Trapezoidal load*: The load is broken down into a uniform central portion

$$\frac{(\ell_y - \ell_x)}{(\ell_y - 0.5\,\ell_x)}\,W_1$$

and two triangular end portions each

$$\frac{0.25\,\ell_x}{(\ell_y - 0.5\,\ell_x)}\,W_1$$

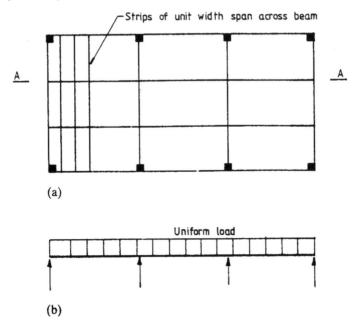

Fig.13.9 (a) Floor plan; (b) beam AA.

where W_1 is the total load on one span of the beam, ℓ_x is the short span of the slab and ℓ_y is the long span of the slab.

The fixed end moments for the two spans in the beam on AA are

$$M_1 = \frac{W_1}{(\ell_y - 0.5\,\ell_x)}\frac{1}{96\,\ell_y}[8\,\ell_y^3 - 4\,\ell_y\,\ell_x^2 + \ell_x^3]$$

(ii) *Triangular load:* The fixed end moments for the two spans in the beam on line BB in Fig.13.10(a) are

$$M_2 = \frac{5\,W_2\,\ell_x^2}{48}$$

where W_2 is the total load on one span of the beam.

13.6.2.5 Alternative distribution of loads from two-way slabs

BS 8110: Part 1, Fig. 3.10, gives the distribution of load on a beam supporting a two-way spanning slab. This distribution is shown in Fig.13.11(a). The design loads on the supporting beams are as follows:

Long span $v_{sy} = \beta_{vy}\,n\,\ell_x$ kN/m

Short span $v_{sx} = \beta_{vx}\,n\,\ell_x$ kN/m

where n is the design load per unit area in kN/m^2. Values of the coefficients β_{vx} and β_{vy} are given in Table 3.15 of the code and are derived in Chapter 8, section 8.9.16.

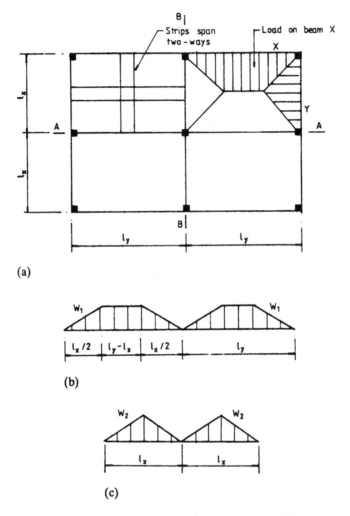

Fig.13.10 (a) Floor plan; (b) beam on AA; (c) beam on BB.

13.6.3 Analysis for Shear and Moment Envelopes

The following methods of analysis can be used to find the shear forces and bending moments for design:

1. Analyses using the matrix stiffness method
2. using coefficients for moments and shear from BS 8110: Part 1, Table 3.5.

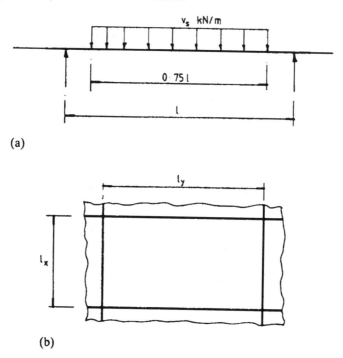

(a)

(b)

Fig.13.11 (a) Load distribution; (b) two-way slab.

In using method 1, the beam is analysed for the various load cases, the shear force and bending moment diagrams are drawn for these cases and the maximum shear and moment envelopes are constructed. Precise values are then available for moments and shear at every point in the beam. This method must be used for two span beams and beams with concentrated loads not covered by Table 3.5 of the code. It is also necessary to use rigorous elastic analysis if moment redistribution is to be made, as set out in section 13.5.

BS 8110: Part 1, clause 3.4.3, gives moments and shear forces in continuous beams with uniform loading. The design ultimate moments and shear forces are given in Table 3.5 in the code which is reproduced as Table 13.2 here.

Table 13.2 Design ultimate bending moments and shear forces for continuous beams

	At outer support	*Near middle of end span*	*At first interior support*	*At middle of interior span*	*At interior supports*
Moment	0	0.09 Fℓ	−0.11 Fℓ	0.07 Fℓ	−0.08 Fℓ
Shear	0.45 F	-	0.6 F	-	0.55 F

ℓ. effective span; F. total design ultimate load on a span equal to $1.4G_k + 1.6Q_k$

The use of coefficients in the table is subject to the following conditions:
1. The characteristic imposed load Q_k may not exceed the characteristic dead load G_k;
2. The loads should be substantially uniformly distributed over three or more spans;
3. Variations in span length should not exceed 15% of the longest span.

The code also states that no redistribution of moments calculated using this table should be made.

13.7 EXAMPLE OF ELASTIC ANALYSIS OF A CONTINUOUS BEAM

(a) Specification
Analyse the continuous beam for the three load cases shown in Fig.13.8 and draw the separate shear force and bending moment diagrams. Construct the maximum shear force and bending moment envelopes. Also calculate the moments and shears using the coefficients from BS 8110: Part 1. Table 3.5.

(b) Analysis by Stiffness method

The fixed end moments are
$$\textbf{Case 1}: \text{All spans: } M = 42.12 \times 8^2/12 = 224.64 \text{ kNm}$$
$$\textbf{Case 2}: \text{Spans AB and CD: } M = 19.8 \times 8^2/12 = 105.60 \text{ kNm}$$
$$\text{Span BC: } M = 42.12 \times 8^2/12 = 224.64 \text{ kNm}$$
$$\textbf{Case 3}: \text{Spans AB and CD: } M = 42.12 \times 8^2/12 = 224.64 \text{ kNm}$$
$$\text{Span BC: } M = 19.8 \times 8^2/12 = 105.60 \text{ kNm}$$

Assuming uniform flexural rigidity EI and equal spans of 8 m, the stiffness matrix K and the load vectors F for the three load cases are

$$K = \frac{EI}{8}\begin{bmatrix} 4 & 2 & 0 & 0 \\ 2 & 8 & 2 & 0 \\ 0 & 2 & 8 & 2 \\ 0 & 0 & 2 & 4 \end{bmatrix}\begin{bmatrix} \theta_A \\ \theta_B \\ \theta_C \\ \theta_D \end{bmatrix}, F = \begin{bmatrix} 224.64 & 105.60 & 224.64 \\ 0 & 119.04 & -119.04 \\ 0 & -119.04 & 119.04 \\ -224.64 & -105.60 & -224.64 \end{bmatrix}$$

The loading and structure are symmetric. Therefore
$$\theta_A = -\theta_D, \theta_B = -\theta_C$$
The stiffness relationship can therefore be condensed to
$$K = \frac{EI}{8}\begin{bmatrix} 8 & 4 \\ 4 & 12 \end{bmatrix}\begin{bmatrix} \theta_A \\ \theta_B \end{bmatrix}, F = \begin{bmatrix} 449.28 & 211.20 & 449.28 \\ 0 & 238.08 & -238.08 \end{bmatrix}$$

The displacements are
$$\frac{EI}{8}\begin{bmatrix} \theta_A = -\theta_D \\ \theta_B = -\theta_C \end{bmatrix} = \begin{bmatrix} 67.392 & 19.776 & 79.296 \\ -22.464 & 13.248 & -46.272 \end{bmatrix}$$

Using clockwise moment as positive, the bending moments at the ends of a span are obtained from the following equations

$$M_{left} = \text{Fixed end moment} + (EI/L) \{4\,\theta_{Left} + 2\,\theta_{Right}\}$$
$$M_{Right} = \text{Fixed end moment} + (EI/L) \{2\,\theta_{Left} + 4\,\theta_{Right}\}$$

The reactions R are given by
$$R_{left} = 0.5\,q\,L + (M_{left} - M_{right})/L$$
$$R_{Right} = 0.5\,q\,L - (M_{left} - M_{right})/L$$

The shear force V and bending moment M at a section x from the left hand support are given by

$$V = R_{left} - q\,x$$
$$M = M_{left} - R_{left}\,x + 0.5\,q\,x^2$$

where L = span (8 m) and q = uniformly distributed loading.

Table 13.3 Summary of elastic analysis

	Beam AB			Beam BC		
	Case 1	Case 2	Case 3	Case 1	Case 2	Case 3
R_{Left}	134.78	54.43	143.71	168.48	168.48	79.2
R_{Right}	202.18	103.97	193.25	168.48	168.48	79.2
M_{Left}	0	0	0	269.57	198.11	198.1
M_{Right}	269.57	198.11	198.11	269.57	198.11	198.1
M_{Max} in Span	215.65	74.82	245.17	67.39	138.85	−39.71
M_{max} at x	3.2	2.75	3.41	4.0	4.0	4.0

Table 13.4 Elastic moment and shear calculations for beam AB

x	Case 1	Case 2	Case 3	Moment		Shear force	
				Max.	Min.	Max.	Min.
0	0.00	0.00	0.00	0.00	143.71	0.00	54.43
1	−113.72	−44.53	−122.65	−44.53	101.59	−122.65	34.63
2	−185.33	−69.26	−203.18	−69.26	59.47	−203.18	14.83
3	−214.81	−74.20	−241.60	−74.20	17.35	−241.60	−4.97
4	−202.18	−59.33	−237.89	−59.33	−24.77	−237.89	−33.70
5	−147.42	−24.66	−192.06	−24.66	−44.57	−192.06	−75.82
6	−50.54	29.81	−104.11	29.81	−64.37	−104.11	−117.94
7	88.45	104.08	25.96	104.08	−84.17	25.96	−160.06
8	269.57	198.14	198.14	269.57	−103.97	198.14	−202.18

Table 13.5 Elastic moment and shear calculations for beam BC

x	Case 1	Case 2	Case 3	Moment		Shear force	
				Max.	Min.	Max.	Min.
0	269.57	198.14	198.14	269.57	198.14	168.48	168.48
1	122.15	50.72	128.81	128.81	50.72	126.36	126.36
2	16.85	−54.58	79.31	79.31	−54.58	84.24	84.24
3	−46.33	−117.76	49.61	49.61	−117.76	42.12	42.12
4	−67.39	−138.82	39.71	39.71	−138.82	0.00	0.00
5	−46.33	−117.76	49.61	49.61	−117.76	−19.8	−42.12
6	16.85	−54.58	79.31	79.31	−54.58	−39.6	−84.24
7	122.15	50.72	128.81	128.81	50.72	−59.4	−126.36
8	269.57	198.14	198.14	269.57	198.14	−79.2	−168.48

The results are summarised in Table 13.3. Detailed calculations for beam AB and beam BC are shown in Tables 13.4 and 13.5 respectively. The bending moment and shear force diagrams for the three load cases are shown in Fig.13.12 (a) and Fig.13.12 (b) respectively.

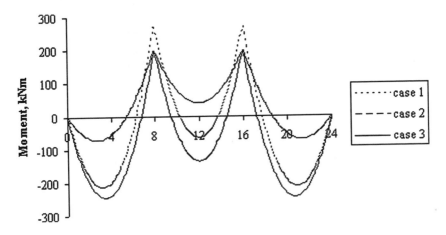

Fig.13.12 (a) Bending moment diagrams.

(c) Analysis using BS 8110: Part 1, Table 3.5

The values of the maximum shear forces and bending moments at appropriate points along the beam are calculated using coefficients from BS 8110: Part 1. Table 3.6. Total design ultimate load per span F

$$F = 42.12 \times 8 = 336.96 \text{ kN, } L = 8 \text{ m}$$

The values of moments and shears are tabulated in Table 13.6.

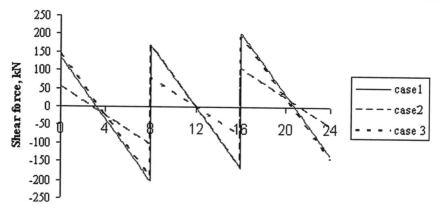

Fig.13.12 (b) Shear force diagrams

Table 13.6 Moments and shears in continuous beam

	Position	BS 8110, Table 3.5 value	Elastic analysis
Shear forces	A	0.45 × 336.96 = 151.63	143.71
	B	0.6 × 336.96 = 202.18	202.18
Bending moments	P	0.09 × 336.96 × 8 = 242.61	245.17
	B	−0.11 × 336.96 × 8 = −296.52	−269.57
	Q	0.07 × 336.96 × 8 = 188.69	138.82

Generally the two results are in reasonable agreement except for the support moment at B. The reason for this is that in the elastic analysis, the loading pattern to give maximum support moment as indicated by the influence line has been ignored. If the beam had been analysed for a load of $1.4G_k + 1.6Q_k$ on spans AB and BC and $1.0G_k$ on span CD so as to give the maximum support moment, the following results would have been obtained.

Load vector F and the displacement vectors are

$$F = \begin{bmatrix} 224.64 \\ 0 \\ -119.04 \\ -105.60 \end{bmatrix}, \frac{EI}{8} \begin{bmatrix} \theta_A \\ \theta_B \\ \theta_C \\ \theta_D \end{bmatrix} = \begin{bmatrix} 63.4240 \\ -14.5280 \\ -5.3119 \\ -23.7443 \end{bmatrix}$$

The support moments are

$M_{AB} = 0$, $M_{BA} = 293.38$, $M_{BC} = -293.38$, $M_{CB} = 174.34$, $M_{CD} = -174.34$, $M_{DC} = 0$

Thus the correct maximum hogging moment at support B would be 293.38 kN m, which agrees with the values from Table 3.5 of the code.

13.8 EXAMPLE OF MOMENT REDISTRIBUTION FOR A CONTINUOUS BEAM

As explained in section 13.4, redistribution gives a more even arrangement for the reinforcement, relieving congestion at supports. It might also lead to a saving in the amount of reinforcement required.

(a) Specification

Referring to the three-span continuous beam analysed in section 13.7 above, redistribute the moments after making a 20% reduction in the maximum hogging moment at the interior support. Draw the envelopes for maximum shear force and bending moment.

(b) Moment redistribution

The maximum elastic hogging moment over the supports B and C in case 1 (Table 13.5), is 269.57. If this is *reduced* by 20%, the hogging moment over the support is $0.8 \times 269.57 = 215.65$. The support section is therefore designed for a moment of 215.65 kNm. Reducing the hogging moment increases the corresponding span moments.

The maximum elastic hogging moment over the supports B and C in case 2 and 3 (Table 13.3) is 198.11. Since the support section is designed to resist a moment of 215.65 kNm, in order to *decrease* the span moments, for redistributed values, the support moment in cases2 and 3 is *increased* to 215.65.

The results are summarised in Tables 13.7 and detailed calculations for beam AB and BC are shown in Tables 13.8 and Table 13.9 respectively. Fig.13.13 (a) and Fig.13.13(b) show the redistributed bending moment and shear force diagrams

Table 13.7 Summary of redistributed Analysis

	Beam AB			Beam BC		
	Case 1	Case 2	Case 3	Case 1	Case 2	Case 3
R_{Left}	141.52	52.24	141.52	168.48	168.48	79.2
R_{Right}	195.44	106.16	195.44	168.48	168.48	79.2
M_{Left}	0	0	0	215.65	215.65	215.65
M_{Right}	215.65	215.65	215.65	215.65	215.65	215.65
M_{Max} in Span	237.76	68.92	237.76	121.31	121.31	−57.25
M_{max} at x	3.36	2.64	3.36	4.0	4.0	4.0

(c) Moment envelop for design

Having calculated the moment distribution for elastic and redistributed cases, the design envelop is constructed. At any section the design moment is larger of

- 70% of Elastic bending moment
- 100% of redistributed moment.

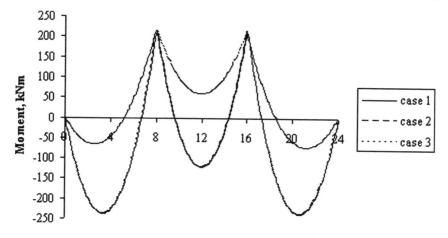

Fig.13.13 (a) Redistributed bending moment diagrams.

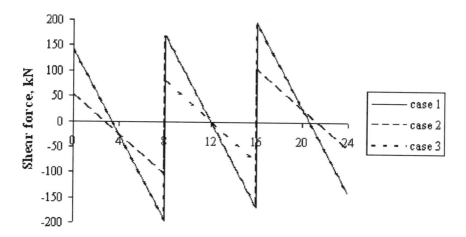

Fig.13.13 (b) Redistributed shear force diagrams.

Table 13.10 and Table 13.11 show detailed calculations for design moment envelops for beam AB and BC respectively. The moment envelops is shown in Fig.13.14(a). When Fig.13.12(a) and Fig.13.13(a) are compared, it is noted that the maximum hogging and sagging moments from the elastic bending moment envelope have both been reduced by the moment redistribution. The redistribution gives a saving in the amount of reinforcement required.

(d) Shear force envelop for design
Redistribution of moments alters the shear force distribution. It has been suggested that when it comes to determining the shear force envelop, at any section one need to take the larger of the elastic and redistributed shear force to guard against the possibility of redistribution not occurring. Detailed calculations are shown in Tables 13.13 and 13.14. Fig.13.14 (b) shows the design shear force envelop.

Table 13.8 Redistributed moment and shear calculations for beam AB

X	Case 1	Case 2	Case 3	Moment		Shear force	
				Max.	Min.	Max.	Min.
0	0.00	0.00	0.00	0.00	0.00	141.52	52.24
1	−120.46	−42.34	−120.46	−42.34	−120.46	99.40	32.44
2	−198.81	−64.89	−198.81	−64.89	−198.81	57.28	12.64
3	−235.03	−67.63	−235.03	−67.63	−235.03	15.16	−7.16
4	−229.13	−50.57	−229.13	−50.57	−229.13	−26.96	−26.96
5	−181.12	−13.72	−181.12	−13.72	−181.12	−46.76	−69.08
6	−90.98	42.94	−90.98	42.94	−90.98	−66.56	−111.20
7	41.28	119.40	41.28	119.40	41.28	−86.36	−153.32
8	215.65	215.65	215.65	215.65	215.65	−106.16	−195.44

Table 13.9 Redistributed moment and shear calculations for beam BC

X	Case 1	Case 2	Case 3	Moment		Shear force	
				Max.	Min.	Max.	Min.
0	215.65	215.65	215.65	215.65	215.65	168.48	79.2
1	68.23	68.23	146.35	146.35	68.23	126.36	59.4
2	−37.07	−37.07	96.85	96.85	−37.07	84.24	39.6
3	−100.25	−100.25	67.15	67.15	−100.25	42.12	19.8
4	−121.31	−121.31	57.25	57.25	−121.31	0	0
5	−100.25	−100.25	67.15	67.15	−100.25	−19.8	−42.12
6	−37.07	−37.07	96.85	96.85	−37.07	−39.6	−84.24
7	68.23	68.23	146.35	146.35	68.23	−59.4	−126.36
8	215.65	215.65	215.65	215.65	215.65	−79.2	−168.48

Table 13.10 Moment envelop for beam AB

X	Redistributed		Elastic		Design	
	Max.	Min.	Max.	Min.	Max.	Min.
0	0.00	0.00	0.00	0.00	0.00	0.00
1	−42.34	−120.46	−44.53	−122.65	−31.17	−120.46
2	−64.89	−198.81	−69.26	−203.18	−48.48	−198.81
3	−67.63	−235.03	−74.20	−241.60	−51.94	−235.03
4	−50.57	−229.13	−59.33	−237.89	−41.53	−229.13
5	−13.72	−181.12	−24.66	−192.06	−13.72	−181.12
6	42.94	−90.98	29.81	−104.11	42.94	−90.98
7	119.40	41.28	104.08	25.96	119.40	18.17
8	215.65	215.65	269.57	198.14	215.65	138.70

Table 13.11 Moment envelop for beam BC

x	Redistributed Max.	Redistributed Min.	Elastic Max.	Elastic Min.	Design Max.	Design Min.
0	215.65	215.65	269.57	198.11	215.65	138.68
1	146.35	68.23	128.81	50.72	146.35	35.51
2	96.85	−37.07	79.31	−54.58	96.85	−38.20
3	67.15	−100.25	49.61	−117.76	67.15	−100.25
4	57.25	−121.31	39.71	−138.82	57.25	−121.31
5	67.15	−100.25	49.61	−117.76	67.15	−100.25
6	96.85	−37.07	79.31	−54.58	96.85	−38.20
7	146.35	68.23	128.81	50.72	146.35	35.51
8	215.65	215.65	269.57	198.11	215.65	138.68

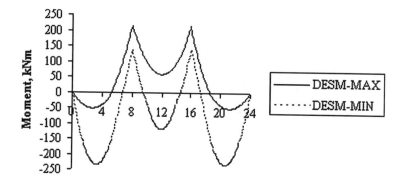

Fig.13.14 (a) Design bending moment envelop.

Table 13.12 Shear force envelop for beam AB

x	Redistributed Max.	Redistributed Min.	Elastic Max.	Elastic Min.	Design Max.	Design Min.
0	141.52	52.24	143.71	54.43	143.71	52.24
1	99.40	32.44	101.59	34.63	101.59	32.44
2	57.28	12.64	59.47	14.83	59.47	12.64
3	15.16	−7.16	17.35	−4.97	17.35	−7.16
4	−26.96	−26.96	−24.77	−33.70	−24.77	−33.70
5	−46.76	−69.08	−44.57	−75.82	−44.57	−75.82
6	−66.56	−111.20	−64.37	−117.94	−64.37	−117.94
7	−86.36	−153.32	−84.17	−160.06	−84.17	−160.06
8	−106.16	−195.44	−103.97	−202.18	−103.97	−202.18

Table 13.13 Shear force envelops for beam BC

x	Redistributed		Elastic		Design	
	Maximum	Minimum	Maximum	Minimum	Maximum	Minimum
0	168.48	79.2	168.48	79.2	168.48	79.2
1	126.36	59.4	126.36	59.4	126.36	59.4
2	84.24	39.6	84.24	39.6	84.24	39.6
3	42.12	19.8	42.12	19.8	42.12	19.8
4	0.00	0.00	0	0	0	0
5	−19.8	−42.12	−19.8	−42.12	−19.8	−42.12
6	−39.6	−84.24	−39.6	−84.24	−39.6	−84.24
7	−59.4	−126.36	−59.4	−126.36	−59.4	−126.36
8	−79.2	−168.48	−79.2	−168.48	−79.2	−168.48

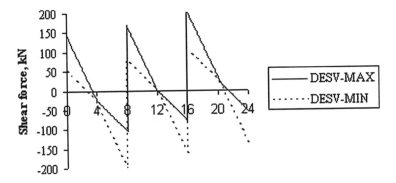

Fig.13.14 (b) Design shear force envelop.

13.9 CURTAILMENT OF BARS

The curtailment of bars may be carried out in accordance with the detailed provisions set out in BS 8110: Part 1, clause 3.12.9.1. The anchorage of tension bars at the simply supported ends is dealt with in clause 3.12.9.4 of the code.

Simplified rules for curtailment of bars in continuous beams are given in clause 3.12.10.2 and Fig. 3.24(a) of the code. The clause states that these rules may be used when the following provisions are satisfied:

1. The beams are designed for predominantly uniformly distributed loads;

2. The spans are approximately equal in the case of continuous beams.
The simplified rules for curtailment of bars in continuous beams are shown in Fig13.15.

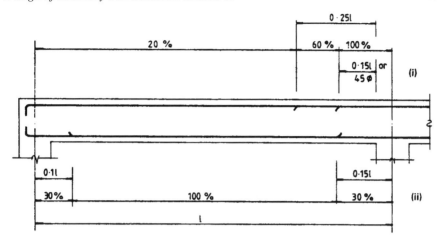

Fig.13.15 Reinforcement as percentage of that required for (i) maximum hogging moment over support and (ii) maximum sagging moment in span.

13.10 EXAMPLE OF DESIGN FOR THE END SPAN OF A CONTINUOUS BEAM

(a) Specification
Design the end span of the continuous beam analysed in section 13.7. The design is to be made for the shear forces and moments obtained after 20% redistribution from the elastic analysis has been made. The shear force and moment envelopes are shown in Fig.13.14. The materials are grade C30 concrete and grade 460 reinforcement.

(b) Design of moment steel
The assumed beam sections for mid-span and over the interior support are shown in Figs 13.16(a) and 13.16 (b) respectively. The cover for mild exposure from Table 3.3 in the code is 25 mm. The cover for a fire resistance period of 2 h from Table 3.4 is 30 mm for continuous beams. Cover of 30 mm is provided to the links.

(i) Section near the centre of the span: The beam acts as a T-beam at this section.
The design moment M = 237.7 kNm.
Effective breadth of flange = (0.7 × 8000/5) + 250 = 1370 mm
The moment of resistance of the section when the entire flange is in compression is
$$M_{Flange} = 0.45\ f_{cu}\ b_f\ h_f\ (d - h_f/2)$$
$$= 0.45 \times 30 \times 1370 \times 125(385 - 0.5 \times 125) \times 10^{-6} = 745.6\ kNm$$
$$M < M_{flange}$$

The neutral axis lies in the flange. The beam can be designed as a rectangular beam.

$$k = 237.7 \times 10^6 / (30 \times 1370 \times 385^2) = 0.039 < 0.156$$
$$z/d = 0.5 + \sqrt{(0.25 - 0.039/0.9)} = 0.956 > 0.95 = 367.5\text{mm}$$
$$A_s = 237.7 \times 10^6 / (0.95 \times 385 \times 0.95 \times 460) = 1487 \text{ mm}^2$$

Provide 4T25, $A_s = 1963$ mm^2.

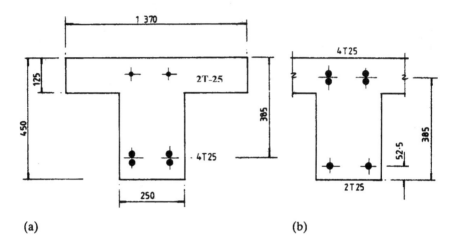

Fig.13.16 (a) T-beam at mid-span; (b) rectangular beam over support.

Note that in this case the amount of redistribution from the elastic moment of 245.17 kNm to redistributed value of 237.76 kNm is just 3% i.e. $\beta_b = 0.97$, and the depth x to the neutral axis must not exceed $(0.97 - 0.4)d = 0.57d$. But the maximum value of $x/d \leq 0.5$. In fact the moment from the elastic analysis at the position where redistributed moment is a maximum is in fact 215.12 kNm. In other words redistribution *increases* the moment at that section rather than decrease it.

The moment of resistance after cutting off two 25 mm diameter bars is calculated where

$$d = 397.5\text{mm}, z = 0.95d \text{ and } A_s = 981 \text{ mm}^2.$$

Note that $z = 0.95d$ when the moment is 237.7 kN m and so the beam will have the same limiting value for $z/d = 0.95$ at a section where the moment is less. The moment of resistance is

$$M_R = (0.95 \times 460) \times (0.95 \times 397.5) \times 981 \times 10^{-6} = 161.89 \text{ kNm}$$

From the design moment envelop this moment of resistance occurs at 1.46 m and 2.74 m from the ends. The anchorage length for type 2 deformed bars is calculated. Refer to clause 3.12.8.3 and Table 3.27 in the code and Chapter 5, section 5.2.1. The anchorage length is 40 bar diameters which is equal to 1000 mm. To comply with the detailed provisions for curtailment of bars given in clause 3.12.9.1 of the code the two 25mm diameter bars will be stopped off at 1000 mm beyond the theoretical cut–off points. This also satisfies the condition that bars extend a distance equal to the greater of the effective depth or 12 bar diameters beyond the theoretical cut–off points. At the support tension bars must be anchored 12 bar diameters past the centre line of the support. The cut–off points are shown in Fig.13.20. These are at $(1460 - 1000) = 460$ mm from the centre of end support and $(2740 - 1000) = 1740$ mm from the interior support respectively.

(ii) Section at the interior support: The beam acts as a rectangular beam at the support. The section is shown in Fig.13.16 (b). The redistribution of 20% has been carried out and so the depth to the neutral axis should not exceed

$$x = (\beta_b - 0.4) d = (0.8 - 0.4) d = 0.4d$$

The design moment is 215.65 kN m. The maximum moment of resistance with no compression steel is calculated from the expressions given in clause 3.4.4.4 of the code. Refer to section 4.7.

$$k` = [(0.402 \times 0.4) - 0.18 \times 0.4^2] = 0.133$$
$$0.133 \, bd^2 \, f_{cu} = 0.133 \times (250 \times 385^2 \times 30) \times 10^{-6} = 147.85 \text{ kNm} < 215.65 \text{ kNm}$$

Compression reinforcement is required.

$$d`/x = 52.5/(0.4 \times 385) = 0.34 < 0.376$$

The stress in the compression steel is $0.95f_y$.

$$A_s = 147.85 \times 10^6/ (0.775 \times 385 \times 0.95 \times 460)$$
$$+ (215.65 - 147.85) \times 10^6/ (0.95 \times 460 \times (385 - 52.5)$$
$$A_s = 1134 + 467 = 1601 \text{ mm}^2$$
$$A_s' = 467 \text{ mm}^2$$

The compression reinforcement will be provided by carrying two 25 mm diameter mid-span bars through the support. For tension reinforcement, provide four 25 mm diameter bars providing an area of 1963 mm².

The theoretical and actual cut–off points for two of the four top bars are determined. The moment of resistance of the section with two 25 mm diameter bars and an effective depth $d = 397.5$ mm is calculated. Assuming that steel yields, equate the total tensile force T and the total compressive force C.

$$T = 0.95 \times 460 \times 981 \times 10^{-3} = 428.70 \text{ kN}$$
$$C = 0.45 \times 30 \times (0.9x) \times 250 \times 10^{-3} = 3038x \text{ kN}$$
$$\text{Equating } T = C, 428.70 = 3038x.$$
$$x = 141 \text{ mm}, x/d = 141/397.5 = 0.36 < 0.5$$
$$\text{Lever arm: } z/d = 1 - 0.45 \, x/d = 0.84 < 0.95$$

The moment of resistance is

$$M_R = T z = 428.7 \times (0.84 \times 397.5) \times 10^{-3} = 143.1 \text{ kNm}$$

From the design moment envelop this moment occurs at 0.73 m from the support. The actual cut–off point after continuing the bars from an anchorage length is $(730 + 1000) = 1730$ mm from support B. The bar cut–offs are shown in Fig.13.17.

(c) Design of shear reinforcement
The design will take account of enhancement of shear strength near the support (BS8110: Part 1, clause 3.4.5.10).

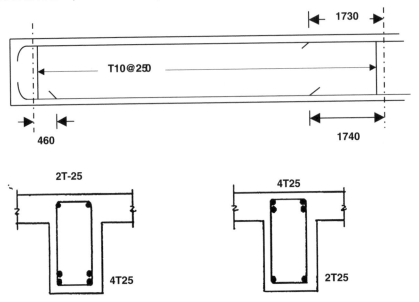

Fig.13.17 Continuous beam.

(i) Simply supported end: From the design shear force envelop

Check for maximum shear stress:
$$V = 143.71 \text{ kN}$$
$$v = 143.71 \times 10^3 / (250 \times 397.5) = 1.45 \text{ N/mm}^2 < \{0.8\sqrt(f_{cu} = 30) = 4.38 \text{ N/mm}^2\}$$
Section is satisfactory and shear links can be designed.

Shear V at d from support
$$V = 143.71 - 42.12x \ (d = 397.5) \times 10^{-3} = 126.97 \text{ kN}$$
$$v = 126.97 \times 10^3 / (250 \times 397.5) = 1.28 \text{ N/mm}^2$$
$$A_s = 2T25 = 982 \text{ mm}^2$$
$$100 \ A_s / (bd) = 100 \times 982 / (250 \times 397.5) = 0.99 < 3.0$$
$$400/d = 400/397.5 = 1.01 > 1.0$$
$$v_c = 0.79 \times (0.99)^{1/3} \ (1.006)^{1/4} \ (30/25)^{1/3}/1.25 = 0.67 \text{ N/mm}^2$$
$$v - v_c = 1.28 - 0.67 = 0.61 > 0.4$$
Provide 10 mm diameter grade 460 links:
$A_{sv} = 157 \text{ mm}^2$ for two–legs.

Calculate the spacing of links
$$157 \geq (1.28 - 0.67) \times 250 \times s_v / (0.95 \times 460), \ s_v \leq 450 \text{ mm}$$

Maximum spacing = 0.75 d = 298 mm.

Calculate the region over which only minimum links are required

Minimum links are required when $v = v_c + 0.4$
For the section with four 25 mm diameter bars, $d = 385$ mm, $A_s = 1964$ mm^2.
$$100 \, A_s / (bd) = 100 \times 1964/ (250 \times 385) = 2.05 < 3.0$$
$$400/d = 400/385 = 1.04 > 1.0$$
$$v_c = 0.79 \times (2.05)^{1/3} (1.04)^{1/4} (30/25)^{1/3}/1.25 = 0.86 \text{ N/mm}^2$$
$$v = 0.86 + 0.40 = 1.26 \text{ N/mm}^2$$
$$V = 1.26 - 250 \times 385 \times 10^{-3} = 121.28 \text{ kN}$$
From the design shear force envelop,
$$121.28 = 143.71 - 42.12 \, a, \, a = 0.53 \text{ m}$$
For minimum links, the spacing is
$$157 \geq (0.4) \times 250 \times s_v / (0.95 \times 460), \, s_v \leq 686 \text{ mm}$$
Rationalize the results from the above calculations and space links at 250 mm centres.

(ii) Near the internal support: From the shear force envelop the maximum shear is V = 202.18 kN
$$v = 202.18 \times 10^3/ (250 \times 385) = 2.10 \text{ N/mm}^2 < (0.8\sqrt{30} = 4.38 \text{ N/mm}^2)$$
The shear at $d = 385$ mm from the support is
$$V = 202.18 - 42.12 \times 385 \times 10^{-3} = 185.96 \text{ kN}$$
$$v = 185.96 \times 10^3/ (250 \times 385) = 1.93 \text{ N/mm}^2$$
$$A_s = 4T25 = 1964 \text{ mm}^2, \, 100 \, A_s / (bd) = 100 \times 1964/ (250 \times 385) = 2.05 < 3.0,$$
$$400/d = 400/385 = 1.04 > 1.0$$
$$v_c = 0.79 \times (2.05)^{1/3} (1.04)^{1/4} (30/25)^{1/3}/1.25 = 0.86 \text{ N/mm}^2$$
Provide 10 mm diameter grade 460 links. $A_{sv} = 157$ mm^2 for two-legs.
$$157 \geq (1.93 - 0.86) \times 250 \times s_v / (0.95 \times 460), \, s_v \leq 256 \text{ mm}$$
On the bottom face where the reinforcement is in compression the link spacing must not exceed $12 \times 25 = 300$ mm.
Space links at 250 mm centres along the full length of the beam. The arrangement of links is shown in Fig.13.20.

(d) Deflection
$$b_w/b = 250/1370 = 0.18 < 0.3$$
From BS8110: Part 1, Table 3.9, the basic span–to–effective depth ratio is 20.8
Modification factor for tension steel:
$$M / (bd^2) = 237.76 \times 10^6 / (1370 \times 385^2) = 1.17$$
$$A_{s \, required} = 1487 \text{ mm}^2, \, A_{s \, provided} = 1963 \text{ mm}^2, \, \beta_b = 0.8$$
$$f_s = (2/3) \times (1487/1963) \times 460 \times (1/ 0.8) = 290 \text{ N/mm}^2$$
$$0.55 + (477 - 290)/ \{120 \times (0.9 + 1.17)\} = 1.30 < 2.0$$
Modification factor for compression steel:
Two 20 mm diameter bars, $A_s' = 625$ mm^2 are provided in the top of the beam which can act as compression reinforcement.

$$100\ A_s'/(bd) = 100 \times 625/(1370 \times 385) = 0.12$$
$$1 + 0.12/(3 + 0.12) = 1.04$$
$$\text{allowable span/d ration } 20.8 \times 1.3 \times 1.04 = 28.12$$
$$\text{actual span/d ratio } 8000/385 = 20.8$$

The beam is satisfactory with respect to deflection.

(e) Cracking

From Fig.13.16, the clear distance between bars on the tension faces at mid-span and over the support is 120 mm. This does not exceed the 125 mm permitted in Table 3.28 of the code for 20% redistribution. The distance from the corner to the nearest longitudinal bar is (cover = 30 mm, links 10 mm, bar diameter = 25 mm)

$$\sqrt{\{(30{+}10 + 25/2)^2 + (30{+}10 + 25/2)^2\}} - 25/2 = 62\ \text{mm}$$

It should not exceed 125/2 = 62.5 mm. The beam is satisfactory with respect to crack control.

(f) Sketch of the beam

A sketch of the beam with the moment and shear reinforcement and curtailment of bars is shown in Fig.13.17.

13.11 EXAMPLE OF DESIGN OF A NON–SWAY FRAME

(a) Specification

Fig.13.18 shows a typical frame supporting a loading bay. The frames are spaced at 4 m centres. The floor consists of 250 mm thick precast slabs simply supported on top of beams. The imposed load is 10 kN/m². The beams are 300 × 600 mm and columns are 300 mm × 300 mm. The materials are grade C30 concrete and grade 460 reinforcement.

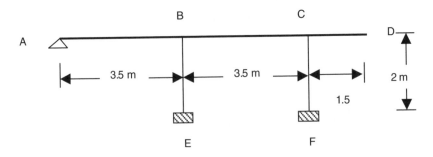

Fig.13.18 Non–sway rigid–jointed frame.

(b) Loads
Dead load:

$$\text{Beam self weight} = 0.3 \times 0.6 \times 24 = 4.32\ \text{kN/m}$$
$$\text{Precast planks: } 0.125 \times 4.0 \times 24 = 24.00\ \text{kN/m}$$
$$G_k = 4.32 + 24.0 = 28.32\ \text{kN/m}$$
$$\text{Imposed load } Q_k = 10 \times 4 = 40\ \text{kN/m}$$

$$(1.4G_k + 1.6\,Q_k) = 1.4 \times 28.32 + 1.6 \times 40 = 103.65 \text{ kN/m}$$
$$1.0\,G_k = 28.32 \text{ kN/m}$$

(c) Elastic analysis

$$\text{I beams} = 0.3 \times 0.6^3/12 = 5.4 \times 10^{-3} \text{ m}^4$$
$$\text{I columns} = 0.3 \times 0.3^3/12 = 0.675 \times 10^{-3} \text{ m}^4$$
$$\text{Beams, I/L: } = 5.4 \times 10^{-3}/3.5 = 1.5429 \times 10^{-3} \text{ m}^3$$
$$\text{Columns, I/L: } = 0.675 \times 10^{-3}/2.0 = 0.3375 \times 10^{-3} \text{ m}^3$$

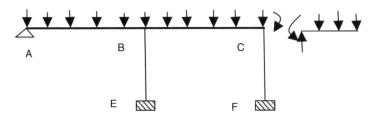

Fig.13.19 Simplified frame used in analysis.

In order to simplify the computation, the structure analysed is as shown in Fig.13.19. The cantilever CD is not included in the stiffness matrix but the moment induced by the cantilever on the rest of the frame in taken into account when computing the load vector and rotations at the joints A, B and C.
The simplified stiffness matrix K is given by

$$K = E \times 10^{-3} \begin{bmatrix} 6.1716 & 3.0858 & 0 \\ 3.0858 & 13.6932 & 3.0858 \\ 0 & 3.0858 & 7.5216 \end{bmatrix} \begin{bmatrix} \theta_A \\ \theta_B \\ \theta_C \end{bmatrix}$$

The structure is analysed for the following four load cases:

Case 1: $(1.4G_k + 1.6\,Q_k)$ on AB and BC, $1.0\,G_k$ on CD
  Fixed end moments AB and BC $= 103.65 \times 3.5^2/12 = 105.8073$ kNm
  Fixed end moment in CD $= 28.32 \times 1.5^2/2 = 31.86$ kNm

Case 2: $1.0\,G_k$ on AB, $(1.4G_k + 1.6\,Q_k)$ on BC and CD
  Fixed end moments AB $= 28.32 \times 3.5^2/12 = 28.91$ kNm
  Fixed end moments BC $= 103.65 \times 3.5^2/12 = 105.8073$ kNm
  Fixed end moment in CD $= 103.65 \times 1.5^2/2 = 116.606$ kNm

Case 3: $(1.4G_k + 1.6\,Q_k)$ on AB and CD, $1.0G_k$ on BC
  Fixed end moments AB $= 103.65 \times 3.5^2/12 = 105.8073$ kNm
  Fixed end moments BC $= 28.32 \times 3.5^2/12 = 28.91$ kNm
  Fixed end moment in CD $= 103.65 \times 1.5^2/2 = 116.606$ kNm

Case 4: 1.0 G_k on AB and CD, $(1.4G_k + 1.6\,Q_k)$ BC
   Fixed end moments AB = $28.32 \times 3.5^2/12 = 28.91$ kNm
  Fixed end moments BC = $103.65 \times 3.5^2/12 = 105.8073$ kNm
   Fixed end moment in CD = $28.32 \times 1.5^2/2 = 31.86$

The load vectors for the four cases are

$$F = \begin{bmatrix} 105.8073 & 28.91 & 105.8073 & 28.91 \\ 0 & 76.8973 & -76.8973 & 76.8973 \\ -73.9473 & 10.7990 & 87.6963 & -73.9473 \end{bmatrix}$$

The displacement vectors for the four cases are

$$E \times 10^{-3} \begin{bmatrix} \theta_A \\ \theta_B \\ \theta_C \end{bmatrix} = \begin{bmatrix} 18.1809 & 2.0194 & 24.7597 & 0.4233 \\ -2.0733 & 5.3299 & -15.2310 & 8.5242 \\ -8.9808 & -0.7509 & 17.9079 & -13.3284 \end{bmatrix}$$

The results are summarised in Table 13.14. The elastic bending moment diagrams for beams AB and BC are shown in Fig.13.20 (a) and Fig.13.20(b).

Table 13.14 Summary of elastic analysis

	Case 1	Case 2	Case 3	Case 4
M_{BA}	149.11	68.04	88.21	82.82
M_{BC}	−146.32	−75.23	−67.65	−94.33
M_{CB}	43.98	117.62	92.43	49.86
M_{CD}	−31.86	−116.61	−116.61	−31.86
M_{AE}	−2.80	7.20	−20.56	11.51
M_{BF}	−12.12	−1.01	24.18	−17.99
Axial: BE	434.62	238.28	249.07	267.32
Axial: CF	194.63	348.98	211.12	211.16

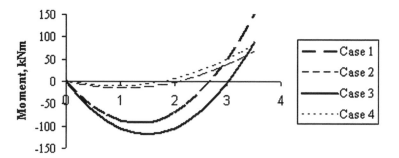

Fig.13.20(a) Elastic bending moment diagrams for beam AB.

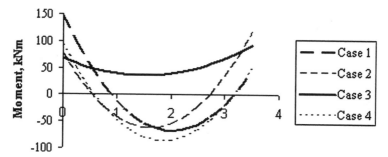

Fig.13.20(b) Elastic bending moment diagrams for beam BC.

(c) Redistribution

(i) Beam AB

The support moment $M_{BA} = 149.11$ from case 1 is the largest value of hogging moment considering all load cases and can be reduced by 30% to
$$M_{BA} = 0.7 \times 149.11 = 104.38$$
The redistributed support moments are as follows.

   Case 1: $M_{BA} = 104.38$ (changed from elastic value of 149.11)
   Case 2: $M_{BA} = 68.04$ (Unchanged from elastic value)
   Case 3: $M_{BA} = 104.38$ (changed from elastic value of 88.21)
   Case 4: $M_{BA} = 104.38$ (changed from elastic value of 82.82)

(ii) Beam BC

Hogging moment $M_{BC} = 146.32$ from case 1 can also be reduced to 104.38 so that the same top steel over the column BE serves for both moments M_{BA} and M_{BC}.
The maximum moment M_{CD} at the root of the cantilever is 116.61. Since CD is a cantilever, this moment cannot be reduced. Moment M_{CB} from case 2 is 117.62 and this can be reduced also to 116.61 so that the same top steel over the column CF for both moments M_{CB} and M_{CD}. The redistributed moments are as follows.

   Case 1: $M_{BC} = 104.38$ (changed from elastic values of 146.32), $M_{CB} = 43.98$
   Case 2: $M_{BC} = 104.38$, $M_{CB} = 116.61$ (changed from elastic values of 75.23 and 117.62 respectively)
   Case 3: $M_{BC} = 104.38$, $M_{CB} = 116.61$ (changed from elastic values of 67.65 and 92.43 respectively)
   Case 4: $M_{BC} = 104.38$, $M_{CB} = 49.86$ (changed from elastic values of 94.33 and 49.86 respectively)

Note that at support C, in cases 1 and 4, no redistribution has been done. The reason for this is that the moment in the cantilever is small. If the support moment M_{CB} is raised to 116.61, then it will result in a very large moment in column CF. The redistributed bending moment diagrams for beams AB and BC are shown in Fig.13.21(a) and Fig.13.21(b).

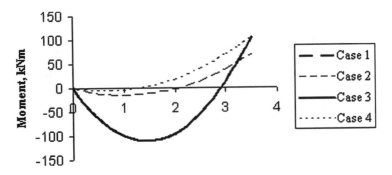

Fig.13.21(a) Redistributed bending moments for beam AB.

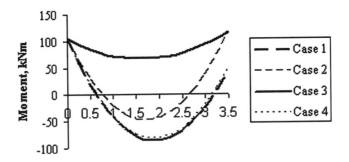

Fig.13.21(b) Redistributed bending moments for beam BC.

(d) Design moment envelops

The design maximum (or minimum) value of bending moment at a section is obtained as follows.

The maximum (or minimum) moment value at a section is obtained by considering all load cases. For a particular load case the values to be considered are:

- If there is no redistribution, then the elastic moment value
- If there is redistribution then the larger of:
a.   0.7 × elastic value
b.   the corresponding redistributed moment value

The design moment envelops for Beams AB and BC are shown in Fig.13.22(a) and Fig.13.22(b) respectively.

(e) Design of moment steel

(i) Section near the centre of span AB

Assuming 20 mm bars and a cover of 30 mm and 8 mm diameter shear links

$$d = 600 - 20/2 - 8 - 30 = 552 \text{ mm}$$

Maximum redistributed moment from the design moment envelop is 110.74 kNm at 1.5 m from left hand support from Case 3.

The corresponding moment before redistribution is 117.67 kNm.
$$\beta_b = 110.74/117.67 = 0.94 > 0.9$$
Redistribution is less than 10%. Therefore K` = 0.156
$$k = 110.74 \times 10^6/ (300 \times 552^2 \times 30) = 0.04 < 0.156$$
No compression steel is required.
$$z/d = 0.5 + \sqrt{(0.25 - 0.04/0.9)} = 0.953 > 0.95$$
$$A_s = 110.74 \times 10^6/ (0.95 \times 552 \times 0.95 \times 460) = 483 \text{ mm}^2$$
Provide 3T16 giving an area of 603 mm².

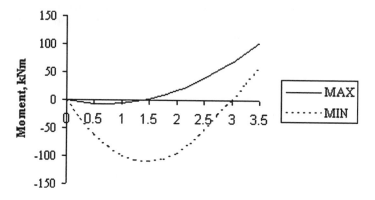

Fig.13.22(a) Design bending moment envelop for beam AB.

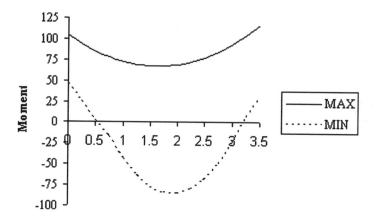

Fig.13.22(b) Design bending moment envelop for beam BC.

(ii) Section over support B
Maximum redistributed moment from the design moment envelop is 104.38 kNm from Case 1. The corresponding moment before redistribution is 149.11 kNm.

$$\beta_b = 104.38/149.11 = 0.7 < 0.9$$
$$K' = 0.402\ (0.7 - 0.4) - 0.18\ (0.7 - 0.4)^2 = 0.104$$
$$k = 104.38 \times 10^6/\ (300 \times 552^2 \times 30) = 0.038 < 0.104$$

No compression steel is required.

$$z/d = 0.5 + \sqrt{\ (0.25 - 0.038/0.9)} = 0.96 > 0.95$$
$$A_s = 104.38 \times 10^6/\ (0.95 \times 552 \times 0.95 \times 460) = 456\ \text{mm}^2$$

Provide 3T16 giving an area of 603 mm².

(iii) Section near the centre of span BC

Maximum redistributed moment from the design moment envelop from Case 1 is 85.61 kNm at 2.0 m from left hand support. The corresponding moment before redistribution is 67.64 kNm. The moment after redistribution has *increased*.

$$k = 81.60 \times 10^6/\ (300 \times 552^2 \times 30) = 0.030 < 0.156$$

No compression steel is required.

$$z/d = 0.5 + \sqrt{(0.25 - 0.03/0.9)} = 0.97 > 0.95$$
$$A_s = 85.61 \times 10^6/\ (0.95 \times 552 \times 0.95 \times 460) = 374\ \text{mm}^2$$

Provide 3T16 giving an area of 603 mm².

(iv) Section over support C

Maximum redistributed moment from the design moment envelop is 116.61 kNm for Case 2. The corresponding moment before redistribution is 117.62 kNm.

$$\beta_b = 116.61/117.62 = 0.99 > 0.9$$
$$k = 116.61 \times 10^6/\ (300 \times 552^2 \times 30) = 0.043 < 0.156$$

No compression steel is required.

$$z/d = 0.5 + \sqrt{(0.25 - 0.043/0.9)} = 0.95$$
$$A_s = 116.61 \times 10^6/\ (0.95 \times 552 \times 0.95 \times 460) = 509\ \text{mm}^2$$

Provide 3T16 giving an area of 603 mm².

By rationalizing the steel area calculations, for simplicity, provide 3T16 at both top and bottom for the beams including the cantilever.

(f) Design shear envelops

The design maximum (or minimum) value is obtained as follows.

The maximum (or minimum) shear force at a section is obtained by considering for all load cases both the elastic analysis shear values and the redistributed shear values. The design shear force envelops for Beams AB and BC are shown in Fig.13.23(a) and Fig.13.23(b).

(g) Design of shear reinforcement

The design will take account of enhancement of shear strength near the support (BS8110: Part 1, clause 3.4.5.10).

As the tension and compression steel is 3T-16 over the entire span, a common value of v_c can be calculated which is applicable over the entire span.

$$100\ A_s/\ (bd) = 100 \times 603/\ (300 \times 552) = 0.36 < 3.0$$
$$400/d = 400/552 = 0.73 < 1.0, \text{take as } 1.0$$
$$v_c = 0.79 \times (0.36)^{1/3}\ (1.0)^{1/4}\ (30/25)^{1/3}/1.25 = 0.48\ \text{N/mm}^2$$

Use 10 mm diameter 2–leg grade 460 links through out. $A_{sv} = 157\ \text{mm}^2$.

Nominal link spacing is
$$157 \geq 0.4 \times 300 \times s_v/ (0.95 \times 460), s_v \leq 572 \text{ mm}$$
Maximum spacing $= 0.75 \text{ d} = 414 \text{ mm}$.

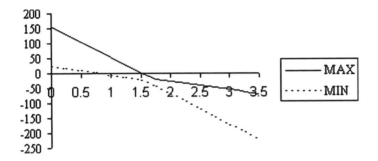

Fig.13.23(a) Design shear force envelop for beam AB.

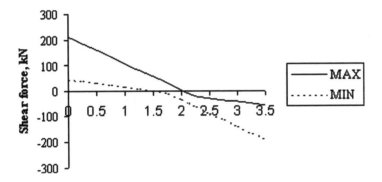

Fig.13.23(b) Design shear force envelop for beam BC.

(i) Simply supported end A
From the design shear force envelop
$$V = 156.18 \text{ kN (From elastic, case 3)}.$$
$$v = 156.18 \times 10^3/ (300 \times 552) = 0.94 \text{ N/mm}^2 < \{0.8\sqrt{(f_{cu} = 30)} = 4.38 \text{ N/mm}^2\}$$
Shear V at d from support
$$V = 156.18 - 103.65 \times (d = 552) \text{ x } 10^{-3} = 98.97 \text{ kN}$$
$$v = 98.97 \times 10^3/ (300 \times 552) = 0.60 \text{ N/mm}^2$$
$$v - v_c = 0.60m - 0.48 = 0.12 < 0.4$$
Nominal links required.

(ii) Continuous end B of span AB over column BE
From the design shear force envelop
$$V = 224.0 \text{ kN (From elastic, case 1)}.$$
$$v = 224.0 \times 10^3/ (300 \times 552) = 1.35 \text{ N/mm}^2 < \{0.8\sqrt{(f_{cu} = 30)} = 4.38 \text{ N/mm}^2\}$$
Shear V at d from support

$$V = 224.0 - 103.65 \times (d = 552) \times 10^{-3} = 166.79 \text{ kN}$$
$$v = 166.79 \times 10^3 / (300 \times 552) = 1.01 \text{ N/mm}^2$$
$$v - v_c = 1.01 - 0.48 = 0.53 > 0.4$$

Design links required. Provide 10 mm diameter grade 460 links.
$A_{sv} = 157 \text{ mm}^2$ for two–legs.

$$157 \geq (0.53) \times 300 \times s_v / (0.95 \times 460), \; s_v \leq 432 \text{ mm}$$

Maximum spacing $= 0.75 \text{ d} = 414 \text{ mm}$.

(iii) Continuous end of span BC over column BE
From the design shear force envelop
$V = 210.63$ kN (From elastic, case 1).
This shear force is less than 224.0 kN for continuous end B for span AB. Use same design as for span AB.

(iv) Continuous end over column CF
From the design shear force envelop $V = 193.50$ kN (from elastic case 2).
This shear force is less than 224.0 kN. Use the same design as for support B.

By rationalizing the link spacings calculated, for simplicity, provide 8 mm diameter two leg links at 400 mm c/c through out the beams including the cantilever.

(h) Deflection
The basic span–to–effective depth ratio is 26 for continuous beam (BS8110: Part 1, Table 3.9).
Over the entire span, compression steel is provided by 3T-16. $A_s' = 603 \text{ mm}^2$.

$$100 \, A_s' / (bd) = 100 \times 603 / (300 \times 552) = 0.36$$

The modification factor for compression steel is

$$1 + 0.36 / (3 + 0.36) = 1.11$$

Calculate the modification factor for tension steel and the allowable span/depth ratio.

(i) Beam AB
Moment at mid-span from the design moment envelop $= 106.52$ kNm

$$M / (bd^2) = 106.52 \times 10^6 / (300 \times 552^2) = 1.17$$
$$A_{s \, required} = 483 \text{ mm}^2, \; A_{s \, provided} = 603 \text{ mm}^2, \; \beta_b = 106.52/114.61 = 0.93$$
$$f_s = (2/3) \times (483/603) \times 460 \times (1/ 0.93) = 264 \text{ N/mm}^2$$
$$0.55 + (477 - 264)/ (120 \times (0.9 + 1.17) = 1.41 < 2.0$$
$$\text{allowable span/d ration } 26 \times 1.41 \text{ x } 1.11 = 40.69$$
$$\text{actual span/d ratio } 3500/552 = 6.34$$

The beam is satisfactory with respect to deflection.

(ii) Beam BC
Moment at mid-span from the design moment envelop $= 81.60$ kNm

$$M / (bd^2) = 85.61 \times 10^6 / (300 \times 552^2) = 0.94$$
$$A_{s \, required} = 374 \text{ mm}^2, \; A_{s \, provided} = 603 \text{ mm}^2, \; \beta_b = 1.0$$

$$f_s = (2/3) \times (374/603) \times 460 \times (1/0.94) = 190 \text{ N/mm}^2$$
$$0.55 + (477 - 190)/\{120 \times (0.9 + 0.94)\} = 1.85 < 2.0$$
$$\text{allowable span/d ration } 26 \times 1.85 \times 1.11 = 53.39$$
$$\text{actual span/d ratio } 3500/552 = 6.34$$

The beam is satisfactory with respect to deflection.

(iii) Cantilever CD

Moment at support = 116.61 kNm
$$M/(bd^2) = 116.61 \times 10^6/(300 \times 552^2) = 1.28$$
$$A_{s\,required} = 509 \text{ mm}^2, A_{s\,provided} = 603 \text{ mm}^2, \beta_b = 1.0$$
$$f_s = (2/3) \times (509/603) \times 460 \times (1) = 259 \text{ N/mm}^2$$
$$0.55 + (477 - 259)/\{120 \times (0.9 + 1.28)\} = 1.38 < 2.0$$
$$\text{allowable span/d ration } 7 \times 1.38 \times 1.11 = 10.7$$
$$\text{actual span/d ratio } 1500/552 = 2.72$$

The cantilever is satisfactory with respect to deflection.

(j) Cracking

The clear distance between bars on the tension faces at mid-span and over the support is
$$(300 - 2 \times 30 \text{ for cover} - 2 \times 8 \text{ for link} - 16)/2 - 16 = 88 \text{ mm}$$
This does not exceed the 110 mm permitted in Table 3.28 of the code for 30% redistribution. The distance from the corner to the nearest longitudinal bar is
$$\sqrt{\{(30 + 8 + 16/2)^2 + (30 + 8 + 16/2)^2\}} - 16/2 = 57 \text{ mm}$$
which should not exceed $110/2 = 55$ mm. The difference between the actual and permitted is very small and the beam is satisfactory with respect to crack control.

Table 13.15 Axial force and moments in columns

		Case 1		Case 2		Case 3		Case 4	
		Elas.	Redis	Elas.	Redis	Elas.	Redis	Elas.	Redis
BE	N	435	410	238	247	249	68	267	276
		(4.8)	(4.6)	(2.7)	(2.7)	(2.8)		(3.0)	(3.1)
	M	2.8	0	7.2	36	21	0	12	0
		(0.1)		(0.3)	(1.4)	(0.8)		(0.4)	
CF	N	195	207	349	340	212	209	211	208
		(2.2)	(2.3)	(3.9)	(3.8)	(2.4)		(2.4)	(2.3)
	M	12	12	1.0	0	24	0	18	18
		(0.5)	(0.5)	(0.04)		(0.9)		(0.7)	(0.7)

(k) Column Design

Table 13.15 shows the axial force and moments at the top of the columns. The figures in brackets are $M/(bh^2)$ and $N/(bh)$ using $b = h = 300$ mm. The column design chart (Fig. 9.11, Chapter 9) for $f_{cu} = 30$ N/mm^2, $f_y = 460$ N/mm^2 and $d/h = 0.90$ shows that only minimum steel equal to $A_{sc}/(bh) = 0.4\%$ is required.
$A_{sc} = (0.4/100) \times 300^2 = 360$ mm^2. Provide one 12 mm bar in each corner.
$A_{sc} = 452$ mm^2. Provide 6 mm diameter links spaced at 125 mm c/c.

13.12 APPROXIMATE METHODS OF ANALYSIS

In the examples of continuous beam and non–sway frame analysed in the previous sections, the relative flexural rigidity EI was assumed in order to carry out the elastic analysis. In the case of statically indeterminate structures, information about the relative stiffness of members is required before analysis can be carried out. In many cases experience can be used to guess at the relative size of members. However it is convenient to have approximate methods of analysis which allow a designer to estimate the relative stiffness of members. Approximate methods of analysis convert a statically indeterminate structure into a statically determinate structure by assuming the position of points of contra-flexure. This enables the determination of inevitably approximate values of bending moment and shear forces in the structure without the need to know the relative stiffness of members.

13.12.1 Analysis for Gravity Loads

Analysis for gravity loads is done by assuming the points of contra-flexure in the individual beams.

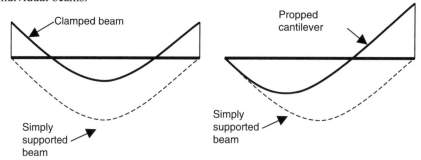

Fig.13.24 (a) Clamped beam; (b) Propped cantilever.

If a beam is continuous at both ends, its behaviour will be between the behaviour of a clamped beam at one extreme and that of a simply supported beam at the other extreme. As shown in Fig.13.24(a), in a clamped beam of span L subjected to a uniformly distributed load q, the contra flexure points are at 0.21 L from the ends. The support and mid-span moments are respectively, $qL^2/12$ and $qL^2/24$. In the corresponding simply supported beam, the points of zero moment are at the support and the moment at mid-span is $qL^2/8$. Assuming the position of contra-flexure at approximately at 0.1L from the ends, the support and mid-span moments are respectively, 0.36 $(qL^2/8)$ and 0.64 $(qL^2/8)$.

If a beam is continuous at one end only, its behaviour will be between the behaviour of a propped cantilever at one extreme and that of a simply supported beam at the other extreme. As shown in Fig.13.24(b), in a propped cantilever of span L clamped at the right hand end and subjected to a uniformly distributed load q, the contra-flexure point is at 0.25 L from the clamped end. The support and mid-span moments are respectively, $qL^2/8$ and $qL^2/16$. In the corresponding simply supported beam, the points of zero moment are at the support and the moment at mid-span is $qL^2/8$. Assuming the position of contra flexure at approximately 0.12L from the ends, the support and mid-span moments are respectively, 0.48 $(qL^2/8)$ and 0.76 $(qL^2/8)$.

The following two examples show how these values can be used to analyse beams subjected to uniformly distributed loading.

13.12.2 Analysis of a Continuous Beam for Gravity Loads

Fig.13.25 shows a three span continuous beam. The beam spans are 8 m each. Assuming the position of contra flexure points, the beams are analysed for the It is given that $1.4G_k+1.6Q_k = 42.12$ kN/m and $1.0\ G_k = 19.8$ kN/m.

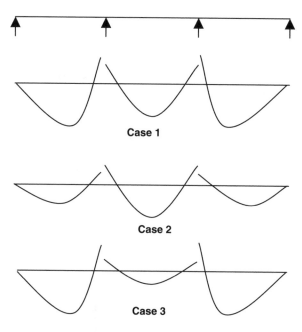

Fig.13.25 Approximate bending moment distribution.

Case 1: All beams are subjected to a uniformly distributed load of 42.12 kN/m.
End spans:
$$\text{support moment} = 0.48 \times 42.12 \times 8^2/8 = 162 \text{ kNm}$$
$$\text{span moment} = 0.76 \times 42.12 \times 8^2/8 = 235 \text{ kNm.}$$
Central span:
$$\text{support moment} = 0.36 \times 42.12 \times 8^2/8 = 121 \text{ kNm}$$
$$\text{span moment} = 0.64 \times 42.12 \times 8^2/8 = 216 \text{ kNm}$$

Case 2: End spans carry 19.8 kN/m and central span carries 42.12 kN/m.
End spans:

support moment = $0.48 \times 19.8 \times 8^2/8 = 76$ kNm
span moment = $0.76 \times 19.8 \times 8^2/8 = 120$ kNm

Central span:

support moment = $0.36 \times 42.12 \times 8^2/8 = 121$ kNm
span moment = $0.64 \times 42.12 \times 8^2/8 = 216$ kNm

Case 3: End beams carry 42.12 kN/m and central span carries 19.8 kN/m.
End spans:

support moment = $0.48 \times 42.12 \times 8^2/8 = \mathbf{162}$ kNm
span moment = $0.76 \times 42.12 \times 8^2/8 = \mathbf{235}$ kNm.

Central span:

support moment = $0.36 \times 19.8 \times 8^2/8 = 57$ kNm
span moment = $0.64 \times 19.2 \times 8^2/8 = 101$ kNm

From the above three analyses, the support needs to be designed for approximately 162 kNm and the mid-span for 235 kNm. Exact elastic analysis assuming uniform flexural rigidity shows that the maximum support moment is 270 kNm and span moment in end spans is 245 kNm. The support moment is quite poorly predicted. The reason for this is that when all the three spans are loaded by the same load, it almost produces a clamped condition at the supports which is a more severe condition than the assumption of position of contra-flexure at 0.1L from supports.

13.12.3 Analysis of a Rectangular Portal Frame for Gravity Loads

Fig.13.26 shows a single by portal frame subjected to gravity loads on the beam. The bending moment distribution can be obtained by assuming the points of contra-flexure in the beam. The moment at the tops of the columns will be the same as in the beam. If the columns are fixed at the base, then the moment at the base of columns is half that of the moment at the top of columns.

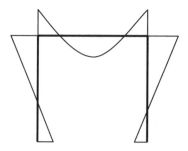

Fig.13.26 Rectangular portal frame.

13.12. 4 Analysis for Wind Loads by Portal Method

Analysis of portal frames for the wind load is made for the whole frame assuming points of contra-flexure at the mid–height of columns and at mid-span of beams. In the portal method the horizontal shear force in each storey is assumed to be divided between the bays in proportion to their spans. The shear force in each bay is then divided equally between the columns. The column end moments are the column shear force multiplied by one–half the column height. Beam moments balance the column moments. The method is considered to be applicable to building frames of regular geometry up to 25 stories high with a height–to–width ratio of less than five. Variations in beam spans and column heights should be small. The application of the method is shown by the analyses of the frame for wind loads shown in Fig.13.27.

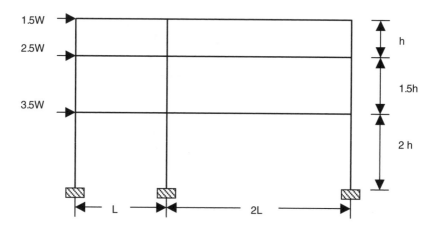

Fig.13.27 Rigid–jointed frame subjected to lateral loads.

(a) Shear force and bending moment in columns

Bay 1: width = L, **Bay 2**: width = 2L
Column heights: 2h, 1.5 h and h.
Total shear force Q in any storey is shared by the two bays in proportion to their widths.

Bay 1 = Q/3, Bay 2: = 2Q/3.
shear force in left column = 0.5 (Q/3) = Q/6
shear force in middle column = 0.5(Q/3 + 2Q/3) = 0.5Q
shear force in right column = 0.5(2Q/3) = Q/3
The bending moment at the top and bottom of columns
= shear force in the column x (height of column /2)
Table 13.16 shows the shear force and bending moments in the columns.

Table 13.16 Shear forces and bending moment in columns

	Q/W	Storey height	Shear in columns/W			Moment in columns/(Wh)		
			Left	Middle	Right	Left	Middle	Right
Top	1.5	h	0.25	0.75	0.50	0.125	0.375	0.25
Middle	1.5+2.5 = 4.0	1.5h	0.67	2.0	1.33	0.50	1.50	1.0
Bottom	4.0 + 3.5 = 7.5	2h	1.25	3.75	2.5	1.25	3.75	2.5

Table 13.17 Moments in beams

Location	Left beam	Right beam
Top	0.125 Wh	0.25 Wh
Middle	0.625 Wh	1.25 Wh
Bottom	1.75 Wh	3.50 Wh

(b) Bending moments in beams

As shown in Fig.13.28, the bending moments at the ends of the left beam are equal to sum of the bending moments at the ends of the columns on the left of the connecting the beam. Similarly the bending moments at the ends of the right beam are equal to sum of the bending moments at the ends of the columns on the right of the connecting the beam. Table 13.17 shows the bending moments in the beams. Fig.13.29 shows the bending moment distribution in the frame.

(c) Axial forces in columns

As shown in Fig.13.28, from the bending moments in the beams, reactions R at the ends of the beam is given by

$$R = 2 \times \text{Bending moments/ Span}.$$

From the reactions in the beam, axial forces in the columns can be determined.

Table 13.18 shows the axial forces in columns. Note that the axial force in the middle column is zero and the axial force in the left column is tensile while in the right column it is compressive.

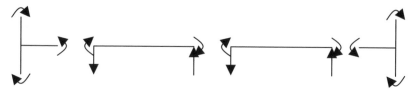

Fig.13.28 Forces at joints and beams.

Table 13.18 Axial forces in columns

Level	Beam moments/(Wh)		Reactions*(L/Wh)		Axial force in column*(L/Wh)
	Beam - left	Beam - Right	Beam - left	Beam - Right	Column-Left and right
Top	0.125	0.25	0.25	0.25	0.25
Middle	0.625	1.25	1.25	1.25	0.25+1.25 = 1.50
Bottom	1.75	3.50	3.50	3.50	0.25+1.25+3.50 =5.0

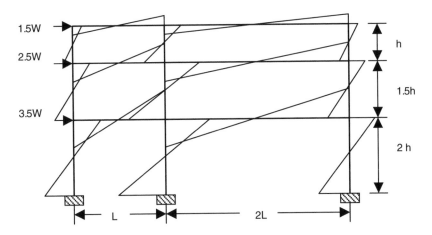

Fig.13.29 Bending moment distribution in the frame.

Table 13.19 Axial forces in beams

	Load at joint/W	Shear columns/W		Axial force in beam/W	
		Left	Right	Beam-Left	Beam-right
Top	1.5	0.25	0.5	1.5 – 0.25 = 1.25	0.5
Middle	2.5	0.67	1.33	2.5 + 0.25 – 0.67 = 2.08	1.33 – 0.5 = 0.83
Bottom	3.5	1.25	2.5	3.5 + 0.67 – 1.25 = 2.92	2.5 – 1.33 = 1.17

Fig.13.30 Axial force in beams.

(d) Axial forces in beams

As shown in Fig.13.30, by considering at joints equilibrium in the horizontal direction of shear forces in the columns and the axial forces in the beams, axial forces in beams can be determined. Table 13.19 shows the axial forces in beams.

CHAPTER 14

REINFORCED CONCRETE FRAMED BUILDINGS

14.1 TYPES AND STRUCTURAL ACTION

Commonly used single-storey and medium-rise reinforced concrete framed structures are shown in Fig.14.1. Tall multi-storey buildings are discussed in Chapter 15. Only *cast-in-situ* rigid jointed frames are dealt with, but the structures shown in the figure could also be precast.

The loads are transmitted by roof and floor slabs and walls to beams and to rigid frames and through the columns to the foundations. In *cast-in-situ* buildings with monolithic floor slabs, the frame consists of flanged beams and rectangular columns. However, it is common practice to base the analysis on the rectangular beam section, but in the design for sagging moments the flanged section is used. If precast slabs are used the beam sections are rectangular.

Depending on the floor system and framing arrangement adopted, the structure may be idealized into a series of plane frames in each direction for analysis and design. Such a system where two-way floor slabs are used is shown in Fig.14.2; the frames in each direction carry part of the load. In the complete three-dimensional frame, torsion occurs in the beams and biaxial bending in the columns. These effects are small and it is stated in BS 8110: Part 1, clause 3.8.4.3, that it is usually only necessary to design for the maximum moment about the critical axis. In rectangular buildings with a one-way floor system, the transverse rigid frame across the shorter plan dimension carries the load. Such a frame is shown in the design example in section 14.5.

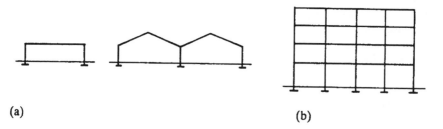

(a)   (b)

Fig.14.1 (a) Single storey; (b) multi-storey.

Resistance to horizontal wind loads is provided by

1. braced structures-shear walls, lift shaft and stairs
2. un-braced structures-bending in the rigid-jointed frames

The analysis for combined shear wall, rigid-jointed frame systems is discussed in Chapter 15.

In multi-storey buildings, the most stable arrangement is obtained by bracing with shear walls in two directions. Stairwells, lift shafts, permanent partition walls as well as specially designed external shear walls can be used to resist the horizontal loading. Shear walls should be placed symmetrically with respect to the building axes. If this is not done the shear walls must also be designed to resist the resulting torque. The concrete floor slabs act as large horizontal diaphragms to transfer loads at floor levels to the shear walls. BS 8110: Part 1, clause 3.9.2.2, should be noted: it is stated that the overall stability of a multi-storey building should not depend on unbraced shear walls alone. Shear walls in a multi-storey building are shown in Fig.14.2.

Foundations for multi-storey buildings may be separate pad or of strip type. However, rafts or composite raft and basement foundations are more usual. For raft type foundations the column base may be taken as fixed for frame analysis. The stability of the whole building must be considered and the stabilizing moment from dead loads should prevent the structure from overturning.

Separate pad type foundations should only be used for multi-storey buildings if foundation conditions are good and differential settlement will not occur. For single-storey buildings, separate foundations are usually provided and, in poor soil conditions, pinned bases can be more economical than fixed bases. The designer must be satisfied that the restraint conditions assumed for analysis can be achieved in practice. If a fixed base settles or rotates, a redistribution of moments occurs in the frame.

14.2 BUILDING LOADS

The load on buildings is due to dead, imposed, wind, dynamic, seismic and accidental loads. In the UK, multi-storey buildings for office or residential purposes are designed for dead, imposed and wind loads. The design is checked and adjusted to allow for the effects of accidental loads. The types of load are discussed briefly.

14.2.1 Dead Load

Dead load is due to the weight of roofs, floors, beams, walls, columns, floor finishes, partitions, ceilings, services etc. The load is estimated from assumed section sizes and allowances are made for further dead loads that are additional to the structural concrete.

14.2.2 Imposed Load

Imposed load depends on the occupancy or use of the building and includes distributed loads, concentrated loads, impact, inertia and snow. Loads for all types of buildings are given in BS 6399: Part 1: 1996

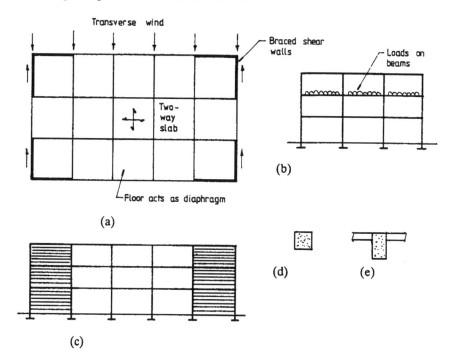

Fig.14.2 (a) Plan; (b) rigid transverse frame; (c) side elevation; (d) column; (e) T-beam.

14.2.3 Wind Loads

Wind load on buildings is estimated in accordance with BS 6399: Part 2: 1997. The following factors are taken into consideration:

1. The basic wind speed V_b depends on the location in the country.
2. The design wind speed V_s

$$V_s = V_b \times S_a \times S_d \times S_s \times S_p$$

where

S_a is an altitude factor depending on the site altitude above mean sea level

S_d is a direction factor normally taken as 1,

S_s is a seasonal factor normally taken as 1

S_p is a probability factor, normally taken as 1.

3. Effective wind speed V_e

$$V_e = V_s \times S_b$$

where S_b is terrain and building factor which depends on location of the building, whether close to the sea or the site is in town and also on the effective height. The height may refer to the total height of the building or the height of the part under consideration. In a multi-storey building the wind load increases with height and the factor S_b should be increased at every floor or every three or four floors (Fig.14.3). The factor varies from a minimum of 1.07 to a maximum of 2.12.

4. The dynamic pressure q_s

$$q_s = 0.613 V_e^2 \; N/m^2$$

q_s is the pressure on a surface normal to the wind and is modified by the dimensions of the building and by openings in the building.

5. The pressure p_e acting on an external surface of a building is given by

$$p_e = q_s \, C_{pe} \, C_a$$

where

C_{pe} is the external pressure coefficient. Values of C_{pe} are given for rectangular plan buildings and for different types of roofs.

C_a is the size effect factor for external pressures. In calculating the size effect factor C_a, the diagonal dimension of the loaded area above the level being considered is used.

6. Similar to the external pressure, the pressure p_i acting on an internal surface of a building is given by

$$p_i = q_s \, C_{pi} \, C_a$$

where depending on whether openings occur on the windward or leeward sides, internal pressure or suction exists inside the building. Tables and guidance are given in the code for evaluating internal pressure coefficient C_{pi}.

C_a is the size effect factor for internal pressures.

7. For enclosed buildings, the net pressure p across a surface is given by

$$p = (p_e - p_i)$$

8. The net load P on an area A of a building surface is given by

$$P = 0.85 \, (1 + C_r) \, pA$$

where C_r = dynamic augmentation factor

Wind loads should be calculated for lateral and longitudinal directions to obtain loads on frames or shear walls to provide stability in each direction. In asymmetrical buildings it may be necessary to investigate wind from all directions.

14.2.4 Load Combinations

Separate loads must be applied to the structure in appropriate directions and various types of loading combined with partial safety factors selected to cause the most severe design condition for the member under consideration. In general the following load combinations should be investigated.

(a) Dead load G_k + imposed load Q_k

1. All spans are loaded with the maximum design load of $1.4G_k + 1.6Q_k$;
2. Alternate spans are loaded with the maximum design load of $1.4G_k + 1.6Q_k$ and all other spans are loaded with the minimum design load of $1.0G_k$.

(b) Dead load G_k + wind load W_k

If dead load and wind load effects are additive the load combination is $1.4(G_k + W_k)$. However, if the effects are in opposite directions the critical load combination is $(1.0G_k - 1.4W_k)$.

(c) Dead load G_k + imposed load Q_k + wind load W_k

The structure is to be loaded with $1.2(G_k + Q_k + W_k)$.

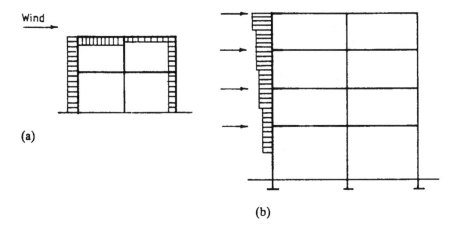

Fig.14.3 Wind loading (a) Loads distributed on surfaces; (b) loads applied at floor levels.

14.2.4.1 Example on load Combinations

Fig.14.4 shows a building supported on pinned base columns spaced 10 m in both directions. Calculate the maximum bending moment and axial force (compression and tension) in the columns. It is given that

$$G_k = 15.0 \text{ kN/m}^2, Q_k = 12.5 \text{ kN/m}^2, W_k = 1.0 \text{ kN/m}^2$$

Wind acts on the face AB or GD and can act from left to right or vice versa.

In considering the load combinations to be considered for a particular force, it is necessary to be aware of the effect of the load on a particular part of the structure on the force under consideration in order that appropriate load factors can be applied.

In this example the following effects can be noted.

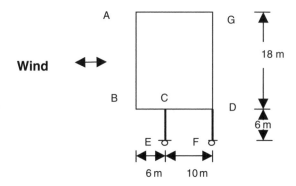

Fig.14.4 Building supported by pinned base columns.

1. Left hand column:
- Vertical load acting on the entire plan area BCD will cause compression in the column.
- Wind blowing from right to left causes compression in the column
- Wind blowing from left to right causes tension in the column

2. Right hand column:
- Vertical load acting on a plan area CD will cause compression
- Vertical load on plan area BC will cause tension.
- Wind blowing from right to left causes tension in the column
- Wind blowing from left to right causes compression in the column

Load combinations:

(i) Dead load G_k + imposed load Q_k
There are four cases to be considered. They are

(a) $(1.4\ G_k + 1.6\ Q_k)$ on the plan area BCD:
    Total vertical load = $(1.4 \times 15 + 1.6 \times 12.5) \times 16 \times 10 = 6560$ kN
Taking moments about F,
    axial force in CE = $\{6560 \times 16/2\}/10 = 5248$ kN
    axial force in DF = $6560 - 5248 = 1312$ kN

(b) $(1.4\ G_k + 1.6\ Q_k)$ on the plan area CD and $1.0\ G_k$ on plan area BC
Total vertical load on plan area CD
    $= (1.4 \times 15 + 1.6 \times 12.5) \times 10 \times 10 = 4100$ kN
Total vertical load on plan area BC
    $= (1.0 \times 15) \times 6 \times 10 = 900$ kN
Taking moments about F,

$$\text{axial force in CE} = \{4100 \times 10/2 + 900 \times (10+6/2)\}/10 = 3220 \text{ kN}$$
$$\text{axial force in DF} = 4100+900 - 3220 = 1780 \text{ kN}$$

(c) 1.0 G_k on the plan area CD and (1.4 G_k + 1.6 Q_k) on plan area BC
Total vertical load on plan area CD
$$= (1.0 \times 15) \times 10 \times 10 = 1500 \text{ kN}$$
Total vertical load on plan area BC
$$= (1.4 \times 15 + 1.6 \times 12.5) \times 6 \times 10 = 2460 \text{ kN}$$
Taking moments about F,
$$\text{axial force in CE} = \{1500 \times 10/2 + 2460 \times (10+6/2)\}/10 = 3948 \text{ kN}$$
$$\text{axial force in DF} = 1500+2460 - 3948 = 12 \text{ kN}$$

(d) 1.0 G_k on the plan area BCD:
$$\text{Total vertical load} = (1.0 \times 15) \times 16 \times 10 = 2400 \text{ kN}$$
Taking moments about F,
$$\text{axial force in CE} = \{2400 \times 16/2\}/10 = 1920 \text{ kN}$$
$$\text{axial force in DF} = 2400 - 1920 = 480 \text{ kN}$$

Column forces: From the above four combinations the maximum and minimum forces in the columns are:
Column CE
Maximum compressive force = 5248 kN (from (a))
Minimum compressive force = 1920 kN (from (d))
Column DF
Maximum compressive force = 1780 kN (from (b))
Minimum compressive force = 12 kN (from (c))

(ii) Dead load and wind load: [(1.4 or 1.0) $G_k \pm 1.4\ W_k$]
For convenience, the calculations are done by considering wind and dead loads separately.

(a) Wind load 1.4 W_k only
$$\text{wind load} = 1.4 \times 1.0 \times 18 \times 10 = 252 \text{ kN}$$
$$\text{Axial force in columns} = \pm 252 \times (6+ 18/2)/10 = \pm 378 \text{ kN}$$
$$\text{Shear force in columns} = \pm 252/2 = 126 \text{ kN}$$
$$\text{Moment in columns} = 126 \times 6 = 756 \text{ kNm}$$

(b) 1.4 G_k on the plan area BCD:
$$\text{Total vertical load} = (1.4 \times 15) \times 16 \times 10 = 3360 \text{ kN}$$
Taking moments about F,
$$\text{axial force in CE} = \{3360 \times 16/2\}/10 = 2688 \text{ kN}$$
$$\text{axial force in DF} = 3360 - 2688 = 672 \text{ kN}$$

(c) 1.4 G_k on the plan area CD and 1.0 G_k on plan area BC
$$\text{Total vertical load on plan area CD} = (1.4 \times 15) \times 10 \times 10 = 2100 \text{ kN}$$
$$\text{Total vertical load on plan area BC} = (1.0 \times 15) \times 6 \times 10 = 900 \text{ kN}$$

Taking moments about F,

axial force in CE = {2100 × 10/2 + 900 × (10+6/2)}/10 = 2220 kN

axial force in DF = 2100+900 – 2220 = 780 kN

(d) 1.0 G_k on the plan area CD and 1.4 G_k on plan area BC

Total vertical load on plan area CD = (1.0 × 15) × 10 × 10 = 1500 kN

Total vertical load on plan area BC = (1.4 × 15) × 6 × 10 = 1260 kN

Taking moments about F,

axial force in CE = {1500x 10/2 + 1260 x (10+6/2)}/10 = 2388 kN

axial force in DF = 1500+1260 – 2388 = 372 kN

(e) 1.0 G_k on the plan area BCD:

Total vertical load = (1.0 × 15) × 16 × 10 = 2400 kN

Taking moments about F,

axial force in CE = {2400 × 16/2}/10 = 1920 kN

axial force in DF = 2400 – 1920 = 480 kN

Column forces: From the above four cases the critical axial force and moment combinations are:

Column CE

Maximum compressive force

= 2688 (from (b)) + 378 (from (a)) = 3066 kN

Moment in column = 756 kNm

Minimum compressive force

= 1920 (from (e)) – 378 (from (a)) = 1542 kN

Moment in column = 756 kNm

Column DF

Maximum compressive force

= 780 (from (c)) + 378 (from (a)) = 1158 kN

Moment in column = 756 kNm

Minimum compressive force

= 372 (from (e)) – 378 (from (a)) = –6 kN

Moment in column = 7564 kNm

(iii) [Dead + imposed + wind] load: 1.2(G_k + Q_k + W_k)

For convenience, the calculations are done by considering wind, dead and imposed loads separately.

(a) Wind load

wind load = 1.2 × 1.0 × 18 × 10 = 216 kN

Axial force in columns = ± 216 × (6 + 18/2)/10 = ± 324 kN

Shear force in columns = ± 216/2 = 108 kN

Moment in columns = 108 × 6 = 648 kNm

(b) Vertical load on the plan area BCD:

Total vertical load = $(1.2 \times 15 + 1.2 \times 12.5) \times 16 \times 10 = 5280$ kN

Taking moments about F,

axial force in CE = $\{5280 \times 16/2\}/10 = 4224$ kN

axial force in DF = $5280 - 4224 = 1056$ kN

Column forces: From the above four cases the critical axial force and moment combinations are:

Column CE

Maximum compressive force

= 4224 (from (b)) + 324 (from (a)) = 4548 kN

Moment in column = 648 kNm

Minimum compressive force

= 4224 (from (b)) – 324 (from (a)) = 3900 kN

Moment in column = 648 kNm

Column DF

Maximum compressive force

= 1056 (from (b)) + 324 (from (a)) = 1380 kN

Moment in column = 648 kNm

Minimum compressive force

= 1056 (from (e)) – 324 (from (a)) = 732 kN

Moment in column = 648 kNm

Design values: From a design point of view, the critical combinations *probably* are

Column CE: N = 1542 kN, M = 756 kNm

Column DF: N = –6 kN, M = 756 kNm

Both arise from $1.0\ G_k + 1.4\ W_k$ combination.

This example shows that even in a simple structure, the number of load cases to be considered may become quite large. In large scale structures, a good understanding of the behaviour of the structure under consideration is necessary in order to limit the number of load cases to be considered, by eliminating load cases clearly not critical.

14.3 ROBUSTNESS AND DESIGN OF TIES

Clause 2.2.2.2 of the code states that situations should be avoided where damage to a small area or failure of a single element could lead to collapse of major parts of the structure. The clause states that provision of effective ties is one of the precautions necessary to prevent progressive collapse. The layout also must be such as to give a stable and robust structure.

The design of ties set out in section 3.12.3 of the code is summarized below.

14.3.1 Types of Tie

The types of tie are
1. peripheral ties
2. internal ties
3. horizontal ties to columns and walls
4. vertical ties

The types and location of ties are shown in Fig.14.5.

14.3.2 Design of Ties

Steel reinforcement provided for a tie can be designed to act at its characteristic strength. Reinforcement provided for other purposes may form the whole or part of the ties. Ties must be properly anchored and a tie is considered anchored to another at right angles if it extends 12 diameters or an equivalent anchorage length beyond the bar forming the other tie.

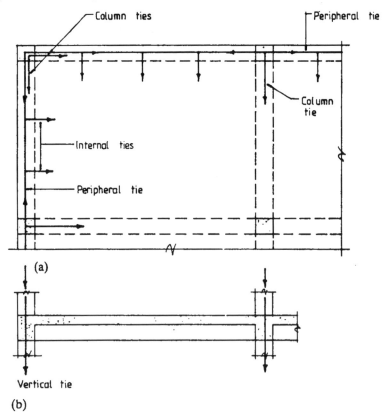

Fig.14.5 Building ties (a) Plan; (b) section.

14.3.3 Internal Ties

Internal ties are to be provided at the roof and all floors in two directions at right angles. They are to be continuous throughout their length and anchored to peripheral ties. The ties may be spread evenly in slabs or be grouped in beams or walls at spacing not greater than $1.5l_r$ where l_r is defined below. Ties in walls are to be within 0.5 m of the top or bottom of the floor slab.

The ties should be capable of resisting a tensile force which is the greater of

$$\frac{g_k + q_k}{7.5}\frac{\ell_r}{5}F_t \text{ or } 1.0F_t$$

where
$g_k + q_k$ = the characteristic dead plus imposed floor load (kN/m^2),
F_t = lesser of $(20 + 4n_o)$ or 60 kN
n_o = the number of storeys
l_r = is the greater of the distance in metres between the centres of columns, frames or walls supporting any two adjacent floor spans in the direction of the tie.

14.3.4 Peripheral Ties

A continuous peripheral tie is to be provided at each floor and at the roof. This tie is to resist F_t as defined above and is to be located within 1.2 m of the edge of the building or within the perimeter wall.

14.3.5 Horizontal Ties to Columns and Walls

Each external column and, if a peripheral tie is not located within the wall, every metre length of external wall carrying vertical load should be tied horizontally into the structure at each floor and at roof level. The tie capacity is to be equal to
- the greater of $2F_t$ or $(l_s/2.5)$ F_t if less , or
- 3% of the ultimate vertical load carried by the column or wall.

where l_s is the floor-to-ceiling height in metres.

Where the peripheral tie is located within the walls the internal ties are to be anchored to it.

14.3.6 Corner Column Ties

Corner columns are to be anchored in two directions at right angles. The tie capacity is the same as specified in section 14.3.5 above.

14.3.7 Vertical Ties

Vertical ties are required in buildings of five or more storeys. Each column and load bearing wall is to be tied continuously from foundation to roof. The tie is to be capable of carrying a tensile force equal to the ultimate dead and imposed load carried by the column or wall from one floor.

14.4 FRAME ANALYSIS

14.4.1 Methods of Analysis

The methods of frame analysis that are used may be classified as
1. using solutions for standard frames
2. simplified methods of analysing sub-frames given in section 3.2.1 of the code (Chapter 3, section 3.4.2, Fig.3.3)
3. plane frame computer programs based on the matrix stiffness method of analysis

All methods are based on elastic theory. For the justification for using internal forces determined from elastic analysis for design at ultimate loads, see Chapter 13. BS 8110 permits redistribution of up to 30% of the peak elastic moment to be made in frames up to four storeys. In frames over four storeys in height where the frame provides the lateral stability, redistribution is limited to 10%.

In rigid frame analysis the sizes for members must be chosen from experience; or established by an approximate design before the analysis can be carried out. Ratios of stiffness of the final member sections should be checked against those estimated and the frame should be reanalysed if it is found necessary to change the sizes of members significantly.

Although BS8110: Part 1, clause 2.5.2, permits the calculation of relative stiffness of members by different methods, it is usual to calculate the relative stiffnesses using gross section sizes. As noted previously in beam-slab floor construction it is normal practice to base the beam stiffness on a uniform rectangular section consisting of the beam depth by the beam rib width. The flanged beam section is taken into account in the beam design for sagging moments near the centre of spans.

14.4.2 Example of Simplified Analysis of Concrete Framed Building under Vertical Load

The application of the various simplified methods of analysis given in section 3.2 of the code is shown in the following example.

(a) Specification
The cross-section of a reinforced concrete building is shown in Fig.14.6(a). The frames are at 4.5 m centres, the length of the building is 36 m and the column bases are fixed. Preliminary sections for the beams and columns are shown in Fig.14.6(b). The floor and roof slabs are designed to span one way between the

frames. Longitudinal beams are provided between external columns of the roof and floor levels only.

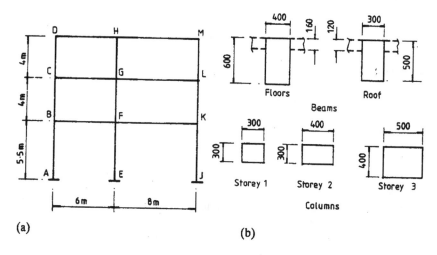

Fig.14.6 (a) Cross-section; (b) assumed member sections.

The dead and imposed loading is as follows:
Roof:

$$\text{total dead load} = 4.3 \text{ kN/m}^2$$
$$\text{imposed load} = 1.5 \text{ kN/m}^2$$

Floor:

$$\text{total dead load} = 6.2 \text{ kN/m}^2$$
$$\text{imposed load} = 3.0 \text{ kN/m}^2$$

Wind load: The wind load is according to BS 6399–2: 1997, Part 2: Code of practice for wind loads. The location is on the outskirts of a city in the Northeast of the UK.

The materials are grade C30 concrete and grade 460 reinforcement.

Determine the design actions for the beam BFK and column length FE for an internal frame for the two cases where the frame is braced and unbraced. Results for selected cases using only the simplified method of analysis from BS 8110: Part 1, section 3.2, are given.

(b) Loading

Braced frame:
The following load cases are required for beam BFK for the braced frame.
Case 1: $(1.4 \, G_k + 1.6 \, Q_k)$ on the whole beam
Case 2: $(1.4 \, G_k + 1.6 \, Q_k)$ on BF and $(1.0 \, G_k)$ on FK
Case 3: $(1.0 \, G_k)$ on BF and $(1.4 \, G_k + 1.6 \, Q_k)$ on FK

Unbraced frame:
For the unbraced frame, an additional load case is required:
Case 4: $1.2(G_k + Q_k)$ on the whole beam

Using a frame spacing of 4.5 m c/c, the characteristic loads are as follows.

Dead load:
$$\text{Roof } 4.3 \times 4.5 = 19.4 \text{ kN/m}$$
$$\text{Floors } 6.2 \times 4.5 = 27.9 \text{ kN/m}$$

Imposed load:
$$\text{Roof } 1.5 \times 4.5 = 6.8 \text{ kN/m}$$
$$\text{Floors } 3.0 \times 4.5 = 14.5 \text{ kN/m}$$

(c) Section properties
The beam and column properties are given in Table 14.1.

Table 14.1 Section properties

Member	b x d	L (mm)	I (mm^4)	I/L
Columns FE, AB, KJ	400 × 500	5500	4.17×10^9	7.58×10^5
Columns GF, BC, KL	300 × 400	4000	1.60×10^9	4.0×10^5
Beam FK	400 × 600	8000	7.20×10^9	9.0×10^5
Beam BF	400 × 600	6000	7.20×10^9	12.0×10^5

(d) Sub-frame analysis for braced frame
Although the correct way to analyse the structure is to analyse the entire frame, for simplicity, very often instead of analysing the entire frame, only parts of the structure called sub-frames are analysed. BS 8110 in clause 3.2.1.2 allows the use of sub-frame consisting of the beams at one level together with the columns above and below. The sub-frame used in this example is shown in Fig.14.7.

The frame is analysed for the three load cases using the Stiffness Method of analysis. From characteristic dead and imposed loads,

$$(1.4 \ G_k + 1.6 \ Q_k) = (1.4 \times 27.9 + 1.6 \times 14.5) = 60.66 \text{ kN/m},$$
$$1.0 \ G_k = = 27.9 \text{ kN/m}$$

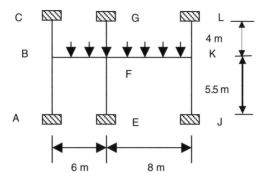

Fig.14.7 Simplified subframe for beam BFK.

The fixed end moments are:
Case 1:

$$\text{span BF} = 60.66 \times 6^2/12 = 181.98 \text{ kNm,}$$
$$\text{span FK} = 60.66 \times 8^2/12 = 323.2 \text{ kN m}$$

Case 2:

$$\text{span BF} = 60.66 \times 6^2/12 = 181.98 \text{ kNm}$$
$$\text{span FK} = 27.9 \times 8^2/12 = 148.80 \text{ kN m}$$

Case 3:

$$\text{span BF} = 27.9 \times 6^2/12 = 83.70 \text{ kNm}$$
$$\text{span FK} = 60.66 \times 8^2/12 = 323.2 \text{ kN m}$$

The stiffness matrix K and load vector F for the three load cases are

$$K = E \times 10^5 \begin{bmatrix} 94.32 & 24.0 & 0 \\ 24.0 & 130.32 & 18.0 \\ 0 & 18.0 & 82.32 \end{bmatrix} \begin{bmatrix} \theta_B \\ \theta_F \\ \theta_K \end{bmatrix}, \; F = \begin{bmatrix} 181.98 & 181.98 & 83.70 \\ 141.54 & -33.18 & 239.82 \\ -323.52 & -148.80 & -323.52 \end{bmatrix}$$

The displacement vectors for the three load cases are

$$E \times 10^5 \begin{bmatrix} \theta_B \\ \theta_F \\ \theta_K \end{bmatrix} = \begin{bmatrix} 1.5783 & 2.0287 & 0.2755 \\ 1.3799 & -0.3903 & 2.4050 \\ -4.2318 & -1.7222 & -4.4559 \end{bmatrix}$$

where E = Young's modulus.

The shear force and bending moment diagrams for beam BFK for are shown in Fig.14.8 and Fig.14.9 and the results of analysis are shown in Table 14.2.

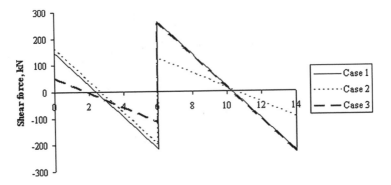

Fig.14.8 Shear force diagrams for beam BFK in sub-frame.

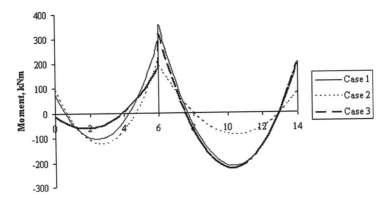

Fig.14.9 Bending moment diagrams for beam BFK in sub-frame.

(e) Sub-sub-frames

As an alternative to the use of sub-frame discussed above, the code in clause 3.2.1.2.3 allows the use of sub-frames which are part of sub-frames. The bending moment in an individual beam may be found by considering a sub-frame formed by consisting of only that beam, the columns attached to the ends of the beam and the beams on either side with *half of their actual stiffness*. Fig 14.10 shows the two sub-frames which can be analysed to determine the bending moment in the beam BF and FK respectively.

The stiffness matrix K and load vector F for the three load cases for sub-sub-frame for beam BF are

$$ K = E \times 10^5 \begin{bmatrix} 94.32 & 24.0 \\ 24.0 & 112.32 \end{bmatrix} \begin{bmatrix} \theta_B \\ \theta_F \end{bmatrix} , F = \begin{bmatrix} 181.98 & 181.98 & 83.70 \\ 141.54 & -33.18 & 239.82 \end{bmatrix} $$

The displacement vectors for the three load cases are

$$E \times 10^5 \begin{bmatrix} \theta_B \\ \theta_F \end{bmatrix} = \begin{bmatrix} 1.7013 & 2.1198 & 0.3639 \\ 0.8966 & -0.7484 & 2.0574 \end{bmatrix}$$

The stiffness matrix K and load vector F for the three load cases for sub-frame for beam FK are

$$K = E \times 10^5 \begin{bmatrix} 106.32 & 18.0 \\ 18.0 & 82.32 \end{bmatrix} \begin{bmatrix} \theta_F \\ \theta_K \end{bmatrix}, F = \begin{bmatrix} 141.54 & -33.18 & 239.82 \\ -323.52 & -148.80 & -323.52 \end{bmatrix}$$

The resulting displacement vectors for the three load cases are

$$E \times 10^5 \begin{bmatrix} \theta_F \\ \theta_K \end{bmatrix} = \begin{bmatrix} 2.0734 & -0.0063 & -4.5933 \\ -4.3834 & -1.8062 & 2.0574 \end{bmatrix}$$

Table 14.2 shows the bending moments in the members from the subframe analysis

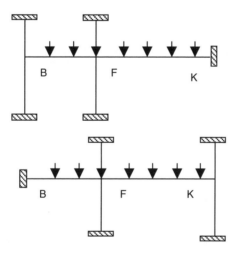

Fig.14.10 Sub-sub-frame for beam BF and beam FK.

(f) Continuous beam simplification

According to clause 3.2.1.2.4, an approach which is even more conservative than the sub-sub-frames considered above is to assume that the beams at any one level is a continuous beam over simple supports. The beam BFK can be taken as a continuous beam over supports that provide no restraint to rotation. The load cases are the same as for the sub-frame analysis above.

The stiffness matrix K and load vector F for the three load cases are

$$K = E \times 10^5 \begin{bmatrix} 48.0 & 24.0 & 0 \\ 24.0 & 84.0 & 18.0 \\ 0 & 18.0 & 36.0 \end{bmatrix} \begin{bmatrix} \theta_B \\ \theta_F \\ \theta_K \end{bmatrix}, F = \begin{bmatrix} 181.98 & 181.98 & 83.70 \\ 141.54 & -33.18 & 239.82 \\ -323.52 & -148.80 & -323.52 \end{bmatrix}$$

The resulting displacement vectors for the three load cases are

$$E \times 10^5 \begin{bmatrix} \theta_B \\ \theta_F \\ \theta_K \end{bmatrix} = \begin{bmatrix} 2.1063 & 4.1863 & -1.1113 \\ 3.3700 & -0.7900 & 5.7100 \\ -10.6717 & -3.7383 & -11.8417 \end{bmatrix}$$

The bending moments from the three simplified idealizations viz. continuous beam idealization, the two sub-sub-frames and the full sub-frame are shown in Table 14.2. The results indicate that for moments M_{BF} and M_{FB}, the full sub-frame and the sub-sub-frame for beam BF produce fairly close results. Similarly, the results indicate that for moments M_{FK} and M_{KF}, the full sub-frame and the sub-sub-frame for beam FK produce fairly close results. The continuous beam idealization is not particularly accurate.

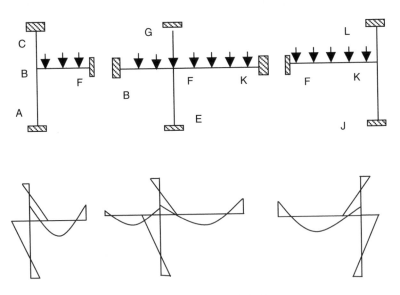

Fig.14.11 Asymmetrically loaded columns.

(g) Asymmetrically loaded columns

Where the beam moments have been obtained from the continuous beam idealization, the moments for a column are calculated assuming that column and beam ends remote from the junction considered are fixed and that beams have one-half their actual stiffness. If the continuous beam idealization is used, then the moments M_{BF} and M_{KF} should be taken to be equal to those found from the sub

Table 14.2 Comparison of bending moments for different idealizations

M	Case 1				Case 2				Case 3			
	Sub-frame-BFK	Sub-frame-BF	Sub-frame-FK	Cont-beam	Sub-frame-BFK	Sub-frame-BF	Sub-frame-FK	Cont-beam	Sub-frame-BFK	Sub-frame-BF	Sub-frame-FK	Cont-beam
M_{BF}	−73	−79	−157	−120*	−94	−98	−182	−120*	−13	−17	−47	−49*
M_{FB}	286	266	232	394	212	197	182	245	206	191	157	331
M_{BA}	48	52			62	64			8	11		
M_{BC}	25	27			33	34			4	6		
M_{FE}	42	27	63		−12	−23	0		73	62	92	
M_{FG}	22	14	33		−6	−12	0		39	33	49	
M_{FK}	−350	−307	−328	−394	−194	−162	−182	−245	−317	−287	−297	−331
M_{KF}	196	332	203	233#	80		87	107#	206	342	213	233#
M_{KJ}	−123		−133		−52		−29		−135		−139	
M_{KL}	−68		−70		−28		−55		−71		−74	

*These values have been obtained from asymmetrical column AC analysis.

These values have been obtained from asymmetrical column JL analysis.

Note: Moment at the fixed end of the columns is one half of the moments at the deformable end

Table 14.3 Analysis of asymmetrically loaded columns

Moment	Case 1			Case 2			Case 3		
	Column ABC	Column EFG	Column JKL	Column ABC	Column EFG	Column JKL	Column ABC	Column EFG	Column JKL
M_{BF}	−120	−163		−120	−187		−59	−116	
M_{FB}	213	220		213	173		96	149	
M_{FK}		−295	−369		−156	−170		−275	−369
M_{KF}		338	233		145	107		348	233
M_{BA}	79			79			32		
M_{FE}		49			−11			82	
M_{KJ}			−152			−70			−152
M_{BC}	41			41			17		
M_{FG}		26			−6			43	
M_{KL}			−81			−37			−81

Note: Moment at the fixed end of the columns is one half of the moments at the deformable end

frame analyses for the asymmetrically loaded outer columns. Fig.14.11 shows the three asymmetrically loaded columns.

The stiffness matrix and the load vectors and the corresponding joint rotations for the three cases are:

Asymmetrical Column ABC:

$$K = E \times 10^5 [70.32][\theta_B], F = [181.98 \quad 181.98 \quad 83.70]$$

$$E \times 10^5 \theta_B = [2.5879 \quad 2.5879 \quad 1.0481]$$

Asymmetrical Column EFG:

$$K = E \times 10^5 [88.32][\theta_F], F = [141.54 \quad -33.18 \quad 239.82]$$

$$E \times 10^5 \theta_F [1.6026 \quad -0.3757 \quad 2.7154]$$

Asymmetrical **Column JKL:**

$$K = E \times 10^5 [64.16][\theta_K], F = [-323.52 \quad -148.80 \quad -323.52]$$

$$E \times 10^5 \theta_K = [-5.0424 \quad -2.3192 \quad -5.0424]$$

The results of analysis are shown in Table 14.3.

(h) Un-braced frame analysis for vertical loads
The analysis for vertical loads can be made in the same way as for the braced frame. The load in this case is $1.2(G_k + Q_k)$.

14.4.3 Example of Simplified Analysis of Concrete Framed Building for Wind Load by Portal Frame Method

(a) Wind loads
The wind loads are calculated using BS 6399: 1997, Part 2: Code of practice for wind loads. For normal calculations the so called Standard Method is used.
The case for which wind load is calculated is wind acting normal to the 40 m width.

The maximum height H = 14.5 m above the ground.
Reference height H_r = 14.5 m
Effective height $H_e = H_r$ = 14.5 m
Building type factor K_b = 1.0 (Table 1 of the Code)
Dynamic augmentation factor C_r, (Fig. 3 of the Code)
$$C_r \approx 0.04$$
The code rules apply.
Basic wind speed: V_b = 20 m/s (Assumed)
Altitude factor, $S_a \approx 1.0$
Direction factor, S_d = 1.0
Seasonal factor, S_s = 1.0

Probability factor, $S_p = 1.0$

Site wind speed, $V_s = V_b \times S_a \times S_d \times S_s \times S_p = 20$ m/s

Terrain and building factor, S_b: Site in town, with the closest distance to sea upwind from the site is greater than 100 kM. Using Table 4 of code and interpolating between 1.85 for 30 m and 1.95 for 50 m,

$$S_b = 1.88$$

Effective wind speed:

$$V_e = V_s \times S_b = 20 \times 1.88 = 37.6 \text{ m/s}$$

Dynamic pressure:

$$q_s = 0.613\ V_e^2 = 0.613 \times 37.6^2 = 866.64 \text{ N/m}^2 = 0.87 \text{ kN/m}^2$$

External and internal surface pressure coefficient, C_{pe} and C_{pi} (Table 5 of the Code). Smaller dimension of the building, $D = 14$ m

$$H = 14.5 \text{ m}$$
$$D/H = 1.04$$

C_{pe} (Windward (front) face)
$$= 0.85 - (0.85 - 0.6)\ (1.04 - 1.0)/\ (4.0 - 1.0) = 0.847$$
C_{pe} (leeward (rear) face) $= -0.5$

Size factor, C_a: (Fig.4 of the Code)

Diagonal dimension of the area on which wind acts,
$$a = \sqrt{(36^2 + 14.5^2)} = 38.45 \text{ m}$$

Using the factors for site in town,
$$C_a \approx 0.83$$

External surface pressure:
$$p_e = q_s \times C_{pe} \times C_a = 0.87 \times 0.847 \times 0.83 = 0.612 \text{ kN/m}^2 \text{ (front face)}$$
$$= 0.87 \times (-0.5) \times 0.83 = -0.361 \text{ kN/m}^2 \text{ (rear face)}$$

Total pressure p on the building:
$$p = 0.612 - (-0.361) = 0.973 \text{ kN/m}^2$$

Fig.14.12 shows the wind load acting as a uniformly distributed load over the height of the building.

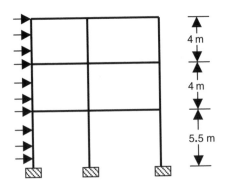

Fig.14.12 Rigid frame subjected to wind load.

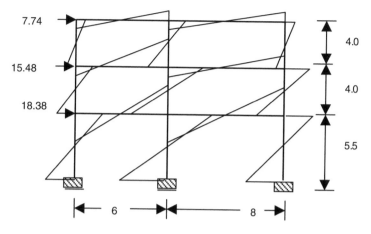

Fig.14.13 Loads at joints and bending moment diagram.

Total horizontal load per frame
$$= 0.85 \times (1 + C_r) \times p \times A$$
$$= 0.85 \times (1 + 0.04) \times 0.973 \times (4.5 \times 13.5) = 52.25 \text{ KN}$$

The total load is distributed at the storey levels in proportion as follows:

Roof level $= 52.25 \times (0.5 \times 4)/13.5 = 7.74$ KN

Second floor level $= 52.25 \times 4.0/13.5 = 15.48$ KN

First floor level $= 52.25 \times \{0.5(4.0 + 5.5)\}/13.5 = 18.38$ kN

Fig.14.13 shows the concentrated loads acting at roof and storey levels.

(b) Analysis by Portal method

The moments and shear forces due to wind load are calculated using the portal method explained in Chapter 13, section 13.12.4.

(i) Shear in bays

Total shear Q in each storey is divided between the bays in proportion to their spans.

Bay 1: Span = 6 m

Bay 2: Span = 8 m

Shear in bay 1 $= Q \times 6/(6 + 8) = 0.4286$ Q

Shear in Bay 2 $= Q \times 8/(6 + 8) = 0.5714$ Q

(ii) Shear in columns

The shear in each bay is divided equally between the columns.

Shear in left column $= 0.5 \times 0.4286$ Q $= 0.2143$ Q

Shear in the middle column $= 0.5 \times (0.4286 + 0.5714)$ Q $= 0.5Q$

Shear in right column $= 0.5 \times 0.5714$ Q $= 0.2857$ Q

(iii) Bending moment in columns

Bending moment at the top and bottom of a column is equal to product of shear in the column and storey height/2.

Table 14.4a summarises the storey shear forces and shear forces in columns. Table 14.4b shows bending moment in columns.

Table 14.4a Shear forces in kN in columns

Location	Q (kN)	Storey height	Shear in columns (kN)		
			Left	Middle	Right
Top	7.74	4.0	1.66	3.87	2.21
Middle	7.74+15.48 = 23.22	4.0	4.97	11.61	6.63
Bottom	23.22 + 18.38 = 41.60	5.5	8.92	20.80	11.89

Table 14.4b Bending moments in kNm in columns

Location	Storey height	Shear in columns			Moment in columns		
		Left	Middle	Right	Left	Middle	Right
Top	4.0	1.66	3.87	2.21	3.32	7.74	4.42
Middle	4.0	4.97	11.61	6.63	9.94	23.22	14.26
Bottom	5.5	8.92	20.80	11.89	24.53	57.2	32.70

(iv) Bending moment in beams

As shown in Fig.13.28, the bending moments at the ends of the left beam are equal to sum of the bending moments at the ends of the columns on the left connecting the beam. Similarly the bending moments at the ends of the right beam are equal to sum of the bending moments at the ends of the columns on the right connecting the beam. Table 14.5 shows the bending moments in the beams. Fig.14.13 shows the bending moment distribution in the frame.

Table 14.5 Moments in beams

Location	Left beam	Right beam
Top	3.32	4.42
Middle	14.26	17.68
Bottom	34.47	45.96

(v) Axial force in columns

Since there is no distributed load on the beams and the contra–flexure point is at mid–span, reactions R at the ends of the beam is given by

R = 2 × Bending moments at the ends/ Span

From the reactions in the beam, axial forces in the columns can be determined. Table 14.6 shows the axial forces in columns. Forces reverse in sign if the wind blows from right to left.

(vi) Axial forces in beams

As shown in Fig.13.33, Chapter 13, from the shear forces in the columns and by considering the horizontal force equilibrium at the joints, axial forces in beams can be determined. Table 14.7 shows the axial forces in beams.

Table 14.6 Axial forces in columns

Level	Beam moments		Reactions in beams	Axial force in columns
	Beam-left	Beam-Right		
Top	3.32	4.42	1.11	1.11
Middle	14.26	17.68	4.42	$1.11 + 4.42 = 5.53$
Bottom	34.47	45.96	11.49	$5.53 + 11.49 = 17.02$

Table 14.7 Axial forces in beams

	Load at joint	Shear columns		Axial force in beam	
		Left	Right	Beam-Left	Beam-right
Top	7.74	1.66	2.21	$7.74 - 1.66 = 6.08$	1.66
Middle	15.48	4.97	6.63	$15.48 + 1.66 - 4.97 = 12.17$	$6.63 - 2.21 = 4.42$
Bottom	18.38	8.92	11.89	$18.38 + 4.97 - 8.92 = 14.43$	$11.89 - 6.63 = 5.26$

14.5 BUILDING DESIGN EXAMPLE

14.5.1 Example of Design of Multi-Storey Reinforced Concrete Framed Buildings

Specification

The framing plans for a multi-storey building are shown in Fig.14.14. The main dimensions, structural features, loads, materials etc. are set out below.

(a) Overall dimensions

The overall dimensions are 36 m × 22 m in plan × 36 m high
Length: six bays at 6 m each; total 36 m
Breadth: three bays, 8 m, 6 m, 8 m; total 22 m
Height: ten storeys, nine at 3.5 m + one at 4.5 m

(b) Roof and floors

The floors and roof are constructed in one-way ribbed slabs spanning along the length of the building. Slabs are made solid for 300 mm on either side of the beam supports.

(c) Stability
Stability is provided by shear walls at the lift shafts and staircases in the end bays.

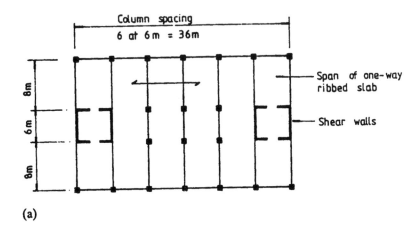

(a)

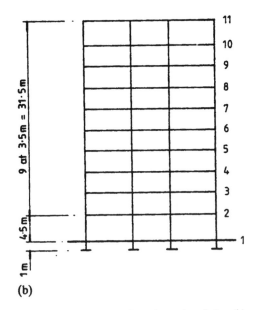

(b)

Fig.14.14 (a) Floor plan and roof plan; (b) end elevation.

(d) Fire resistance
All elements are to have a fire resistance period of 2h.

(e) Loading condition
Roof imposed load: 1.5 kN/m^2

Floors imposed load: 3.0 kN/m^2
Finishes, roof: 1.5 kN/m^2
Finishes, floors, partitions, ceilings, services: 3.0 kN/m^2
Parapet: 2.0 kN/m
External walls at each floor: 6.0 kN/m

The load due to self–weight is estimated from preliminary sizing of members. The imposed load contributing to axial load in the columns is reduced by 50% for a building with ten floors including the roof as permitted by Table 2 of BS 6399–1, 1996.

(f) Exposure conditions
External moderate
Internal mild

(g) Materials
Concrete grade 30 Reinforcement grade 460

(h) Foundations
Pile foundations are provided under each column and under the shear walls.

Scope of the work
The work carried out covers analysis and design for
1. transverse frame members at floor 2 outer span only
2. an internal column between floors 1 and 2
The design is to meet requirements for robustness. In this design, the frame is taken as completely braced by the shear walls in both directions. A link–frame analysis can be carried out to determine the share of wind load carried .by the rigid frames (Chapter 15). The design for dead and imposed load will be the critical design load case.

Preliminary sizes and self–weights of members:

(a) Floor and roof slab
The one-way ribbed slab is designed first. The size is shown in Fig.14.15. The weight of the ribbed slab is 0.5 m wide × 1 m is
$$= 24[(0.5 \times 0.275) - (0.375 \times 0.215)] = 1.365 \text{ kN/m}$$
$$= 1.365/ (0.5) = 2.73 \text{ kN/m}^2$$

(b) Beam sizes
Beam sizes are specified from experience:
$$\text{depth} = \text{span}/15 = 500 \text{ mm,}$$
$$\text{width} = 0.6 \times \text{depth} = 400 \text{ mm, say}$$
Preliminary beam sizes for roof and floors are shown in Fig.14.15. The weights of the beams including the solid part of the slab are:
roof beams, 24[(0.3 × 0.45) + (0.6 × 0.275)] = 7.2 kN/m
floor beams, 24[(0.4 x 0.5) + (0.6 x 0.275)] = 8.8 kN/m.

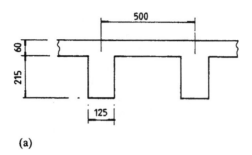

(a)

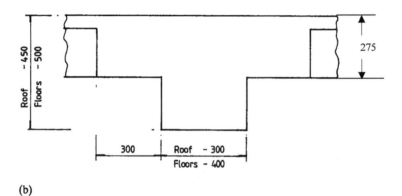

(b)

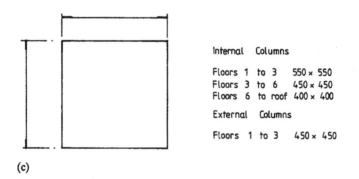

(c)

Fig 14.15 (a) Roof and floor slab; (b) roof and floor beams; (c) columns

(c) Column sizes

Preliminary sizes are shown in Fig.14.15. The self–weights are as follows.

$$\text{Floors 1 to 3: } 0.55^2 \times 24 = 7.3 \text{ kN/m}$$
$$\text{Floors 3 to 7: } 0.45^2 \times 24 = 4.9 \text{ kN/m}$$
$$\text{Floor 7 to roof: } 0.4^2 \times 24 = 3.8 \text{ kN/m}$$

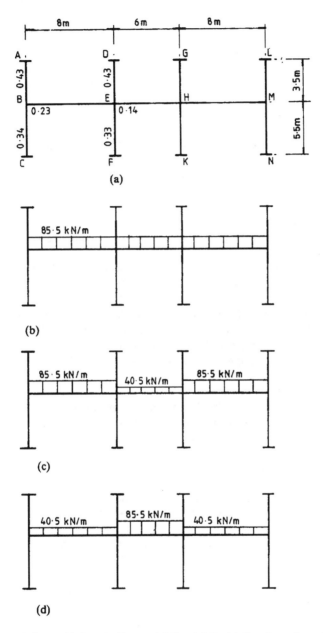

Fig 14.16 (a) Sub-frame; (b) Case 1, all spans $1.4\,G_k + 1.6\,Q_k$; (c) Case 2, maximum load on outer spans; (d) Case 3, maximum load on centre span.

Vertical loads
(a) Roof beam
The floor slab extends to only $(6000 - 2 \times 300 - 300) = 5100$ mm
$$\text{dead load (slab, beam, finishes)} = (2.73 \times 5.1) \text{ floor slab}$$
$$+ \ 7.2 \text{ beam self weight} + (1.5 \times 6) \text{ finishes}= 35 \text{ kN/m}$$
$$\text{imposed load} = 6 \times 1.5 = 9 \text{ kN/m}$$

(b) Floor beams
The floor slab extends to only $(6000 - 2 \times 300 - 400) = 5000$ mm
$$\text{dead load} = (2.73 \times 5) \text{ floor slab} + 8.8 \text{ beam self weight}$$
$$+ \ (3 \times 6) \text{ finishes} = 40.5 \text{kN/m}$$
$$\text{imposed load} = 6 \times 3 = 18 \text{ kN/m}$$

(c) Internal column below floor 2
The entire load over a width $(8+6)/2 = 7$ m is carried by the internal column.
The dead load is:
$$\text{Beam: } 7[35 \text{ roof} + (40.5 \times 9 \text{ floors})] = 2796.5 \text{ kN}$$
$$\text{Column self weight: } 3.5[7.3 + 4(4.9 + 3.8)] = 147.35 \text{ kN}$$
$$\text{Total: } 2943.85 \text{ kN}$$
$$\text{imposed load} = 7 \times [9 \text{ roof} + (18 \times 9 \text{ floors})] = 1197 \text{ kN}$$
Table 2 of BS 6399, Part 1 allows for 40% reduction in total distributed imposed floor loads carried by a member for the number of floors including roof up 10.
$$\text{reduced imposed load} = (1 - 0.4) \times 1197 = 718.2 \text{ kN}$$
Sub-frame analysis
(a) Sub-frame
The sub-frame consisting of the beams and columns above and below the floor level 1 is shown in Fig.14.16. The properties of members are shown in Table 14.8.

(c) Loads and load combinations

(i) Case 1: All spans are loaded with $(1.4G_k + 1.6Q_k)$
$$(1.4G_k + 1.\ Q_k) = (1.4 \times 40.5) + (1.6 \times 18) = 85.5 \text{ kN/m}$$
The fixed end moments are:
$$\text{Spans BE and HM: } 85.5 \times 8^2/12 = 456 \text{ kNm}$$
$$\text{Span EH: } 85.5 \times 6^2/12 = 256.5 \text{ kNm}$$

(ii) Case 2: Alternate spans are loaded with $(1.4G_k + 1.6Q_k)$ and the other spans are loaded with $1.0G_k$
$$1.0G_k = 40.5 \text{ kN/m}$$
The fixed end moments are:
$$\text{Spans BE and HM: } 85.5 \times 8^2/12 = 456 \text{ kNm}$$
$$\text{Span EH: } 40.5 \times 6^2/12 = 121.5 \text{ kNm}$$

(iii) Case 3: Alternate spans are loaded with $(1.4G_k + 1.6Q_k)$ and the other spans are loaded with $1.0G_k$

The fixed end moments are

Spans BE and HM: $40.5 \times 8^2/12 = 216$ kNm

Span EH: $85.5 \times 6^2/12 = 256.5$ kNm

The design load cases are shown in Fig.14.16.

Table 14.8 Section properties

Member	Length (mm)	Second Moment of area, I (mm^4)	Stiffness, I/L
BC, MN	4500	3.417×10^9	7.5938×10^5
EF, HK	4500	7.626×10^9	16.9456×10^5
AB, LM	3500	3.417×10^9	9.7634×10^5
DE, GH	3500	7.626×10^9	21.787×10^5
BE, HM	8000	4.167×10^9	5.208×10^5
EH	6000	4.167×10^9	6.945×10^5

Note: Columns BC, EF and MN, the length is taken as 4.5 m, although it is shown as 5.5 m in Fig.14.14 and Fig.14.16. It is assumed that they are deformable only over 4.5 m.

(d) Analysis
The stiffness matrix relationship is given by

$$E \times 10^6 \begin{bmatrix} 9.026 & 1.0416 & 0 & 0 \\ 1.0416 & 20.3540 & 1.3888 & 0 \\ 0 & 1.3888 & 20.3540 & 1.0416 \\ 0 & 0 & 1.0416 & 9.026 \end{bmatrix} \begin{bmatrix} \theta_B \\ \theta_E \\ \theta_H \\ \theta_M \end{bmatrix} = \begin{bmatrix} M_B \\ M_E \\ M_H \\ M_M \end{bmatrix}$$

Since the structure and the loading are both symmetrical, indicating that

$$\theta_B = -\theta_M, \theta_E = -\theta_H$$
$$M_B = -M_M, M_E = -M_H$$

The stiffness matrix and the load vector can be condensed to

$$E \times 10^6 \begin{bmatrix} 18.052 & 2.0832 \\ 2.0832 & 37.9304 \end{bmatrix} \begin{bmatrix} \theta_B \\ \theta_E \end{bmatrix} = \begin{bmatrix} 2M_B \\ 2M_E \end{bmatrix}$$

The condensed load vector for the three cases and the corresponding displacement vector are given by

$$\text{Load vector} = \begin{bmatrix} 912 & 912 & 432 \\ -399 & -669 & 81 \end{bmatrix}$$

$$E \times 10^6 \begin{bmatrix} \theta_B \\ \theta_E \end{bmatrix} = \begin{bmatrix} 52.0646 & 52.8913 & 23.8355 \\ -13.3787 & -20.5425 & 0.8264 \end{bmatrix}$$

Table 14.9 shows the values of the bending moments in the symmetrical half of the sub-frame. The shear force and bending moment diagrams for the load cases are shown in Fig.14.17 and Fig.14.18 respectively

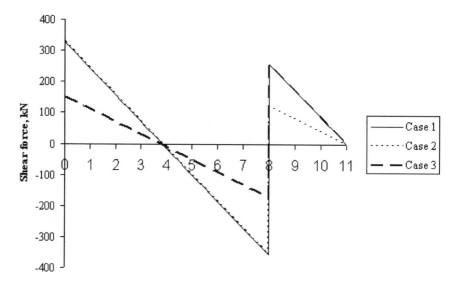

Fig.14.17 Shear force diagram for a symmetrical half of beam.

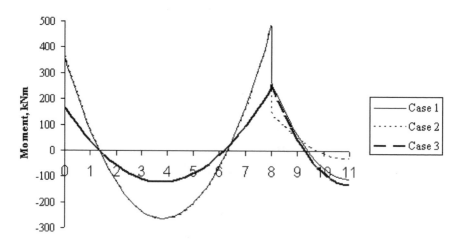

Fig.14.18 Bending moment diagram for a symmetrical half of beam.

Table 14.9 Bending moments in kNm in the sub-frame

Moment	Case 1	Case 2	Case 3
M_{BE}	−361.47	−367.21	−165.49
M_{EB}	482.36	468.30	242.55
M_{EH}	−275.08	−150.03	−255.35
M_{BA}	203.32	206.55	93.08
M_{BC}	158.15	160.66	72.40
M_{ED}	−116.59	−179.02	7.20
M_{EF}	−90.69	−139.25	5.60

Design of the outer span of beam BEH

(a) Design of moment reinforcement
(i) Section at mid–span: The exposure is mild, and the fire resistance 2 h. Cover is 30 mm for a continuous beam. Refer to BS8110: Part 1, Tables 3.4 and 3.5.
Assume 25 mm diameter bars and 10 mm diameter links:
$$d = 500 - 30 - 10 - 12.5 = 447.5 \text{mm}$$
Maximum span moment occurs for case 2 loading. The support moments from Table 14.8 are
$$M_{BE} = 367.21, M_{EB} = 468.3 \text{ kNm, load q} = 85.5 \text{ kN/m, Span} = 8 \text{ m}$$
Reaction at left support
$$V_{left} = 0.5 \times 85.5 \times 8 + (367.21 - 468.3)/8 = 329.36 \text{ kN}$$
Shear force is zero when
$$329.36 - 85.5 \text{ a} = 0, \text{ a} = 3.85 \text{ m}$$
Maximum span moment
$$M_{span} = -367.21 + 329.36 \times 3.85 - 85.5 \times 3.85^2/2 = 267.16 \text{ kNm}$$
$$k = M/ (f_{cu} \text{ b d}^2) = 267.16 \times 10^6 / (30 \times 1000 \times 447.5^2) = 0.044 < 0.156$$
$$z/d = 0.5 + \sqrt{(0.25 - 0.044/0.9)} = 0.948 < 0.95$$
$$A_s = 267.16 \times 10^6/ (0.948 \times 447.5 \times 0.95 \times 460) = 1440 \text{ mm}^2$$
Provide 4T25 to give an area of 1963 mm² (Fig.14.19). This will provide for tie reinforcement.

(ii) Section at outer support: Maximum support moment occurs for case 2 loading. From Table 14.8,
$$M_{BE} = 367.21 \text{ kN m.}$$
At support tension is at the top. The flange will be in tension. The beam section is therefore rectangular.
$$b = 400 \text{ mm. Provide for 25 mm bars; } d = 447.5 \text{ mm.}$$
$$k = M/ (f_{cu} \text{ b d}^2) = 367.21 \times 10^6 / (30 \times 400 \times 447.5^2) = 0.153 < 0.156$$
$$z/d = 0.5 + \sqrt{(0.25 - 0.153/0.9)} = 0.783 < 0.95$$
$$A_s = 367.21 \times 10^6/ (0.783 \times 447.5 \times 0.95 \times 460) = 2398 \text{ mm}^2$$
Provide 5T25 = 2454 mm².

(iii) Section at inner support: Maximum support moment occurs for case 1 loading. From Table 14.8, $M_{EB} = 482.36 \text{ kN m.}$

Assuming 32 mm bars,
$$d = 500 - 30 - 10 - 32/2 = 444 \text{ mm}$$
$$k = M/ (f_{cu} \, b \, d^2) = 482.36 \times 10^6 / (400 \times 444^2 \times 30) = 0.203 > 0.156$$
Need compression reinforcement.
Design as a doubly reinforced beam.
$$d' = 52.5 \text{ mm}; \; d' /d = 0.121 < 0.171.$$
Compression steel yields.
$$0.156 \, bd^2 \, f_{cu} = 0.156 \times 400 \times 444^2 \times 30 \times 10^{-6} = 369.04 \text{ kNm}$$
$$A_s = 369.04 \times 10^6 / (0.775 \times 444 \times 0.95 \times 460)$$
$$+ (482.36 - 369.04) \times 10^6 / \{(444 - 52.5) \times 0.95 \times 460\}$$
$$A_s = (2454 + 663) = 3117 \text{ mm}^2$$
$$A_s' = 663 \text{ mm}^2$$

For the compression steel, carry 2T25 through from the centre span.
$A_s' = 982 \text{ mm}^2$. For the tension steel, provide 4T32 to give an area of 3217 mm^2.

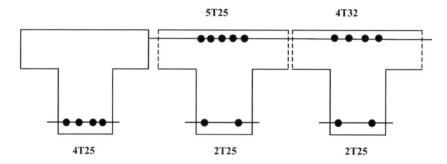

Fig.14.19 Beam BE (a) Mid–span; (b) outer column; (c) inner column.
Note: Links not shown for clarity.

(b) Curtailment and anchorage

As moments have been calculated by detailed analysis, the cut–off points will be calculated in accordance with section 3.12.9.1 in the code.

(i) Top steel at outer support: Refer to Fig.14.19. The section has 5T25 bars at the top and 2T25 bars at the bottom. Determine the positions along the beam where the three bars can be cut off.

The moment of resistance of the section with 2T25 bars ($A_s = 981$ mm^2) at top in tension is calculated.

Assuming that tension steel yields and the stress in the steel is 0.95 f_y,
Equate the forces in the section:
$$0.95 \times 460 \times 981 = 0.45 \times 30 \times 400(0.9x)$$
$$x = 88 \text{ mm}$$
Check strain in tension steel
$$\varepsilon_s = 0.0035 \times (447.5 - x)/x$$
$$= 14.3 \text{ x } 10^{-3} > \text{yield strain} (0.95 \times 460/E = 2.185 \times 10^{-3})$$

Tension steel yields, $f_s = 0.95 \times 460 = 437$ N/mm^2

Taking moment about the tension steel, moment of resistance is given by

$$M_R = 0.45\, f_{cu}\, b\, (0.9\, x)\, (d - 0.9x/2)$$
$$= (0.45 \times 30 \times 400 \times (0.9 \times 88)\, (447.5 - 0.45 \times 88) \times 10^{-6}$$
$$= 174.45 \text{ kNm}$$

Using case 2 loads, the theoretical cut–off point for two bars is given by the solution of the equation

$$-174.45 = -367.21 + 329.36 \times a - 85.5 \times a^2/2, \text{ giving } a = 0.64 \text{ m}$$

In accordance with the general rules for curtailment of bars in the tension zone given in clause 3.12.9.1, using the code designation, the code states that one of the following requirements should be satisfied:

(a) effective depth of the member, $d = 447.5$ mm

(c) The bar must continue for an anchorage length (40 diameters = 1000 mm) beyond the point where it is no longer required to resist bending moment.

3T25 will therefore extend a distance of $(0.64 + \text{anchorage length}) = 1.64$ m from the centre line of support.

The position of point of contra–flexure is given by

$$0 = -367.21 + 329.36\, a - 85.5\, a^2/2, \text{ giving } a = 1.35 \text{ m}$$

The remaining 2T25 bars will extend to lap with 2T32 top bars running from the inner support.

(ii) Top steel at inner support: The section has 4T32 bars at the top and 2T25 bars at the bottom. Determine where 2T32 can be stopped.

Equate the forces in the section with 2T32 at top and 2T25 at bottom. Assume that tension steel yields but compression steel remains elastic.

Strain in compression steel is

$$\varepsilon_{sc} = 0.0035 \times (x - 52.5)/x$$

stress in compression steel is

$$f_{sc} = \varepsilon_{sc\,x}\, 200 \times 10^3 = 700\, (x - 52.5)/x$$

Equate the total tension and total compression forces.

$$0.95 \times 460 \times 1608 = 0.45 \times 30 \times 400(0.9x) + 981 \times 700\, (x - 52.5)/x$$

Simplifying

$$x^2 - 3.2914\, x - 7418.0558 = 0, \text{ giving } x = 88 \text{ mm}$$

Check strains in compression and tension steel to see if the assumptions were justified.

Compression steel:

$$\varepsilon_{sc} = 0.0035 \times (x - 52.5)/x$$
$$\varepsilon_{sc} = 1.412 \times 10^{-3} < \text{yield strain } (0.95 \times 460/E = 2.185 \times 10^{-3})$$
$$f_{sc} = 1.412 \times 10^{-3} \times 200 \times 10^3 = 282 \text{ N/mm}^2$$

Tension steel:

$$\varepsilon_s = 0.0035 \times (444 - x)/x$$
$$\varepsilon_s = 14.66 \times 10^{-3} > \text{yield strain } (0.95 \times 460/E = 2.185 \times 10^{-3})$$

Tension steel yields, $f_s = 0.95 \times 460 = 437$ N/mm^2
Taking moment about the tension steel, moment of resistance is given by
$$M_R = 0.45\ f_{cu}\ b\ (0.9\ x)\ (d - 0.9x/2) + f_{sc}\ A_s\ (d - d')$$
$$= [(0.45 \times 30 \times 400 \times (0.9 \times 88)\ (444 - 0.45 \times 88) + 981 \times 282\ (444 - 52.5)] \times 10^{-6}$$
$$M_R = 172.95 + 108.31 = 281.26\ kNm$$
As shown in Fig.14.18, maximum support moment occurs for case 1 loading. From Table 14.8,
$$M_{BE} = 361.47\ kNm,\ M_{EB} = 482.36\ kNm,\ q = 85.5\ kN/m,\ L = 8\ m$$
Reaction on the right
$$V = 85.5 \times 8/2 - (361.47 - 482.63)/8 = 357.15\ kN$$
Determine the position a from the right hand support where the moment is equal to moment of resistance due to 2T32 at top and 2T25 at bottom.
$$-281.26 = -482.26 + 357.15\ a - 85.5 \times a^2/2,\ giving\ a = 0.61\ m$$
The 2T32 at top therefore will continue to a distance of (0.61 m + anchorage length of 40 × bar diameter) = 1.89 m from the right hand support.
The position of contra–flexure is given by
$$0 = -482.26 + 357.15\ a - 85.5 \times a^2/2,\ giving\ a = 1.69\ m$$
The remaining 2T32 at top therefore will continue to a distance of (1.69 m + anchorage length of 40 × bar diameter) = 2.93 m from the right hand support.

(iii) Bottom steel at outer support: The bottom steel consists of 4T25 bars. The point where two bars can be cut off will be determined. Assume that at the theoretical cut off points the effective bottom steel consists of 2T25 bars. The beam section is shown in Fig.14.19 with a flange breadth of 1000 mm.
Calculate the moment of resistance due to 2T25 only. Neglect the 'compression steel' at the top and assuming that tension steel yields, equate forces in the section:
$$0.95 \times 460 \times 981 = 0.45 \times 30 \times 1000 \times 0.9x,\ giving\ x = 35.3\ mm$$
$$z = 447.5 - 0.5 \times 0.9 \times 35.3 = 432\ mm > (0.95d = 425\ mm)$$
The moment of resistance of the section is
$$M_R = (0.95 \times 460 \times 981 \times 425) \times 10^{-6} = 182.2\ kNm$$
Consider case 2 loads and solve the following equation to give the theoretical cut off points:
$$182.2 = -367.21 + 329.36\ a - 85.5\ a^2/2$$
$$Solving,\ a = 2.44\ and\ 5.26\ m$$

Left hand support: From the left hand support, extend 4T25 to a distance of
$$(2.44 - anchorage\ length\ of\ 40\ bar\ diameters) = 1.44\ m$$

Right hand support: From the right hand support, extend 4T25 to a distance of
$$= (8.0 - 5.26) - anchorage\ length\ of\ 40\ bar\ diameters = 1.74\ m$$

The position of point of contra–flexure from left hand support is given by
$$0 = -367.21 + 329.36\ a - 85.5\ a^2/2,$$
$$Solving,\ a = 1.35\ and\ 6.35$$

(iv) Anchorage of top bars at outer support: 5T25 bars are to be anchored. The arrangement for anchorage is shown in Fig.14.20. The anchorage length is calculated for pairs of bars.

A larger steel area has been provided than is required (section 14.5.6(a) (ii)). The stress in the bars at the start of the bend is
$$0.95 \times 460 \times 2398/2454 = 427 \text{ N/mm}^2$$

From Table 3.28 of the code, $\beta = 0.5$ for type 2 deformed bars in tension.
Anchorage bond stress:
$$f_{bu} = 0.5 \sqrt{f_{cu}} = 0.5 \sqrt{30} = 2.74 \text{ N/mm}^2$$

Anchorage length
$$= 427 \times (\phi = 25 \text{ mm}) / (2.74 \times 4) = 974 \text{ mm}$$
$$\text{Assume bend radius} = 275 \text{ mm}$$
$$\text{Area of 25 mm bar} = 490 \text{ mm}^2$$
$$F_{bt} = 427 \times 490 = 209.23 \text{ kN}$$

centre to centre distance between bars:
$$a_b = (400 - 2x \ 30 - 2 \times 10 - 25)/4 = 74 \text{ mm}$$
$$F_{bt}/(r\phi) = 209.23 \times 10^3/ (275 \times 25) = 30.4 \text{ N/mm}^2$$
$$2 f_{cu}/ (1 + 2\phi/a_b) = 2 \times 30/ (1 + 2 \times 25/74) = 35.8 \text{ N/mm}^2$$
$$30.4 < 35.8 \text{ N/mm}^2$$

For outer bar:
$$a_b = (\text{cover} + \text{link} + \text{bar diameter}) = 65 \text{ mm}$$
$$\text{permissible stress is } 33.9 \text{ N/mm}^2$$
$$30.4 < 33.9 \text{ N/mm}^2$$

(v) Arrangement of longitudinal bar: The arrangement of the longitudinal bars is shown in Fig.14.21.

(c) Design of Shear Reinforcement

The shear force envelope constructed from the shear force diagrams is shown in Fig.14.17. Take account of the enhanced shear strength near the support using the simplified approach set out in clause 3.4.5.9 of the code.

(i) Inner support: From case 1 loading, shear force at support
$$V = 357.11 \text{ kN}$$
Distance from centre line of support (column = 550 mm wide) to a distance (d = 444) from the support
$$= 550/2 + 444 = 719 \text{ mm}$$
The shear at d from the face of the support is
$$V = 357.11 - 85.5 \times (0.719) = 295.6 \text{ kN.}$$
$$v = 295.6 \times 10^3/ (400 \times 444) = 1.66 \text{ N/mm}^2$$
The effective tension steel is 4T32 bars, $A_s = 3217 \text{ mm}^2$, d = 444 m (Fig.14.20). The bars continue for at least a distance d past the section.
$$100 \ A_s/ (b_v \ d) = 100 \times 3217/ (400 \times 444) = 1.81 < 3.0$$
$$400/d = 400/444 < 1, \text{ take as } 1.0$$

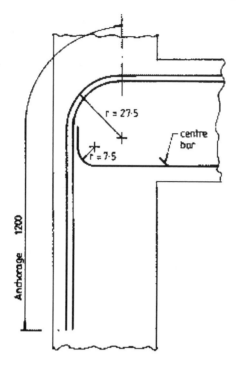

Fig.14.20 Anchorage arrangement for top bars.

$$v_c = (0.79/1.25) \times 1.81^{1/3} \times (1)^{1/4} \times (30/25)^{1/3} = 0.82 \text{ N/mm}^2$$

Use T10 links, $A_{sv} = 157 \text{ mm}^2$: Spacing of links,

$$157 > 400 \, s_v \, (1.66 - 0.82)/ \, (0.95 \times 460)$$
$$s_v < 204 \text{ mm}.$$

Maximum spacing $< 0.75d = 326$ mm
Provide links at 200 mm centres.

The spacing for minimum links is

$$157 > 400 \, s_v \, (0.4)/ \, (0.95 \times 460)$$
$$s_v < 429 \text{ mm}$$

Maximum spacing $< 0.75d = 326$ mm
Adopt a minimum spacing of 300 mm.
Determine the distance from the centre of the inner support where minimum links at 300 mm centres can be used. At this location the effective tension steel is 2T32 bars.

$$A_s = 1608 \text{ mm}^2, d = 444 \text{ mm}$$

Calculate v_c for this value.

$$100 \, A_s/ \, (b_v \, d) = 100 \times 1608/ \, (400 \times 444) = 0.91 < 3$$
$$400/d = 400/444 < 1, \text{ take as } 1.0$$
$$v_c = (0.79/1.25) \times 0.91^{1/3} \times (1)^{1/4} \times (30/25)^{1/3} = 0.65 \text{ N/mm}^2$$

Taking account of the shear resistance of the T10 links at 300 mm centres, the average shear stress v in the concrete can be found:
$$v - 0.65 = 0.95 \times 460 \times 157 / (400 \times 300) = 0.57 \text{ N/mm}^2$$
$$v = 1.22 \text{ N/mm}^2$$
The shear V at the section is
$$V = 1.22 \times 444 \times 400 \times 10^{-3} = 216.7 \text{ kN}$$
The distance a from the centre support where the shear force is 216.7 kN is given by
$$216.7 = 357.1 - 85.5 \times a, \text{ giving } a = 1.64 \text{ m}$$
where 85.5 kN/m is the uniformly distributed load on the beam.

(ii) Outer support: From case 1 and case 2 loading, shear force V at support is
$$V = 329.36 \text{ kN}$$
Distance from centre line of support (column = 450 mm wide) to a distance d from the support
$$= 450/2 + 447.5 = 672.5 \text{ mm}$$
The shear at d from the face of the support is
$$V = 329.36 - 85.5 \times (0.673) = 271.9 \text{ kN}.$$
The shear stress
$$v = 271.9 \times 10^3 / (400 \times 444) = 1.53 \text{ N/mm}^2$$
For 4T25 bars, $A_s = 1963 \text{ mm}^2$, at the top gives
$$100 \, A_s / (b_v \, d) = 100 \times 1963 / (400 \times 444) = 1.11 < 3.0$$
$$400/d = 400/444 < 1, \text{ Take as } 1.0$$
$$v_c = (0.79/1.25) \times 1.11^{1/3} \times (1)^{1/4} \times (30/25)^{1/3} = 0.70 \text{ N/mm}^2$$
The spacing s_y for the T10 links is
$$157 > 400 \, s_v \, (1.53 - 0.70)/ (0.95 \times 460)$$
$$s_v = 207 \text{ mm}.$$
For 2T25 bars, $A_s = 982 \text{ mm}^2$,
$$100 \, A_s / (b_v \, d) = 100 \times 982 / (400 \times 444) = 0.55 < 3$$
$$400/d = 400/444 < 1, \text{ take as } 1.0$$
$$v_c = (0.79/1.25) \times 0.55^{1/3} \times (1)^{1/4} \times (30/25)^{1/3} = 0.55 \text{ N/mm}^2$$

For a minimum spacing of 300mm with $v_c = 0.55 \text{ N/mm}^2$, calculate the permissible value of v.
$$157 > 400 \text{ x } 300(v - 0.55)/ (0.95 \text{ x } 460), v = 0.57 \text{ N/mm}^2$$
$$v = 0.57 + 0.55 = 1.12 \text{ N/mm}^2$$
$$V = 1.12 \times 400 \times 444 \times 10^{-3} = 198.9 \text{ kN}$$
This shear occurs at a from the left hand support.
$$198.9 = 329.36 - 85.5 \times a, \text{ giving } a = 1.53 \text{ m}$$

(iii) Rationalization of link spacings: The following rationalization of link spacings will be adopted:
1. From face of outer support to a distance of 1400mm, provide 9T10-175mm c/c
2. From face of centre (inner) support to a distance of 1575 mm, provide 10T10-175 mm c/c
3. Centre portion over a distance of 5025 mm provide 18T10-280 mm c/c

The link spacing is shown in Fig.14.21.

(d) Deflection
Refer to Fig.14.15
$$b_w / b = 400/1000 = 0.4 > 0.3$$
Interpolating from Table 3.10 of the code the basic span/d ratio is
$$20.8 + (26 - 20.8) \times (0.4 - 0.3)/(1.0 - 0.3) = 21.5$$

Modification factor for tension steel:
$$M/ (bd^2) = 267.16 \times 10^6/(400 \times 447.5^2) = 3.33$$
$$f_s = (2/3) \times 460 \times (1569/1963) = 245 \text{ N/mm}^2$$
$$0.55 + (477 - 245)/ \{120 \times (0.9 + 3.33)\} = 1.01$$

The modification factor for compression reinforcement with 2T25 bars supporting the links is
$$100 \times A_s` \text{ provided } / (bd) = 100 \times 981/ (1000 \times 447.5) = 0.219$$
$$1 + 0.219/ (3 + 0.219) = 1.07$$
$$\text{allowable span/d ratio} = 21.5 \times 1.01 \times 1.07 = 23.2$$
$$\text{actual span/d ratio} = 8000/447.5 = 17.9$$
The beam is satisfactory with respect to deflection.

(e) Cracking
Referring to Table 3.28 in the code, the clear spacing between bars in the tension zone for grade 460 steel and no redistribution should not exceed 155 mm

(i). *Top steel at outer support*: 5T25 bars with 30 mm cover and 10 mm links. Spacing between bars is
$$= (400 - 30 - 10 - 30 - 10 - 25)/4 = 49 \text{ mm} < 155 \text{ mm}$$
If the inner 3T25 bars are curtailed, spacing between bars is
$$= (400 - 30 - 10 - 30 - 10 - 25) = 295 \text{ mm} > 155 \text{ mm}$$
In the interests of crack control, curtail only the inner 2T25 bars.

(ii) *Top steel at Inner support*: 4T32 bars with 30 mm cover and 10 mm links. Spacing between bars is
$$= (400 - 30 - 10 - 30 - 10 - 32)/3 = 96 \text{ mm} < 155 \text{ mm}$$
If the inner 2T32 bars are curtailed, spacing between bars is
$$= (400 - 30 - 10 - 30 - 10 - 32) = 288 \text{ mm} > 155 \text{ mm}$$
In the interests of crack control, add an additional 25 mm bar to link with middle 25 bar from outer support steel.

(iii) Bottom steel: Curtailing 2T25 bars will again make the bar spacing too wide. The simple solution is to add an extra 20 mm bar in the interests of crack control.

(f) Arrangement of reinforcement
The final arrangement of the reinforcement is shown in Fig.14.21.

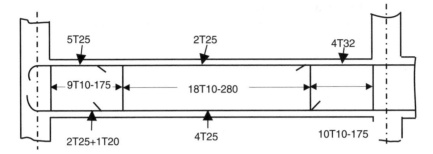

Fig.14.21 Reinforcement detail.

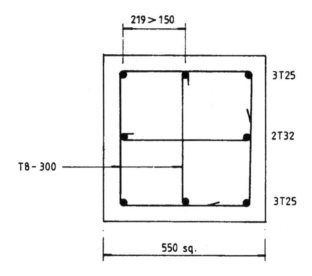

Fig.14.22 Column reinforcement.

Design of lower length of centre column

(a) Design loads and moments
The axial load and moment at the column top are as follows.

Case 1:
$$\text{axial load } N = (1.4 \times 2943.85) + (1.6 \times 718.2) = 5270.5 \text{ kN}$$
$$\text{moment } M = M_{EF} = 90.7 \text{ kNm}$$

Case 2: At the first floor the centre beam carries dead load only. The axial load can be calculated from the axial load in case 1 by deducting $(0.4 \, G_k + 1.6 \, Q_k)$
$$\text{axial load } N = 5270.5 - [(0.4 \times 40.5) + 1.6 \times 18] \times 6/2 = 5135.5 \text{ kN}$$
$$\text{moment } M = M_{EF} = 139.25 \text{ kNm}$$

(b) Effective length and slenderness
Refer to BS8110: Part 1, section 3.8.1.6. The column is square with assumed dimensions 550 mm × 550 mm. The restraining members are as follows:
Transverse direction beam 500 mm deep
Longitudinal direction ribbed slab 275 mm deep
Check the slenderness in the longitudinal direction. The end conditions for a braced column are (Table 3.19 of the code)
Top Condition 3, ribbed slab
Bottom Condition 1, moment connection to base
$$\beta = 0.9$$
$$\text{effective length } l_c = 0.9(5500 - 250) = 4725 \text{ mm}$$
$$\text{slenderness} = 4725/550 = 8.59 < 15$$
The column is short.

(c) Column reinforcement
Use column design chart, $f_{cu} = 30$, $f_y = 460$, $d/h = 0.85$
Case 1:
$$N/(bh) = 5270.5 \times 10^3/550^2 = 17.42$$
$$M/(bh^2) = 90.7 \times 10^6/550^3 = 0.55$$
$$100A_{sc}/bh = 1.4.$$
Calculations show that for $x/h = 1.22$, $N/(bh) = 18.05$, $M/(bh^2) = 0.55$,

Case 2:
$$N/(bh) = 5135.5 \times 10^3/550^2 = 16.98$$
$$M/(bh^2) = 139.25 \times 10^6/550^3 = 0.84$$
$$100A_{sc}/bh = 1.4.$$
Calculations indicate that for $x/h = 1.09$, $N/(bh) = 17.33$, $M/(bh^2) = 0.84$.

$$A_{sc} = 1.4 \times 550^2/100 = 4235 \text{ mm}^2$$
Provide 6T25 + 2T32 to give an area of 4552 mm².
The links required are 8 mm in diameter at 300 mm centres. The reinforcement is shown in Fig.14.22. Note that no bar must be more than 150mm from a restrained bar. So centre links are provided.

Robustness -design of ties
The design must comply with the requirements of sections 2.2.2.2 and 3.12.3 of the code regarding robustness and the design of ties. These requirements are examined.

(a) Internal ties
(i) Transverse direction: The ties must be able to resist a tensile force in kN/m width that is the greater of
$$[\{(g_k + q_k)/7.5\} (l_r/5) F_t = (7 + 3)/7.5\} (8/5) F_t$$
$$= 2.13F_t \text{ or}$$
$$F_t$$
where

$g_k = 2.73$ (self weight of ribbed slab) $+ 8.8/6$ (Self weight of floor beam)
$+ 3.0$ (Finishes) $= 7kN/m^2$
$q_k = 3.0kN/m^2$, $l_r = 8.0m$ (transverse direction)
$F_t =$ lesser of $(20 + 4n_o = 80$ kN) or 60 kN
$n_o = 10$ is the number of storeys.
tie force $= 2.13 \times 60 = 127.8$ kN/m
steel area $= 127.8 \times 10^3 / (0.95 \times 460) = 293$ mm^2/m

Provide 3T12 bars, $A_s = 339$ mm^2, in the topping of the ribbed slab per metre width.

(ii) Longitudinal direction: The area of ties must be added to the area of steel in the ribs.

(b) Peripheral ties
The peripheral ties must resist a force F_t of 60 kN. This will be provided by an extra steel area in the edge L–beams running around the building.

(c) External column tie
The force to be resisted is the greater of
$$2.0 \, F_t = 120 \text{ kN}$$
or
$$(\ell_s/2.5)F_t = 3 \times 60/2.5 = 72 \text{ kN}$$
or
3% of the design ultimate vertical load carried by the column at that level, which is,
$$(3/100) \times 4887.5 = 146.6 \text{ kN}$$
where ℓ_s is the floor to ceiling height (3.0 m)
$$\text{steel area} = 146.6 \times 10^3 / (0.95 \times 460) = 336 \text{ mm}^2$$
At the centre of the beam 4T25 bars equal to 1963 mm^2 are provided whereas 1440.4 mm^2 are required. The extra moment reinforcement provided at the bottom of the beam is $(1963 - 1440) = 523$ mm^2 is adequate to resist this force. The top reinforcement will also provide resistance. The bars are anchored at the external column.
The corner columns must be anchored in two directions at right angles.

(d) Vertical ties
The building is over five storeys and so each column must be tied continuously from foundation to roof. The tie must support in tension the design load of one floor section on vertical loads.
$$\text{design load} = 7 [1.4 \times 40.5 + 1.6 \times 18] = 598.5 \text{ kN}$$
The steel area required is $598.5 \times 10^3 / (0.95 \times 460) = 1370$ mm^2.
The column reinforcement is lapped above floor level with a compression lap of 40 times the bar diameter (Table 3.27 of the code). This reinforcement is more than adequate to resist the code load.

CHAPTER 15

TALL BUILDINGS

*Modified version of initial contribution by
J.C.D. Hoenderkamp, formerly of Nanyang Technological
Institute, Singapore*

15.1 INTRODUCTION

For the structural engineer the major difference between low and tall buildings is the influence of the horizontal loads due to wind and earthquake on the design of the structure. Lateral deflection of a tall concrete building is generally limited to H/1000 to H/200 of the total height H of the building. In the case of tall buildings, in addition to limiting this so called lateral drift, attention has to be focussed on the comfort of the occupants because vibratory motion could induce mild discomfort to acute nausea.

Another aspect that needs to be addressed in tall buildings is the vertical movement due to creep and shrinkage in addition to that due to elastic shortening. These movements can cause distress in non-structural elements and must be allowed for in detailing.

This chapter is mainly concerned with the elastic static analysis of tall structures subject to lateral loads. An attempt is made to explain the complex behaviour of such structures and to suggest simplified methods of analysis of those types of structures which do not require full 3-D analysis. The behaviour of individual planar bents and the interaction between shear walls and rigid-jointed frames will be examined in detail as it highlights the complexity involved in the analysis of three dimensional structures subjected to horizontal forces.

15.2 ASSUMPTIONS FOR ANALYSIS

The structural form of a building is inherently three dimensional. The development of efficient methods of analysis for tall structures is possible only if the usual complex combination of many different types of structural members can be reduced or simplified whilst still representing accurately the overall behaviour of the structure. A necessary first step is therefore the selection of an idealized structure that includes only the significant structural elements with their dominant modes of behaviour. It is often possible to ignore the asymmetry in a structural floor plan of a building, thereby making a three dimensional analysis unnecessary. One common assumption made is to assume that floor slabs are fully rigid in their own plane. Consequently, all vertical members at any level are subject to the same components of translation and rotation in the horizontal plane. This does not hold true for very long narrow buildings and for slabs which have their widths drastically reduced at one or more locations. Similarly contributions from the out-of-plane stiffness of floor slabs and structural bents can be neglected because of their low stiffness compared with inplane stiffness.

15.3 PLANAR LATERAL LOAD RESISTING ELEMENTS

15.3.1 Rigid-jointed Frames

The most common type of planar bent used for medium height structures is the rigid-jointed frame. They are economic for buildings up to about 25 stories. Beyond that height, control of drift becomes problematic and requires uneconomically large members.

15.3.2 Braced Frames

The lateral stiffness of a rigid frame can be improved significantly by providing diagonal members. In fact such structures could be economic in case of very tall structures. Bracing can be either in storey height-bay width module or they could extend over many bays and stories. Fig.15.1 shows rigid frame and braced frames.

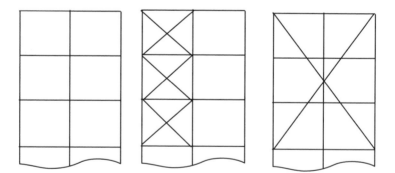

Fig.15.1 (a). Rigid-jointed frame; (b). braced frame; (c) braced frame with large diagonal bracing.

15.3.3 Shear walls

The simplest form of bracing against horizontal loading is the plane cantilevered shear wall. The main difficulty with shear walls is their solid form which tends to restrict planning where wide open internal spaces are required. They are particularly suitable for hotel and residential buildings requiring repetitive floor plans. This allows the walls to be vertically continuous and they can serve both as room dividers and also provide sound and fire insulation.

15.3.4 Coupled Shear walls

As shown in Fig.15.2, a coupled shear wall structure is a shear wall with openings. The two halves of the wall could be connected by beams or slabs at each floor level. For analysis purposes coupled shear walls are treated as rigid frames. However compared to the width of a column in a rigid frame, the width of the wall is very large. To allow for the large width of the walls, the beams connecting the walls are assumed to be rigid over half the width of the walls as shown in Fig.15.3.

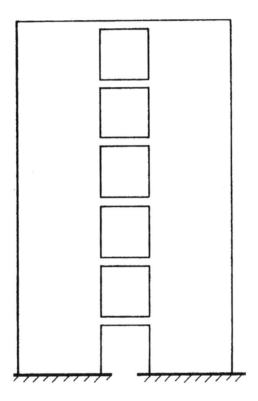

Fig 15.2 Coupled shear wall

15.3.5 Wall-frame Structures

When rigid-jointed structures which deflect in a shear mode as shown in Fig.15.4(a) are combined with shear walls which deflect in a flexural mode as shown in Fig.15.4(b), they are constrained to adopt a common deflected shape because of the horizontally stiff girders and slabs. As a consequence, the two horizontal load resisting structural forms interact, especially at the top to produce a stiffer and stronger structure than a simple addition of the stiffnesses of the two elements would indicate. This combined form has been found to be suitable for structures in the 40–60 storey range.

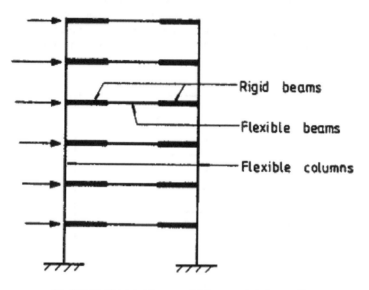

Fig.15.3 Rigid-jointed frame model for a coupled shear wall

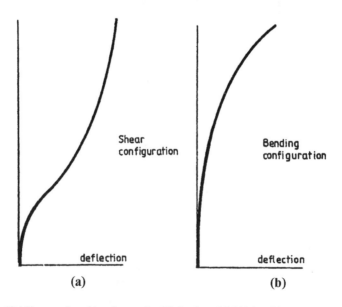

Fig.15.4 Shear mode and bending mode of deflection of rigid-jointed frames and walls.

15.3.6 Framed-tube Structures

In this type of structure the lateral load resisting system consists of moment resisting rigid-jointed frames in two orthogonal directions which form a closed tube around the perimeter of the building plan as shown in Fig.15.5. The frame consists of closely spaced columns at around 2-4 m centres joined by deep girders. The lateral load is carried by the perimeter frames but gravity load is shared between internal columns and perimeter frames. Perimeter frames aligned in the direction of the lateral load act as the webs and the frames normal to the direction of loading act as the flanges of the massive box cantilever. Inevitably, with a wide flange, shear lag effect as shown in Fig.15.5 makes the flanges much less efficient.

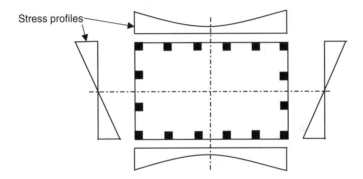

Fig.15.5 Framed-tube structure

15.3.7 Tube-in-Tube Structures

This is similar to framed tube structures except that apart from the perimeter tube there is an internal tube formed of a service core and lift cores as shown in Fig.15.6. Both tubes participate in resisting lateral loads.

15.3.8 Outrigger-braced Structures

Fig.15.7 shows an outrigger structure. This consists of a central core which could be shear walls which form part of the elevator and service cores or a braced frame. The core is connected to the perimeter columns by horizontal cantilevers or 'outriggers'. Under horizontal loads the core bends and rotation of the core is restrained by the outrigger trusses through tension and compression in the perimeter columns. In effect the outriggers considerably increase the effective depth of the building and provide a very stiff structure. The number of outriggers up the height is generally limited to a maximum of about four. This type of structure has been found to be efficient in the design of buildings in the 40-70 storeys range.

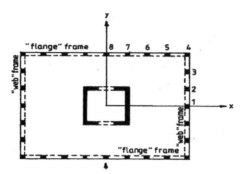

Fig.15.6 Tube-in tube structure.

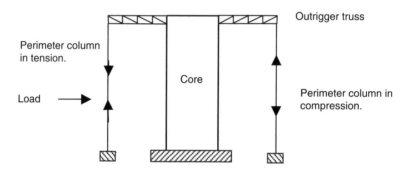

Fig.15.7 Outrigger braced structure.

15.4 INTERACTION BETWEEN BENTS

The analysis of a tall building structure subject to horizontal and vertical loads is a three dimensional problem. In many cases, however, it is possible to simplify and reduce this to a 2-D problem by splitting the structure into several smaller two-dimensional components which then allow a less complicated planar analysis to be carried out. The procedure for subdividing a three dimensional structure requires some knowledge of the sway behaviour of individual bents subjected to lateral loads. As stated in section 15.3.5, rigid frames subject to lateral load will mainly deflect in a shear configuration and shear walls will adopt a flexural configuration under identical loading conditions. These types of behaviour describe extreme cases of deflected shapes along the height of the structures. Other bents such as coupled walls will show a combination of the two deflection curves. In general they behave as flexural bents in the lower region of the structure and show some degree of shear behaviour in the upper storeys. Combining several bents with characteristically different types of behaviour in a single three dimensional structure will inevitably complicate the lateral load analysis. It would be incorrect to isolate one of the bents and subject it to a percentage of the horizontal loading.

Fig.15.8 (a) shows the structural floor plan of a multi-storey building that consists of a single one-bay frame combined with a shear wall. The symmetrically applied lateral load will cause the structure to rotate owing to the distinctly different characteristics of the two bents. A side view of the deflections of both cantilevers is shown in Fig.15.8 (b). Fig.15.8 (c) shows rotation of sections taken at different levels. It shows that it cannot be assumed not only that the rotation of the floor plans continuously increase along the height in one direction but also that the structure has a single centre of rotation. To deal with these complications a more sophisticated three dimensional analysis will be necessary.

15.5 THREE DIMENSIONAL STRUCTURES

15.5.1 Classification of Structures for Computer Modelling

In many cases it is possible to simplify the analysis of a three dimensional tall building structure subject to lateral load by considering only small parts which can be analysed as two-dimensional structures. This type of reduction in the size of the problem can be applied to many different kinds of building. The degree of reduction that can be achieved depends mainly on the layout of the structural floor plan and the location, in plan, of the horizontal load resultant. The analysis of tall structures as presented here is divided into three main categories on the basis of the characteristics of the structural floor plan.

15.5.1.1 Category I: Symmetric floor plan with identical parallel bents subject to a symmetrically applied lateral load q

The structure shown in Fig.15.9 (a) comprises six rigid-jointed frames, four in the y-direction and two in the x-direction. Because of symmetry about the y axis, all beams and columns at a particular floor level will have identical translations in the y-direction when subjected to load q. There will be no deflections in the x-direction. For the analysis of this model consisting of four identical rigid frames parallel to the applied load, it will be sufficient to analyse only one frame subjected to a quarter of the total load.

15.5.1.2 Category II: Symmetric structural floor plan with non-identical bents subject to a symmetric horizontal load q

The lateral load-resisting component of the structure shown in Fig.15.10(a) comprises two rigid frames and two shear walls orientated parallel to the direction of the horizontal load q. The behaviour of the structure is similar to Category I structures except that for the analysis a symmetrical half of the structure needs to be analysed. In addition the shear wall and the rigid-jointed frame need to be connected in line such that the horizontal deflections of the two elements at any level are identical. The two structures can be linked by members of high axial stiffness to achieve the required compatibility of deflections.

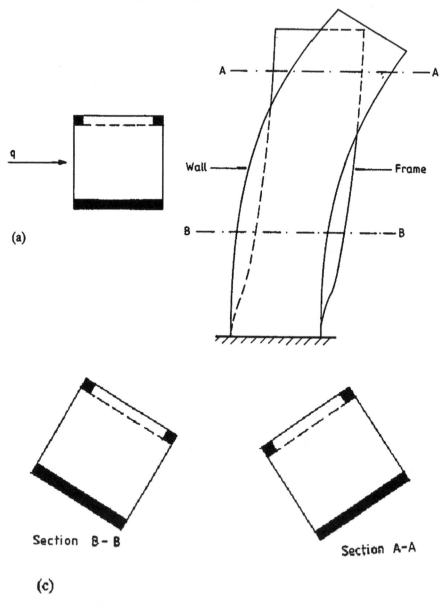

Fig 15.8 (a) Structural floor plan; (b) deflected profiles; (c) floor rotations.

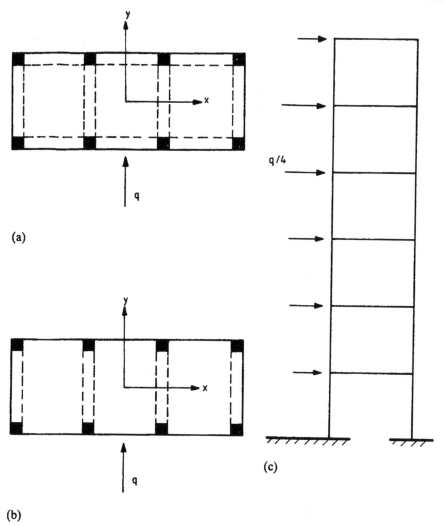

Fig.15.9 (a) Structural floor plan of tall rigid frame building; (b) simplified floor plan; (c) one-bay rigid frame computer model.

Note that as long as symmetry about the y-axis is maintained, it is possible to cope with any variation in geometry with height of different frames/walls. A setback in the upper storeys for all exterior bays in the floor plan shown in Fig.15.11(a) will still allow a plane frame analysis for the linked bents shown in Fig.15.11(b). If the setback causes a loss of symmetry about the *y* axis, however, the structure will rotate in the horizontal plane and a full three dimensional analysis will be necessary.

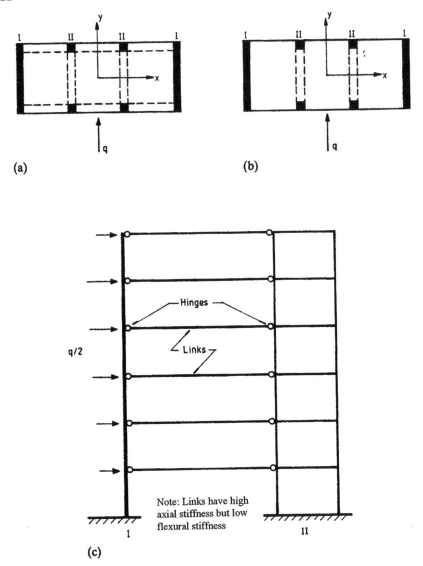

Fig.15.10 (a) Structural floor plan of frame-wall building; (b) simplified floor plan;
(c) computer model of linked bents in a single plane.

15.5.1.3 Category III: Non-symmetric structural floor plan with identical or non-identical bents subject to a lateral load q

A category III structure, of which an example floor plan is shown in Fig.15.12, will rotate in the horizontal plane regardless of the location of the lateral load. It cannot be reduced to a plane frame problem and a complete three dimensional analysis is required.

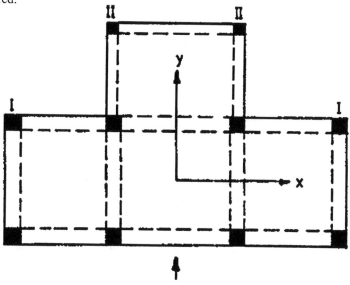

Fig 15.11 (a) Structural floor plan of rigid frame building;

15.6 ANALYSIS OF FRAMED-TUBE STRUCTURES

Framed tube structures shown in Fig.15.5 can be analysed as a pair of cantilevers lying in the same plane. However it is necessary to allow for the shear lag effect in columns in the 'flange' frame. This can be allowed for by treating the 'web' frame and the 'flange' frame as two cantilevers as shown in Fig.15.13. The lateral load is applied to the web frame. At the junction between the two frames, at each storey level the web frame is connected to the flange frame through a set of `rigid` vertical springs. This ensures that at the junction between the two frames, both frames move in the vertical z-direction by the same amount. However the displacement in the y-direction of web frame is different from the deflection of the flange frame in the x-direction, although in the analysis they lie in the same plane. The compatibility of deflections is valid only in the vertical direction.

15.7 ANALYSIS OF TUBE-IN-TUBE STRUCTURES

The distribution of horizontal load between the inner core and the perimeter tube of a tube-in-tube structure (Fig.15.14) depends on the characteristics of the floor system connecting the vertical elements. Two assumptions about these connections can be made, resulting in different computer models.

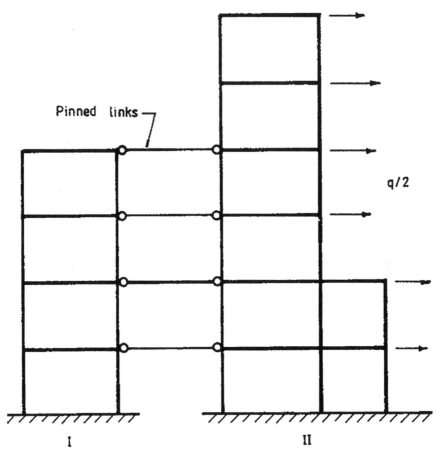

Fig 15.11 (b) linked rigid frames in a single plane.

(a) The interior beams and/or floors are effectively pin connected to the cores and columns: If the structural floor plan is symmetric about the y axis as shown in Fig.15.14 (a), the structure can be classified under category II and a plane frame analysis is possible. Only half the structure needs to be considered. As shown in Fig.15.14 (b), half the core is bent B, i.e. one channel-shaped cantilever wall is bundled with its exterior columns of the two exterior frames perpendicular to the direction of the load. In calculating the second moment of area of the channel section, allowance has to be made for shear lag effect by assuming a reduced effective width for flange width. Together they can be modelled as a single flexural cantilever with a combined bending stiffness represented by the wall and columns 1 and 2. One rigid frame parallel to the direction of the load, bent A, is then connected to it in a single plane by means of rigid links at each floor level. The two dimensional model is to be subjected to half the lateral load.

(b) Beams spanning from the exterior columns to the cores can be considered rigidly connected: The channel-shaped shear wall which is parallel to the load and rigidly connected to floor beams will behave as a wide column and must be

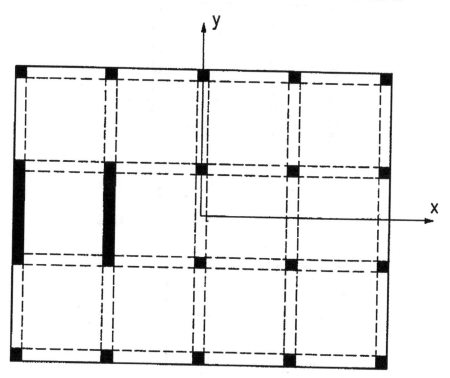

Fig.15.12 Non-symmetric structural floor plan.

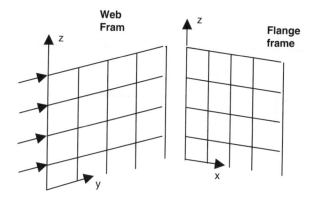

Fig.15.13 Web and flange frames

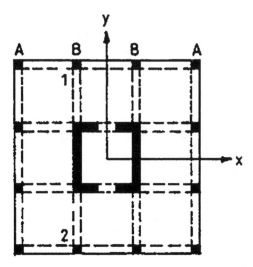

Fig.15.14 (a) Structural floor plan of a tube-in-tube building.

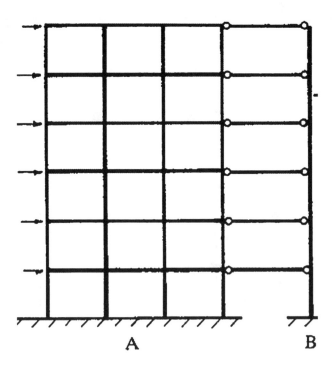

Fig.15.14 (b) rigid frame linked to core and columns.

modelled as such. Flexural column elements are located on the neutral axis of the wall but in the plane of the bent. The moment of inertia of these members should represent the full section of the channel-shaped wall. Rigid arms are then attached in two directions at each floor level. Floor beams are rigidly connected to these arms and the columns of the perpendicular frames. The plane frame model of half the structure subjected to half the horizontal loading is shown in Fig.15.14(c). The short deep beams connecting the shear walls at each floor level will not influence the deflection behaviour of the structure in the y direction since both walls adopt exactly the same deflection profile when subjected to lateral load.

When the core is turned through 90°, without loss of symmetry, a wide arm column model is still possible. The flexible column elements are to be placed on the neutral axis of the channel-shaped section but in the plane of the bent to be analysed. The second moment of area of this element should represent only one-half of one channel-shaped section. The two unequal rigid arms at each floor level add up to the width of the 'flange' of the channel-shaped cantilever. Beams connecting the wide column to other walls or columns can then be rigidly jointed to the arms.

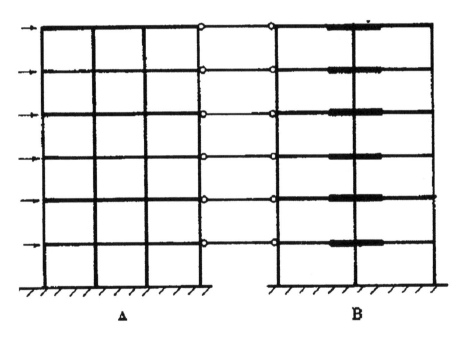

Fig.15.14 (c). Linked bents

15.8 REFERENCES

Stafford Smith, Bryan and Coull, Alex. 1991, *Tall building structures: Analysis and Design*, (Wiley).
Schueller, Wolfgang. 1977, *High-rise building structures*, (Wiley).
Taranath, Bungale S.,1997, *Steel, concrete & composite design of tall buildings*, (McGraw-Hill)

CHAPTER 16

PRESTRESSED CONCRETE

16.1 INTRODUCTION

Prestressed concrete structure can be defined as a concrete structure where external compressive forces are applied to overcome tensile stresses caused by unavoidable loads due to gravity, wind, etc. In other words, it is pre–compressed concrete meaning that compressive stresses are introduced into areas where tensile stresses might develop under working load and this pre–compression is introduced even before the structure begins its working life.

One of the disadvantages of reinforced concrete is that tensile cracks due to bending occur even under working loads. This has four major disadvantages.

- Cracks encourage corrosion of steel.
- A cracked concrete beam is much more flexible than an uncracked beam. This means that when using a reinforced concrete beam, one could have serviceability problems due to deflection or even due to cracking if too slender a beam is used.
- Cracked concrete is not, on the whole, contributing to strength but rather it is simply adding to dead weight.
- The width of the cracks is to a large extent governed by the strain in reinforcing steel. If high tensile steel is used as reinforcement, then the resulting width of the cracks at working loads would be unacceptable. Ordinary reinforced concrete precludes the utilization of high strength steel and the resulting possible economies.

Clearly the above problems can be overcome if we can apply external compressive forces to the beam to prevent it from cracking or even better if the external compressive forces can be applied so as to neutralize the stresses created by applied loads under serviceability conditions, a very efficient structure can be designed.

Consider the simply supported beam supporting loads as shown in Fig.16.1. Bending moment at a section XX produces tensile and compressive stresses at bottom and top fibres respectively. If a compressive force is applied at the centroidal axis, it sets up uniform compression throughout the beam cross–section. It does neutralize the tensile stresses at the bottom portion of the beam caused by bending but it has the disadvantage of increasing the total compressive stresses at the top face. If however the compressive force is applied towards the bottom face at an eccentricity of 'e' from the centroidal axis, then in addition to an axial force of P, a bending moment equal to Pe of a nature *opposite* to that caused by external loads is created. The total stresses due to the bending moment M and the prestress P at an eccentricity e are

$$f_{top} = -\frac{P}{A} + \frac{Pe}{z_t} - \frac{M}{z_t}, \quad f_{bottom} = -\frac{P}{A} - \frac{Pe}{z_b} + \frac{M}{z_b}$$

where z_t and z_b are the section moduli of the cross section for top and bottom fibres respectively and A is the cross sectional area of the beam.

As shown in Fig.16.1, by proper manipulation of the values of P and e, it is possible to ensure that at working loads, the entire cross–section is in compression. It is often stated that one tonne of prestressing steel can result in up to 15 times the amount of building that is made possible by one tonne of structural steel.

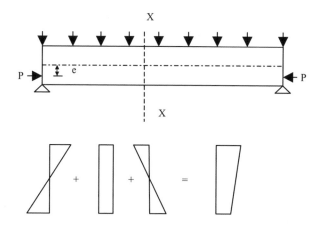

Fig.16.1 Stresses due to prestress and external loads.

16.2 HOW TO APPLY PRESTRESS?

There are two main methods of pre–stressing. They are called pre–tensioning and post–tensioning.

16.2.1 Pre–tensioning

This is used for producing precast prestressed concrete products such as bridge beams, double T beams for floors, floor slabs, railway sleepers, etc. In this method, as shown in Fig.16.2, the process consists of the following steps.

- any reinforcing steel such as links etc. are threaded onto the high tensile steel 'cables'. The cables are tensioned or 'jacked' to the desired force between abutments. The cable is anchored using a simple barrel and wedge device.

Because of the fact that the cables are tensioned before concrete is cast, the name pre–tensioning is used for this process.

- the formwork is assembled round the steel cables
- concrete is placed in the moulds around it and is allowed to cure to gain desired level of strength. This is often speeded up using steam curing. This also enables the prestressing bed to be reused quickly for another job.
- steel is released from the abutments, transferring the compressive force to the concrete through the bond between steel and concrete.

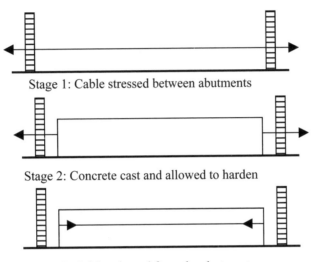

Stage 1: Cable stressed between abutments

Stage 2: Concrete cast and allowed to harden

Stage 3: Cable released from the abutments

Fig.16.2 Basic stages of pre–tensioning.

In practice a large number of identical units are cast at the same time using what is known as Long–Line production method.

It is worth noting that when the force in the cable is transferred to concrete, it contracts. Because of the full bond between concrete and steel, steel also suffers the same contraction leading to a certain loss of stress from the stress at the time of jacking. This is known as Loss of prestress at Transfer and is generally of the order of 10%. Thus

$$P_{Transfer} \approx 0.9 \, P_{jack}$$

where

$P_{Transfer}$ = Total force in the cable after initial loss of stress due to compression of concrete.

P_{jack} = Total force used at the time of initial jacking the cable between the abutments.

16.2.1.1 Debonding

One of the disadvantages of having the same prestressing force P and eccentricity e at all sections is that while normally the external loads produce large bending stresses at the mid–span of the simply supported beam but small stresses towards the supports, the stresses due to prestressing remains constant at all sections. This clearly defeats the idea of tailoring stresses due to pre–stressing to match the stresses due to external loads. This disadvantage can be overcome to a great extent by two methods as follows.

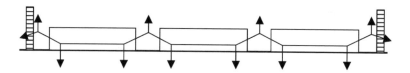

Fig.16.3 Deflected tendons.

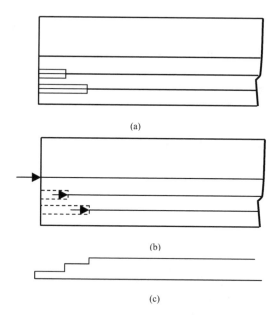

Fig.16.4 Debonding (a) Plastic sleeves around the cables; (b) Effective position of where prestress starts; (c) variation of prestress and eccentricity along span.

(a) Deflected Tendons: As shown in Fig.16.3, the cable is deflected along its length by pulling the cable up or down as necessary. However this is generally not preferred because of extra cost.

(b) De–bonding: In this method by preventing bond from developing between concrete and steel by sheathing some of the cables in plastic tubing as shown in Fig.16.4, both the prestressing force and eccentricity can be varied in a stepwise fashion along the span. This is generally the preferred option due to low cost.

The transfer of force between concrete and steel takes place gradually. The force transfer takes place due to two basic actions.

- Bond between concrete and steel plays an important part. It is therefore essential to ensure that the 'cable' is clean and free from loose rust and that concrete is well compacted.
- The cable is stretched and therefore has a very slightly reduced diameter due to the Poisson effect. However when the cable is released from the abutments, the wire regains its original diameter at the ends. This creates a certain amount of wedging action and in addition frictional forces also come into play.

16.2.1.2 Transmission length

As shown in Fig.16.2, once the cable is released from the abutments, the force in the cable becomes zero at the ends of the cable. However away from the ends, 'bond' between cable and concrete prevents the cable from regaining its original length. As shown in Fig.16.5, the force in the cable gradually builds up to its full value over a certain length. This is known as Transmission Length. This varies depending on the surface characteristics of the cables and the strength of concrete. It is generally of the order of about 50 diameters for 7-wire strand.

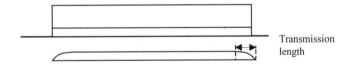

Transmission length

Fig.16.5 Gradual build up of force in the cable.

16.2.2 Post–tensioning

One of the limitations of Pre–tensioning is that normally the cables need to remain straight because the cable is pre–tensioned before concrete hardens. This limitation can be overcome if as shown in Fig.16.6, the cable is laid to any desired profile inside a metal ducting fixed to the required profile to the reinforcement cage with the permanent anchorages also positioned at the ends of the duct. Afterwards, concrete is cast and once it has hardened the cable is tensioned and anchored to the concrete using permanent external anchors rather than relying on bond between the cable and the concrete as in the case of Pre–tensioning. This is the basic idea of Post–tensioning. Because of the fact that the cables are tensioned after the concrete has hardened, the system is known as post–tensioning. Finally, the duct is filled with a colloidal grout under pressure in order to establish bond between concrete and steel and also as protection against corrosion.

There are various types of anchors used in practice but they are generally of two types.
*   The bar is threaded at the ends and anchoring is by a nut bearing on concrete. The threads are rolled rather than cut to reduce stress concentration. The main advantage of this system is that prestress can be applied in stages to suit design considerations or losses can be compensated at any time prior to grouting. The anchorage is completely positive and there is no loss of pre–stress at the transfer stages.
*   Anchoring is done using a system of cones and wedges. In this case, there is loss of pre–stress at transfer stage because of the slip between the cable and the wedge before the wedges 'bite in'.

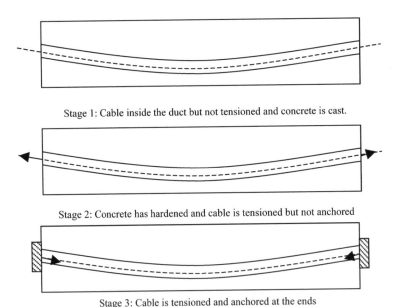

Stage 1: Cable inside the duct but not tensioned and concrete is cast.

Stage 2: Concrete has hardened and cable is tensioned but not anchored

Stage 3: Cable is tensioned and anchored at the ends

Fig.16.6 Different stages in post–tensioning.

16.2.3 External Prestressing

One of the disadvantages of traditional Post–tensioning is that there is no guarantee that the ducts are properly filled with grout to prevent corrosion and if the steel corrodes, cables cannot be replaced. In order to overcome these problems, external prestressing as shown diagrammatically in Fig.16.7 is used. The cables are on the *outside* of the beam and the eccentricity is varied using saddles at appropriate places to obtain the required profile. This is similar to the use of deflected tendons in a pre–tensioned system. This system allows replacement of cables as required and also allows the use of additional cables at a later stage in order to strengthen the structure. Although the term external prestressing is used, it is not necessary for the cables to be 'outside' the structure. For example in the case of box girders, cables can be placed inside the void of the box girder.

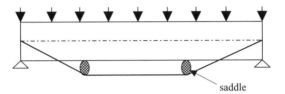

saddle

Fig.16.7 External prestressing.

16.2.4 Un–bonded Construction

Because of the relative unreliability of grouting to prevent corrosion of the cables and also because of the fact grouting is a time consuming job and sufficient time has to elapse for the bond between the cable (also called tendons or strands) and concrete to become effective, a common form of construction used in practice is to dispense altogether with the bond between the concrete and steel. Cables used in this form of construction are manufactured with the cables coated by grease and encased in a plastic sleeve. The plastic sleeve acts as the duct and the construction process is similar to normal post–tensioning. The main advantage of this 'unbonded' system is speed of construction as no grouting is done. However this is not a particularly structurally efficient system because the ultimate bending capacity tends to be only about 70% of a corresponding beam using bonded construction. Nevertheless, un–bonded post–tensioned slabs are a very common form of construction.

16.2.5 Statically Indeterminate Structures

Because the bending moment due to external loading in a simply supported beam is parabolic, the cable profile is also parabolic. One of the advantages of post–tensioning is that the cable profile can be varied so that the bending moment due to external loading can be neutralized by varying the eccentricity to match the shape of the bending moment diagram. A Post–tensioning system is essential when constructing prestressed statically indeterminate structures. Fig.16.8 shows the cable profile in a two span continuous beam.

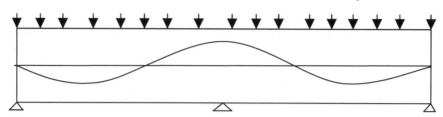

FIG.16.8 Cable profile in a two span continuous beam

16.2.6 End–block

One important aspect of post–tensioning that needs special attention is the area where cables are anchored. Because of the fact that many cables are anchored to the same anchorage block of a relatively small size, large compressive forces are transferred at the anchorage block. Depending on the number of cables anchored and their diameter, the force at an anchor block can vary from 100 kN to 12000 kN. This large transfer of force has the same effect as driving a wedge into a block of wood and has the tendency to burst the concrete transversely near the anchorage. The bursting forces have to be resisted using a large number of links near the anchors. This area of beam is known as an End–block.

16.3 MATERIALS

16.3.1 Concrete

Concrete used for prestressing work is generally of much higher quality than that used for reinforced concrete work. Concrete of grade C50 or over is common. Certain deformational properties of concrete affect the design of prestressed concrete structures and it is necessary to understand them. One of the important properties of concrete is Creep. Creep is defined as the increase of strain with time when the stress is held constant. Under the action of compressive stress due to prestress, concrete continues to deform. Because of the bond between steel and concrete, steel also experiences compressive strain which reduces the tension in the cables. In addition to creep, shrinkage of concrete also contributes to the loss of prestress. This long–term loss can be as high as 25% of the initial stress. Thus

$$P_{Service} \approx 0.75 \, P_{jack}$$

where

$P_{Service}$ = Total prestress remaining in the long term under working load conditions, after all the losses have taken place.

P_{jack} = Total load in the cables at the time of jacking.

It should be noted that these long term losses of prestress is common to both pre– and post–tensioning systems.

One important effect of creep is increased deflections. Because in a prestressed concrete member a greater part of the cross section is in compression compared to the corresponding reinforced concrete section, long term creep movements are increased.

16.3.2 Steel

Compared to normal high yield steel bars used in reinforced concrete which has an ultimate tensile stress of about 460 N/mm^2, prestressing steel is usually cold drawn high tensile steel wires or alloy steel bars with an ultimate tensile stress of about 1800 N/mm^2. Apart from the fact that steel used in prestressing work is of higher strength, it is also much less ductile compared with reinforcing bars. Steel used in prestressed concrete is available in the form of:

- Wires from 7 mm to 3 mm diameter. In order to improve bond, wires are often indented. This is called crimping.
- Tendons used today are almost always 7-wire strand made from six wires spun round a straight central wire. The overall nominal diameter varies from 12.5 mm to 18 mm. Two basic shapes of cables are available. In Standard cables the individual wires maintain their circular cross section. In order to reduce the overall diameter, the Standard strand can be passed through a die to compress the cable and reduce the voids. The final shape of the individual wires is trapezoidal rather than circular. This type of cable is called 'Drawn' and has, for the same nominal diameter, a higher amount of steel in the cross–section.
- Bars: 25 mm to 50 mm diameter. Two types are common:

(i) Dividag Bbar: This is a bar with ribs along its entire length. The ribs are rolled rather than cut to reduce stress concentration problems. The ribs act as threads for coupling purposes.
(ii) Macalloy Bar: This is a smooth bar with threads rolled only at the ends for coupling or for anchorage purposes.

16.3.2 .1 Relaxation of steel

Just as concrete exhibits time dependent deformation due to creep, steel exhibits a property called Relaxation. If the strain in steel is maintained constant, then the stress required to maintain that strain reduces with time. This property is known as Relaxation. Relaxation is thus loss of stress under constant strain. Generally tests are conducted for duration of 1000 hours (about 42 days) to determine Relaxation properties. Final long–term relaxation loss is expressed as a multiple of the 1000 hour loss. Relaxation also contributes to the loss of prestress in the long term.

Heat treatment is used to improve the elastic and 'yield' properties of strands. Two types of strands are available. They are:

- Relaxation Class 1: This is also called as Normal Relaxation or Stress relieved strand. In order to remove residual stresses due to cold drawing, the strand is heated to about $350°$ C and allowed to cool slowly.
- Relaxation Class 2: This is also called Low relaxation or Strain tempered strand. The strand is heated to about $350°$ C *while the strand is under tension* and allowed to cool slowly.

This process is known as Strain tempering.

16.4 DESIGN OF PRESTRESSED CONCRETE STRUCTURES

Although design of prestressed concrete structures has to satisfy both serviceability and ultimate limit state criteria, there is a fundamental difference in the approach to the design of reinforced and prestressed concrete structures. The normal design procedure for a reinforced concrete structure is to design the structure for ultimate limit state and then check that the structure behaves satisfactorily at serviceability limit state. On the other hand, the normal design procedure for a prestressed concrete structure is to design the structure for the serviceability limit state and then check that the structure behaves satisfactorily at ultimate limit state. The reason for this difference is that generally speaking in prestressed concrete structures, serviceability limit state conditions are much more critical than the conditions at ultimate limit state. Thus generally structures designed for serviceability limit state also satisfy the ultimate limit state criteria, but not the other way round.

16.5 LIMITS ON PERMISSIBLE STRESSES IN CONCRETE

Since prestressed concrete structures are primarily designed to satisfy the serviceability limit state, it is necessary to limit the stresses in concrete and steel. The structure is assumed to behave elastically and the two critical conditions to be considered are

- Stress state at transfer of prestress: At this stage the loads acting are the self weight of the structure and prestress with only elastic shortening during transfer having taken place.
- Stress state at serviceability limit state when the loads acting are the dead and live loads along with the prestress with all the long term losses assumed to have taken place.

16.5.1 Definition of Class

In the design of prestressed concrete structures, a structure is designed to the criterion of a certain Class. Class is simply dependent on the amount of tensile stress f_{st} permitted in concrete at serviceability limit state.

There are three classes defined in the code BS8110. These are

Class 1: $f_{st} = 0$
Class 2: No visible cracking, $f_{st} > 0$
Class 3: Surface crack width not greater than 0.1 mm in members exposed to severe environment (such as alternate wetting and drying, occasional freezing, severe condensation, severe rain, etc.) and limited to 0.2 mm in other cases.
It has been found in practice that for design purposes, Class 1 and Class 2 are generally governed by SLS criteria based on stress limitation whereas Class 3 can be governed by either Limit State of Deflection or Ultimate limit state condition.

16.5.1.1 Partial prestressing

Class 1 structures are described as being fully prestressed as the whole cross section is maintained in compression at working loads.

Class 3 structures are termed as partially prestressed because like reinforced concrete cracking is allowed at working loads. In Class 3 structures part of the steel present is tensioned and in addition there is also untensioned steel as in ordinary reinforced concrete. The main purpose of unstressed steel is to improve ultimate limit capacity. The main advantage of partial prestressing is that of economy of materials, especially prestressing steel. Partial prestressing has the advantage over reinforced concrete in that the prestress present aids the recovery of deflections after an overload and also reduces the crack widths or even closes the cracks altogether. Such structures also tend be less brittle compared to Class 1 structures.

Class 2 structures fall between Class 1 and Class 3 structures.

16.5.2 Permissible Compressive Stress in Concrete at Transfer

Permissible stress in compression at transfer, f_{tc} is given in Clause 4.3.5.1. The *numerical* value of $f_{tc} \leq 0.5 \, f_{ci}$ at the extreme fibre, where f_{ci} = Cube strength at transfer of prestress. If the distribution of prestress is near uniform, then this should be limited to $0.4 \, f_{ci}$.

16.5.3 Permissible Tensile Stress in Concrete at Transfer

Permissible stress in tension, f_{tt} at transfer are given in Clause 4.3.5.2.

Class 1: 1 N/mm^2

Class 2 and Class 3:

$0.45\sqrt{f_{ci}}$ N/mm^2 for pre–tensioned structures.

$0.36\sqrt{f_{ci}}$ N/mm^2 for post–tensioned structures

16.5.4 Permissible Compressive Stress in Concrete at Serviceability Limit State

Permissible stresses in compression at serviceability limit state, f_{sc} are given in Clause 4.3.4.2. The *numerical* value of $f_{sc} \leq 0.33 \, f_{cu}$ at the extreme fibre, where f_{cu} is the cube strength at 28 daysis. In continuous beams and other statically indeterminate structures, near support moments, this may be increased to $0.40 \, f_{cu}$. If the distribution of prestress is near uniform, then this should be limited to $0.25 \, f_{cu}$.

Note that although f_{cu} is greater than f_{ci}, still $f_{tc} \leq 0.5 \, f_{ci}$ is numerically larger than $f_{sc} \leq 0.33 \, f_{cu}$. The reason for this is that the transfer state lasts only for a short time as the stresses beginto reduce due to creep.

16.5.4 Permissible Tensile Stress in Concrete at Serviceability Limit State

Permissible stress in tension, f_{st} at serviceability limit state are given in Clause 4.3.4.3.

Class 1: $f_{st} = 0$; ie. No tensile stress is permitted.

Class 2:

$0.45\sqrt{f_{cu}}$ N/mm^2 for pre–tensioned structures.

$0.36\sqrt{f_{cu}}$ N/mm^2 for post–tensioned structures.

f_{cu} = Cube strength at 28 days.

Class 3: Although crack width is the limitation, for design purposes, hypothetical values of tensile stresses are given as shown in Table 16.1.

Table 16.1 Permissible values of f_{st} (hypothetical)

Group	Crack width Limit in mm	C30	C40	C50 and over
Pre–tensioned	0.1		4.1	4.8
	0.2		5.0	5.8
Grouted Post–tensioned Tendons	0.1	3.2	4.1	4.8
	0.2	3.8	5.0	5.8

16.6 LIMITS ON PERMISSIBLE STRESSES IN STEEL

As described in section 16.3.2, many different types of prestressing steel are available in the form of wires, bars, tendons, etc. The ultimate tensile stress varies from about 1030 N/mm² for hot rolled bars to 1860 N/mm² for Super or Drawn strands. The value of Young's modulus E = (195 ±10) kN/mm².

16.6.1 Maximum Stress at Jacking and at Transfer

The permissible stresses are given in Clause 4.7.1. They are:

*   Jacking force should not normally exceed 75% of the characteristic strength of the tendon but may be increased to 80% provided additional consideration is given to safety.
*   At transfer, the initial prestress should not normally exceed 70% of the characteristic strength of the tendon, and in no case should it exceed 75%.

It is worth noting that with the usual long term loss of about 20 to 25% of the jacking stress, the stress in steel at service, f_{pe} is of the order of 50 to 60% of the of the characteristic tensile stress f_{pu} of the tendon. Essentially, for ultimate strength purposes, tendons act as reinforcement with a characteristic strength of $(f_{pu} - f_{pe})$.

16.7 EQUATIONS FOR STRESS CALCULATION

In a statically determinate structure, the stresses at top and bottom fibres are given by the following equations. The sign convention used is as follows.
Eccentricity e is *positive below* the neutral axis.
Tensile stresses are positive and compressive stresses are negative.
Bending moment causing sagging is considered positive.

16.7.1 Transfer State

At transfer, the external load acting is normally only the self weight. With a large prestress force applied below the neutral axis, the beam will hog up. The critical stress conditions are

*   tension is critical at the top of the beam
*   compression is critical at the bottom of the beam.

The expressions for the stress at top and bottom fibres are given by

$$f_{top} = [-\frac{P_t}{A} + \frac{P_t e}{z_t} - \frac{M_{self\,weight.}}{z_t}] \le f_{tt} \tag{16.1}$$

$$f_{bottom} = [-\frac{P_t}{A} - \frac{P_t e}{z_b} + \frac{M_{self\,weight.}}{z_b}] \ge f_{tc} \tag{16.2}$$

where f_{tt} and f_{tc} are the permissible stresses at transfer in tension and compression respectively. The first subscript, t stands for transfer and the second subscript, (t or c) stands for tension and compression respectively. Note that f_{tc} is a compressive stress, and it takes on a negative number.

P_t = Prestress at transfer. $P_t = \alpha\, P_{jack}$, where $\alpha \approx 0.90$ because at transfer, there is generally loss of prestress of about 10%.

z_t and z_b are the section moduli of the beam for top and bottom fibres respectively.

16.7.2 Serviceability Limit State

At working loads, external loads on the beam increase due to live loads and other dead loads. In addition due to long term loss the prestress in the cables also decreases. These effects cause the beam to sag. The critical stress conditions are:

- compression is critical at the top of the beam,
- tension is critical at the bottom of the beam.

The expressions for the stress at top and bottom fibres are given by

$$f_{top} = [-\frac{P_s}{A} + \frac{P_s e}{z_t} - \frac{M_{service}}{z_t}] \ge f_{sc} \tag{16.3}$$

$$f_{bottom} = [-\frac{P_s}{A} - \frac{P_s e}{z_b} + \frac{M_{service}}{z_b}] \le f_{st} \tag{16.4}$$

where

f_{st} and f_{sc} are the permissible stresses at service in tension and compression respectively. The first subscript, s stands for service and the second subscript (t or c) stands for tension and compression respectively.

$M_{service}$ = Moment at Serviceability limit state. It includes the effect of self–weight, live loads, etc.

$P_s = \beta\, P_{jack}$ where $\beta \approx 0.75$ because of about 25% prestress is lost due to elastic shortening, creep, shrinkage and relaxation.

16.7.3 Example of Stress Calculation

Fig.16.9 shows a pre–tensioned symmetric double–T–beam.

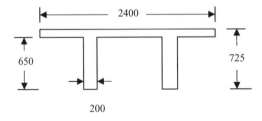

Fig.16.9 A double–T–beam.

It is used as a simply supported beam to support a total characteristic load (excluding self weight) of 45 kN/m over a span of 10 m. It is prestressed by a total force of $P_{jack} = 1450$ kN. The constant eccentricity is 390 mm. Calculate the stresses at mid–span and support at transfer and serviceability limit state. It is given that the structure has been designed for Class 2 criterion. $f_{ci} = 35$ N/mm² and $f_{cu} = 50$ N/mm². Assume loss at transfer and serviceability limit state as respectively 10% and 25% of the jacking force.

(i) Section Properties
Area of cross section
$$A = 2400 \times 75 + 2 \times (200 \times 650) = 44 \times 10^4 \text{ mm}^2$$
Position of the centroid from the bottom: Taking moment about the bottom of the webs,
$$A \, y_b = 2400 \times 75 \times (725 - 75/2) + 2 \times (200 \times 650 \times 650/2) = 2.083 \times 10^8 \text{ mm}^3$$
Distance from the centroidal axis to bottom and top fibres are:
$$y_b = 473 \text{ mm}, \, y_t = 725 - y_b = 252 \text{ mm}$$
Second moment of area, I:
$$I = [2400 \times 75^3/12 + 2400 \times 75 \times (y_t - 75/2)^2] + 2 \times [200 \times 650^3/12 + 200 \times 650 \times (650/2 - y_b)^2] = 2.322 \times 10^{10} \text{ mm}^4$$
$$z_t = I/y_t = 92.12 \times 10^6 \text{ mm}^3$$
$$z_b = I/y_b = 49.08 \times 10^6 \text{ mm}^3$$

(ii) Calculation of moments
Unit weight of concrete = 24 kN/m³
$$\text{Self weight} = (44 \times 10^4) \times 10^{-6} \times 24 = 10.56 \text{ kN/m}$$
$$M_{self \, weight} = 10.56 \times 10^2/8 = 132.0 \text{ kNm at mid–span}$$
$$\text{Total load on the beam (including self weight)}$$
$$= 10.56 + 45.0 = 55.56 \text{ kN/m}$$
$$M_{service} = 55.56 \times 10^2/8 = 694.5 \text{ kNm at mid–span}$$

(iii) Prestress
$$P_t = 0.90 \, P_{jack} = 0.9 \times 1450 = 1305 \text{ kN}$$
$$P_s = 0.75 \, P_{jack} = 0.9 \times 1450 = 1087.5 \text{ kN}$$

(iv) Permissible stresses

(a) Transfer
$$f_{tt} = 0.45 \sqrt{f_{ci}} = 0.45 \times \sqrt{35} = 2.7 \text{ N/mm}^2$$
$$f_{tc} = -0.5 f_{ci} = -0.5 \times 35 = -17.5 \text{ N/mm}^2$$

(b) Service
$$f_{st} = 0.45 \sqrt{f_{cu}} = 0.45 \times \sqrt{50} = 3.2 \text{ N/mm}^2$$
$$f_{sc} = -0.33 f_{cu} = -0.33 \times 50 = -16.7 \text{ N/mm}^2$$

(v) Stress calculation at transfer
$$P_t = 1305 \text{ kN}, \, e = 390 \text{ mm}$$

Expressions for stresses at top and bottom fibres are given by equations 16.1 and 16.2.

(a) Support: At support self weight moment is zero because of the simply supported condition.

$$f_{top} = -1305 \times 10^3 / (44 \times 10^4) + 1305 \times 10^3 \times 390 / (92.12 \times 10^6)$$
$$= -2.97 + 5.53 = 2.56 < 2.70 \text{ N/mm}^2$$

$$f_{bottom} = -1305 \times 10^3 / (44 \times 10^4) - 1305 \times 10^3 \times 390 / (49.08 \times 10^6)$$
$$= -2.97 - 10.37 = -13.33 > -17.50 \text{ N/mm}^2$$

(b) Mid–span:

$$\text{Self weight moment} = 132.0 \text{ kNm}$$
$$f_{top} = -1305 \times 10^3 / (44 \times 10^4) + 1305 \times 10^3 \times 370 / (92.12 \times 10^6)$$
$$- 132.0 \times 10^6 / (92.12 \times 10^6)$$
$$= -2.97 + 5.53 - 1.43 = 1.13 < 2.70 \text{ N/mm}^2$$

$$f_{bottom} = -1305 \text{ x } 10^3 / (44 \times 10^4) - 1305 \times 10^3 \times 370 / (49.08 \times 10^6)$$
$$+ 132.0 \times 10^6 / (49.08 \times 10^6)$$
$$f_{bottom} = -2.97 - 10.37 + 2.69 = -10.65 > -17.50 \text{ N/mm}^2$$

Note: Stresses at the supports are larger than at mid–span.

(vi) Stress calculation at serviceability limit state
$$P_s = 1087.5 \text{ kN, } e = 390 \text{ mm}$$
Expressions for stresses at top and bottom fibres are given by equations 16.3 and 16.4.

(a) Support: At support moment is zero because of the simply supported condition.
$$f_{top} = -1087.5 \times 10^3 / (44 \times 10^4) + 1087.5 \times 10^3 \times 390 / (92.12 \times 10^6)$$
$$= -2.47 + 4.60 = 2.13 < 3.2 \text{ N/mm}^2$$

$$f_{bottom} = -1087.5 \times 10^3 / (44 \times 10^4) - 1087.5 \times 10^3 \times 390 / (49.08 \times 10^6)$$
$$= -2.47 - 8.64 = -11.11 > -16.70 \text{ N/mm}^2$$

(b) Mid–span:
Serviceability limit state moment = 694.5 kNm
$$f_{top} = -1087.5 \times 10^3 / (44 \times 10^4) + 1087.5 \times 10^3 \text{ x } 390 / (92.12 \times 10^6)$$
$$- 694.5 \times 10^6 / (92.12 \times 10^6)$$
$$= -2.47 + 4.60 - 7.54 = -5.41 > -16.7 \text{ N/mm}^2$$

$$f_{bottom} = -1087.5 \times 10^3 / (44 \times 10^4) - 1087.5 \times 10^3 \times 370 / (49.08 \text{ x } 10^6)$$
$$+ 694.5 \times 10^6 / (49.08 \times 10^6)$$
$$= -2.47 - 8.64 + 14.15 = 3.04 < 3.20 \text{ N/mm}^2$$

Note: Stresses at the supports are smaller than at transfer condition. The stresses at mid–span are larger than at transfer condition. In addition the state of stress has reversed from tension at top to tension at bottom and vice versa.

16.8 DESIGN FOR SERVICEABILITY LIMIT STATE

For a given structural configuration and loads, design in prestressed concrete for serviceability limit state requirements involves two things.
- A suitable section
- Choice of prestress and corresponding eccentricity

16.8.1 Initial Sizing of Section

Consider the four equations 16.1 to 16.4 associated with the calculation of stresses at top and bottom fibres at a cross section under transfer and serviceability conditions. In these equations

$$P_{transfer} = \alpha\, P_{Jack}, \alpha \approx 0.90$$
$$P_{Service} = \beta\, P_{Jack}, \beta \approx 0.75$$
$$P_{transfer} = \eta\, P_{Service,}$$
$$\eta = \beta/\alpha \approx 0.83$$

Expressing P_t in terms of P_s, equations 16.1 and 16.2 can be expressed in terms of P_s as

$$[-\frac{P_s}{A} + \frac{P_s e}{z_t} - \eta\frac{M_{self\,weight.}}{z_t}] \leq \eta\, f_{tt} \tag{16.5}$$

$$[-\frac{P_s}{A} - \frac{P_s e}{z_b} + \eta\frac{M_{self\,weight.}}{z_b}] \geq \eta\, f_{tc} \tag{16.6}$$

Eliminating P_s and e from equations 16.5 and 16.3,

$$[\frac{M_{service}}{z_t} - \frac{\eta\, M_{self\,weight.}}{z_t}] \leq \{\eta\, f_{tt} - f_{sc}\}$$

$$z_t \geq \frac{M_{service} - \eta M_{self\,weight}}{\eta\, f_{tt} - f_{sc}}$$

$$z_t \geq \frac{(M_{service} - M_{self\,weight}) + (1-\eta)M_{self\,weight}}{\eta\, f_{tt} - f_{sc}} \tag{16.7}$$

Similarly, eliminating P_s and e from equations 16.6 and 16.4,

$$\frac{M_{service}}{z_b} - \frac{\eta\, M_{self\,weight.}}{z_b} \leq f_{st} - \eta f_{tc}$$

$$z_b \geq \frac{M_{service} - \eta M_{self\,weight}}{f_{st} - \eta\, f_{tc}}$$

$$z_b \geq \frac{(M_{Service} - M_{self\,weight}) + (1-\eta)M_{self\,weight}}{f_{st} - \eta\, f_{tc}} \tag{16.8}$$

Initially the self weight moment is not known. However, $(M_{service} - M_{Selfweight})$ represents the moment due to external loads which are known. As $\eta \approx 0.83$, the effect of including $M_{self\ weight}$ has a small effect on the required section moduli. Therefore for an initial estimate, it is reasonable to take $M_{self\ weight}$ as zero. Once an initial section has been decided upon, if necessary the required value of section modulus can be recalculated.

16.8.1.1 Example of initial sizing

Calculate the section moduli required for a simply supported beam to support a characteristic load of 45 kN/m (excluding self weight) over a span of 10 m. It is given that the allowable stresses are:

$$f_{tt} = 2.7\ N/mm^2,\ f_{tc} = -17.5\ N/mm^2$$
$$f_{st} = 3.2\ N/mm^2,\ f_{sc} = -16.7\ N/mm^2$$

The loss of prestress at transfer and service can be taken as 10% and 25% of the force at jacking.

$$M_{service} - M_{self\ weight} = 45 \times 10^2/8 = 562.5\ kNm\ at\ mid–span.$$
$$\eta = 0.9/0.75 = 0.83$$
$$\eta\ f_{tt} - f_{sc} = 0.83 \times 2.7 - (-16.7) = 18.9\ N/mm^2$$
$$f_{st} - \eta\ f_{tc} = 3.2 - 0.83\ x\ (-17.5) = 17.7\ N/mm^2$$

Ignoring $M_{self\ weight}$ for an initial estimate of moduli, from equations 16.7 and 16.8,
$$z_t \geq 562.5 \times 10^6/18.9 = 29.76 \times 10^6\ mm^3$$
$$z_b \geq 562.5 \times 10^6/17.7 = 31.78 \times 10^6\ mm^3$$

If it is decided to choose a T–section shown in Fig.16.10, the section properties can be expressed as functions of the two parameters (h_f/h) and (b_w/b) as follows. Table 16.2 gives the section properties.

$$\alpha = \frac{h_f}{h}, \beta = \frac{b_w}{b}$$
$$A = bh[\alpha + \beta(1-\alpha)]$$
$$y_b = h\frac{[0.5\beta(1-\alpha)^2 + \alpha(1-0.5\alpha)]}{[\alpha + \beta(1-\alpha)]}$$
$$y_t = h\frac{0.5[\beta(1-\alpha)^2 + 0.5\alpha^2]}{[\alpha + \beta(1-\alpha)]}$$
$$I = \frac{bh^3}{12}[\alpha^3 + 12\alpha(\frac{y_t}{h} - 0.5\alpha)^2 + \beta(1-\alpha)^3 + 12\beta(1-\alpha)(0.5 - 0.5\alpha - \frac{y_b}{h})^2]$$
$$z_t = \frac{I}{y_t}, z_b = \frac{I}{y_b}$$

Table 16.2 Section properties of T–beams

h_f/h	b_w/b	$A/(bh)$	y_b/h	y_t/h	$I/(bh^3)$	$z_t/(bh^2)$	$z_b/(bh^2)$
0.1	0.1	0.19	0.713	0.287	0.018	0.063	0.025
0.2	0.1	0.28	0.757	0.243	0.019	0.079	0.025
0.3	0.1	0.37	0.755	0.245	0.019	0.079	0.026
0.1	0.2	0.28	0.629	0.371	0.028	0.076	0.045
0.2	0.2	0.36	0.678	0.322	0.031	0.098	0.046
0.3	0.2	0.44	0.691	0.309	0.032	0.103	0.046
0.1	0.3	0.37	0.585	0.415	0.037	0.088	0.062
0.2	0.3	0.44	0.627	0.373	0.041	0.109	0.065
0.3	0.3	0.51	0.644	0.356	0.042	0.117	0.065
0.1	0.4	0.46	0.559	0.441	0.044	0.100	0.079
0.2	0.4	0.52	0.592	0.408	0.049	0.119	0.082
0.3	0.4	0.58	0.609	0.391	0.050	0.127	0.082

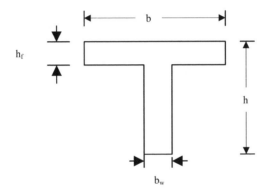

Fig.16.10 T–section.

Choosing
$$h_f/h = 0.1,\ b_w/b = 0.2,$$
A variety of section sizes is possible, for example:

(i) b = 2000 mm, h = 595 say 600 mm, b_w = 400 mm, h_f = 60 mm,
$$z_t = 0.076 \times 2000 \times 600^2 = 54.72 \times 10^6 \text{ mm}^3$$
$$z_b = 0.045 \times 2000 \times 600^2 = 32.40 \times 10^6 \text{ mm}^3$$

(ii) b = 1500 mm, h = 686 say 700 mm, b_w = 300 mm, h_f = 70 mm,
$$z_t = 0.076 \times 1500 \times 700^2 = 55.86 \times 10^6 \text{ mm}^3$$
$$z_b = 0.045 \times 1500 \times 700^2 = 33.08 \times 10^6 \text{ mm}^3$$

If a double T–section (Fig.16.9) is desired, the web width b_w can be shared between two webs with the width of each web equal to $0.5b_w$.

Having chosen a section, its self weight can be calculated. For example for the section:

$$b = 1500 \text{ mm}, \ h = 700 \text{ mm}, \ b_w = 300 \text{ mm}, \ h_f = 70 \text{ mm},$$
$$A = 294.0 \times 10^3 \text{ mm}^2$$
$$\text{Self weight} = 7.056 \text{ kN/m},$$
$$M_{\text{self weight}} = 88.2 \text{ kNm}$$
$$(1 - \eta) M_{\text{self weight}} = 15.00 \text{ kNm}$$

Using the self weight moment, the revised required section moduli become

$$z_t \geq (562.5 + 15.00) \times 10^6/18.9 = 30.56 \times 10^6 \text{ mm}^3$$
$$z_b \geq (562.5 + 15.00) \times 10^6/17.7 = 32.63 \times 10^6 \text{ mm}^3$$

The section modulus z_b of the chosen section is $33.08 \times 10^6 \text{ mm}^3$ which is only slightly greater than the required value of $32.63 \times 10^6 \text{ mm}^3$. The chosen section is adequate.

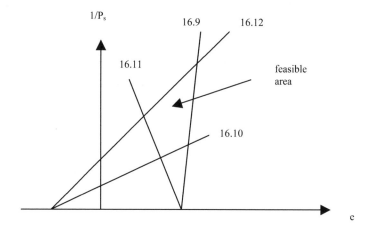

Fig.16.11 Magnel diagram (The numbers against the lines correspond to equation numbers).

16.8.2 Choice of Prestress and Eccentricity

Having chosen a section, the next step is to choose the required value of prestress and eccentricity such that none of the stress criteria are violated. By dividing throughout by $1/P_s$, equations 16.3 to 16.6 can be rewritten as follows

$$-\frac{1}{A} + \frac{e}{z_t} \leq \eta \langle \frac{M_{\text{self weight.}}}{z_t} + f_{tt} \rangle \frac{1}{P_s} \tag{16.9}$$

$$-\frac{1}{A} - \frac{e}{z_b} \geq \eta \langle -\frac{M_{\text{self weight.}}}{z_b} + f_{tc} \rangle \frac{1}{P_s} \tag{16.10}$$

$$-\frac{1}{A}+\frac{e}{z_t} \geq \langle\frac{M_{service}}{z_t}+f_{sc}\rangle\frac{1}{P_s} \tag{16.11}$$

$$-\frac{1}{A}-\frac{e}{z_b} \leq \langle-\frac{M_{service}}{z_b}+f_{st}\rangle\frac{1}{P_s} \tag{16.12}$$

If the inequality signs are replaced by an equality sign, a plot of e vs. $1/P_s$ of each equation is a straight line and the plot of all the four equations encloses a quadrilateral as shown in Fig.16.11. Any choice of e and P_s inside the quadrilateral satisfies all the four stress criteria. This plot is known as a Magnel diagram.

16.8.2.1 Example of construction of Magnel diagram

Fig.16.9 shows a pre–tensioned symmetric double–T–beam. It is used as a simply supported beam to support a total characteristic load (excluding self weight) of 45 kN/m over a span of 10 m. Construct the Magnel diagram for the mid–span section using the following data. Assume loss at transfer and serviceability limit state as respectively 10% and 25% of the jacking force.

(i) Section properties:

$$A = 44 \times 10^4 \text{ mm}^2$$
$$z_t = 92.12 \times 10^6 \text{ mm}^3$$
$$z_b = 49.08 \times 10^6 \text{ mm}^3$$
$$1/A = 227.27 \times 10^{-8},$$
$$1/z_b = 2.035 \times 10^{-8},$$
$$1/z_t = 1.086 \times 10^{-8}$$

(ii) Moments at mid–span:

$$M_{self\,weight} = 132.0 \text{ kNm}$$
$$M_{self\,weight}/z_t = 132.0 \times 10^6/(92.12 \times 10^6) = 1.43 \text{ N/mm}^2$$
$$M_{self\,weight}/z_b = 132.0 \times 10^6/(49.08 \times 10^6) = 2.69 \text{ N/mm}^2$$

$$M_{service} = 694.5 \text{ kNm}$$
$$M_{service}/z_t = 694.5 \times 10^6/(92.12 \times 10^6) = 7.54 \text{ N/mm}^2$$
$$M_{service}/z_b = 694.5 \times 10^6/(49.08 \times 10^6) = 14.15 \text{ N/mm}^2$$

(iii) Permissible stresses:

$$f_{tt} = 2.7 \text{ N/mm}^2,$$
$$f_{tc} = -17.5 \text{ N/mm}^2$$
$$f_{st} = 3.2 \text{ N/mm}^2$$
$$f_{sc} = -16.7 \text{ N/mm}^2$$

(iv) Prestress losses:

$$\eta = 0.75/0.9 = 0.83$$

(v) Solution:
Substituting the above values in equations 16.9 to 16.11, the following four linear equations are obtained.

$$-227.27 + 1.086\ e = 0.83(1.43 + 2.7)\ 10^8/P_s = 3.43\ (10^8/P_s)$$
$$-227.27 - 2.035\ e = 0.83(-2.69 - 17.5)\ 10^8/P_s = -16.76\ (10^8/P_s)$$
$$-227.27 + 1.086\ e = (7.54 - 16.7)\ 10^8/P_s = -9.16\ (10^8/P_s)$$
$$-227.27 - 2.035\ e = (-14.15 + 3.2)\ 10^8/P_s = -10.95\ (10^8/P_s)$$

Fig.16.12 shows the Magnel diagram.

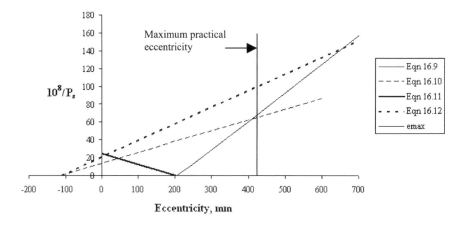

Fig.16.12 Magnel diagram for mid–span.

16.8.2.2 Example of choice of prestress and eccentricity

Fig.16.12 shows the feasible region. Any combination of P_s and e within the feasible region will satisfy all the four stress criteria. Unfortunately practical limitations of cover, etc, reduce the extent of the feasible area. In the above example $y_b = 473$ mm. Assuming that cables require a cover of approximately 50 mm, the maximum eccentricity e_{max} allowable is only $(y_b - 50) = 423$ mm. This limitation is shown in Fig.16.12 by the vertical line.

In choosing a value of P_s and e two important points to keep in mind are:

- Choose as small a value of P_s (ie. as large a value of $10^8/P_s$) and as large a value of e as possible. This will keep the costs down. In this example this is approximately,
$$10^8/P_s \approx 100 \text{ and } e \approx 420 \text{ mm}.$$

- It is not advisable to work right at the edge of the feasible region as it does not allow for any flexibility in the arrangement of cables in the cross section.

(i) Determination of number of cables(strands,tendons) required

A value of P_s can be chosen which is near the top right hand part of the feasible region. Choosing

$$10^8/P_s \approx 85,$$
$$P_s \approx 1180 \text{ kN},$$
$$P_{jack} = P_s/0.75 \approx 1570 \text{ kN}$$

Having calculated the value of P_{jack}, the next stage is to choose the type and number of cables required. Table 16.3 gives the strengths of various types of 7-wire strands. If for example 7-wire drawn strand of 18.0 mm nominal diameter is chosen, its breaking load is 380 kN. However jacking force should not normally exceed 75% of the characteristic strength of the tendon but may be increased to 80% provided additional consideration is given to safety (see 16.6.1).

Force per cable at jacking = $0.75 \times 380 = 285$ kN
Number of cables required = $P_{jack}/285 = 1570/285 = 5.5$ say 6 cables.

Table 16.3 Properties of strands

Type of strand	Nominal diameter, mm	Net area of cross section, mm^2	Characteristic breaking load, kN	Nominal tensile strength, N/mm^2
Standard	12.5	92.66	164	1770
	15.2	138.92	232	1670
Super	12.9	100.00	186	1860
	15.7	149.72	265	1770
Drawn	12.7	112.36	209	1860
	15.2	164.84	300	1820
	18.0	223.53	380	1700

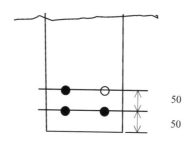

50

50

Fig.16.13 Arrangement of cables in the web.

(ii) Determination of eccentricity

Assuming that cables can be placed in the webs in horizontal layers at 50 mm intervals vertically with two cables per layer, 6 cables can be accommodated with three cables in each web with two cables at the lowest level and one cable at the next level as shown in Fig.16.13.

The resultant eccentricity
$$e = y_b - (2 \times 50 + 1 \times 100)/3 = 473 - 67 = 406 \text{ mm}$$

The point corresponding to
$$10^8/P_s = 85, e = 406 \text{ mm}$$

is inside the feasible region. Therefore the arrangement and force in the cables is satisfactory.

16.8.2.3 Example of debonding

If it is decided to debond some cables towards the support, then a Magnel diagram has to be drawn for the support section as well. At a support section, the critical condition is at transfer. Conditions at service are not critical because of the long term losses in the prestress. Since there are no moments acting at supports, the two critical equations are:

$$-227.27 + 1.086\, e = 0.83(0 + 2.7)\, 10^8/P_s = 2.24\, (10^8/P_s)$$
$$-227.27 - 2.035\, e = 0.83(0 - 17.5)\, 10^8/P_s = -14.53\, (10^8/P_s)$$

Fig.16.14 shows the Magnel diagram at the support. The feasible area is *not* a closed polygon.

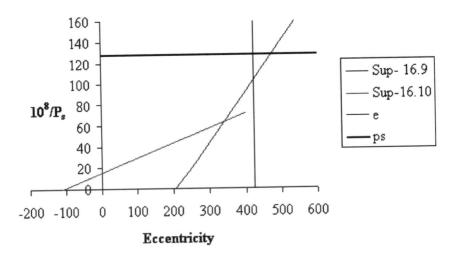

Fig.16.14 Magnel diagram at support.

The point corresponding to $10^8/P_s = 85$, e = 406 mm is outside the feasible region and cannot be accepted. In order to bring the point inside the feasible region, remove two cables, one from each web. The number of strands is reduced from six to four. The corresponding $10^8/P_s$ is given by

$$10^8/P_s = 85 \times (6/4) = 127.5$$
$$e = 473 - 50 = 423 \text{ mm.}$$

This point now lies inside the feasible region and can be accepted. Thus one cable in each web needs to be debonded towards the support.

16.9 COMPOSITE BEAMS

In a very common form of bridge construction, precast pre–tensioned beams are erected first and the in–situ concrete is cast on top of them using form–work which is supported on the precast beams. The form–work is just left in place. This type of formwork is called sacrificial formwork. Beams are placed at approximately 1 m apart. Once the in–situ concrete has hardened, the beam and the deck slab act as a composite structure. Fig.16.15 shows a typical bridge superstructure using inverted T–beams.

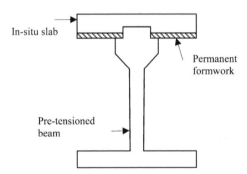

Fig.16.15 Composite beam.

In this type of beams, the weight of the slab and associated permanent formwork is carried wholly by the precast beam. However once the slab hardens, then all subsequent loads acting on the slab will be resisted by the pre–tensioned beam acting in conjunction with the cast–in–situ slab. The cast–in–situ slab acts as the compression flange of the composite I–beam.

Since the object is to calculate the value of P_s and e so that the stresses in the precast section are within permissible limits, the stresses are calculated in the precast section only.

(i) Transfer stage: Prestress acts on the precast beam. The only external moment is due to the self weight of the beam. The governing equations are 16.1 and 16.2 (repeated here for completeness)

$$f_{top} = [-\frac{P_t}{A} + \frac{P_t e}{z_t} - \frac{M_{self\,weight.}}{z_t}] \le f_{tt} \tag{16.1}$$

$$f_{bottom} = [-\frac{P_t}{A} - \frac{P_t e}{z_b} + \frac{M_{self\,weight.}}{z_b}] \ge f_{tc} \tag{16.2}$$

(ii) Serviceability limit: The weights of slab and precast beam are supported by the precast beam. The 'Live' loads are supported by composite beam. In addition to the stresses caused by the loads, one needs to include the stresses caused by the shrinkage of the insitu slab. Stresses due to shrinkage occur because as the cast–in–situ slab dries, it shrinks and forces the precast beam to bend.

$$f_{top} = [-\frac{P_s}{A} + \frac{P_s e}{z_t} - \frac{M_{Dead}}{z_t} - \frac{M_{live}}{Comp.z_{top\,of\,precast}} - Shrinkage\ stress] \ge f_{sc} \tag{16.13}$$

$$f_{bottom} = [-\frac{P_s}{A} - \frac{P_s e}{z_b} + \frac{M_{Dead}}{z_b} + \frac{M_{live}}{Comp.z_b} + Shrinkage\ stress] \le f_{st} \tag{16.14}$$

16.9.1 Magnel Equations for a Composite Beam

Fig.16.16 shows a composite beam. The precast pre–tensioned inverted T–beam is made composite with an cast–in–situ slab acting as the top flange of the composite beam. It is used as a simply supported beam over a span of 24 m.

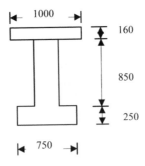

Fig 16.16 Composite beam section.

The section properties on precast and composite beam are as follows.

(a) Precast beam:

$$\text{Area} = 4.425 \times 10^5 \text{ mm}^2$$
$$y_b = 442 \text{ mm}$$
$$y_t = 658 \text{ mm}$$
$$I = 4.90 \times 10^{10} \text{ mm}^4,$$
$$z_b = 111.0 \times 10^6 \text{ mm}^3$$
$$z_t = 74.5 \times 10^6 \text{ mm}^3.$$
$$\text{self weight} = 10.62 \text{ kN/m}$$

(b) Composite beam:

$$A_{\text{composite}} = 6.025 \times 10^5 \text{ mm}^2$$
$$y_{b \text{ Composite}} = 638 \text{ mm},$$
$$y_t \text{ to top of precast} = 1100 - 638 = 462 \text{ mm}$$
$$I_{\text{Composite}} = 11.33 \times 10^{10} \text{ mm}^4$$
$$z_{b \text{ Composite}} = 177.6 \times 10^6 \text{ mm}^3$$
$$\text{Composite } z_{t \text{ to top of precast}} = 245.2 \times 10^6 \text{ mm}^3$$
$$\text{self weight} = 14.46 \text{ kN/m}$$

(c) Loads:

$$\text{self weight of precast} = 10.62 \text{ kN/m}$$
$$q_{\text{dead}} = \text{Weight of (Composite beam + permanent formwork)}$$
$$q_{\text{dead}} = 14.46 + \text{say } 1.2 = 15.66 \text{ kN/m}$$
$$q_{\text{live}} = 18.2 \text{ kN/m}$$

(d) Moments and stresses at mid–span:

Self weight:

$$q_{\text{Self weight}} = 10.62 \text{ kN/m}$$
$$M_{\text{self weight}} = 10.62 \times 24^2/8 = 764.64 \text{ kNm}$$
$$\frac{M_{\text{Self weight}}}{z_b} = \frac{764.64 \times 10^6}{111.0 \times 10^6} = 6.9 \text{ N/mm}^2$$
$$\frac{M_{\text{Self weight}}}{z_t} = \frac{764.64 \times 10^6}{74.5 \times 10^6} = 10.3 \text{ N/mm}^2$$

Total dead load:

$$q_{\text{dead}} = 15.66 \text{ kN/m}$$
$$M_{\text{Dead}} = 15.66 \times 24^2/8 = 1127.57 \text{ kNm}$$
$$\frac{M_{\text{Dead}}}{z_b} = \frac{1127.57 \times 10^6}{111.0 \times 10^6} = 10.2 \text{ N/mm}^2$$
$$\frac{M_{\text{Dead}}}{z_t} = \frac{1127.57 \times 10^6}{74.5 \times 10^6} = 15.1 \text{ N/mm}^2$$

Live load:

$$q_{Live} = 18.2 \text{ kN/m}$$
$$M_{Live} = 18.2 \times 24^2/8 = 1310.4 \text{ kNm}$$

$$\frac{M_{Live}}{\text{Comp} z_b} = \frac{1310.4 \times 10^6}{177.6 \times 10^6} = 7.4 \text{ N/mm}^2$$

$$\frac{M_{Live}}{\text{Comp} z_t \text{ to top of precast}} = \frac{1310.4 \times 10^6}{245.0 \times 10^6} = 5.3 \text{ N/mm}^2$$

Shrinkage stresses: Assume:

$$\text{Top} = -1.7 \text{ N/mm}^2$$
$$\text{Bottom} = 0.6 \text{ N/mm}^2.$$

(e) Magnel Equations: Magnel equations consider the stresses in precast section only. Using precast beam properties,

$$1/A = 1/ (4.425 \times 10^5) = 226.0 \times 10^{-8}$$
$$1/z_b = 1/ (111.0 \times 10^6) = 0.90 \times 10^{-8},$$
$$1/z_t = 1/ (74.5 \times 10^6) = 1.34 \times 10^{-8}$$

(f) Losses
Take 10% loss at transfer and 25% long term.

$$\eta = (1 - 0.1)/ (1 - 0.25) = 0.83$$

(g) Permissible stresses: Assume

$$f_{tt} = 3.0 \text{ N/mm}^2$$
$$f_{tc} = -20\ 0 \text{ N/mm}^2$$
$$f_{st} = 3.2 \text{ N/mm}^2$$
$$f_{sc} = -21\ 0 \text{ N/mm}^2$$

(h) Stress conditions and Magnel equations:

At transfer:
Top: (Equation 16.1)

$$(-226.0 \times 10^{-8} + 1.34e \times 10^{-8})P_s - 0.83 \times 10.3 \leq 0.83 \times 3.0$$
$$-226.0 + 1.34\ e \leq 11.0 \times (10^8/P_s)$$

Bottom: (Equation 16.2)

$$(-226.0 \times 10^{-8} - 0.90e \times 10^{-8})P_s + 0.83 \times 6.9 \geq 0.83 \times (-20.0)$$
$$-226.0 - 0.90\ e \geq -22.3 \times (10^8/P_s)$$

At Service:

Top: (Equation 16.13)

$$(-226.0 \times 10^{-8} + 1.34e \times 10^{-8})P_s - 15.1 - 5.3 - 1.7 \geq -21.0$$
$$-226.0 + 1.34 \text{ e} \geq 1.1 \times (10^8/P_s)$$

Bottom: (Equation 16.14)

$$(-226.0 \times 10^{-8} - 0.90e \times 10^{-8})P_s + 10.2 + 7.4 + 0.6 \leq 3.2$$
$$-226.0 - 0.90 \text{ e} \leq -15.0 \times (10^8/P_s)$$

The Magnel diagram for the above set of four equations can be drawn.

16.10 POST–TENSIONED BEAMS: CABLE ZONE

In pre–tensioned beams, the strands are straight (except when cables are deflected) and due to debonding, prestress and eccentricity vary along the span in a stepwise manner. In post–tensioned beams, cables take a curved profile. Thus the *eccentricity can vary* along the span but the *prestress remains constant* (if losses in prestress along the span can be ignored). The permissible eccentricity at any section can be calculated by rearranging equations 16.9 to 16.12 as follows.

$$e \leq \frac{z_t}{A} + \eta(M_{self\,weight.} + z_t f_{tt}) \frac{1}{P_s} \tag{16.15}$$

$$e \leq -\frac{z_b}{A} + \eta(M_{self\,weight.} - z_b f_{tc}) \frac{1}{P_s} \tag{16.16}$$

$$e \geq \frac{z_t}{A} + (M_{service} + z_t f_{sc}) \frac{1}{P_s} \tag{16.17}$$

$$e \geq -\frac{z_b}{A} + (M_{service} - z_b f_{st}) \frac{1}{P_s} \tag{16.18}$$

16.10.1 Example of a Post–tensioned Beam

Fig.16.9 shows a post–tensioned symmetric double–T–beam. It is used as a simply supported beam to support a total load (excluding self weight) of 45 kN/m over a span of 10 m. From the Magnel diagram at mid–span, the value of $10^8/P_s = 85$, giving $P_s = 1176.5$ kN. Assume loss at transfer and serviceability limit state as respectively 10% and 25% of the jacking force.

(i) Section properties:

$$A = 44 \times 10^4 \text{ mm}^2$$
$$z_t = 92.12 \times 10^6 \text{ mm}$$
$$z_b = 49.08 \times 10^6 \text{ mm}^3$$
$$z_t/A = 209.36 \text{ mm}$$
$$z_b/A = 111.56 \text{ mm}$$

(ii) Moments (At position x from left hand support):
$$q_{\text{self weight}} = 10.56 \text{ kN/m}$$
$$M_{\text{self weight}} = 52.8 \, x - 5.28 \, x^2$$
$$q_{\text{service}} = 55.56 \text{ kN/m}$$
$$M_{\text{service}} = 277.8 \, x - 27.78 \, x^2$$

(iii) Permissible stresses:
$$f_{tt} = 2.7 \text{ N/mm}^2$$
$$f_{tc} = -17.5 \text{ N/mm}^2$$
$$f_{st} = 3.2 \text{ N/mm}^2$$
$$f_{sc} = -16.7 \text{ N/mm}^2$$

(iv) Loss of prestress
$$\eta = 0.75/0.9 = 0.83$$

(v). Prestress
$$P_s = 1176.5 \text{ kN}$$

(v) Limits on eccentricity:

Substituting the above values in equations 16.15 to 16.18 and simplifying
$$e \leq \{384.83 + 37.25 \, x - 3.725 \, x^2\}$$
$$e \leq \{717.50 + 37.25 \, x - 3.725 \, x^2\}$$
$$e \geq \{-1098.25 + 236.12 \, x - 23.612 \, x^2\}$$
$$e \geq \{-518.32 + 236.12 \, x - 23.612 \, x^2\}$$

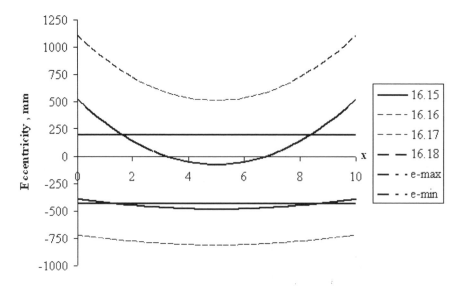

Fig.16.17 Feasible cable zone.

Lower limits for e (e $\geq$) are governed by equations 16.15 and 16.16. Clearly equation 16.15 gives a lower value than equation 16.16. Similarly the upper limits for e (e $\leq$) are governed by equations 16.17 and 16.18. Clearly equation 16.18 gives a larger value than equation 16.17. Thus the feasible cable zone lies between the curves corresponding to equations 16.15 and 16.18. Cables placed inside the feasible zone thus satisfy all the stress criteria. Assuming a minimum cover to the cables of 50 mm, the maximum and minimum values of e attainable are equal to

$$y_b - 50 = 423 \text{ mm}$$
$$y_t - 50 = 202 \text{ mm}.$$

Fig.16.17 shows the feasible region. The range of e at the support and mid–span are:

$$\text{Ends: } -202 \leq e \leq 384.83$$
$$\text{Mid–span: } 71.98 \leq e \leq 423$$

16.11 ULTIMATE MOMENT CAPACITY

One aspect of design of prestressed sections which is different from the procedure used in the case of a reinforced concrete section is that the designs are carried out for SLS and the designed section is checked to ensure that ULS conditions are also satisfied. The calculations for determining the ultimate moment capacity are similar to the ultimate moment capacity calculation in the case of reinforced concrete section as explained in Chapter 4, section 4.6.2. As in the case of reinforced concrete sections, the compressive stress distribution in concrete is assumed to be that given by rectangular stress block assumption with the maximum stress equal to $0.445f_{cu}$ and the maximum strain equal to 0.0035. The main difference from the calculations for a reinforced concrete section is in calculating the strains in steel. In the case of reinforced concrete sections, the strain in steel is due to bending. However, in the case of prestressed concrete sections, because the cables are pre–tensioned before the application of load, the total strain in the cable is the sum of the prestrain due to prestress $P_{service}$ and the strain due to applied bending.

16.11.1 Example of Ultimate Moment Capacity Calculation

Fig.16.18 shows the cross section of a precast pre–tensioned inverted T–beam made composite with an cast–in–situ slab. The beam is used to carry a total *factored* uniformly distributed dead load of 20 kN/m and 30 kN/m live load over a simply supported span of 24 m. Calculate the ultimate moment capacity of the section.

(a) Specification
The properties of the beam are as follows.
Total prestressing force P_s at service is 3712 kN applied at an eccentricity of 283 mm. The prestress is applied by 32 number 15.2 mm diameter 7-wire standard strands with an ultimate breaking load of 232 kN.

The 32 strands are positioned as follows:

10 cables at 60 mm from the soffit

14 cables at 110 mm from the soffit

6 cables at 160 mm from the soffit

2 cables at 1000 mm from the soffit.

The cross sectional area A_{ps} of cable

$$A_{Ps} = 138.92 \text{ mm}^2$$
$$f_{cu} \text{ for precast beam} = 50 \text{ N/mm}^2$$
$$f_{cu} \text{ for insitu slab} = 35 \text{ N/mm}^2$$

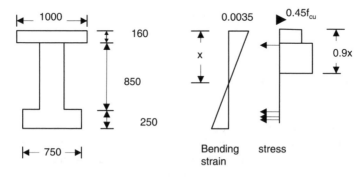

Fig.16.18 Composite beam.

(b) Stress–strain relationship

The stress–strain relationship for prestressing steel is given in Fig. 2.3 of the code BS8110. It is a trilinear curve as shown in Fig.16.19.

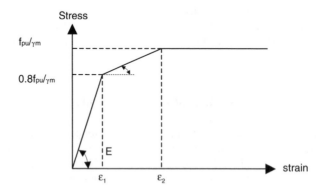

Fig.16.19 Stress–strain relationship for prestressing steel.

The ultimate tensile strength f_{pu} of steel in 15.2 mm diameter 7-wire strand 'standard' cable is

$$f_{pu} = 1670 \text{ N/mm}^2$$

$$\text{Young's modulus } E = 195 \text{ kN/mm}^2$$
$$\gamma_m = 1.05 \text{ for steel.}$$
$$0.8 \, f_{pu}/ \gamma_m = 1272 \text{ N/mm}^2$$
$$\varepsilon_1 = 1272/E = 1272/ (195 \times 10^3) = 6.523 \times 10^{-3}$$
$$f_{pu}/ \gamma_m = 1591 \text{ N/mm}^2$$
$$\varepsilon_2 = 0.005 + 1591/E$$
$$\varepsilon_2 = 0.005 + 1591/ (195 \times 10^3) = 13.154 \times 10^{-3}$$
$$E_2 = (1591 - 1272)/ (\varepsilon_2 - \varepsilon_1) = 48.11 \text{ kN/mm}^2$$

(c) Pre–strain calculation

$$f_{pe} = \text{prestress in the cables} = P_s/\text{Total area of cables}$$
$$= 3712 \times 10^3/ (32 \times 138.92) = 835 \text{ N/mm}^2$$

$$\varepsilon_{pe} = \text{prestrain in the cables} = f_{pe} /E = 835/(195 \times 10^3) = 4.28 \times 10^{-3}$$

(d) Stress and strain in cables

For a given depth x of neutral axis, at a depth a from the compression face,
Strain ε_b due to bending
$$\varepsilon_b = 0.0035 \times (a - x)/x = 3.5 \times 10^{-3} (a/x - 1.0)$$
$$\text{Total depth of composite beam} = 1100 + 160 = 1260 \text{ mm}$$
$$a = 1260 - \text{distance to the layer from soffit}$$
Table 16.4 summarises the data for all the cables.

Table 16.4 Data for cables

Layer	c = Depth from soffit, mm	No. in the layer	a =1260 – c
1	60	10	1200
2	110	14	1150
3	160	6	1100
4	1000	2	260

Total strain ε at a depth a from the compression face
$$\varepsilon = \varepsilon_{pe} + \varepsilon_b = \{4.28 + 3.5(a/x - 1)\} \, 10^{-3}$$

From Fig.16.19, for a given strain ε, the corresponding stress σ is given by the following equations.
$$0 < \varepsilon \le \varepsilon_1, \sigma = \varepsilon E$$
$$\varepsilon_1 < \varepsilon \le \varepsilon_2, \sigma = 1272 + (\varepsilon - \varepsilon_1) E_2$$
$$\varepsilon > \varepsilon_2, \sigma = 1591 \text{ N/mm}^2$$

(e) Compressive stress in concrete

Using the rectangular stress block, the depth of the stress block is 0.9x. The compressive stress in concrete is $0.45 \, f_{cu}$.

(f) Determination of neutral axis depth x

The determination of the neutral axis depth is a trial and error process. The steps involved are as follows.

(i) Assume a value for neutral axis depth, x

(ii) Calculate the bending strain in ε_b in the cables at different levels a:
$$\varepsilon_b = 3.5 \times 10^{-3} \, (a/x - 1.0)$$

(iii) Calculate the total strain $\varepsilon = \varepsilon_{pe} + \varepsilon_b$, $\varepsilon_{pe} = 4.28 \times 10^{-3}$

(iv) Calculate the stress σ in the cables

(v) Calculate the total tensile force F in each layer
$$F = \sigma \times (\text{Area of cable}= 138.92 \text{ mm}^2) \times \text{No. of cables in the layer}$$

(vi) Total tensile force $T = \Sigma F$

(vi) Calculate the total compressive force C:

(a) If $x <$ depth of slab (= 160 mm),
$$C = 0.447 \times f_{cu \, Slab} \times 1000 \times (0.9 \, x)$$

(b) If $x >$ (depth of slab = 160 mm)
$$C_{Slab} = 0.447 \times f_{cu \, Slab} \times (1000 \times 160)$$
$$C_{Beam} = 0.447 \, f_{cu \, beam} \times (0.9x - 160) \times 300$$
$$C = C_{Slab} + C_{Beam}$$

(vii) Check if $T = C$. If not go back to step (i) and repeat.

(g) Trial 1:

Assume $x = 600$ mm

Table 16.5 summarizes the calculation of forces in layers.
$$F = \sigma \times \text{No. of cables in the layer} \times 138.92 \times 10^{-3} \text{ kN}$$

Table 16.5 Force calculation in the cables for Trial 1.

Layer	$\varepsilon_b \times 10^3$	$(\varepsilon = \varepsilon_{pe} + \varepsilon_b) \times 10^3$	σ N/mm^2	F, kN
1	3.5	7.78	1333	1852
2	3.208	7.488	1318	2563
3	2.917	7.197	1304	1087
4	−1.983	2.297	448	125
				$T = \Sigma F = 5657$

$$C_{slab} = 0.447 \times 35 \times 1000 \times 160 \times 10^{-3} = 2503.2 \text{ kN}$$
$$C_{beam} = 0.447 \times 50 \times (0.9x - 160) \times 300 \times 10^{-3} = 2547.9 \text{ kN}$$

$$C = C_{slab} + C_{beam} = 2503.2 + 2547.9 = 5051.1 \text{ kN}$$
$$T - C = 605.9 \text{ kN}$$

Since T > C, increase the value of x and repeat.

(h) Trial 2:

Assume $x = 700$ mm

Table 16.6 summarises the calculation of forces in layers.

Table 16.6 Force calculation in the cables for Trial 2.

Layer	$\varepsilon_b \times 10^3$	$(\varepsilon = \varepsilon_{pe} + \varepsilon_b) \times 10^3$	σ N/mm^2	F, kN
1	2.5	6.78	1284	1784
2	2.25	6.53	1272	2474
3	2.0	6.28	1225	1021
4	−2.2	2.08	406	113
				$T = \Sigma F = 5392$

$$C_{slab} = 0.447 \times 35 \times 1000 \times 160 \times 10^{-3} = 2503.2 \text{ kN}$$
$$C_{beam} = 0.447 \times 50 \times (0.9x - 160) \times 300 \times x \, 10^{-3} = 3151.4 \text{ kN}$$
$$C = C_{slab} + C_{beam} = 2503.2 + 3151.4 = 5654.6 \text{ kN}$$
$$T - C = -262.6 \text{ kN}$$

(i) Linear interpolation

Since there are two values of neutral axis depth for which values of (T − C) are known, linear interpolation between x = 600 and x = 700 can be done to determine the value of x for which T − C = 0.

From Fig.16.20,

$$x = 600 + \frac{(700 - 600)}{\{606 - (-263)\}} 606 = 670$$

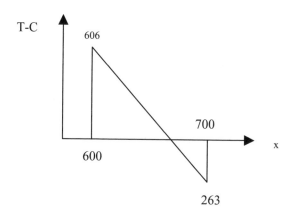

Fig.16.20 Linear interpolation.

(j) Calculation of tensile and compressive forces at $x = 670$ mm.

Table 16.7 shows the force calculation in the cables.

Table 16.7 Force calculation in the cables for interpolated value of x.

Layer	$\varepsilon_b \times 10^3$	$(\varepsilon = \varepsilon_{pe} + \varepsilon_b) \times 10^3$	σ N/mm^2	F, kN
1	2.769	7.049	1297	1802
2	2.508	6.788	1285	2499
3	2.246	6.526	1272	1060
4	−2.142	2.138	417	116
				T = ΣF = 5477

$$C_{slab} = 0.447 \times 35 \times 1000 \times 160 \times 10^{-3} = 2503.2 \text{ kN}$$
$$C_{beam} = 0.447 \times 50 \times (0.9x - 160) \times 300 \times 10^{-3} = 2970.0 \text{ kN}$$
$$C = C_{slab} + C_{beam} = 2503.2 + 2970.0 = 5473.2 \text{ kN}$$
$$T - C = 3.8 \text{ kN which is small enough to be ignored.}$$

(k) Calculation of ultimate moment M_u

Since the total T and C form a couple, the ultimate moment is calculated by taking moments about any convenient point of the tensile and compressive forces. Taking moments about the top of the cross section of the forces shown in Fig.16.21, the ultimate moment capacity is equal to 4897.53 kNm.

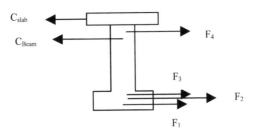

Fig.16.21 Forces in the cross section

Table 16.8 Calculation of ultimate moment capacity, M_u

	Force, kN	Lever arm = Distance from top, m	Moment, kNm
C_{slab}	−2503.2	$(160/2 = 80) \times 10^{-3}$	−200.26
C_{Beam}	−2970.0	$\{160 + (0.9x - 160)/2 = 381.5\} \times 10^{-3}$	−1133.06
F_1	1802	1200×10^{-3}	2162.40
F_2	2499	1150×10^{-3}	2873.85
F_3	1060	1100×10^{-3}	1166.00
F_4	110	260×10^{-3}	28.6
			$\Sigma = 4897.5$

The applied bending moment at ultimate load is

$$M \text{ at ULS} = 20 \times 24^2/8 \text{ due to dead load} + 30 \times 24^2/8 \text{ due to live load}$$
$$= 3600 \text{ kNm} < 4897.53 \text{ kNm}.$$

The applied moment is less than the ultimate moment capacity M_u. The beam has sufficient capacity to resist the applied bending moment at ULS.

16.11.2 Ultimate Moment Capacity Calculation using Tables in BS8110

The code BS8110 provides Table 4.4 from which the ultimate moment can be calculated for rectangular sections and for flanged sections where the stress block is inside the top flange. The Table is based on strain compatibility method applied to a *rectangular beam* with all the steel concentrated at the effective depth d. Calculations have been done for $f_{pu} = 1860$ N/mm^2 and $E = 195$ kN/mm^2. The given values are slightly conservative for lower strength tendons. In using the table, it has to be remembered that only those tendons, which are in the tensile zone, should be included. In other words the stress in the tendons, which are included, should not be vastly different from the 'average' stress.

Assuming a neutral axis depth of x, the basic equations used in the development of the tabular values are as follows.

Total compression :

$$C = 0.445 f_{cu} \times b \times \{0.9x\}$$

If the area of prestressing cables is A_{ps} and the stress in the cables is f_{pb}, total tension T is,

$$T = f_{pb} A_{ps}$$

For equilibrium C must be equal to T.

$$0.445 f_{cu} \times b \times \{0.9x\} = f_{pb} A_{ps}$$

Rewriting the above equation as

$$0.4 \{x/d\} [f_{cu} b d] = \{f_{pb}/f_{pu}\} [f_{pu} A_{ps}]$$
$$0.4 \{x/d\} = [(f_{pu} A_{ps})/ (f_{cu} b d)] \{f_{pb}/f_{pu}\}$$

For a given section and material strengths, $[(f_{pu} A_{ps})/ (f_{cu} b d)]$ is a non–dimensional parameter.

Strain in steel:

$$\varepsilon = f_{pe}/E + 0.0035(d/x - 1)$$
$$= [f_{pe}/f_{pu}] \{f_{pu} /E\} + 0.0035(d/x - 1)$$

For a given value of prestress in the cables, $[f_{pe}/f_{pu}]$ is aother non–dimensional parameter.

For prescribed values of the non–dimensional parameters, assuming values of (x/d), stress in steel can be obtained from the stress–strain curve for steel and equilibrium can be checked. The value of (x/d) which satisfies equilibrium is the correct value. Once the value of x is known, the value of f_{pb} can be calculated.

The ultimate moment is then given by

$$M_u = A_{ps} f_{pb} (d - 0.45 x)$$

16.11.2 .1 Example of ultimate moment capacity calculation using tables in BS8110

Calculate the ultimate moment capacity of the composite beam shown in Fig.16.19. Use data as given in section 16.11.1.

Assuming that the stress block depth is inside the depth of the slab, the cross section will be treated as a rectangular with a width b equal to width of slab. The beam has 32 cables each of cross sectional area equal to 138.92 mm^2. Since the cables at 1000 mm from the soffit are too far away from the tension zone, excluding these two cables, the number of cables in the tension zone is 30. This is made up of 10 cables at 60 mm from the soffit, 14 cables at 110 mm from the soffit, and 6 cables at 160 mm. The resultant effective depth is

$$d = \{10 \times (1260 - 60) + 14 \times (1260 - 110) + 6 \times (1260 - 160)\}/30 = 1157 \text{ mm}$$
$$b = \text{width of slab} = 1000 \text{ mm}$$

Area of prestressing steel A_{ps}
$$A_{ps} = 30 \times 138.92 = 4167.6 \text{ mm}^2$$
$$f_{cu} = f_{cu\ Slab} = 35 \text{ N/mm}^2$$
$$f_{pu} = 1670 \text{ N/mm}^2, f_{pe} = 835 \text{ N/mm}^2$$
$$\frac{f_{pu}}{f_{cu}} \frac{A_{ps}}{bd} = \frac{1670}{35} \frac{4167.6}{1000 \times 1157} = 0.17$$
$$\frac{f_{pu}}{f_{cu}} = \frac{835}{1670} = 0.5$$

From Table 4.4 of the code, using linear interpolation
$$\frac{f_{pb}}{0.95 f_{pu}} = 0.92 + \frac{(0.84 - 0.92)}{(0.20 - 0.15)}(0.17 - 0.15) = 0.888$$
$$\frac{x}{d} = 0.32 + \frac{(0.40 - 0.32)}{(0.20 - 0.15)}(0.17 - 0.15) = 0.352$$
$$f_{pb} = 0.888 \times (0.95 \times 1670) = 1409 \text{ N/mm}^2$$
$$x = 0.532 \times 1157 = 407 \text{ mm},$$
$$0.9x = 366 \text{ mm} > \text{depth of slab}$$

The tabular values are inapplicable for calculating the ultimate moment capacity of the section. Strain compatibility method is the valid approach.

16.12 ULTIMATE SHEAR CAPACITY OF SECTIONS CRACKED IN FLEXURE

The ultimate shear capacity depends on whether the section is cracked in flexure or not. A section is said to be cracked in flexure if the applied moments cause a tensile stress at the soffit sufficient to neutralize the compressive stress due to prestress. If a section is cracked in flexure, then the shear capacity V_{cr} due to concrete is given by the following empirical formula.

$$V_{CR} = (1 - 0.55 \frac{f_{pe}}{f_{pu}}) v_c b_v d + \frac{M_o}{M} V \geq 0.1 b_v d \sqrt{f_{cu}}$$

where

M and V are shear force and moment acting on the section

M_o = Moment required to neutralize 0.8 of the compressive stress due to prestress at the soffit.

f_{pe} = stress in prestressing cables due to P_s, prestress at SLS

f_{pu} = ultimate tensile stress in the cables

b_v = width of web

d = effective depth

v_c = permissible shear stress in concrete

$$v_c = \frac{0.79}{1.25} (\frac{100 A_s}{b_v d})^{\frac{1}{3}} (\frac{400}{d})^{\frac{1}{4}} (\frac{f_{cu}}{25})^{\frac{1}{3}}$$

16.12.1 Example of Calculation of V_{cr}

Calculate V_{cr} at the mid–span section of the beam in section 16.11.1.

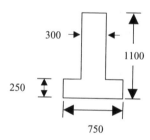

Fig 16.22 Precast beam section

(a) Calculate M_0

The prestress P_s acts on the precast pre–tensioned inverted T–beam shown in Fig.16.22.

The properties of the precast section are:

$A = 4.425 \times 10^5 \text{ mm}^2$

$y_b = 442 \text{ mm}, y_t = 658 \text{ mm}$

$I = 4.90 \times 10^{10} \text{ mm}^4$

$z_b = 111.0 \times 10^6 \text{ mm}^3,$

$z_t = 74.5 \times 10^6 \text{ mm}^3$

$P_s = 3712 \text{ kN}$, eccentricity $e = 283 \text{ mm}$.

Compressive stress σ at soffit due to prestress

$$\sigma = \frac{P_s}{A} + \frac{P_s\,e}{z_b} = \frac{3712 \times 10^3}{4.425 \times 10^5} + \frac{3712 \times 10^3 \times 283}{111.0 \times 10^6} = 8.4 + 9.5 = 17.9 \text{ N/mm}^2$$

$$0.8\,\sigma = 0.8 \times 17.9 = 14.3 \text{ N/mm}^2$$

Dead load acts on the precast beam:
$$\text{Bending moment due to dead load} = 20 \times 24^2/8 = 1440 \text{ kNm}$$
$$\text{Tensile stress at soffit due to dead load} = 1440 \times 10^6/z_b = 13.0 \text{ N/mm}^2$$

Bending stress 13.0 N/mm^2 due to dead load is insufficient to neutralize 0.8 σ. Additional moment $M_{additional}$ acting on the composite section shown in Fig.16.19 is necessary.

The properties of the composite section are:
$$A_{composite} = 6.025 \times 10^5 \text{ mm}^2$$
$$\text{Centroid: } y_b = 638 \text{ mm, } y_t = 622 \text{ mm}$$
$$\text{Second moment of area } I_{composite} = 11.33 \times 10^{10} \text{ mm}^4$$
$$\text{Section moduli: } z_{b\ Composite} = 177.6 \times 10^6 \text{ mm}^3, \ z_{t\ Composite} = 182.2 \times 10^6 \text{ mm}^3$$

The 0.8 of the compressive stress due to prestress is neutralized when
$$-0.8\sigma + \frac{M_{Dead}}{z_b} + \frac{M_{Additional}}{z_{bComposite}} = 0$$
$$-14.3 + 13.00 + M_{Additional} \times 10^6 / (177.6 \times 10^6) = 0$$
$$M_{Additional} = 230.9 \text{ kN}$$
$$M_o = M_{Dead} + M_{Additional} = 1440 + 230.9 = 1670.9 \text{ kNm}$$

(b) Calculate M and V
From the influence line for shear force at a section, it can be shown that in order to obtain maximum shear force at mid–span, the live load will occupy only one half of the span. Dead load being a stationary load will occupy the entire span. The load positions are as shown in Fig.16.23.
The total moment M at mid–span
$$= q_{dead} \times 24^2/8 + q_{live} \times 24^2/16$$
$$M = 20 \times 24^2/8 + 30 \times 24^2/16$$
$$M = 2530 \text{ kNm}$$
$$V = 0 + q_{live} \times 24/8$$
$$V = 0 + 30 \times 24/8 = 90 \text{ kN}$$

(c) Calculate v_c
In calculating v_c, only cables in the tension zone should be included in calculating the effective depth d and the area of cross section A_s. Excluding the two cables at 1000 mm from the bottom, using the data from section 16.11.2.1,
$$d = 1157 \text{ mm}$$
$$A_s = 4167.6 \text{ mm}^2$$

$$b_v = 300 \text{ mm}$$
$$100\, A_s/(b_v d) = 1.2 < 3.0$$
$$400/d < 1.0, \text{ take as } 1.0$$
$$f_{cu} = 50 > 40 \text{ N/mm}^2, \text{ take as } 40 \text{N/mm}^2$$

$$v_c = \frac{0.79}{1.25}(1.2)^{\frac{1}{3}}(1.0)^{\frac{1}{4}}(\frac{40}{25})^{\frac{1}{3}} = 0.79 \text{ N/mm}^2$$

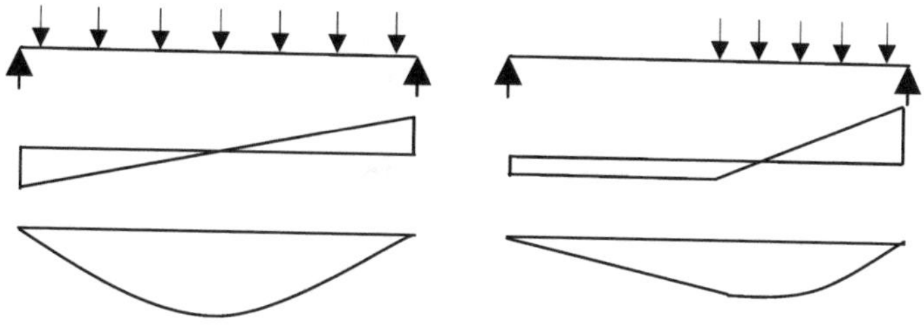

Fig.16.23 Dead and live loads to give maximum shear force at mid–span.

(d) Calculate f_{pe}/f_{pu}

$$f_{pu} = 1670 \text{ N/mm}^2$$
$$f_{pe} = \text{prestress in the cables}$$
$$f_{pe} = P_s/\text{Total area of } all \text{ cables}$$
$$f_{pe} = 3712 \times 10^3/(32 \times 138.92) = 835 \text{ N/mm}^2$$
$$f_{pe}/f_{pu} = 835/1670 = 0.50$$

(e) Calculate V_{cr}

$$V_{CR} = (1 - 0.55\frac{f_{pe}}{f_{pu}})v_c b_v d + \frac{M_o}{M}V \ge 0.1 b_v d\sqrt{f_{cu}}$$

$$V_{CR} = (1 - 0.55 \times 0.5)0.79 \times 300 \times 1157 \times 10^{-3}$$
$$+ \frac{1671}{2520}90 \ge 0.1 \times 300 \times 1157\sqrt{50} \times 10^{-3}$$

$$V_{CR} = 258.5 > 245.4 \text{ kN}$$
$$V_{CR} = 258.5 \text{ kN}$$

Applied shear force is V = 90 kN and hence the shear capacity of concrete alone is sufficient. There is no need for any shear links but minimum links will always be provided as in the case of reinforced concrete beam.

16.13 ULTIMATE SHEAR CAPACITY V_{CO} OF SECTIONS UNCRACKED IN FLEXURE

Sections are said to be uncracked in flexure if the applied moment M is less than the moment M_0 required to neutralize 0.8 of the compressive stress at the soffit due to prestress. In sections which are uncracked in flexure, it is necessary to limit the maximum principal tensile stress in the web to a value f_t, where $f_t = 0.24\sqrt{f_{cu}}$.

In the case of a rectangular section b × h, the maximum shear stress τ due to a shear force V occurs at the neutral axis and is given by

$$\tau = 1.5\frac{V}{bh}$$

If f_{cp} is the compressive stress due to prestress at the neutral axis, then the state of stress at the neutral axis is as shown in Fig. 16.24.

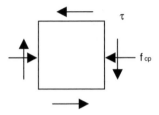

Fig.16.24 Normal and shear stresses at the neutral axis.

The maximum principle tensile stress for the biaxial state of stress shown in Fig.16.24 is given by

$$f_t = -\frac{f_{cp}}{2} + \sqrt{\{(-\frac{f_{cp}}{2})^2 + \tau^2\}}$$

From which the permissible value of τ can be calculated as follows.

$$f_t + \frac{f_{cp}}{2} = \sqrt{\{(-\frac{f_{cp}}{2})^2 + \tau^2\}}$$

$$(f_t + \frac{f_{cp}}{2})^2 = (-\frac{f_{cp}}{2})^2 + \tau^2$$

$$f_t^2 + f_t f_{cp} = \tau^2$$

$$\tau = \sqrt{\{f_t^2 + f_t f_{cp}\}} = 1.5\frac{V}{bh}$$

$$V = 0.67\,bh\sqrt{\{f_t^2 + f_t f_{cp}\}}$$

The code BS8110 gives the following formula for calculating V_{CO}, the shear capacity due to concrete alone in sections uncracked in flexure, where the compressive stress is taken as 0.8 f_{cp}.

$$V_{CO} = 0.67\,bh\sqrt{\{f_t^2 + 0.8 f_t f_{cp}\}}$$

Although the formula is derived for the case of a rectangular section, it can be applied for flanged sections also with little loss of accuracy because the shear force is mainly resisted by the web.

16.13.1 Example of Calculating Ultimate Shear Capacity V_{CO}

Calculate V_{co} at the support section of the beam in section 16.11.1. The number of cables at the support has been reduced by 50% by debonding all cables at 60 mm from soffit and six cables at 110 mm from the soffit.

P_s = Prestress force from 16 cables only

$$P_s = 0.5 \times 3712 = 1856 \text{ kN}$$
$$e = y_b \text{ of precast} - \{8 \times 110 + 6 \times 160 + 2 \times 1000)/16$$
$$e = 442 - 240 = 202 \text{ mm}$$

f_{cp} = compressive stress (taken as positive value) due to prestress at centroidal axis of the composite beam.

$$f_{cp} = -[-\frac{P_s}{A_{Precast\,beam}} + \frac{P_s \times e}{I_{Precast\,beam}}(y_{bComposite\,beam} - y_{bPrecast\,beam})]$$

$$f_{cp} = -[-\frac{1856 \times 10^3}{4.425 \times 10^5} + \frac{1856 \times 10^3 \times 202}{4.90 \times 10^{10}}(638 - 442)]$$

$$f_{cp} = 4.2 - 1.5 = 2.7 \text{N/mm}^2$$
$$f_t = 0.24\sqrt{f_{cu}} = 0.24\sqrt{50} = 1.7 \text{ N/mm}^2$$
$$b = b_v = 300 \text{ mm}, h = 1260 \text{ mm}$$
$$V_{CO} = 0.67 \times 300 \times 1260 \times \sqrt{\{1.7^2 + 0.8 \times 1.7 \times 2.7\}} \times 10^{-3} = 648.8 \text{kN}$$

Applied shear force V

$$V = 20 \times 24/2 \text{ from dead load} + 30 \times 24/2 \text{ from live load}$$
$$V = 600 \text{ kN}$$

Since $V_{CO} > V$, theoretically no shear reinforcement is required but but minimum links will always be provided as in the case of reinforced concrete beam

16.13.1.1 Calculation of V_{co} from First Principles

The formula given in the code does not distinguish between the loads carried by the precast beam section and the loads carried by the composite section. If this fact is taken into account, the calculation of V_{CO} can be refined as follows.

1. Permissible shear stress
The permissible shear stress τ at the centroid of the composite beam from Section 16.13.1 is

$$\tau = \sqrt{\{1.7^2 + 0.8 \times 1.7 \times 2.7\}} = 2.6 \text{ N/mm}^2$$

2. Shear stress due to dead load
The shear stress τ_{dead} due to dead load is given by

$$\tau_{dead} = \frac{V_{dead} \times \int ydA}{I_{Precast\,beam} \times b_v}$$

where

$$V \text{ dead} = 20 \times 24/2 = 240 \text{ kN}$$
$$b_v = 300 \text{ mm}$$

$\int y \, dA$ = First moment of area of the part of the section above the level considered for calculating the shear stress about the centroidal axis of the beam.

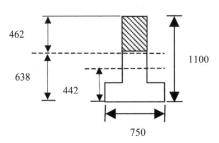

FIG.16.25 Calculation of first moment of area of precast section.

From Fig.16.25, for the hatched area
$$\int y \, dA = 462 \times 300 \times \{638 - 442 + 462/2\} = 59.2 \times 10^6 \text{ mm}^3$$
$$\tau_{dead} = \frac{240 \times 10^3 \times 59.2 \times 10^6}{4.90 \times 10^{10} \times 300} = 1.0 \text{ N/mm}^2$$

3. Shear stress $\tau_{Additional}$ from live load
The permissible value of shear stress carried by the composite section without exceeding the total permissible shear stress is $\tau_{Additional}$

$$\tau = \tau_{dead} + \tau_{Additional}$$
$$2.6 = 1.0 + \tau_{Additional}$$
$$\tau_{Additional} = 1.6 \text{ N/mm}^2$$

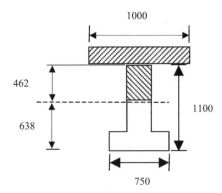

Fig.16.26 Calculation of first moment of area of composite section.

The $\tau_{Additional}$ arises from any live loads acting on the composite beam.

$$\tau_{Additional} = \frac{V_{Additional} \times \int y\,dA}{I_{Composite\ beam} \times b_v}$$

From Fig.16.26, for the hatched area

$$\int y\ dA = 462 \times 300 \times \{462/2\} + 1000 \times 160 \times (462 + 160/2)$$
$$\int y\ dA = 118.7 \times 10^6\ mm^3$$
$$1.6 = \frac{V_{Additional} \times 10^3 \times 118.7 \times 10^6}{11.33 \times 10^{10} \times 300}$$
$$V_{Additional} = 458.2\ kN$$
$$V_{CO} = V_{dead} + V_{Additional} = 240 + 458.2 = 698.2\ kN$$

The approximate value of V_{CO} from code formula = 648.8 kN. In this example, the code formula under estimates V_{CO} by about 7%.

16.14 DESIGN OF SHEAR REINFORCEMENT

Design of shear links is carried out in a manner similar to the design in reinforced concrete (See Chapter 5, section 5.1.3).

1. Check if section is adequate

$$\frac{V}{b_v d} \le\ lesser\ of\,(0.8\sqrt{f_{cu}}\ or\ 5)\,N/mm^2$$

2. $0.5\ V_c < V \le (V_c + 0.4\ b_v\ d)$,
Provide minimum shear reinforcement A_{sv} at a spacing of s_v.

$$A_{sv} \ge \frac{0.4\,b_v\,s_v}{0.95\,f_{yv}}$$

3. $V > (V_c + 0.4\ b_v\ d)$
Provide shear reinforcement A_{sv} at a spacing of s_v.

$$A_{sv} \ge (V - V_c)\frac{s_v}{d}\frac{1}{0.95\,f_{yv}}$$

where

$$s_v = spacing\ of\ links$$
$$A_{sv} = Area\ of\ all\ legs\ of\ a\ link\ in\ a\ cross\ section$$
$$f_{yv} = yield\ stress\ of\ link\ steel$$
$$d = effective\ depth\ to\ steel\ in\ the\ tension\ zone$$
$$b_v = width\ of\ the\ web.$$
$$V_c = V_{CO}\ at\ a\ section\ uncracked\ in\ flexure$$
$$V_c = lesser\ of\ (V_{CO}\ and\ V_{CR})\ at\ a\ section\ cracked\ in\ flexure.$$
Maximum spacing s_v of links is limited to 0.75 d.

16.14.1 Example of Shear Link Design

Design the shear reinforcement for the beam in section 16.13.1 using 8 mm diameter two legged high yield steel links.
$$V = 600 \text{ kN}$$

$$V_{CO} = 648.8 \text{ kN (using code formula)}$$
$$b_v = 300 \text{ mm}$$
Ignoring the steel at 1000 mm from the soffit,
$$d = 1129 \text{ mm.}$$

Check adequacy of section:
$$V/ (b_v d) = 600 \times 10^3/ (300 \times 1129)$$
$$= 1.8 \text{ N/mm}^2 < \text{lesser of } \{(0.8\sqrt{50} = 5.7 \text{ and } 5 \text{ N/mm}^2\}$$
Section size is adequate and shear links can be provided.

Check if minimum or design links are required.

$$0.4 \, b_v \, d = 0.4 \times 300 \times 1129 \times 10^{-3} = 135.5 \text{ kN}$$
$$V_{co} + 0.4 \, b_v \, d = 648.8 + 135.5 = 784.3 \text{ kN}$$
$$(0.5 \, V_c = 300) < (V = 600) \le \{(V_c + 0.4 \, b_v \, d) = 784.3)\}$$
Only minimum steel is required to resist shear.

Minimum links:
$$f_{yv} = 460 \text{ N/mm}^2, A_{sv} = 2 \times (\pi/4) \times 8^2 = 100.5 \text{ mm}^2.$$
$$100.5 \ge \frac{0.4 \times 300 \times s_v}{0.95 \times 460}$$

giving $s_v = 366$ mm. $0.75 \, d = 847$ mm. Spacing of links at say 350 mm can be used.

16.15 HORIZONTAL SHEAR

In the case of composite beams, it is necessary to ensure that the horizontal shear stress between the precast beam and the cast–in–situ slab as shown in Fig.16.27 can be safely resisted. If there there were to be a shear failure at the slab–beam junction, then composite beam action will be destroyed. BS8110 gives guidance on horizontal shear in clause 5.4.7.

   If the compressive force in slab due to ultimate moment at mid–span is C_{slab}, the average shear stress $\tau_{average}$ is given by
$$\tau_{average} = C_{slab}/(0.5 \times \text{Span} \times \text{Contact width w})$$

In continuous beams, instead of half span, the length between the maximum positive or negative moment and the point of zero moment should be used.

The average design shear stress should be distributed in proportion to the vertical design shear force diagram. The design shear stress v_h should be less than the appropriate value in Table 5.5 of the code.

16.15.1 Shear Reinforcement to Resist Horizontal Shear Stress

In clauses 5.4.7.3 and 5.4.7.4, code gives guidance on the required shear reinforcement in the form of shear links projecting from the precast beam into the cast–in–situ slab.

(a) Nominal links

If $v_h \leq$ permitted value from Table 5.5, links should be provided such that

$$A_h \geq \frac{0.15}{100} \times w \times s_h$$

(b) Design links

If $v_h >$ permitted value from Table 5.5, links should be provided such that

$$A_h \geq s_h \times w \times \frac{v_h}{0.95 f_{yv}}$$

where A_h = area of two–legged link, s_h = spacing of links, w = contact width. s_h in T–beam ribs with cast–in–situ slab acting compositely with the precast beam should not exceed four times the minimum thickness of slab nor 600 mm which ever is *greater.*

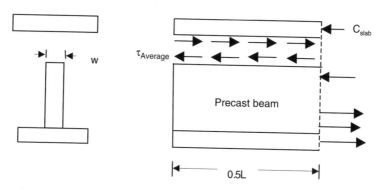

Fig.16.27 Horizontal shear stress.

16.15.2 Example of Design for Horizontal Shear

Design the shear reinforcement for the beam in section 16.11.1. At ULS, the neutral axis depth x = 670 mm. The cast–in–situ slab is 1000 × 160 mm, $f_{cu\ slab} = 35$ N/mm^2.
Since the neutral axis depth x > depth of slab,
$$C_{slab} = 0.447 \times 35 \times 1000 \times 160 \times 10^{-3} = 2503.2 \text{ kN}$$

$$\text{Width of contact } w = 300 \text{ mm}$$
$$0.5 \text{ span} = 24/2 = 12 \text{ m} \qquad 1.4 \text{ N/mm}^2$$
$$\tau_{average} = C_{slab}/(0.5 \times \text{Span} \times \text{Contact width } w)$$
$$\tau_{average} = 2503.2 \times 10^3/(12 \times 10^3 \times 300) = 0.7 \text{ N/mm}^2$$

At ULS to cause maximum horizontal shear both dead and live loads cover the entire span. Shear force at support is 600 kN (Section 16.13.1) and zero at mid–span.

$$\frac{\tau_{Support} + 0}{2} = \tau_{Average}$$

$$0.5 \times \tau_{Support} = \tau_{Average} = 0.7$$

$$v_h = \tau_{Support} = 1.4 \text{ N/mm}^2$$

From Table 5.5 of code, minimum permissible interface shear stress for grade 30 concrete is 1.8 N/mm^2. Therefore only nominal links are required. Using 8 mm diameter two legged links,

$$(A_h = 100.5) \geq \frac{0.15}{100} \times (w = 300) \times s_h$$

s_h = 233 mm < greater of (4 × 160 or 600 mm)

Note that this spacing is *smaller* than 366 mm which was required for resisting vertical shear.

16.16 LOSS OF PRESTRESS IN PRE–TENSIONED BEAMS

In sections 16.2.1 and 16.3.1 it was stated that that although at the time of stressing the cables, the total force is P_{jack}, due to losses that occur during transfer of prestress to concrete and also due to long term deformation of steel and concrete, there is considerable reduction in prestress at the long term SLS stage.

16.16.1 Loss at Transfer

The loss at transfer occurs because of the fact that when the force is transferred to concrete, it contracts. Because of the full bond between steel and concrete, steel also suffers the same contraction. The loss in prestress can be calculated by using a simple model where the prestress and eccentricity are constant over the whole length and all the prestressing steel A_{ps} can be assumed to be concentrated at an eccentricity e. If P_t is the force in the cables, then at the centroid of steel the stress σ_c in concrete is given by

$$\sigma_c = \frac{P_t}{A} + \frac{P_t e}{I} e$$

where A and I are respectively the area of cross section and second moment area of pre–tensioned beam.

The strain ε_c in concrete is given by

$$\varepsilon_c = \sigma_c / E_c$$

where E_c = Young's modulus for concrete considering immediate contraction. Because of full bond, the strain ε_s in steel is same as strain in concrete.

$$\varepsilon_s = \varepsilon_c = \sigma_c / E_c$$

The stress σ_s in steel corresponding to ε_s is

$$\sigma_s = E_s \varepsilon_s = E_s \sigma_c / E_c$$

The loss of prestress is given by

$$\text{Loss} = A_{ps} \sigma_s = A_{ps} E_s \sigma_c / E_c$$
$$P_t = P_{jack} - \text{Loss}$$

16.16.1.1 Example on calculation of loss at transfer

Calculate the loss at transfer for the pre-tensioned beam in section 16.11.1. The properties of the precast section are:

Area $A = 4.425 \times 10^5$ mm^2
Second moment of area $I = 4.90 \times 10^{10}$ mm^4
Prestressing force used to stress 32 cables each of area 138.92 mm^2 is
$P_{Jack} = 4700$ kN
eccentricity $e = 283$ mm.
$E_s = 195$ kN/mm^2

The cables are normally released after one or two days or earlier using steam curing to speed up the gain in strength. Assuming that at the time of transfer of prestress, f_{cu} for concrete is approximately 30 N/mm^2, from Table 7.2, BS8110, Part 2,

$$E_c = (20 \text{ to } 32 \text{ kN/mm}^2)$$
$$\text{Let } E_c = 30 \text{ kN/mm}^2.$$

Assuming that P_t is in kN, compressive stress σ_c due to prestress at centroid of steel

$$\sigma_c = \frac{P_t \times 10^3}{4.425 \times 10^5} + \frac{P_t \times 10^3 \times 283}{4.90 \times 10^{10}} \times 283 = 3.894 \times 10^{-3} \times P_t$$

$$\text{Loss} = A_{ps} E_s \sigma_c / E_c$$
$$\text{Loss} = (32 \times 138.92) \times (195/30) \times 3.894 \times 10^{-3} \times P_t \times 10^{-3} \text{ kN}$$
$$\text{Loss} = 0.113 \, P_t \text{ kN}$$
$$P_t = P_{jack} - \text{Loss} = 4700 - 0.113 \, P_t$$
$$P_t = 4223 \text{ kN, } P_t/P_{jack} = 0.90$$

There is 10 % loss of prestress at the time of transfer.

16.16.2 Long Term Loss of Prestress

After the force has been transferred, concrete continues to contract due to creep. In addition concrete also suffers shrinkage due to loss of moisture. Because the steel is under stress there is reduction in stress due to the relaxation effect. These losses can be calculated using the simple model used in section 16.15.1

Reinforced concrete

(i) Loss of prestress due to creep
Code gives guidance on calculating the loss due to creep in clause 4.8.5.2. The long term Young's modulus of concrete is obtained by dividing the short term E_c at transfer by creep coefficient which varies between 1.8 for transfer at 3 days and 1.4 for transfer at 28 days. The stress in concrete should be taken as the value at transfer.

Assuming that $P_t = 4223$ kN, compressive stress σ_c due to prestress at centroid of steel

$$\sigma_c = \frac{4223 \times 10^3}{4.425 \times 10^5} + \frac{4223 \times 10^3 \times 283}{4.90 \times 10^{10}} \times 283 = 16.4\,\text{N/mm}^2$$

Using Creep coefficient = 1.8, E_c including creep deformation
$$E_c = 30/1.8 = 16.7\,\text{kN/mm}^2$$
$$\text{Loss due to creep} = A_{ps}\,E_s\,\sigma_c\,/E_c$$
$$\text{Loss due to creep} = (32 \times 138.92) \times 195/16.7 \times 16.4 \times 10^{-3} = 851.3\,\text{kN}$$

(ii) Loss of prestress due to shrinkage of concrete
In clause 4.8.4 of BS8110, shrinkage strain ε_{sh} of 100×10^{-6} for UK outdoor exposure and 300×10^{-6} for indoor exposure are suggested.
$$\text{Loss of prestress} = A_{ps}\,E_s\,\varepsilon_{sh}$$
$$\text{Using } \varepsilon_{sh} = 100 \times 10^{-6},$$
$$\text{Loss due to shrinkage} = (32 \times 138.92) \times 195 \times 100 \times 10^{-6}$$
$$\text{Loss due to shrinkage} = 86.7\,\text{kN}$$

(iii) Loss of prestress due to relaxation of steel
In clause 4.8.2 of BS8110 gives guidance on calculation of prestress loss due to relaxation of steel. Tests are normally conducted for 1000 hour duration and the long term relaxation loss is a multiple of 1000 hour loss by relaxation by factors given in Table 4.6 of the code.
For 7-wire strand, if the initial stress is 60%, 70% and 80% of f_{pu}, then the corresponding 1000 hour loss is approximately 1%, 2.5% and 4.5% respectively.
Taking 2.5% as the 1000 hour loss and a relaxation factor equal to 1.5, then the total loss due to relaxation is
$$\text{Loss due to relaxation} = 2.5 \times 1.5 = 3.75\% \text{ of } P_t.$$
$$\text{Taking } P_t = 4141\,\text{kN},$$
$$\text{Loss due to relaxation} = (3.75/100) \times 4141 = 155.3\,\text{kN}$$

(iv) Total long term loss
Total long term loss = Loss due to (Creep + Shrinkage + Relaxation)
Total long term loss = 851.3 + 86.7 + 155.3 = 1093.3 kN
$P_s = P_t - \text{long term loss} = 4223 - 1093.3 = 3129.7$ kN
% Loss = $(1 - P_s/P_{jack}) \times 100 = 33\%$

As can be seen the greatest part of the long term loss is due to creep. Creep can also substantially increase long term deformation leading to unacceptable deflection. It is very important to make realistic estimation of the effects of creep.

16.17 LOSS OF PRESTRESS IN POST–TENSIONED BEAMS

The difference between the losses in pre–tensioned and post–tensioned occurs only due to losses during jacking and transfer. The long term loss calculations are identical.

(i) Transfer loss

In post–tensioned beams, because concrete contracts while the cables are being stressed, any loss due to elastic contraction of concrete can be compensated to a certain extent. In clause 4.8.3.3, the code suggests that this loss is approximately 50% of that in a corresponding pre–tensioned beam. However once the jacking is done and the wedges are driven into the anchors, a certain amount of slip takes place before the wedges bite in. This is known as 'Draw–in' during anchorage. As the amount of slip is same whatever the length of the member, the loss of prestress due to 'draw–in' is particularly important in short members.

$$\text{Loss} = A_{ps} \times (\text{Slip/Length of member}) \times E_s$$

(ii) Loss due to friction between the cable and the duct and curvature of the tendons

Friction between the duct and the cable reduces the force in the cable away from the jacking end. Loss also occurs because of the curvature of the cables. The loss may be minimized by jacking from both ends of the beam. If P_o is the prestress at the jacking end, then prestress P_x at a distance x from the anchorage is given by the equation

$$P_x = P_0 \, e^{-(k + \mu \frac{1}{r_{ps}})x}$$

where
k = profile coefficient. k is dependent on many factors such as the type of duct, how well it is supported while concrete is being cast, degree of vibration used. In clause 4.9.3.3, the code suggests $k = 33 \times 10^{-4}$ per meter length in normal use.
μ = friction coefficient. In clause 4.9.4.3, the code suggests $\mu = 0.20$ to 0.30 for strands running in steel ducts, the value to be used depending on the surface characteristics of the duct and the strand.
r_{ps} = radius of curvature. In the case of symmetric parabolic profile with the dip at mid–span of Δ and span of L, then $r_{ps} = L^2 / (8\, \Delta)$
As an example, in a simply supported post–tensioned beam of span 15 m assuming

$$k = 33 \times 10^{-4} \,/\text{m}$$
$$\mu = 0.30$$
$$\Delta = 200\text{mm}$$
$$r_{ps} = L^2/(8\, \Delta) = 15^2/(8 \times 0.2) = 140.6 \text{ m}$$
$$k + \mu/\, r_{ps} = 5.43 \times 10^{-3}/\text{m}$$

Loss from prestress from jacking end to mid span is for x = 15/2 = 7.5 m

$$(k + \frac{\mu}{r_{ps}})x = (33 \times 10^{-4} + \frac{0.3}{140.6}) \times 7.5 = 0.041$$

$$\frac{P_x}{P_0} = e^{-(k+\mu\frac{1}{r_{ps}})x} = e^{-0.041} = 0.96$$

There is 4% loss of prestress from anchorage to mid–span.

16.18 DESIGN OF END–BLOCK IN POST–TENSIONED BEAMS

In post–tensioned members, after stressing the cables which are inside ducts fixed to the reinforcement cage, they are anchored at the ends using proprietary anchorages. After anchoring, the ducts are grouted to prevent corrosion of the cables and also to bond the cables to concrete.

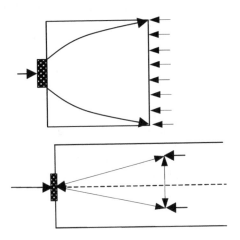

Fig.16.28 Bursting forces in an End block.

When the cables are anchored, a very high force is transferred to the concrete over a small area. As shown in Fig.16.28, if an axial force representing the force applied to the anchor acts at the end face, the load gradually diffuses into concrete along curved paths and after a certain distance from end, the stresses normal to the cross section become uniform. A simplified force system can be visualized by replacing the curved stress path by straight inclined struts. In order to maintain equilibrium, a vertical tie is needed. This represents the bursting or splitting force which can cause tensile failure of the end block.

The bursting stresses are local to the anchorage and generally there is little of an interference effect from neighbouring anchorages. The code uses a simplified approach to the determination of the bursting force. Considering an end block of square cross section with a load on the anchor of P_0 over a square area as shown in Fig.16.29, the ratio of bursting force F_{bst} to P_0 is given in Table 4.7 of the code. The smaller the ratio of y_{po}/y_0, the larger will be the ratio of F_{bst}/P_0.

Since in practice the ratio y_{po}/y_0 will not be the same in both vertical and horizontal directions, the smaller of the two ratios will be used for determining the bursting force to be restrained.

The total bursting force F_{bst} is resisted by closed links extending from $0.2y_0$ to $2y_0$ from the loaded face. The stress in the reinforcement must be limited to 200 N/mm^2.

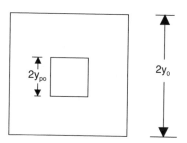

Fig.16.29 Idealized End block used for calculation.

It is not unusual for multiple anchors to occur as shown in Fig.16.30. Normally the anchors are so spaced that there is very little interference between the end blocks of individual anchors. Therefore in case of multiple anchors, each anchor will be treated on its own with the associated idealized square end block. For the end block shown in Fig.16.29, in the vertical direction the value of y_0 is calculated as follows.

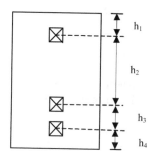

Fig.16.30 End block with multiple anchors.

Top anchor: y_0 = smaller of (h_1 and $0.5\ h_2$)
Middle anchor: y_0 = smaller of ($0.5h_2$ and $0.5\ h_3$)
Bottom anchor: y_0 = smaller of (h_4 and $0.5\ h_3$)

16.18.1 Example of End–Block Design

Fig.16.31 shows the cross section at the support of a post–tensioned T–beam with three anchors. Each tendon is prestresssed with a force at jacking of 1200 kN. The anchorages are all 150 mm square. Design the necessary reinforcement to prevent bursting.

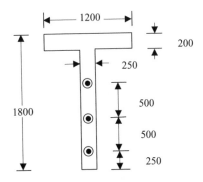

Fig.16.31 End–block with multiple anchors

1. Horizontal direction:
In the horizontal direction for all the three anchors:
$$y_o = 250/2 = 125 \text{ mm},$$
$$y_{Po} = 150/2 = 75 \text{ mm}$$
$$y_{Po}/ y_o = 0.6$$

2. Vertical direction:
$$y_{Po} = 150/2 = 75 \text{ mm}$$
Bottom anchor: y_o = lesser of (250 and 0.5 × 500) = 250 mm
Middle anchor: y_o = lesser of (0.5 × 500 and 0.5 × 500) = 250 mm
Top anchor: y_o = lesser of (0.5 × 500 and 550) = 250 mm
Therefore for all the three anchors:
$$y_o = 250 \text{ mm}$$
$$y_{Po} = 150/2 = 75 \text{ mm}$$
$$y_{Po}/ y_o = 0.3$$

3. Bursting force:
The smaller of two ratios viz. 0.6 and 0.3 governs design. From Table 4.7 of the code,
$$\text{for } y_{Po}/ y_o = 0.3, \ F_{bst}/P_0 = 0.23$$
$$P_0 = \text{Force at anchor} = 1200 \text{ kN}$$
$$F_{bst} = 0.23 \times P_0 = 0.23 \times 1200 = 276 \text{ kN}$$

4. Reinforcement to resist bursting force:

$$F_{bst} = A_{sv} \times 200 \times N$$

where A_{sv} = Area of 2-legs of a link, N = Number of links required.
Using 8 mm links, $A_{sv} = 100.5$ mm^2.

$$276 \times 10^3 = 100.5 \times 200 \times N, N = 14 \text{ links.}$$
$$\text{Spacing of links} = (2y_0 - 0.2y_0)/(N-1)$$
$$= (2 \times 250 - 0.2 \times 250)/(14-1) = 34 \text{ mm}$$

16.19 REFERENCES

Nawy, Edward G. 2000, *Prestressed Concrete*, 3rd Edition, (Prentice Hall).
Nicholson, B.A. 1997, *Simple bridge design using prestressed beams*, (Prestressed Concrete Association).
Hurst, M.K. 1998, *Prestressed Concrete Design*, 2nd Edition (E & F.N. Spon), (uses Euro code EC2).
Allen, A.H.1992, *An Introduction to Prestressed concrete*, (British Cement Association).

CHAPTER 17

DESIGN OF STRUCTURES RETAINING AQUEOUS LIQUIDS

17.1 INTRODUCTION

Structures such as tanks for retaining water or effluents in sewage treatment works are designed using the code

BS 8007: 1987: *Code of practice for Design of concrete structures for retaining aqueous liquids.*

The code (in this chapter this refers to BS8007 rather than to BS8110) generally adopts the relevant clauses of code BS 8110: 1997 with additional clauses as required. The following is a brief summary of the relevant clauses. The reader should always refer to the complete texts in the codes.

The design is normally carried out according to the limit state principles. However, unlike normal reinforced concrete structures, design is often governed by the serviceability limit state considerations of limiting the crack width rather than by ultimate limit state considerations.

17.1.1 Load Factors

Although the load factors for various load combinations at ultimate limit state are as given by Table 2.1 of BS 8110, in the case of tanks located below ground, the possibility of floatation of the tank when empty due to ground water pressure should be considered. The uplift is normally resisted by the dead weight of the structure. The required factor of against floatation during construction and in service should not be less than 1.1.

17.1.2 Crack Width

At serviceability limit state in a reinforced concrete structure, the maximum crack width due to direct tension and flexure or restrained temperature and moisture effects should be limited to 0.2 mm in case of severe or very severe exposure (see Table 3.2 of BS 8110) or to 0.1 mm when aesthetic appearance is critical.

In the case of prestressed concrete structure, the crack width is governed by the permissible tensile stress according to the Class to which it is designed (see Chapter 16, section 16.5.1)

17.1.3 Span/Effective Depth Ratios

Limitations on the permissible deflection in cantilever walls is satisfied by ensuring that span/effective depth ratio is limited to 7 as given in Table 3.9 of BS 8110. However because of the fact that the loading on the walls is hydrostatic rather than uniform and also that the thickness of the wall at top is smaller than at the base, the following reduction is suggested. Linear interpolation is permissible between quoted figures.

Table 17.1 Reduction factor for tapered cantilever

d_{top}/d_{base}	Reduction factor	L/d
1.0	1.0	7
0.6	1.0	7
0.3	0.78	5.46

17.1.4 Cover

Clause 2.7.6 specifies that nominal cover should not be less than 40 mm. Care has to be taken to take account of the fact that especially in sections less than 300 mm thick, increasing the cover will lead to increased crack width.

17.1.5 Mix Proportions

Clause 6.3 specifies that the 28 day characteristic cube strength should not be less than 35 N/mm^2 and classed as grade C35A as this is not in accordance with BS 8110. Additional advice is given on the maximum cement content which for ordinary Portland cement should not exceed 400 kgs/m^3.

17.1.6 Minimum Reinforcement

The minimum reinforcement is 0.64% for 250 grade steel and 0.35% for 460 grade steel. The total reinforcement is provided as follows:
  (a) Walls and suspended slabs: Total depth h ≤ 500 mm, the required reinforcement is calculated for the whole area of concrete and on each face half the required reinforcement is provided. If h > 500 mm, the required reinforcement is calculated for the outer 250 mm depth of concrete and on each face half the required reinforcement is provided.

(b) Ground slabs:

(i) h < 300 mm: Minimum reinforcement calculated on the basis of top half of the slab only. Provide this area of reinforcement in the top half of the slab. There is no reinforcement in the bottom part of the slab in contact with the ground.

(ii) 300 < h ≤ 500 mm: Provide reinforcement for the upper half of slab as in (i) above. In addition calculate the reinforcement for the 100 mm depth of the slab in contact with ground and provide the same.

(iii) h > 500 mm: Calculate and provide reinforcement as for (ii) above, except that the depth of the upper half is limited to 250 mm only.

17.2 BENDING ANALYSIS FOR SERVICEABILITY LIMIT STATE

Consider a rectangular beam of width b and effective depth d shown in Fig.17.1.

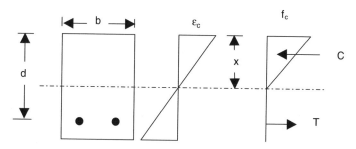

Fig 17.1 Elastic strain and stress distribution.

Let the area of tension steel be A_s. Under serviceability limit loads, concrete and steel are assumed to remain elastic. Assuming that plane sections remain plane, and ignoring the contribution from the concrete in the tensile zone, the stress distribution in the compression zone is linear. The maximum fibre strain and stress are respectively ε_c and f_c.

$$f_c = E_c \, \varepsilon_c$$

where E_c = Young's modulus for concrete.

Since the average stress in the compressive zone is 0.5 f_c, the total compressive force C is

$$C = bx\frac{f_c}{2}$$

The strain in tension steel is ε_s, where

$$\varepsilon_s = \varepsilon_c \frac{d-x}{x}$$

The stress in steel is

$$f_s = E_s \, \varepsilon_s,$$

where E_s = Young's modulus for steel

$$f_s = E_s \varepsilon_s = \frac{E_s}{E_c} E_c \varepsilon_c \frac{d-x}{x} = \alpha_e f_c \frac{d-x}{x}$$

where $\alpha_e = E_s/E_c$ is the modular ratio.
The total tensile force T is

$$T = A_s f_s = A_s \alpha_e f_c \frac{d-x}{x}$$

Equating C = T, for equilibrium

$$bx\frac{f_c}{2} = A_s \alpha_e f_c \frac{d-x}{x}$$

Dividing throughout by f_c and simplifying

$$\frac{1}{2} bx^2 = \alpha_e A_s (d-x)$$

The lever arm z is

$$z = d - x/3$$

For a given moment M, the stresses in steel and concrete can be determined as follows.

$$M = T z = A_s f_s z = C z = 0.5 f_c bx z$$

$$f_s = \frac{M}{A_s z}, \quad f_c = \frac{2M}{b x z}$$

17.2.1 Example of Stress Calculation At SLS

Fig 17.2 shows a cantilever wall which is part of a water tank. The tank retains water to a depth of 3.5 m. The base is 400 mm thick overall and is reinforced with 12 mm diameter bars at 100 mm c/c. Calculate the stresses in concrete and steel at serviceability limit state. Also check if the moment of resistance is sufficient at ultimate limit state. Assume $f_{cu} = 35$ N/mm^2 , $f_y = 460$ N/mm^2 and modular ratio $\alpha_e = 15$.

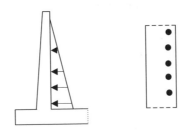

Fig.17.2 A cantilever wall.

Consider 1 m length of wall. The steel area $A_s = 1131$ mm^2/m, b = 1000 mm
h = 400 mm. Assuming cover = 40 mm,

$$d = 400 - 40 - 12/2 = 354 \text{ mm}$$

(i) Calculate the neutral axis depth:

$$\frac{1}{2}bx^2 = \alpha_e A_s (d-x)$$

$$\frac{1}{2}1000x^2 = 15 \times 1131 \times (354 - x)$$

Simplifying:

$$x^2 + 33.93x - 12011.2 = 0, \text{ giving } x = 94 \text{ mm}$$
$$z = d - x/3 = 354 - 94/3 = 323 \text{ mm}$$

(ii) Moment at base at serviceability limit state:

Assuming γ = unit weight of water = 10 kN/m³,

$$M = \gamma H^3/6$$
$$M = 10 \times 3.5^3/6 = 71.46 \text{ kNm/m}$$

(iii) Stresses in steel and concrete

$$f_s = \frac{M}{A_s z} = \frac{71.46 \times 10^6}{1131 \times 323} = 196 \text{ N/mm}^2$$

$$f_c = \frac{2M}{bxz} = \frac{2 \times 71.46 \times 10^6}{1000 \times 94 \times 323} = 4.7 \text{ N/mm}^2$$

(iv) Moment at ultimate limit state:

Using a load factor of 1.4,

$$M = 1.4 \times 71.46 = 100.04 \text{ kNm/m}$$

(v) Neutral axis depth at ULS:

Assuming that steel yields, equating total tension and total compression,

$$0.45 f_{cu} b\, 0.9x = 0.95 f_y A_s$$
$$0.45 \times 35 \times 1000 \times 0.9x = 0.95 \times 460 \times 1131, \text{ giving } x = 34.9 \text{ mm}$$

Check the strain in tension steel:

$$\varepsilon_s = \frac{0.0035}{x}(d-x)$$

$$\varepsilon_s = \frac{0.0035}{34.9}(354 - 34.9) = 0.032 > (\frac{0.95 \times 460}{200 \times 10^3} = 0.0022)$$

Steel yields and the assumption is justified.

(vi) Moment of resistance at ULS:

$$M_u = 0.95 f_y A_s (d - 0.45x)$$
$$= 0.95 \times 460 \times 1131 \times (354 - 0.45 \times 34.9) \times 10^{-6}$$
$$= 167.2 \text{ kNm/m} > 100.04 \text{ kNm/m}$$

The design is satisfactory at ULS.

17.2.2 Crack Width Calculation in a Section Subjected to Flexure Only

Provided that the stress in steel $f_s \leq 0.8f_y$ and the stress in concrete $f_c \leq 0.45f_{cu}$, the design crack width w is given by

$$w = \frac{3a_{cr}\,\varepsilon_m}{1 + 2\dfrac{(a_{cr} - c_{min})}{(h - x)}}$$

$$\varepsilon_m = \varepsilon_1 - \varepsilon_2$$

$$\varepsilon_1 = \varepsilon_s \frac{(h - x)}{(d - x)}, \varepsilon_s = \frac{f_s}{E_s}$$

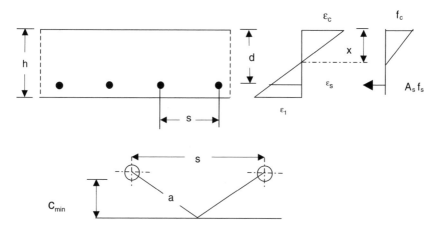

Fig.17.3 Crack width calculation.

If limiting design crack width is 0.2 mm,

$$\varepsilon_2 = \frac{b_t(h - x)(a' - x)}{3E_s A_s(d - x)}$$

If limiting design crack width is 0.1 mm,

$$\varepsilon_2 = 1.5 \frac{b_t(h - x)(a' - x)}{3E_s A_s(d - x)}$$

where as shown in Fig.17.3

w = design crack width, x = neutral axis depth

h = overall depth of member, d = effective depth

c_{min} = minimum cover to tension steel

b_t = width of section at the centroid of steel

ε_1 = apparent strain at the level considered

ε_2 = strain due to the stiffening effect of concrete between cracks.

s = centre to centre distance between bars

a_{cr} = distance from the point considered to the surface of the nearest longitudinal bar.

17.2.2.1 Example of crack width calculation in flexure only

Determine the design crack width using data from Example in 17.2.1:
Steel reinforcement is provided by 12 mm bars at 100 mm c/c,

$$A_s = 1131 \text{ mm}^2/\text{m}$$
$$h = 400 \text{ mm}$$
$$d = 354 \text{ mm},$$
$$\text{cover } c = 40 \text{ mm}$$
$$f_{cu} = 35 \text{ N/mm}^2$$
$$f_y = 460 \text{ N/mm}^2$$
$$x = 94 \text{ mm}$$
$$f_s = 196 \text{ N/mm}^2$$
$$f_c = 4.7 \text{ N/mm}^2$$

(i) Check that the stresses in steel and concrete are within permissible limits:

$$f_c = 4.7 \text{ N/mm}^2 < (0.45 \times 35 = 15.8 \text{ N/mm}^2)$$
$$f_s = 196 \text{ N/mm}^2 < (0.8 \times 460 = 368 \text{ N/mm}^2)$$

(ii) Strain at steel level:

Taking Young's modulus for steel $E_s = 200 \text{ kN/mm}^2$,

$$\varepsilon_s = f_s/E_s$$
$$\varepsilon_s = 196/(200 \times 10^3) = 0.98 \times 10^{-3}$$

(iii) Apparent strain at the surface

$$\varepsilon_1 = \varepsilon_s (h - x)/ (d - x)$$
$$\varepsilon_1 = 0.98 \text{ x } 10^{-3} (400 - 94)/ (354 - 94) = 1.153 \text{ x } 10^{-3}$$

(iv) Tension stiffening effect:

If limiting design crack width is 0.2 mm,

$$\varepsilon_2 = \frac{b_t(h - x)(a' - x)}{3E_s A_s (d - x)}$$

$b_t = 1000$ mm, $a' = h$, because the crack is calculated at the surface of the wall.

$$\varepsilon_2 = \frac{1000(400 - 94)(400 - 94)}{3 \times 200 \times 10^3 \times 1131 \times (354 - 94)} = 0.531 \times 10^{-3}$$

(v) Average strain at the surface:

$$\varepsilon_m = \varepsilon_1 - \varepsilon_2$$
$$\varepsilon_m = 1.153 \times 10^{-3} - 0.531 \times 10^{-3} = 0.622 \times 10^{-3}$$

(vi) Calculate a_cr

If ϕ = diameter of bar, c = cover and s = bar spacing, from Fig.17.3,

$$a = \sqrt{[(s/2)^2 + (c+\phi/2)^2]}$$

$$a = \sqrt{[(100/2)^2 + (40+12/2)^2]} = 68 \text{ mm}$$

$$a_{cr} = a - \phi/2 = 68 - 12/2 = 62 \text{ mm}$$

(vii) Width of crack, w:

$$w = \frac{3\,a_{cr}\,\varepsilon_m}{1+2\dfrac{(a_{cr}-c_{min})}{(h-x)}}$$

$$w = \frac{3\times 62\times 0.622\times 10^{-3}}{1+2\dfrac{(62-40)}{(400-94)}} = 0.10 \text{ mm}$$

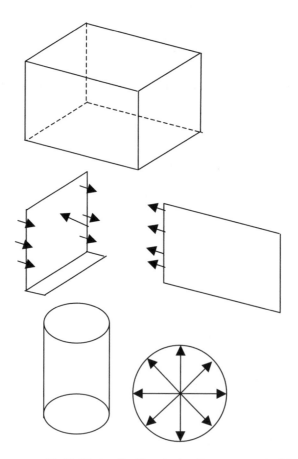

Fig.17.4 Tank wall subjected to bending moment and axial tension.

17.2.3 Crack Width Calculation in a Section Subjected to Bending Moment and Direct Tension

Fig.17.4 shows rectangular and circular water tanks. Normal load acting on the side wall of the rectangular tank is resisted by shear forces at the base and at the sides. The shear forces on the sides act as tensile forces on the front and back walls.

Similarly, in the case of circular water tanks, the radial pressure causes circumferential tension and depending on the fixity at the base of the wall may result in bending moment.

If the tensile force is not so large as to crack the whole section, the equations for calculating the neutral axis depth x and the stresses in concrete and steel can be developed in a manner is similar to the calculations for the case of pure flexure.

From the equations in section 17.2, the total compressive force C is

$$C = bx\frac{f_c}{2} + A_s^{'}\,\alpha_e f_c \frac{(x-d^{'})}{x}$$

where $A_s^{'}$ is the area per unit length of compression steel.
The total tensile force T is

$$T = A_s\,f_s = A_s\,\alpha_e f_c \frac{d-x}{x}$$

For equilibrium in the axial direction,

$$T - C = \text{Applied tensile force N}$$

$$A_s\,\alpha_e f_c \frac{d-x}{x} - bx\frac{f_c}{2} - A_s^{'}\alpha_e f_c \frac{(x-d^{'})}{x} = N \tag{17.1}$$

Taking moments about the tension steel,

$$bx\frac{f_c}{2}(d-\frac{x}{3}) + A_s^{'}\alpha_e f_c \frac{(x-d^{'})}{x}(d-d^{'}) = M - N\frac{h}{2} \tag{17.2}$$

where M = applied moment , h = overall depth of the section.

There are two unknowns viz. f_c and x in the two equations. Eliminating f_c from the two equations,

$$(M - N\frac{h}{2})A_s\alpha_e(d-x) - A_s^{'}\alpha_e(x-d^{'})\{M + N(d-d^{'}-\frac{h}{2})\}$$

$$= b\frac{x^2}{2}\{N(d-\frac{x}{3}-\frac{h}{2})+M\} \tag{17.3}$$

Solution of the above cubic equation gives the value of neutral axis depth, x. Compressive stress in concrete f_c can be obtained from equation 17.1 or 17.2.

The equations are valid only if $x \ge d^{'}$ because it is assumed that the stress in 'compression' steel is actually compressive.

17.2.3.1 Example of calculation of crack width under bending moment and axial tension

Calculate the crack width using the following data.
$$h = 400 \text{ mm}$$
$$\text{cover} = 40 \text{ mm}$$
Tension steel: 16 mm diameter bars at 100 mm c/c giving
$$A_s = 2011 \text{ mm}^2/\text{m}$$
$$f_{cu} = 35 \text{ N/mm}^2, f_y = 460 \text{ N/mm}^2, \alpha_e = 15$$
Applied actions at serviceability limit state:
$$M = 100 \text{ kNm/m}, N = 60 \text{ kN/m}$$
$$b = 1000 \text{ mm},$$
$$d = 400 - 40 - 16/2 = 352 \text{ mm}$$
$$A_s' = 0$$

1. Neutral axis depth:
From equation 17.3,

$$(M - N\frac{h}{2})A_s\alpha_e(d - x) = b\frac{x^2}{2}\{N(d - \frac{x}{3} - \frac{h}{2}) + M\}$$

$$(100 \times 10^6 - 60 \times 10^3 \frac{400}{2}) \times 2011 \times 15(352 - x)$$

$$= 1000\frac{x^2}{2}\{60 \times 10^3(352 - \frac{x}{3} - \frac{400}{2}) + 100 \times 10^6\}$$

Simplifying:
$$x^3 - 5456 \, x^2 - 265452 \, x + 93.4391 \text{x} 10^6 = 0$$
Solving the cubic equation by trial and error gives $x = 109.7$ mm

2. Compressive stress in concrete
Determine the compressive stress f_c in concrete using equation 17.1.
$$bx\frac{f_c}{2}(d - \frac{x}{3}) = M - N\frac{h}{2}$$
$$1000 \times 109.7 \times \frac{f_c}{2}(352 - \frac{109.7}{3}) = 100 \times 10^6 - 60 \times 10^3\frac{400}{2}$$
$$f_c = 5.09 \text{ N/mm}^2$$

3. Tensile stress in steel
Calculate tensile stress in steel:
$$f_s = \alpha_e f_c \frac{d - x}{x} = 5 \times 5.09 \times \frac{352 - 109.7}{109.7} = 168.7 \text{ N/mm}^2$$

4. Check that the stresses in steel and concrete are with in permissible limits:

$$f_c = 5.09 \text{ N/mm}^2 < (0.45 \times 35 = 15.8 \text{ N/mm}^2)$$
$$f_s = 168.7 \text{ N/mm}^2 < (0.8 \times 460 = 368 \text{ N/mm}^2)$$

5. Strain at steel level:

Taking Young's modulus for steel $E_s = 200$ kN/mm^2,
$$\varepsilon_s = f_s/E_s = 168.7/(200 \times 10^3) = 0.84 \times 10^{-3}$$

6. Apparent strain at the surface

$$\varepsilon_1 = \varepsilon_s \, (h - x)/ (d - x)$$
$$\varepsilon_1 = 0.84 \times 10^{-3} \, (400 - 109.7)/ (352 - 109.7) = 1.006 \times 10^{-3}$$

7. Tension stiffening effect:

If limiting design crack width is 0.2 mm,
$$\varepsilon_2 = \frac{b_t(h-x)(a'-x)}{3E_s A_s (d-x)}$$

Substituting $b_t = 1000$ mm, $a' = h$,
$$\varepsilon_2 = \frac{1000(400-109.7)(400-109.7)}{3 \times 200 \times 10^3 \times 2011 \times (352-109.7)} = 0.288 \times 10^{-3}$$

8. Average strain at the surface:

$$\varepsilon_m = \varepsilon_1 - \varepsilon_2 = 1.006 \times 10^{-3} - 0.288 \times 10^{-3} = 0.718 \times 10^{-3}$$

9. Calculate a_{cr}

If ϕ = diameter of bar, c = cover and s = bar spacing, from Fig.17.3,
$$a = \sqrt{[(s/2)^2 + (c+\phi/2)^2]}$$
$$a = \sqrt{[(100/2)^2 + (40+16/2)^2]} = 69.3 \text{ mm}$$
$$a_{cr} = a - \phi/2 = 69.3 - 16/2 = 61.3 \text{ mm}$$

10. Width of crack, w:

$$w = \frac{3a_{cr}\varepsilon_m}{1+2\dfrac{(a_{cr}-c_{min})}{(h-x)}} = \frac{3 \times 61.3 \times 0.718 \times 10^{-3}}{1+2\dfrac{(61.3-40)}{(400-109.7)}} = 0.12 \text{ mm}$$

17.2.4 Crack Width Calculation in Direct Tension

Provided that the stress in steel $f_s \le 0.8 \, f_y$, according to the code the crack width w is given by

$$w = 3\,a_{cr}\,\varepsilon_m$$
$$\varepsilon_m = \varepsilon_1 - \varepsilon_2$$
$$\varepsilon_1 = \varepsilon_s = \frac{f_s}{E_s}$$

$$[w = 0.2 \text{ mm}, \; \varepsilon_2 = \frac{2b_t h}{3E_s A_s}] \text{ and } [w = 0.1 \text{ mm}, \; \varepsilon_2 = \frac{b_t h}{E_s A_s}]$$

17.2.4.1 Example of crack width calculation in direct tension

A circular water tank 10 m internal diameter and 6.5 m high retains water to a depth of 6.0 m. The base of the tank is designed to be free sliding. Design the reinforcement at the base and calculate the crack width.

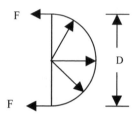

Fig.17.5 Ring tension.

The pressure p at the base of the tank:
$$p = \gamma H = 10 \text{ kN/m}^3 \times 6 = 60 \text{ kN/m}^2$$
As shown in Fig.17.5, the circumferential tension at the base T is given by
$$T = p \, D/2 = 60 \times 10/2 = 300 \text{ kN/m}$$
where D = Diameter of tank = 10 m
Assuming 350 mm thick walls, the minimum steel reinforcement in each face (Clause 2.6.2.3) should not be less than 0.35%
$$A_s = (0.35/100) \times (350/2) \times 1000 = 613 \text{ mm}^2/\text{m}$$
$$\text{Cover} = 40 \text{ mm}$$
Provide 12 mm bars at 175 c/c on each face. $A_s = 646 \text{ mm}^2/\text{m}$ on each face.

(i) Stress and strain in steel:
$$f_s = T/(2 \, A_s)$$
$$f_s = 300 \times 10^3/ (2 \times 646) = 232 \text{ N/mm}^2 < (0.8 \times 460 = 368 \text{ N/mm}^2)$$
$$\varepsilon_1 = \varepsilon_s = \frac{f_s}{E_s} = \frac{232}{200 \times 10^3} = 1.16 \times 10^{-3}$$

(ii) Effect of tension stiffening:
If limiting design crack width is 0.2 mm,
$$\varepsilon_2 = \frac{2b_t h}{3E_s A_s} = \frac{2 \times 1000 \times 350}{3 \times 200 \times 10^3 \times (2 \times 646)} = 0.903 \times 10^{-3}$$

(iii) Average surface strain:
$$\varepsilon_m = \varepsilon_1 - \varepsilon_2 = (1.16 - 0.903) \times 10^{-3} = 0.257 \times 10^{-3}$$

(iv) Calculate a_{cr}
If ϕ = diameter of bar, c = cover and s = bar spacing, from Fig.17.3,
$$a = \sqrt{[(s/2)^2 + (c + \phi/2)^2]} = \sqrt{[(125/2)^2 + (40 + 12/2)^2]} = 77.6 \text{ mm}$$

$$a_{cr} = a - 12/2 = 77.6 - 12/2 = 71.6 \text{ mm}$$

(v) Crack width

$$w = 3\,a_{cr}\,\varepsilon_m = 3 \times 71.6 \times 0.257 \times 10^{-3} = 0.06 \text{ mm}$$

17.2.4 Deemed to Satisfy Clause

Instead of calculating the width of cracks, BS 8007 allows a simpler approach. If the stress at serviceability limit state is less than shown in Table 17.2 (corresponds to Table 3.1 in the code), then crack widths can be assumed to be satisfactory.

It is generally found that these limitations are quite restrictive and designs which do not satisfy the above criterion are often found to be quite satisfactory when crack widths are calculated as illustrated by the three examples from previous sections as shown in Table 17.3.

17.2.5 Design Tables

Since it is not possible to design directly a section for a fixed value of design crack width, design tables can be developed for specific sections with specified bar diameter and spacing. Tables 17.4 and 17.5 are typical design aids. Notice that only those cases where the ratio of ultimate moment capacity to moment capacity at serviceability limit state is greater than 1.4 are retained.

17.3 CONTROL OF RESTRAINED SHRINKAGE AND THERMAL MOVEMENT CRACKING

Changes in temperature and moisture content of the concrete cause movements. If these movements are restrained they lead to tensile stresses in concrete and possibility of cracks. During hydration of cement heat is generated and as the concrete cools, it contracts. Similarly loss of moisture leads to drying shrinkage. In most structures these effects are of no significance compared with the stresses due to external loads. However in thin sections such as walls, these effects are important and must be taken into account if the structure is not to be rendered unserviceable due to wide cracks. It is necessary to reinforce the structures to ensure that a number of well distributed cracks of acceptable width occur rather than a few wide cracks.

The restraint to movement can be reduced by proper sequence of construction. Fig.17.6 (a) shows the preferred sequence because after each bay is cast, the slab is unrestrained at one edge and can contract during cooling. On the other hand a sequence of construction shown in Fig.17.6 (b) is not recommended because the middle slab is restrained on both sides.

Table 17.2 Allowable stress in steel at SLS

Design crack width mm	Allowable stress N/mm^2	
	$f_y = 250$ N/mm^2	$f_y = 460$ N/mm^2
0.1	85	100
0.2	115	130

Table 17.3 Calculated f_s and w

Example in section	f_s, N/mm^2	w, mm
17.2.2.1	196	0.1
17.2.3.1	169	0.12
17.2.4.1	169	0.06

Table 17.4 Design table (Crack width = 0.1 mm)

h=450 mm, cover =52 mm, w = 0.1 mm					
bar dia	Spacing, mm	M-service, kNm/m	f_s, service N/mm^2	M-ultimate, kNm/m	f_s, ultimate N/mm^2
12	100	98.24	242	184.05	437
	150	87.56	319	122.70	437
16	100	116.60	167	314.01	437
	150	96.75	204	215.71	437
	200	87.99	244	162.77	437
	250	83.58	288	130.21	437
20	100	140.88	133	462.69	437
	150	110.01	152	324.01	437
	200	95.84	174	248.84	437
	250	88.40	199	201.87	437
	300	84.15	226	168.68	437
25	100	178.00	111	656.09	437
	150	131.57	120	475.36	437
	200	109.45	131	370.76	437
	250	97.45	144	303.44	437
	300	90.34	159	256.66	437
32	100	239.17	95	894.25	376
	150	169.29	98	693.66	437
	200	134.57	102	556.62	437
	250	115.07	108	463.64	437
	300	103.17	115	396.56	437

Table 17.5 Design table (Crack width = 0.2 mm)

	Spacing, mm	M-service, kNm/m	f_s, service N/mm^2	M-ultimate, kNm/m	f_s, ultimate N/mm^2
bar dia					
16	100	176.87	253	314.01	437
	150	132.67	279	215.71	437
	200	112.23	311	162.77	437
20	100	231.03	218	462.69	437
	150	164.40	227	324.01	437
	200	132.81	242	248.84	437
	250	115.57	260	201.87	437
	300	105.24	282	168.68	437
25	100	311.16	194	656.09	437
	150	213.15	194	475.36	437
	200	165.41	198	370.76	437
	250	138.82	205	303.44	437
	300	122.34	215	256.66	437
32	100	440.13	175	894.25	376
	150	295.16	171	693.66	437
	200	222.04	168	556.62	437
	250	180.28	169	463.64	437
	300	154.27	172	396.56	437

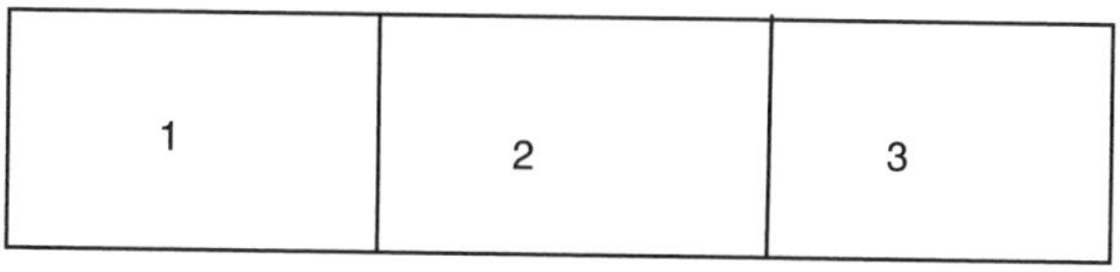

(h = 450 mm, cover =52, w = 0.2 mm)

(a) Preferredoption

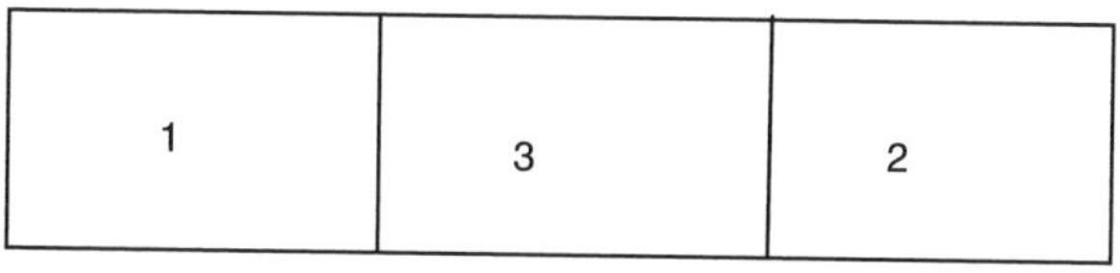

(b) Not recommended

Fig.17.6 Sequence of construction.

The restraint to movement due to friction between the ground and the wall can be reduced by laying a sheet of polythene sheet on a layer of smooth blinding concrete.

17.3.1 Movement Joints

Stresses due to shrinkage and thermal movements are controlled by the provision of movement joints. See Fig.5.1 in BS 8007. There are basically three types of movement joints. They are

- Expansion joints: Fig.17.7 shows a typical expansion joint. This has no restraint to movement. There is no continuity of steel or concrete across the joint. An initial gap is provided for expansion and leakage of water is prevented by using a water stop made from rubber or similar materials.
- Complete contraction joint: This is similar to the expansion joint except that there is no initial gap.
- Partial contraction joint: As shown in Fig.17.8, in these types of joint there is continuity of 50% of steel across the joint. However, there is no concrete continuity across the gap.

Generally the maximum distance between the full contraction joints is about 15 m with a partial contraction joint in between the full contraction joints.

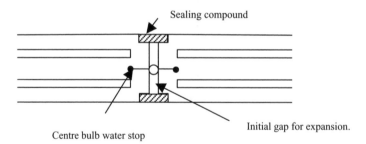

Fig.17.7 Complete expansion joint.

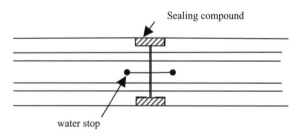

Fig.17.8 Partial contraction joint.

17.3.2 Critical Amount of Reinforcement

Although movement joints reduce the effect of restraint to *overall* contraction in individual bays, the effect of contraction *within bays* can be minimized by ensuring that reinforcement can distribute the contraction between a large number of fine cracks as opposed to a small number of large cracks. The critical amount of reinforcement necessary to control the early thermal and shrinkage cracking is based on the ability of the reinforcement to crack the immature concrete and hence produce a series of fine cracks. Fig.17.9 shows the state of equilibrium adjacent to a crack. The tensile force in steel is resisted by tensile stress in concrete. The stress in the concrete f_{ct} is the early age (about three days) tensile strength which is quite small. The tensile stress in the steel in uncracked concrete is small and is ignored. The maximum allowable stress in steel is f_y, the yield stress in steel. For equilibrium,

$$A_s\, f_y = A_c\, f_{ct}$$
$$\rho_{\text{critical}} = A_s/A_c = f_{ct}/f_y$$

If the steel ratio $\rho = A_s/A_c$ is less than ρ_{critical}, the steel will yield in tension resulting in a few rather wide cracks.

BS 8007 recommends in Table A.1 for C35A concrete,

$$\rho_{\text{critical}} = 0.64\% \text{ for grade 250 steel}$$
$$\rho_{\text{critical}} = 0.35\% \text{ for grade 460 steel.}$$

17.3.3 Crack Spacing

Once the first crack forms, local slipping starts between the concrete and steel. If the steel ratio is greater than ρ_{critical} the stress in steel is less than the yield stress. Additional cracks will form if the bond stress f_b between steel and concrete exceeds the tensile strength f_{ct} of concrete.

$$f_b\, s\, \Sigma u \ge f_{ct}\, A_c$$

where
s = development length for bond stress
Σu = total perimeter of all bars at the section.

$$s \ge \frac{f_{ct}}{f_b} \frac{A_c}{A_s} \frac{A_s}{\Sigma u}$$

The ratio of area to perimeter in the case of bars of diameter ϕ is given by

$$(\pi\, \phi^2/4)/(\pi\, \phi) = \phi/4$$

Because in general only one type of bars will be used at a cross section,

$$\Sigma u / A_s = 4/\phi$$

$$s \ge \frac{f_{ct}}{f_b} \times \frac{A_c}{A_s} \times \frac{\phi}{4} = \frac{f_{ct}}{f_b} \times \frac{\phi}{4\rho}$$

where $\rho = A_s/A_c$

The minimum crack spacing is given by the equality sign and the maximum crack spacing s_{max} is twice the value of development length s immediately prior to the formation of a new crack. Thus

$$2s = s_{max} = \frac{f_{ct}}{f_b} \times \frac{\phi}{2\rho}$$

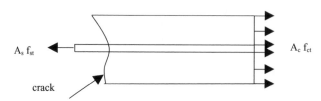

Fig.17.9 Forces adjacent to a crack.

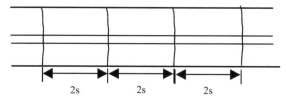

Fig.17.10 Crack spacing.

The crack spacing reduces with decreasing bar size (hence smaller diameter bars are preferred), greater bond strength and larger steel ratio.

The code BS 8007 recommends in Table A.1 for C35A concrete,

$f_{ct}/f_b = 1.0$ and $f_b = 1.6$ N/mm^2 for grade 250 steel
$f_{ct}/f_b = 0.67$ and $f_b = 2.4$ N/mm^2 for grade 460 steel.

17.3.4 Width of Cracks

The estimated crack width w_{max} is given by

$$w_{max} = s_{max} \frac{\alpha}{2}(T_1 + T_2)$$

Substituting for s_{max},

$$w_{max} = \frac{f_{ct}}{f_b} \frac{\phi}{2\rho} \frac{\alpha}{2}(T_1 + T_2)$$

T_1 = the fall in temperature between the hydration peak and the ambient. Table A.2 of BS 8007 gives the value to be used for ordinary Portland concretes. These are shown in Table 17.7.

T_2 = Further fall in temperature because of seasonal variations.

α = Coefficient of thermal expansion. Typical values are given in Table 7.3 of BS 8110, Part 2. It varies from about 8×10^{-6}/°C to 12×10^{-6}/°C.

Table 17.7 Typical values of $T_1°C$ for OPC concretes

Section thickness, mm	Walls						Ground slabs OPC content, kg/m³		
	Steel formwork OPC content, kg/m³			18 mm plywood formwork OPC content, kg/m³					
	325	350	400	325	350	400	325	350	400
300	20	20	20	23	25	31	15	17	21
500	20	22	27	32	35	43	25	28	34
700	28	32	39	38	42	49	-	-	-
1000	38	42	49	42	47	56	-	-	-

17.3.5 Design Options for Control of Thermal Contraction and Restrained Shrinkage

The provision of movement joints and their spacing depends on whether the designer prefers to provide substantial amount of reinforcement (in the form of small diameter bars at close spacing, see section 17.3.3) and no movement joints or closely spaced movement joints with a moderate amount of reinforcement or any combination in between these two extreme options. BS 8007 in Table 5.1 reproduced in Table 17.8 provides three major options with many more sub–options in each major option.

17.3.6 Example of Options for Control of Thermal Contraction and Restrained Shrinkage

A wall slab in a water retaining structure is 350 mm thick. Calculate the required steel area to control cracking due to thermal effects and the joint spacing for the various options in Table 17.8 for a design crack width of 0.2 mm. Concrete is C35A grade and steel is 460 grade.

(a) Calculate ρ_{crit}:
$$\rho_{crit} = 0.35\%, \ h = 350 \ mm < 500 \ mm$$
$$A_s = (0.35/100) \ 350 \times 1000 = 1225 \ mm^2/m$$
Provide T10-125 mm c/c = 628 mm²/m. $A_s = 2 \times 628 = 1256 \ mm^2/m$.

(b) Movement joint spacing:

Option 1: Continuous casting with no movement joints.
Provide T10-125 mm c/c = 628 mm²/m

Option 2: Semi–continuous for partial restraint:
Provide T10-125 mm c/c = 628 mm²/m
The spacing of joints can de adapted to any of the three options as follows:

(a) Complete joints at a spacing ≤ 15 m
(b) Alternate partial and complete joints (by interpolation), ≤ 11.25 m
(c) Partial joints, ≤ 7.5 m

Table 17.8 Design options for control of thermal contraction and restrained shrinkage

Option	Type of construction and method of control	Movement joint spacing	Steel ratio	Comments
1	Continuous: for full restraint	No joints, but expansion joints at wide spacing may be desirable in walls and roofs that are not protected from solar heat gain or where the contained liquid is subjected to a substantial temperature range	Minimum of ρ_{crit}	Use small size bars at close spacing to avoid high steel ratios well in excess of ρ_{crit}
2	Semi continuous: for partial restraint	a) Complete joints, spacing ≤ 15 m b) Alternate partial and complete joints (by interpolation), spacing ≤ 11.25 m c) Partial joints, ≤ 7.5 m	Minimum of ρ_{crit}	Use small size bars but less steel than in option 1
3	Close movement joint spacing: for freedom of movement	(a) Complete joints, in metres $\leq 4.8 + w/\varepsilon$ (b) Alternate partial and complete joints, in metres $\leq 0.5\,s_{max} + 2.4 + w/\varepsilon$ (c) Partial joints, in metres $\leq s_{max} + w/\varepsilon$	$2/3\ \rho_{crit}$	Restrict the joint spacing for options 3 (b) and 3(c).

w = Design crack width, ε = thermal strain = $0.5\ \alpha\ T_1$ (see Table 17.7)

Option 3: Close movement joint spacing:
$A_s = (2/3)1225 = 817$ mm^2/m, $817/2 = 408$ mm^2/m on each face.

Provide T10-175 mm c/c = 448 mm^2/m
$$w = 0.2 \text{ mm}$$
Assume 18 mm plywood formwork and $T_1 = 20°C$, $\alpha = 10^{-6} /°C$
$$\varepsilon = 0.5 \ \alpha \ T_1 = 0.5 \times 10 \times 10^{-6} \times 20 = 0.1 \times 10^{-3} \text{ mm/mm}$$
$$w/ \varepsilon = 2 \times 10^3 \text{ mm} = 2 \text{ m}$$
$$f_{ct}/f_b = 0.67 \text{ (Table A.1 of BS 8007)}$$
$$\phi = 10\text{mm}$$
$$\rho = A_s/ (350 \times 1000) = 2 \times 448/ (350 \times 1000) = 0.256\%$$
$$S_{max} = \frac{f_{ct}}{f_b} \times \frac{\phi}{2\rho}$$
$$s_{max} = 0.67 \times 10/(2 \times 0.00256) = 1309 \text{ mm} = 1.31 \text{ m}$$

The spacing of joints can de adapted to any of the three options as follows:

(a) Complete joints:
  Spacing of joints $\leq (4.8 + w/\varepsilon = 4.8 + 2.0 = 6.8 \text{ m})$
  Spacing of joints ≤ 6.8 m

(b) Alternate partial and complete joints:
  Spacing of joints $\leq (0.5 \ s_{max} + 2.4 + w/\varepsilon)$
  Spacing of joints $\leq = (0.5 \ x \ 1.31 + 2.4 + 2.0 = 5.055 \text{ m})$
  Spacing of joints $\leq = 5.055$ m

(c) Partial joints:
  Spacing of joints $\leq (s_{max} + w/\varepsilon = 1.31 + 2.0 = 3.31 \text{ m})$
  Spacing of joints ≤ 3.31 m

17.4 DESIGN OF A RECTANGULAR COVERED TOP UNDER GROUND WATER TANK

Specification:
Design a rectangular water tank with two equal compartments as shown in Fig.17.11.
$$\text{Soil: Unit weight } \gamma = 18 \text{ kN/m}^3$$
$$\text{Soil: Submerged unit weight } \gamma = (18 - \gamma_w) = 8 \text{ kN/m}^3$$
$$\text{Coefficient of friction } \phi = 30°.$$
$$\text{Surcharge: 12 kN/m}^2.$$
$$\text{Unit weight of water } \gamma_w = 10 \text{ kN/m}^3$$
Consider the possibility of water logging up to 1 m below the ground level.
Design for severe exposure, design crack width = 0.2 mm.
Use C35A concrete and 460 grade steel.
Assume walls and slabs are 400 mm thick. The roof is *not* integrally connected to the walls and is simply supported on the external walls but continuous over the central dividing wall.

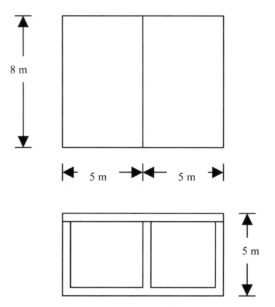

Fig.17.11 Rectangular water tank

(a) Check uplift:
Total weight W of the tank when empty:
$$W = \{5 \times 10 - (5 - 0.4 - 0.4)(10 - 0.4 - 0.4 - 0.4)\} \times 8 \times 24$$
$$W = 2504 \text{ kN}$$

Uplift Pressure of water under the floor due to 4 m head of water
$$\text{Uplift pressure} = 10 \times 4 = 40 \text{ kN/m}^2$$
$$\text{Uplift force} = 8 \times 10 \times 40 = 3200 \text{ kN}$$

Additional weight required to have a factor of safety against floatation of 1.1
Additional weight = $3200 \times 1.1 - 2504 = 1016$ kN
This can be provided by extending the base as shown in Fig.17.12.

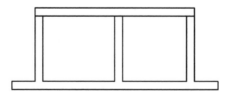

Fig.17.12 New design of base to increase total weight of tank.

The submerged unit weight of the soil = $18 - 10 = 8$ kN/m^3
Pressure due to 1 m high dry soil plus 3.6 m of submerged soil
$$= 1 \times 18 + 3.6 \times 8 = 46.8 \text{ kN/m}^2$$
Submerged weight of slab = $(24 - 10) \times 0.4 = 5.6$ kN/m^2

If b = width of the projecting base slab, then
$$\{(10 + 2b) \times (8+2b) - 10 \times 8\} \times (46.8 + 5.6) = 1016$$
If b = 0.55 m, the additional weight is 1101 kN

(b) Pressure calculation on the walls:

Case 1: Tank empty.
Coefficient of active earth pressure:
$$k_a = (1 - \sin\phi)/ (1 + \sin\phi) = (1 - 0.5)/ (1 + 0.5) = 0.33$$
$$\text{Pressure due to surcharge} = k_a \times 12 = 4 \text{ kN/m}^2$$
The wall is 5000 – 400 – 400 = 4200 mm high.
For the top (1000 – 400) = 600 mm, unit weight of soil = 18 kN/m³
Below this level submerged unit weight of soil = 8 kN/m³
In addition to the soil pressure there is also the pressure due to ground water.
The pressures at different levels are:

(i) At 400 mm below ground:
$$p = 4 \text{ kN/m}^2 \text{ due to surcharge} + k_a \times 18 \times 0.4 = 6.4 \text{ kN/m}^2$$

(ii) At 1000 mm below ground:
$$p = 4 \text{ kN/m}^2 \text{ due to surcharge} + k_a \times 18 \times 1.0 = 10.0 \text{ kN/m}^2$$

(iii) At 4600 mm below ground :
$$p = 10 + k_a \times 8 \times (4.6 - 1.0) + 10 \times (4.6 - 1.0) \text{ due to ground water}$$
$$p = 55.6 \text{ kN/m}^2$$

Case 2: Tank full.
Ignore any passive pressure due to soil and assume that the ground is dry.

(i) At 400 mm below ground
$$p = 10 \times 0.4 = 4.0 \text{ kN/m}^2$$

(ii) At 4600 mm below ground:
$$p = 10 \times 4.6 = 46 \text{ kN/m}^2$$

(c) Check shear capacity:
Effective depth:
$$d = 400 - 40 \text{ mm cover} - 12 \text{ mm bar } /2 = 354 \text{ mm}$$

Case 1: Tank empty:
Total shear force at base is approximately
$$V = 0.5 \times (6.4 + 10.0) \times 0.6 + 0.5 \times (10.0 + 55.6) \times 3.6 = 123.0 \text{ kN/m}$$
$$v = 123.0 \times 10^3/ (1000 \times 354) = 0.35 \text{ N/mm}^2$$
Assuming minimum area of steel $A_s = 0.35\%$

$$v_c = \frac{0.79}{1.25} \times 0.35^{\frac{1}{3}} \times (\frac{400}{354})^{\frac{1}{4}} \times (\frac{35}{25})^{\frac{1}{3}} = 0.51 \text{ N/mm}^2$$

$$v < v_c$$

Section thickness is adequate.

Case 2: Tank full.
Total shear force at base is approximately
$$V = 0.5 \times (4.0 + 46.0) \times 4.2 = 105.0 \text{ kN/m}$$
$$v = 105.0 \times 10^3 / (1000 \times 354) = 0.30 \text{ N/mm}^2$$
$$v < v_c$$

Section thickness is adequate.

(d) Minimum steel:
From Table A1 of 8007, $\rho_{crit} = 0.0035$ for 460 grade steel.
Minimum steel A_s area required $= 0.0035 \times 1000 \times 400 = 1400 \text{ mm}^2/\text{m}$
$$w_{max} = 0.2 \text{ mm}$$
$$\alpha = 12 \times 10^{-6} \text{ from Table 3.2 of BS 8110, Part 2}$$
$$T_1 = 25°C \text{ (Table A.2 of BS 8007)}$$
From Table A1 of 8007, $f_{ct}/f_b = 0.67$ for deformed bars of type 2.
Choose bar diameter $\phi = 12$ mm

$$w_{max} = \frac{f_{ct}}{f_b} \frac{\phi}{2\rho} \frac{\alpha}{2} (T_1 + T_2)$$

$$\rho = 0.0030 < 0.0035$$

Using continuous construction for full restraint (Table 5.1 of BS 8007), minimum steel required is
$$A_s = 0.0035 \times 1000 \times 400 = 1400 \text{ mm}^2/\text{m}$$
Provide T12-150 mm c/c = 755 mm^2/m on each face.
Total steel area = 1510 mm^2/m.

(e) Design of walls for bending at serviceability limit state:
For calculating moments in the walls of the tank, ready made tables of moment coefficients are available. These coefficients have been obtained from elastic analysis of thin plates using analytical methods based on multiple Fourier series or using the finite element method. Typical results are shown in Table 17.9 for the case of side and bottom edges being clamped and the top edge being free as shown in Fig.17.13.

(i) Transverse walls:
The wall is designed as a 7.2 m × 4.2 m slab clamped on three sides and free at top and subjected to a hydrostatic loading giving base pressures of 55.6 kN/m^2 for case 1 and 46.0 kN/m^2 for case 2. Since the pressure difference is not large, design for Case 1 and use the same steel area for case 2.

(1) Vertical bending moment at base

From Table 17.9, interpolating between b/a of 1.5 and 2.0,

$$\text{bending moment coefficient} = (0.084 + 0.058)/2 = 0.071$$

Vertical bending moment M at SL:

$$M = 0.071 \times 55.6 \times 4.2^2 = 69.64 \text{ kNm/m (SLS)}$$

Vertical bending moment at base (ULS)

$$M = 1.4 \times 69.64 = 97.50 \text{ kNm/m (ULS)}$$

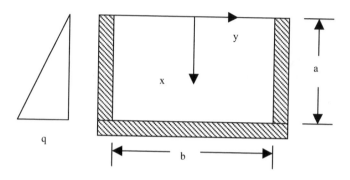

Fig.17.13 Notation for Table 17.9.

Table 17.9 Moment coefficients

b/a	x/a	y = 0		y = b/4		y = b/2
		M_x	M_y	M_x	M_y	M_y
2.0	0	0	0.027	0	0.010	−0.064
	0.25	0.012	0.024	0.006	0.010	−0.060
	0.50	0.016	0.017	0.011	0.010	−0.048
	0.75	−0.007	0.003	−0.002	0.003	−0.024
	1.0	−0.084	-	−0.058	-	-
1.5	0	0	0.021	0	0.006	−0.039
	0.25	0.009	0.020	0.004	0.007	−0.044
	0.50	0.015	0.017	0.009	0.008	−0.041
	0.75	0.003	0.006	0.004	0.004	−0.023
	1.0	−0.058	-	−0.039	-	-
1.0	0	0	0.010	0	0.002	−0.014
	0.25	0.003	0.012	0.001	0.003	−0.023
	0.50	0.009	0.013	0.005	0.005	−0.028
	0.75	0.008	0.008	0.005	0.004	−0.020
	1.0	−0.032	-	−0.021	-	-

$$\text{Moment} = \text{Coefficient} \times q \times a^2$$

Table 17.10 Moment capacity for a fixed crack width.

h =400 mm, cover = 40 mm, crack width w =0.2 mm					
bar dia	Spacing (mm)	SLS Moment (kNm)	SLS f_s (N/mm^2)	ULS Moment (kNm)	ULS f_s (N/mm^2)
	100	125.70	344	165.89	437
	150	96.05	389	110.81	437
	200	83.31	445	83.11	437
	250	76.84	510	66.48	437
12	300	73.16	580	55.40	437
	100	174.43	278	280.62	437
	150	122.83	288	193.45	437
	200	99.97	309	146.91	437
	250	88.05	337	117.53	437
16	300	81.10	370	97.94	437
	100	235.15	247	410.53	437
	150	157.52	243	289.24	437
	200	122.31	248	222.76	437
	250	103.57	260	181.01	437
20	300	92.48	276	152.16	437
	100	324.84	226	577.71	437
	150	211.02	215	421.02	437
	200	157.42	210	330.00	437
	250	128.58	212	270.84	437
25	300	111.22	218	229.49	437
	100	469.02	208	744.74	347
	150	299.83	194	624.66	437
	200	218.16	185	489.85	437
	250	172.83	181	410.22	437
32	300	145.06	180	352.04	437

From the data in Table 17.10, using T12-150 mm c/c gives at SLS and ULS moment of resistances of 96.05 kNm/m and 110.81 kNm/m respectively. Provide on both faces T12-150 mm c/c in the vertical direction.

(2) Horizontal bending moment at fixed vertical edges

From data in Table 17.9, interpolating between b/a of 1.5 and 2.0,

$$\text{bending moment coefficient} = (0.064 + 0.039)/2 = 0.052$$
$$M \text{ at SLS} = 0.052 \times 55.6 \times 4.2^2 = 51.0 \text{ kNm/m}$$

(3) Horizontal bending moment at mid–span

From data in Table 17.9, interpolating between b/a of 1.5 and 2.0,

$$\text{bending moment coefficient} = (0.027 + 0.021)/2 = 0.024$$
$$M \text{ at SLS} = 0.024 \times 55.6 \times 4.2^2 = 23.54 \text{ kNm/m}$$

From the data in Table 17.9, using T12-150 mm c/c gives at SLS and ULS moment of resistances of 96.05 kNm/m and 110.81 kNm/m respectively. Provide on both faces T12-150 mm c/c in the horizontal direction. A_s on each face = 754 mm²/m

(4) Direct tension in walls

In case 2 there is also direct tension in the horizontal direction in the wall due to water pressure on the 10 m long walls. Average pressure p is approximately

$$p = 0.5 \times 46.0 = 23 \text{ kN/m}^2$$

Ignoring the resistance provided by the base, tensile force N per meter is

$$N = 0.5 \times 5.0 \times 23 = 57.5 \text{ kN/m}.$$

The tensile stress in steel due to tensile force is

$$N = 57.5 \times 10^3 / (2 \times 754) = 38 \text{ N/mm}^2$$

The tensile force N is combined with a maximum bending moment of 51.0 kNm/m. Check the crack width using steel area provided by T12 @ 150 mm c/c on each face

(5) Calculate the crack width using the following data

h = 400 mm, cover = 40 mm, Steel: 12 mm diameter bars at 150 mm c/c.
Applied forces at serviceability limit state:

$$M = 51.0 \text{ kNm/m}, N = 57.5 \text{ kN/m (tension)}$$
$$f_{cu} = 35 \text{ N/mm}^2, f_y = 460 \text{ N/mm}^2, \alpha_e = 15$$
$$b = 1000 \text{ mm}, d = 354 \text{ mm}, A_s = 754 \text{ mm}^2/\text{m}$$

(i) Calculate the neutral axis depth including compression steel area:

$$(M - N\frac{h}{2})A_s\alpha_e(d - x) - A_s'\alpha_e(x - d')\{M + N(d - d' - \frac{h}{2})\}$$

$$= b\frac{x^2}{2}\{N(d - \frac{x}{3} - \frac{h}{2}) + M\}$$

$$(51.0 \times 10^6 - 57.5 \times 10^3 \frac{400}{2}) \times 754 \times 15 \times (354 - x)$$

$$-(51.0 \times 10^6 + 57.5 \times 10^3 (354 - 46 - \frac{400}{2}) \times 754 \times 15 \times (x - 46)$$

$$= 1000 \frac{x^2}{2} \{57.5 \times 10^3 (354 - \frac{x}{3} - \frac{400}{2}) + 51.0 \times 10^6\}$$

Simplifying:

$$x^3 - 3122.87 \, x^2 - 114134.61 \, x + 19.2031 \times 10^6 = 0$$
$$\text{Solving, } x = 62.7 \text{ mm}$$

(ii) Calculate the compressive stress in concrete:

$$bx \frac{f_c}{2} (d - \frac{x}{3}) + A_s' \alpha_e f_c \frac{(x - d')}{x} (d - d') = M - N \frac{h}{2} \quad (17.2)$$

$$\{1000 \times 62.7 \times \frac{1}{2} (354 - \frac{62.7}{3}) + 754 \times 15 \times \frac{(62.7 - 46)}{62.7} \times (354 - 46)\} f_c$$

$$= 51.0 \times 10^6 - 57.5 \times 10^3 \times \frac{400}{2}$$

Solving for f_c, $f_c = 3.5$ N/mm^2 < (0.45 × 35 = 15.8 N/mm^2)

(iii) Calculate the tensile stress in steel:

$$f_s = \alpha_e f_c \frac{d - x}{x} = 15 \times 3.5 \times \frac{354 - 62.7}{62.7} = 244 \text{ N/mm}^2$$

$$f_s = 244 \text{ N/mm}^2 < (0.8 \times 460 = 368 \text{ N/mm}^2)$$

(iv) Strain at steel level:
Taking Young's modulus for steel $E_s = 200$ kN/mm^2,
$$\varepsilon_s = f_s/E_s = 244/(200 \times 10^3) = 1.22 \times 10^{-3}$$

(v) Apparent strain at the surface
$$\varepsilon_1 = \varepsilon_s (h - x)/(d - x) = 1.22 \times 10^{-3} (400 - 62.7)/(354 - 62.7)$$
$$\varepsilon_1 = 1.41 \times x \, 10^{-3}$$

(vi) Tension stiffening effect:
If limiting design crack width is 0.2 mm,
$$\varepsilon_2 = \frac{b_t (h - x)(a' - x)}{3 E_s A_s (d - x)}$$

$b_t = 1000$ mm, $a' = h$ because the crack is calculated at the surface of the wall.
$$\varepsilon_2 = \frac{1000(400 - 62.7)(400 - 62.7)}{3 \times 200 \times 10^3 \times 754 \times (354 - 62.7)} = 0.86 \times 10^{-3}$$

(vii) Average strain at the surface:
$$\varepsilon_m = \varepsilon_1 - \varepsilon_2 = 1.41 \times 10^{-3} - 0.86 \times 10^{-3} = 0.55 \times 10^{-3}$$

(viii) Calculate a_{cr}

If ϕ = diameter of bar, c = cover and s = bar spacing, from Fig.17.3,
$$a = \sqrt{[(s/2)^2 + (c+\phi/2)^2]} = \sqrt{[(150/2)^2 + (40+12/2)^2]} = 88.0 \text{ mm}$$
$$a_{cr} = a - \phi/2 = 88.0 - 12/2 = 82.0 \text{ mm}$$

(ix) Width of crack, w:

$$w = \frac{3 a_{cr} \, \varepsilon_m}{1+2\dfrac{(a_{cr} - c_{min})}{(h-x)}} = \frac{3 \times 82 \times 0.55 \times 10^{-3}}{1+2\dfrac{(82-40)}{(400-62.7)}} = 0.11 \text{ mm}$$

(x) Check ultimate conditions:

Applied forces:
$$M = 1.4 \times 51.0 = 71.4 \text{ kNm/m},$$
$$N = 1.4 \times 57.5 = 80.5 \text{ kN/m}$$

Using the method in Chapter 9 on columns for calculating moment–axial force interaction,
$$x/h = 0.034, \, M = 75.0 \text{ kNm/m}, \, N = 468.6 \text{ kN/m}$$
both steels yield in tension.
$$x/h = 0.094, \, M = 236.6 \text{ kNm/m}, \, N = 51.0 \text{ kN/m},$$
Tensile stress in 'compression' steel = 157 N/mm², tension steel yields.

(ii) Longitudinal walls: The wall is designed as a 4.4 m × 4.2 m slab clamped on three sides and free at top and subjected to a hydrostatic loading giving at base pressures of 55.6 kN/m² for case 1 and 46.0 kN/m² for case 2. Since the pressure difference is not large, design for Case 1 and use the same steel area for case 2 as well.

(1) Vertical bending moment at base

From Table 17.9, using the coefficient for b/a = 1.0,
$$\text{Moment at SLS} = 0.032 \times 55.6 \times 4.2^2 = 31.4 \text{ kNm/m (SLS)}$$

Vertical bending moment at base (ULS)
$$\text{Moment at ULS} = 1.4 \times 31.4 = 43.9 \text{ kNm/m (ULS)}$$
From the data in Table 17.10, using minimum steel of T12-150 mm c/c, gives at SLS and ULS moment of resistance of 96.05 kNm/m and 110.81 kNm/m respectively. Provide on both faces T12-150 mm c/c in the vertical direction.

(2) Horizontal bending moment at fixed vertical edges

From Table 17.9, using the coefficient for b/a = 1.0,
$$\text{Moment at SLS} = 0.028 \times 55.6 \times 4.2^2 = 27.5 \text{ kNm/m}$$

(3) Horizontal bending moment at mid–span

From Table 17.9, using the coefficient for b/a = 1.0,
$$\text{Moment at SLS} = 0.013 \times 55.6 \times 4.2^2 = 12.8 \text{ kNm/m}.$$

Provide on both faces T12-150 mm c/c in the horizontal and vertical directions. A_s on each face = 754 mm²/m

In case 2 there is also direct tension in the horizontal direction in the wall due to water pressure on the 8 m long walls. Average pressure p is approximately
$$p = 0.5 \times 46.0 = 23 \text{ kN/m}^2$$
Ignoring the resistance provided by the base, tensile force N per meter is
$$N = 0.5 \times 8.0 \times 23 = 92.0 \text{ kN/m}$$
The tensile stress due to tensile force is
$$= 92.0 \times 10^3 / (2 \times 754) = 61 \text{ N/mm}^2$$
The tensile force is combined with a maximum bending moment of 27.5 kNm/m. Check the crack width using steel area provided by T12-150 mm c/c on each face.

(4) Calculate the crack width using the following data

h = 400 mm, cover = 40 mm, Steel: 12 mm diameter bars at 150 mm c/c.
Applied forces at serviceability limit state: M = 27.5 kNm/m, N = 92.0 kN/m
f_{cu} = 35 N/mm², f_y = 460 N/mm², α_e = 15
b = 1000 mm, d = 354 mm, A_s = 754 mm²/m. Since the tensile force is small, ignore the steel in the compression zone.

(i) Calculate the neutral axis depth ignoring compression steel:

$$(M-N\frac{h}{2})A_s\alpha_e(d-x)=b\frac{x^2}{2}\{N(d-\frac{x}{3}-\frac{h}{2})+M\}$$

$$(27.5\times10^6 - 92.0\times10^3\frac{400}{2})\times754\times15\times(354-x)$$

$$=1000\frac{x^2}{2}\{92.0\times10^3(354-\frac{x}{3}-\frac{400}{2})+27.5\times10^6\}$$

Simplifying:
$$x^3 - 1358.74\,x^2 - 6712.24\,x + 2.3761 \times 10^6 = 0$$
$$\text{Solving,} x = 40 \text{ mm}$$

(ii) Calculate the compressive stress in concrete:

$$bx\frac{f_c}{2}(d-\frac{x}{3})=M-N\frac{h}{2}$$

$$1000\times40\times\frac{f_c}{2}(354-\frac{40}{3})=27.5\times10^6 -92.0\times10^3\times\frac{400}{2}$$

Solving,
$$f_c = 1.34 \text{ N/mm}^2 < (0.45 \times 35 = 15.8 \text{ N/mm}^2)$$

(iii) Calculate the tensile stress in steel:

$$f_s = \alpha_e f_c\frac{d-x}{x}=15\times1.34\times\frac{354-40}{40}=158 \text{ N/mm}^2$$

$$f_s = 158 \text{ N/mm}^2 < (0.8 \times 460 = 368 \text{ N/mm}^2)$$

(iv) Strain at steel level:
Taking Young's modulus for steel $E_s = 200$ kN/mm^2,
$$\varepsilon_s = f_s/E_s = 158/(200 \times 10^3) = 0.79 \times 10^{-3}$$

(v) Apparent strain at the surface
$$\varepsilon_1 = \varepsilon_s \, (h - x)/ (d - -) = 0.79 \times 10^{-3} \, (400 - 40)/ (354 - 40) = 0.91 \times 10^{-3}$$

(vi) Tension stiffening effect:
If limiting design crack width is 0.2 mm,
$$\varepsilon_2 = \frac{b_t (h - x)(a' - x)}{3 E_s A_s (d - x)}$$
$b_t = 1000$ mm, $a' = h$, because the crack is calculated at the surface of the wall.
$$\varepsilon_2 = \frac{1000(400 - 40)(400 - 40)}{3 \times 200 \times 10^3 \times 754 \times (354 - 40)} = 0.91 \times 10^{-3}$$

(vii) Average strain at the surface:
$$\varepsilon_m = \varepsilon_1 - \varepsilon_2 = 0.91 \times 10^{-3} - 0.91 \times 10^{-3} = 0$$
Section is uncracked!

(viii) Check ultimate conditions:
Applied forces:
$$M = 1.4 \times 27.5 = 38.5 \text{ kNm/m}$$
$$N = 1.4 \times 92.0 = 128.8 \text{ kN/m}$$
Using the method in Chapter 9 on columns for calculating moment–axial force interaction,
$$x/h = 0.018, \, M = 40.0 \text{ kNm/m}, \, T = 558 \text{ kN/m},$$
both steels yield in tension.
$$x/h = 0.079, \, M = 184.7 \text{ kNm/m}, \, T = 130.0 \text{ kN/m},$$
tensile stress in 'compression' steel $= 325$ N/mm^2, tension steel yields.

(f) Detailing at corners
Proper detailing of steel at corners is extremely important to realize the full strength of the sections. Fig.17.14 shows the recommended detail for closing and opening corners. The details are taken from
Standard Method of detailing structural concrete, Institution of Structural Engineers, 1989.

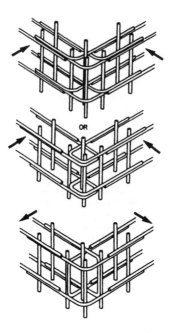

Fig.17.14 Corner details: (a) & (b) closing; (b) opening.

(g) Design of base slab for serviceability limit state
The slab is subjected to concentrated load from the walls and bending moment at the ends from the walls. There is also a small amount of direct tension from the internal pressure in the tanks but this has been ignored in the following design.

(i) Longitudinal direction:

(1) Tank empty:

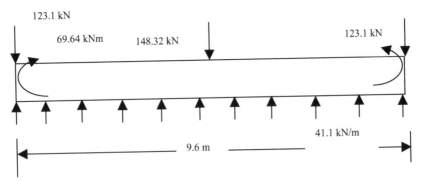

Fig.17.15 Forces on the base slab in the longitudinal direction: Tank empty.

(i) Load on end walls:

Vertical load from roof slab: $= (5.0/2) \times 0.4 \times 24 = 24.0$ kN/m
Surcharge: $(5/2) \times 12 = 30.0$ kN/m
Weight of wall: $4.2 \times 0.4 \times 24 = 40.32$ kN/m
Weight of soil on the 0.66 m projection
$= 0.66 \times (18.0 \times 1.0$ dry soil at top $+ 8.0 \times 3.2$ submerged soil$) = 28.78$ kN/m
Total $= 24.0 + 30.0 + 40.32 + 28.78 = 123.1$ kN/m
From previous calculation of wall design, moment from the external pressure
$= 0.071 \times 55.6 \times 4.2^2 = 69.64$ kNm/m

(ii) Load on central wall:

Vertical load from roof slab: $= 5.0 \times 0.4 \times 24 = 48.0$ kN/m
Surcharge: $5 \times 12 = 60.0$ kN/m
Weight of wall: $4.2 \times 0.4 \times 24 = 40.32$ kN/m
Total $= 48.0 + 60.0 + 40.32 = 148.32$ kN/m
Moment from the external pressure $= 0$

(iii) Uplift pressure: There is an uplift pressure of $10 \times 4.0 = 40$ kN/m^2

(iv) Net pressure p on the ground

$$p = (2 \times 123.1 + 148.32)/9.6 - 40.0 = 1.1 \text{ kN/m}^2$$

Fig.17.15 and Fig.17.16 show respectively the forces on the base slab and the corresponding bending moment distribution.

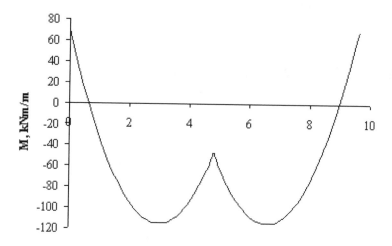

Fig.17.16 Bending moment distribution in base slab in the longitudinal direction: Tank empty.

(2) Both tanks full and no ground water:

(i) Load on end walls:

Vertical load from roof slab: $= (5.0/2) \times 0.4 \times 24 = 24.0$ kN/m
Surcharge: $(5/2) \times 12 = 30.0$ kN/m
Weight of wall: $4.2 \times 0.4 \times 24 = 40.32$ kN/m
Weight of soil on the 0.66 m projection $= 0.66 \times 18.0 \times 4.2 = 49.90$ kN/m
Total $= 24.0 + 30.0 + 40.32 + 49.90 = 144.22$ kN/m
Vertical bending moment at base (SLS)
$M = 0.071 \times (10 \times 4.2) \times 4.2^2 = 52.60$ kNm/m

(ii) Load on central wall:

Vertical load from roof slab: $= 5.0 \times 0.4 \times 24 = 48.0$ kN/m
Surcharge: $5 \times 12 = 60.0$ kN/m
Weight of wall: $4.2 \times 0.4 \times 24 = 40.32$ kN/m
Total $= 48.0 + 60.0 + 40.32 = 148.32$ kN/m
Moment from the external pressure $= 0$

(iii) There is no uplift pressure

(iv) Net pressure p on the ground
$$p = (2 \times 144.2 + 148.32)/9.6 = 45.5 \text{ kN/m}^2$$
Fig.17.17 and Fig.17.18 show respectively the forces on the base slab and the corresponding bending moment distribution.

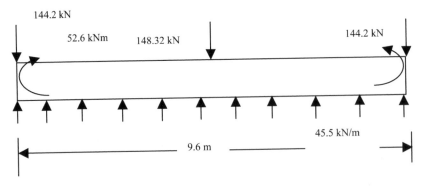

Fig.17.17 Forces on the base slab in the longitudinal direction, tank full.

Design of Reinforcement

The maximum bending moment causing tension at top is 281.0 kNm/m from tank full case and the maximum bending moment causing tension at bottom is

69.64 kNm/m from tank empty case. From Table 17.10, T25-100c/c gives a moment of resistance at SLS of 324.84 kNm and at ULS of 577.71 kNm/m. Similarly, T12-150 gives moment of resistance of 96.05 kNm/m at SLS and 110.81 kNm at ULS. Provide T25@100 at top and T12@150 at bottom.

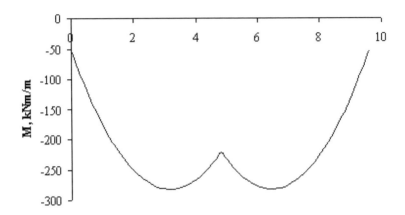

Fig.17.18 Bending moment distribution in base slab in the longitudinal direction, tank empty.

(ii) Transverse direction:
The slab is subjected to concentrated loads from the walls and bending moment at the ends from the walls.

(1) Tank empty:

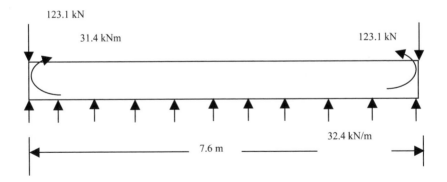

Fig.17.19 Forces on the base slab in the transverse direction, tank empty.

(i) Load on end walls:
Vertical load = 123.1 kN/m (from previous calculation for longitudinal wall)

Moment from the external pressure = $0.032 \times 55.6 \times 4.2^2 = 31.4$ kNm/m

(ii) Uplift pressure = $10 \times 4.0 = 40$ kN/m^2

(iii) Net pressure p on the ground
$$p = 2 \times 123.1/7.6 - 40.0 = -7.6 \text{ kN/m}^2$$
Although calculation indicates that the slab will not be in equilibrium, since the overall stability against floatation of the structure has been established, calculation will be continued. Fig.17.19 shows the forces on the base slab.
Maximum bending moment causing tension at bottom = 31.4 kNm/m
Maximum moment causing tension at top = $32.4 \times 7.6^2/8 - 31.4 = 205.5$ kNm/m

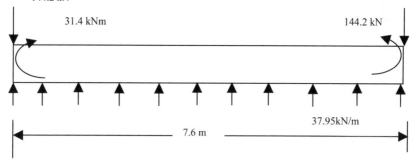

Fig.17.20 Forces on the base slab in the transverse direction, tank full.

(2) Both tanks full and no ground water:

(i) Load on end walls:
Vertical load = 144.2 kN/m (from previous calculation for longitudinal wall)
Moment from the external pressure = $0.032 \times (10 \times 4.2) \times 4.2^2 = 23.7$ kNm/m

(ii) There is no uplift pressure

(iii) Net pressure p on the ground
$$p = 2 \times 144.2/7.6 = 37.95 \text{ kN/m}^2$$
Fig.17.20 shows the forces on the base slab.
Maximum moment causing tension at top
$$= 37.95 \times 7.6^2/8 + 23.7 = 297.7 \text{ kNm/m}$$

Design of Reinforcement
The maximum bending moment causing tension at top is 297.7 kNm/m from tank full case and the maximum bending moment causing tension at bottom is 31.4 kNm/m from tank empty case. From Table 17.10, T25-100c/c gives a moment of resistance at SLS of 324.84 kNm and at ULS of 577.71 kNm/m. Similarly, T12-150 gives moment of resistance of 96.05 kNm/m at SLS and 110.81 at ULS. Provide T25-100c/c at top and T12-150 at bottom.

17.5 DESIGN OF CIRCULAR WATER TANKS

Circular water tanks are commonly employed especially in prestressed concrete. Fig.17.21 shows a circular tank subjected to an internal pressure which can be constant as in the case of gas tanks or increase towards the base as in the case of liquid retaining tanks.

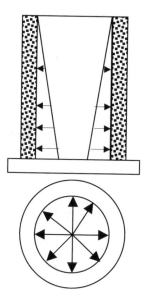

Fig.17.21 A circular tank.

If the tank is not restrained in the radial direction at top and bottom, then considering the tank as a thin walled cylinder, under a constant internal pressure p, the circumferential tension T in the wall is given by

$$T = pR$$

where R = internal radius of the tank.
The displacement w in the radial direction is given by

$$w = \frac{pR^2}{Et}$$

where E = Young's modulus, t = thickness of the wall.

If the pressure variation is hydrostatic and at p any depth y from the top is

$$p = \gamma y$$

where γ = unit weight of the liquid retained.
The circumferential tension T in the wall is given by

$$T = \gamma y \, R.$$

The displacement w in the radial direction at a depth y from the top is given by

$$w = \frac{\gamma R^2}{Et} y$$

If the displacement is constrained at the bottom, then the total pressure p is resisted partly by circumferential tension and partly by bending action in the vertical direction as shown in Fig.17.22. In addition to the bending moment in the vertical direction, there is also a bending moment in the circumferential direction given by vM, where v is the Poisson's ratio.

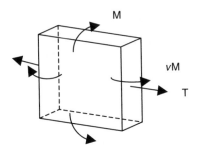

Fig.17.22 Forces on an element of the wall.

The pressure p_t resisted by tension causes a radial displacement w given by

$$w = \frac{p_t R^2}{Et}, \quad \therefore p_t = \frac{Et}{R^2} w$$

The pressure p_b resisted by bending action is given by

$$EI \frac{d^4 w}{dy^4} = p_b$$

$I = t^3/12$ per unit length, t = thickness of the wall.
Because of the Poisson effect,

$$EI = \frac{Et^3}{12(1-v^2)}$$

If p is the internal pressure,

$$p = p_b + p_t = EI \frac{d^4 w}{dy^4} + \frac{Et}{R^2} w$$

The bending moment M and shear force V and circumferential tension T are given by

$$M = -EI \frac{d^2 w}{dy^2}, \quad V = -EI \frac{d^3 w}{dy^3}, \quad T = \frac{Et}{R} w$$

The differential equation is known as the Beam on Elastic Foundation equation and can be solved for given boundary conditions. Ready made tables are available for calculating the circumferential tension T and bending moment M for the two cases of the base of the tank being either fully fixed or pinned. Fig.17.23 shows the base reinforcement details for achieving pinned and fixed joints. Table 17.11 shows typical values for a specific tank of dimensions $h^2/(Rt) = 8.0$

Table 17.11 Vertical bending moment and ring tension coefficients for cylindrical tanks

| y/h | $h^2/(Rt) = 8.0, v = 0.2$ | | | |
| | Fixed base | | Pinned base | |
	M	T	M	T
0: Top	0	0.067	0	0.017
0.1	0.0003	0.163	0.0001	0.136
0.2	0.0013	0.256	0.0006	0.254
0.3	0.0028	0.339	0.0016	0.367
0.4	0.0047	0.402	0.0033	0.468
0.5	0.0066	0.430	0.0056	0.545
0.6	0.0077	0.410	0.0084	0.579
0.7	0.0069	0.334	0.0109	0.552
0.8	0.0023	0.210	0.0118	0.446
0.9	−0.0081	0.073	0.0092	0.255
1.0: Base	−0.0267	0	0	0

Moment M = Coefficient × (γh^3) kNm/m.
Positive moment causes tension on the outer face.
Tension T = Coefficient × $(\gamma h R)$ kN/m

17.5.1 EXAMPLE OF DESIGN OF A CIRCULAR WATER TANK

Design an above ground fixed base water tank for the following specification.

(a) Specification
Internal radius R = 15 m, Height h = 6 m, wall thickness t = 300 mm
Unit weight of water γ = 10 kN/m³
Design crack width = 0.2 mm

(b) Calculation of forces
Parameter $(h^2/Rt) = 6^2/ (15 \times 0.3) = 8.0$
$q = \gamma h = 10 \times 6 = 60$ kN/m²
Fig.17.24, Fig.17.25 and Fig.17.26 show the distribution of vertical bending moment, shear force and circumferential tension.
Maximum shear force V at base at SLS:
$$V = 0.063 \times 60 \times 6 = 22.90 \text{ kN/m}$$
At ULS shear force V:
$$V = 1.4 \times 22.90 = 32.06 \text{ kN/m}$$
Maximum bending moment causing tension on inner face at base at SLS:
$$M = 0.0267 \times 60 \times 6^2 = 57.67 \text{ kNm/m. (SLS)}$$

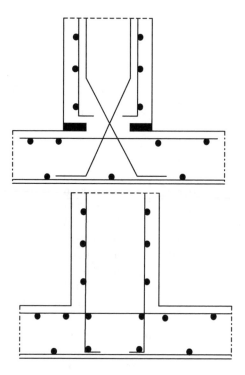

Fig.17.23 Base details for pinned and fixed joints.

M at ULS:
$$M = 1.4 \times 57.67 = 80.74 \text{ kNm/m (ULS)}$$
Maximum bending moment causing tension on the outer face at 0.4h at SLS:
$$M = 0.0077 \times 60 \times 6^2 = 16.63 \text{ kNm/m. (SLS)}$$
M at ULS:
$$M = 1.4 \times 16.63 = 23.29 \text{ kNm/m (ULS)}$$
Maximum ring tension T occurs at mid–height
$$T = 0.43 \times 60 \times 15 = 387 \text{ kN/m}$$
Corresponding moment:
$$M = 0.0066 \times 60 \times 6^2 = 14.26 \text{ kNm/m}$$
Circumferential moment:
$$= v \, M = 0.2 \times 14.17 = 2.83 \text{ kNm/m}$$

(c) Design

(i) Check shear capacity
Effective depth:
$$d = 300 - 40 \text{ cover} - 16/2 = 252 \text{ mm}$$
$$v = 32.06 \times 10^3 / (1000 \times 252) = 0.13 \text{ N/mm}^2$$
Assuming minimum area of steel $A_s = 0.35\%$,

$$v_c = \frac{0.79}{1.25} \times 0.35^{\frac{1}{3}} \times \left(\frac{400}{252}\right)^{\frac{1}{4}} \times \left(\frac{35}{25}\right)^{\frac{1}{3}} = 0.56\,\text{N/mm}^2$$

$$v < v_c$$

Depth is adequate.

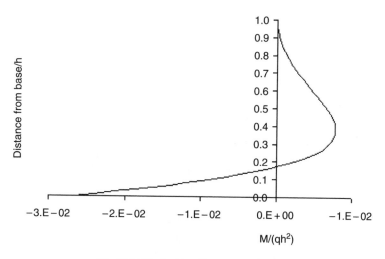

Fig.17.24 Vertical bending moment in the wall.

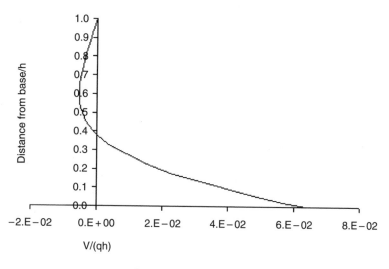

Fig.17.25 Shear force in the wall.

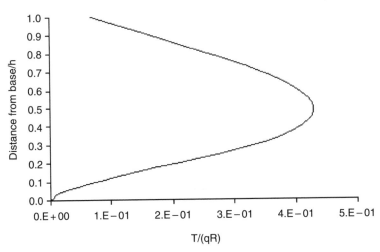

Fig.17.26 Circumferential tension in the wall.

(ii) Steel to control thermal cracking

From Table A1 of BS8007, $\rho_{crit} = 0.0035$ for 460 grade steel.

Minimum steel A_s area required = 0.0035 x 1000 x 300 = 1050 mm^2/m

$$w_{max} = \frac{f_{ct}}{f_b} \frac{\phi}{2\rho} \frac{\alpha}{2}(T_1 + T_2)$$

$w_{max} = 0.2$ mm,

$\alpha = 12 \times 10^{-6}$ from Table 3.2 of BS 8110, Part 2

$T_1 = 25°C$ (Table A.2 of BS 8007)

$T_2 = 0$, from Table A1 of 8007

$f_{ct}/f_b = 0.67$ for deformed bars of type 2.

$\phi = 12$ mm diameter bars,

$\rho = 0.30\% < (\rho_{crit} = 0.35)$

Provide T12-200 mm c/c. $A_s = 565$ mm^2/m on each face. Total steel area = 1130 mm^2/m.

$$\rho = 100 \times 1130/ (1000 \times 300) = 0.38\%.$$

(iii) Design for vertical bending:

(1) Vertical steel on inner face

M at base at SLS = 57.67 kNm/m

M at ULS = 1.4 × 57.67 = 80.74 kNm/m

Using T12-100 gives for a maximum crack width of 0.2 mm, moment capacities at SLS and ULS equal to 78.16 kNm/m and 116.47 kNm/m respectively. This steel is required for only a height of approximately 0.2 h from base. Above this only minimum steel required. Alternate bars can be terminated beyond (0.2h + anchorage length of 38 ϕ) = 1656 mm, say 1700 mm above base.

(2) Vertical steel on outer face:

$$M = 16.63 \text{ kNm/m}$$

Using T12-200 gives for a maximum crack width of 0.2 mm, moment capacities at SLS and ULS equal to 49.51 kNm/m and 59.63 kNm/m respectively.

(iv) Design for ring tension

$$\text{T at mid–height} = 0.043 \times 60 \times 15 = 387 \text{ kN/m}$$
$$\text{Circumferential moment} = 2.83 \text{ kNm/m}$$

Try tensile stress in steel equal to 300 N/mm²
$$A_s = 387 \times 10^3/300 = 1290 \text{ mm}^2/\text{m}$$

Use T16-300 on each face giving total A_s = 1340 mm²/m

(d) Check crack width

Moment is small and can be ignored.

(i) Stress and strain in steel:

$$f_s = T/A_s = 387 \times 10^3/ (1340) = 289 \text{ N/mm}^2 < (0.8 \times 460 = 368 \text{ N/mm}^2)$$

$$\varepsilon_1 = \varepsilon_s = \frac{f_s}{E_s} = \frac{289}{200 \times 10^3} = 1.44 \times 10^{-3}$$

(ii) Effect of tension stiffening:

If limiting design crack width is 0.2 mm,

$$\varepsilon_2 = \frac{2b_t h}{3E_s A_s} = \frac{2 \times 1000 \times 300}{3 \times 200 \times 10^3 \times (1340)} = 0.75 \times 10^{-3}$$

(iii) Average surface strain:

$$\varepsilon_m = \varepsilon_1 - \varepsilon_2 = (1.44 - 0.75) \times 10^{-3} = 0.69 \times 10^{-3}$$

(iv) Calculate a_{cr}

ϕ = diameter of bar, c = cover and s = bar spacing, from Fig.17.3,
$$a = \sqrt{[(s/2)^2 + (c+\phi/2)^2]} = \sqrt{[(150/2)^2 + (40+16/2)^2]} = 89 \text{ mm}$$
$$a_{cr} = a - 16/2 = 89 - 16/2 = 81 \text{ mm}$$

(v) Crack width

$$w = 3 a_{cr}\, \varepsilon_m = 3 \times 81 \times 0.69 \times 10^{-3} = 0.17 \, mm$$

Use T16-300 on each face giving A_s = 1340 mm²/m,
$$\rho = 1340/ (1000 - 300) = 0.45 \% > (\rho_{crit} = 0.35\%)$$

17.6 REFERENCES

Hughes, B.P. 1990, *Limit state theory for reinforced concrete design*, 3rd Edition, pp. 284-310, (Pitman).

Anchor, Robert .D. 1992, *Design of liquid retaining concrete structures*, 2nd Edition, (Edward Arnold).

Batty, Ian and Westbrook, Roger. 1991, *Design of water retaining concrete structures*, (Longman Scientific and Technical).

Green, Keith J. and Perkins, Phillip.H.1980, *Concrete liquid retaining structures*. (Applied Science Publishers).

Ghali, Amin.1979, *Circular storage tanks and silos*, (E. & F.N. Spon).

Anchor, Robert.D., Hill, A.W. and Hughes, B.P.1979, *Handbook on BS 5337*: 1976. (A Viewpoint Publication).

Perkins, Phillip.H.1986, *Repair, Protection and water proofing of concrete structures*, (Elsevier Applied Science).

CHAPTER 18

EUROCODE 2

The search for harmonization of technical standards throughout the European Community (EC) has led to the development of Eurocodes which are intended to replace the national codes. Eurocode 2 is the replacement standard for BS 8110. The provisions of Eurocode2(EC2) are similar to BS8110. This chapter summarises the main clauses with a few examples showing the comparison between the codes. Two useful publications are given at the end.

18.1 LOAD FACTORS

Dead load and imposed loads are designated as Permanent load and Variable loads respectively.

18.1.1 Load Factors for Ultimate Limit State

Table 18.1 Load combinations and values of γ_f for the ultimate limit state EC2 (BS 8110)

Load combinations	Load type					
	Permanent		Variable		Earth and water pressure	Wind
	Adverse	Beneficial	Adverse	Beneficial		
Permanent and variable (and earth and water pressure)	1.35 (1.40)	1.0 (1.0)	1.50 (1.60)	0 (0)	1.35 (1.40)	-
Permanent and wind (and earth and water pressure)	1.35 (1.40)	1.0 -	-	-	1.35 (1.40)	1.50 (1.40)
Permanent and variable and wind (and earth and water pressure)	1.35 (1.2)	1.0 -	1.35 (1.2)	0 -	1.35 (1.2)	1.35 (1.2)

Note that in Table 18.1, figures in brackets refer to BS 8110 values.

Table 18.1 shows the load factors to be used for various load combinations when designing for ultimate limit state.

As can be seen, the main differences are for

- Dead and imposed load: Dead load when adverse instead using a load factor of 1.4, the new value is 1.35 and for the imposed load instead of 1.6 the new load factor is 1.5.
- Dead and wind load: The new factor for dead load is 1.35 instead of 1.40 and for wind 1.5 instead of 1.4.
- Dead + Imposed + wind: The factor has increased to 1.35 from BS8110 value of 1.2.

18.1.2 Load Factors for Serviceability Limit State

Table 18.1 shows the load factors to be used for various load combinations when designing for serviceability limit state. As can be seen, the main difference is

- Dead + Imposed + wind: The factors for imposed and wind loading has decreased from 1.0 to 0.9.

Load combinations	Permanent	Variable	Wind
Permanent and variable	1.0	1.0 (1.0 or 0.25 to 0.75)*	-
Permanent and wind	1.0	-	1.0
Permanent and variable and wind	1.0	0.9 (1.0)	0.9 (1.0)

*Generally 1.0 but for deflection calculations, 0.25 for domestic or office occupancy and 0.75 for storage. (BS 8110, Part 2, clause 3.3.3)

18.2 MATERIAL SAFETY FACTORS

Table 18.2 Material safety factors γ_m

Limit State	Type of 'stress'	Material	
		Concrete	Steel
Ultimate	Flexure and axial load	1.5	1.15 (1.05)
	Shear	1.5 (1.25)	1.15 (1.05)
	Bond	1.5 (1.4)	
Serviceability		1.0	1.0

Note that in Table 18.2, figures in brackets refer to BS 8110 values.

Table 18.2 shows the material safety factors for steel and concrete. The main difference is for steel γ_m is 1.15 instead of 1.05. For concrete slightly larger values of γ_m than BS 8110 are used for shear and bond stresses.

18.3 MATERIALS

Concrete: Properties are specified in terms of characteristic cylinder strength f_{ck}. Table 18.3 shows the relationship between cylinder strength f_{ck} and corresponding cube strength f_{cu}.

Table 18.3 Relationship between cylinder and cube strength

f_{ck}	f_{cu}	f_{ck}/f_{cu}	Lowest class for use as specified
12	**15**	0.80	
16	20	0.80	Plain concrete
20	**25**	0.80	
25	30	0.83	Reinforced concrete
30	**37**	0.81	Prestressed concrete
35	45	0.78	
40	**50**	0.80	
45	55	0.82	
50	**60**	0.83	

Figures in **bold** are the preferred class.

The ratio of f_{ck}/f_{cu} varies from 0.78 to 0.83 with an average value of 0.81.

The stress–strain relationship is made up of parabolic and straight segments. The maximum stress is limited to 0.85 $f_{ck}/$ ($\gamma_c = 1.5$). Or in terms of cube strength this is approximately 0.459 f_{cu}, which corresponds to BS 8110 value of 0.447 f_{cu}. The maximum strain is limited in both codes to 0.0035. In EC2, the strain at the end of parabolic variation is 0.002 for all value of f_{ck}. In BS 8110 this is equal to $2.4 \times 10^{-4}\sqrt{}\,[f_{cu} / (\gamma_c = 1.5)]$. From the formula, for $f_{cu} = 30$ N/mm^2, the value of the strain at the end of parabolic variation is 0.001.

Steel

The specified characteristic strength is designated by f_{yk} and corresponds to f_y in BS 8110

18.4 BENDING ANALYSIS

The basic assumptions such as plane sections remaining plane, full bond between steel and concrete, maximum concrete strain limited to 0.0035 and a material safety factor on concrete of 1.5 are identical in both codes.

18.4.1 Maximum Depth of Neutral Axis x

In the absence of any redistribution of moments, in EC2 the permissible maximum depth of neutral axis depends on the value of f_{cu} and is in addition smaller than the corresponding value in BS 8110.

EC2:
$$x/d \leq 0.45 \text{ for } f_{cu} \leq 45 \text{ N/mm}^2$$
$$x/d \leq 0.35 \text{ for } f_{cu} > 45 \text{ N/mm}^2$$

BS 8110: $x/d \leq 0.5$ for all values of f_{cu}

18.4.2 Stress Block Depth

EC2: The depth of the rectangular stress block and the average compressive stress are respectively $0.8x$ and $(0.567 f_{ck} \approx 0.459 f_{cu})$

BS 8110: The depth of the rectangular stress block and the average compressive stress are respectively $0.9x$ and $0.447 f_{cu}$.

18.4.3 Maximum Moment Permitted in a Rectangular Beam With no Compression Steel

The maximum moment permitted in EC2 is generally smaller than in BS 8110. Therefore in designs based on EC2 compression steel will be required at a smaller value of maximum moment than in designs based on BS 8110.

EC2:
  (i) x/d ≤ 0.45 for f_{cu} ≤ 45 N/mm^2
$$M_{max} = 0.459 f_{cu} \, b \, 0.8 \, (0.45d) \, (d - 0.8 \times 0.45d/2)$$
$$M_{max} = 0.136 \, bd^2 \, f_{cu}$$
  (ii) x/d ≤ 0.35 for f_{cu} > 45 N/mm^2
$$M_{max} = 0.459 f_{cu} \, b \, 0.8 \, (0.35d) \, (d - 0.8 \times 0.35d/2)$$
$$M_{max} = 0.110 \, bd^2 \, f_{cu}$$
BS 8110:
For all values of f_{cu},
$$M_{max} = 0.156 \, bd^2 \, f_{cu}$$

18.4.4 Lever Arm Z

Lever arm in EC 2 will be marginally larger than in BS 8110.

EC 2:
$$M = 0.459 \ f_{cu} \ b \ 0.8x \ (d - 0.8x/2)$$
$$z = d - 0.8x/2$$
$$x = (d - z)/0.4$$

Substituting for x in terms of z
$$M = 0.918 \ b \ f_{cu} \ z \ (d - z)$$
If $k = M/ (bd^2 \ f_{cu})$,
$$z/d = 0.5 + \sqrt{(0.25 - k/0.918)}$$

BS 8110:
$$M = 0.45 \ f_{cu} \ b \ 0.9x \ (d - 0.9x/2)$$
$$z = d - 0.9x/2$$
$$x = (d - z)/0.45$$

Substituting for x in terms of z
$$M = 0.9 \ b \ f_{cu} \ z \ (d - z)$$
If $k = M/ (bd^2 \ f_{cu})$,
$$z/d = 0.5 + \sqrt{(0.25 - k/0.9)}$$

Table 18.4 Neutral axis depth and maximum value of M

% Redistribution	δ (β_b)	$f_{cu} \leq 45$		$f_{cu} > 45$		BS 8110	
		x/d	k	x/d	k	x/d	k
0	1	0.45	0.135	0.35	0.111	0.5	0.156
5	0.95	0.41	0.126	0.310	0.100	0.5	0.156
10	0.9	0.37	0.116	0.270	0.088	0.5	0.156
15	0.85	0.33	0.105	0.230	0.077	0.45	0.144
20	0.8	0.29	0.094	0.190	0.064	0.4	0.132
25	0.75	0.25	0.083	0.15	0.052	0.35	0.119
30	0.7	0.21	0.071	0.11	0.039	0.3	0.104

18.4.5 Moment Redistribution

The maximum permitted value of neutral axis depth in EC 2 depends on the value of f_{cu}.

EC2:
δ = Moment after redistribution/Moment before redistribution
$$x/d \leq (0.8 \ \delta - 0.35), \ f_{cu} \leq 45$$
$$x/d \leq (0.8 \ \delta - 0.45), \ f_{cu} > 45$$

For high ductility steel, $\delta \leq 0.7$

BS 8110:
β_b = Moment after redistribution/Moment before redistribution
$$x/d \leq (\beta_b - 0.4), \text{ for all values of } f_{cu}$$
Table 18.4 shows a comparison between EC2 and BS 8110 values of the maximum permitted values of moment for singly reinforced rectangular sections. Especially when f_{cu} is greater than 45 N/mm^2, the differences between the values are quite large.

18.5 EXAMPLES OF BEAM DESIGN FOR BENDING

Three examples are given with 'parallel' calculations for EC2 and BS 8110 rules.

18.5.1 Singly Reinforced Rectangular Beam

A simply supported rectangular beam of 8 m span carries a uniformly distributed dead load (which includes an allowance for self weight) of 7 kN/m and an imposed load of 5 kN/m. Assuming breadth, b = 250 mm, design the beam. Use $f_{cu} = 30$ N/mm^2 and $f_y = 460$ N/mm^2

EC2:
$$\text{Design load} = 1.35 \times 7 + 1.5 \times 5 = 16.95 \text{ kN/m}$$
$$\text{Ultimate moment} = 16.95 \times 8^2/8 = 135.6 \text{ kNm}$$
Minimum depth required for no compression steel:
$$x/d \leq 0.45 \text{ for } f_{cu} \leq 45 \text{ N/mm}^2, M_{max} = 0.136 \text{ bd}^2 f_{cu}$$
$$d_{min} = \sqrt{[135.6 \times 10^6/ (0.136 \times 250 \times 30)]} = 365 \text{ mm}$$

BS 8110:
$$\text{Design load} = 1.40 \times 7 + 1.6 \times 5 = 17.80 \text{ kN/m}$$
$$\text{Ultimate moment} = 17.80 \times 8^2/8 = 142.4 \text{ kNm}$$
Minimum depth required for no compression steel:
$$x/d \leq 0.5, M_{max} = 0.156 \text{ bd}^2 f_{cu}$$
$$d_{min} = \sqrt{[142.4 \times 10^6/ (0.156 \times 250 \times 30)]} = 349 \text{ mm}$$

Total depth, h:
Assuming 30 mm cover, 25 mm bars and 8 mm links, total depth h is
$$h = 365 + 30 + 8 + 25/2 = 416 \text{ mm}$$
Assume an overall depth of 425 mm
Effective depth, d:
$$d = 425 - 30 - 8 - 25/2 = 375 \text{ mm}$$

EC2:
$$k = M/ (\text{bd}^2 f_{cu}) = 135.6 \times 10^6/ (250 \times 375^2 \times 30)] = 0.129$$
$$z/d = 0.5 + \sqrt{(0.25 - 0.129/0.908)} = 0.83$$
$$A_s = 135.6 \times 10^6/ (0.83 \times 375 \times 0.87 \times 460) = 1089 \text{ mm}^2$$

BS 8110:
$$k = M/ (bd^2 f_{cu}) = 142.4 \times 10^6/ (250 \times 375^2 \times 30)] = 0.135$$
$$z/d = 0.5 + \sqrt{(0.25 - 0.135/0.9)} = 0.82$$
$$A_s = 142.4 \times 10^6/ (0.82 \times 375 \times 0.95 \times 460) = 1060 \text{ mm}^2$$
In both designs, provide 3T25 giving an area of 1472 mm^2

18.5.2 Doubly Reinforced Beam

A simply supported rectangular beam of 6 m span carries a uniformly distributed dead load (which includes an allowance for self weight) of 12.7 kN/m and an imposed load of 6 kN/m. The breadth b = 200 mm and the overall depth h is limited to 400 mm. Design the beam. Use f_{cu} = 30 N/mm^2 and f_y = 460 N/mm^2 Assume 30 mm cover, 8 mm shear links and 25 mm diameter bars. Effective depth
$$d = 400 - 30 - 8 - 25/2 = 350 \text{ mm}$$
$$d' = 30 + 8 + 25/2 = 51 \text{ mm}$$

EC2:
$$\text{Design load} = 1.35 \times 12.7 + 1.5 \times 6.0 = 26.15 \text{ kN/m}$$
$$\text{Ultimate moment} = 26.15 \times 6^2/8 = 117.68 \text{ kNm}$$
Maximum moment allowed with no compression steel:
$$f_{cu} \leq 45 \text{ N/mm}^2$$
$$M_{max} = 0.136 \text{ bd}^2 f_{cu} = 0.136 \times 200 \times 350^2 \times 30 \times 10^{-6} = 99.96 \text{ kNm}$$
$$M_{max} < 117.68$$
Compression steel required.
$$x = 0.45 \text{ d} = 0.45 \times 350 = 157.5 \text{ mm}$$
$$\text{stress block depth} = 0.8 \times 157.5 = 126 \text{ mm}$$
$$\text{lever arm, } z = 350 - 0.5 \times 126 = 287 \text{ mm}$$
$$\text{Strain in compression steel} = 0.0035 \times (157.5 - 51)/157.5 = 2.37 \times 10^{-3}$$
Compression steel yields.
$$A_s = 99.96 \times 10^6/ (287 \times 0.87 \times 460)$$
$$+ (117.68 - 99.96) \times 10^6/ [(350 \times 51) \times 0.87 \times 460]$$
$$A_s = 870 + 148 = 1018 \text{ mm}^2,$$
$$A_s` = 148 \text{ mm}^2$$

BS 8110:
$$\text{Design load} = 1.40 \times 12.7 + 1.6 \times 6.0 = 27.38 \text{ kN/m}$$
$$\text{Ultimate moment} = 27.38 \times 6^2/8 = 123.21 \text{ kNm}$$
Maximum moment allowed with no compression steel:
$$M_{max} = 0.156 \text{ bd}^2 f_{cu} = 0.156 \times 200 \times 350^2 \times 30 \times 10^{-6} = 114.66 \text{ kNm}$$
$$M_{max} < 123.21$$
Compression steel required.
$$x = 0.5 \text{ d} = 0.5 \times 350 = 175 \text{ mm}$$
$$\text{stress block depth} = 0.9 \times 175 = 157.5 \text{ mm},$$
$$\text{lever arm, } z = 350 - 0.5 \times 157.5 = 271 \text{ mm}$$

Strain in compression steel = $0.0035 \times (175 - 51)/175 = 2.48 \times 10^{-3}$ Compression steel yields.

$$A_s = 114.66 \times 10^6/ (271 \times 0.95 \times 460)$$
$$+ (123.21 - 114.66) \times 10^6/ [(350 - 51) \times 0.95 \times 460]$$
$$A_s = 968 + 66 = 1034 \text{ mm}^2$$
$$A_s^` = 66 \text{ mm}^2$$

In both designs, provide

Tension steel: 2T25 + 1T16 = 1182 mm^2
Compression steel: 2T12 = 226 mm^2.

18.5.3 T-Beam Design

Determine the area of reinforcement required for the T-beam section with breadth of flange, b = 600 mm, depth of flange, h_f = 100 mm, width of web, b_w = 250 mm, overall depth = 425 mm.
The beam is subjected to an ultimate moment of 280 kN m. The materials are grade 30 concrete and grade 460 reinforcement.
Assume cover = 30 mm, link diameter = 8 mm, 25 mm bars, effective depth d:

$$d = 425 - 30 - 8 - 25/2 = 375 \text{ mm}$$

EC 2:
Calculate M_{flange} to check if the stress block is inside the flange.
$$M_{flange} = 0.459 \, f_{cu} \, b \, h_f \, (d - h_f/2)$$
$$M_{flange} = 0.459 \times 30 \times 600 \times 100 \times (375 - 0.5 \times 100) \times 10^{-6} = 268.5 \text{ kNm}$$
The design moment of 280 kNm is greater than M_{flange}. Therefore the stress block extends in to the web.
Check if compression steel is required.
$$M_{max} = 0.459 \, f_{cu} \, (b - b_w) \, h_f(d - h_f/2) + 0.136 \, f_{cu} \, b_w \, d^2$$
$$M_{max} = \{0.459 \times 30 \times (600 - 250) \times 100 \times (375 - 100/2)$$
$$+ 0.136 \times 30 \times 250 \times 375^2\} \times 10^{-6}$$
$$M_{max} = 156.63 + 143.44 = 300.1 \text{ kNm}$$
$$M_{max} > (M = 280 \text{ kNm})$$
The beam does not need compression steel.
Determine the depth of the neutral axis from

$$\frac{M}{bd^2 f_{cu}} = 0.459(1 - \frac{b_w}{b})\frac{h_f}{d}(1 - \frac{h_f}{2d}) + 0.459\frac{b_w}{b}0.8\frac{x}{d}(1 - 0.4\frac{x}{d})$$

$$\frac{280 \times 10^6}{600 \times 375^2 \times 30} = 0.459(1 - \frac{250}{600})\frac{100}{375}(1 - \frac{100}{2 \times 375})$$

$$+ 0.459\frac{250}{600}0.8\frac{x}{d}(1 - 0.4\frac{x}{d})$$

Setting $x/d = \alpha$

$$0.1106 = 0.0619 + 0.153 \, \alpha - 0.0612 \, \alpha^2$$

Simplifying

$$\alpha^2 - 2.50\,\alpha + 0.7958 = 0$$

Solving the quadratic in α,

$$\alpha = x/d = (2.50 - 1.7524)/2 = 0.3738$$
$$x = 0.3738 \times 375 = 140 \text{ mm} < (0.45\,d = 169 \text{ mm})$$
$$T = 0.95\,f_y\,A_s = 0.459\,f_{cu}\,(b - b_w)\,h_f + 0.459\,f_{cu}\,b_w\,0.8\,x$$
$$T = (0.459 \times 30 \times (600 - 250) \times 100 + 0.459 \times 30 \times 250 \times 0.8 \times 140 \times 10^{-3}$$
$$T = (472.5 + 385.6) = 858.1 \text{ kN}$$
$$A_s = 858.1 \times 10^3 / (0.87 \times 460) = 2144 \text{ mm}^2$$

BS 8110:

Calculate M_{flange} to check if the stress block is inside the flange.

$$M_{flange} = 0.45\,f_{cu}\,b\,h_f\,(d - h_f/2)$$
$$M_{flange} = 0.45 \times 30 \times 600 \times 100 \times (375 - 0.5 \times 100) \times 10^{-6}$$
$$M_{flange} = 263.25 \text{ kNm}$$

The design moment of 280 kNm is greater than M_{flange}. Therefore the stress block extends in to the web.

Check if compression steel is required.

$$M_{max} = 0.45\,f_{cu}\,(b - b_w)\,h_f\,(d - h_f/2) + 0.156\,f_{cu}\,b_w\,d^2$$
$$M_{max} = \{0.45 \times 30 \times (600 - 250) \times 100 \times (375 - 100/2)$$
$$+ 0.156 \times 30 \times 250 \times 375^2\} \times 10^{-6}$$
$$M_{max} = (153.6 + 164.5) = 318.1 \text{ kNm}$$
$$M_{max} > (M = 280 \text{ kNm})$$

The beam can be designed without any need for compression steel. Two approaches can be used for determining the area of tension steel required.

(a) Exact approach:

Determine the depth of the neutral axis from

$$\frac{M}{bd^2 f_{cu}} = 0.45(1 - \frac{b_w}{b})\frac{h_f}{d}(1 - \frac{h_f}{2d}) + 0.45\frac{b_w}{b}0.9\frac{x}{d}(1 - 0.45\frac{x}{d})$$

$$\frac{280 \times 10^6}{600 \times 375^2 \times 30} = 0.45(1 - \frac{250}{600})\frac{100}{375}(1 - \frac{100}{2 \times 375}) + 0.45\frac{250}{600}0.9\frac{x}{d}(1 - 0.45\frac{x}{d})$$

Setting $x/d = \alpha$

$$0.1106 = 0.0607 + 0.1688\,\alpha - 0.0759\,\alpha^2$$

Simplifying

$$\alpha^2 - 2.22\,\alpha + 0.657 = 0$$

Solving the quadratic in α,

$$\alpha = x/d = (2.22 - 1.5167)/2 = 0.352$$
$$x = 0.352 \times 375 = 132 \text{ mm} < (0.5\,d = 188 \text{ mm})$$
$$T = 0.95\,f_y\,A_s = 0.45\,f_{cu}\,(b - b_w)\,h_f + 0.45\,f_{cu}\,b_w\,0.9\,x$$
$$T = (0.45 \times 30 \times (600 - 250) \times 100 + 0.45 \times 30 \times 250 \times 0.9 \times 132) \times 10^{-3}$$
$$T = (472.5 + 400.95) = 873.5 \text{ kN}$$
$$A_s = 873.5 \times 10^3 / (0.95 \times 460) = 1999 \text{ mm}^2$$

(b) Calculation of A_s using simplified code formula which uses x = 0.5

$$A_s = \frac{280\times10^6 + 0.1\times30\times250\times375\times(0.45\times375-100)}{0.95\times460\times(375-0.5\times100)} = 2108 \text{ mm}^2$$

In both designs, provide 3T32, $A_s = 2412$ mm^2

18.6 SHEAR DESIGN: STANDARD METHOD

EC2 allows two methods for shear design, Standard method and Variable inclination method. Only the former is considered here.

18.6.1 Maximum Permissible Shear Stress

EC2:

$$v_{max} = 0.3 \text{ v } f_{ck}$$
$$v = (0.7 - f_{ck}/200) \geq 0.5 \text{ i.e. } f_{ck} \leq 40 \text{ N/mm}^2$$
$$\text{Taking } f_{ck} \approx 0.81 f_{cu},$$
$$v_{max} = 0.243 (0.7 - f_{cu}/247) f_{cu}, f_{cu} \leq 50 \text{ N/mm}^2$$

BS 8110:

$$v_{max} = 0.8\sqrt{f_{cu}}$$

Table 18.5 shows a comparison between the v_{max} permitted by the two codes.

Table 18.5 Maximum permissible shear stress

f_{cu}	f_{ck}	v_{max}	
		EC2	BS 8110
25	20.25(20)	3.64	4.00
30	24.30 (25)	4.22	4.38
35	28.35	4.75	4.73
40	32.40	5.23	5.06
50	40.50 (40)	6.05	5.66

Figures in brackets correspond to values in Table 18.2.

18.6.2 Permissible Shear Stress in Reinforced Concrete

EC2:

$$v_c = 0.035 f_{ck}^{(2/3)} (1.6 - d/1000) \{1.2+ 0.4 (100 A_s/b_v d)\}$$
$$\text{(A material safety factor of } \gamma_m = 1.5 \text{ is included)}$$
$$100 A_s/ (b_v d) \leq 2, (1.6 - d/1000) > 1.0, \text{ i.e. } d < 600 \text{ mm}$$
$$\text{Taking } f_{ck} \approx 0.81 f_{cu}, 0.035 f_{ck}^{(2/3)} = 0.030 f_{cu}^{(2/3)}$$
$$v_c = 0.03 f_{cu}^{(2/3)} (1.6 - d/1000) \{1.2+ 0.4 (100 A_s/b_v d)\}$$

BS 8110

$$v_c = (0.79/1.25) \, (f_{cu}/25)^{(1/3)} \, (400/d)^{(1/4)} \, (100 \, A_s/b_v d)^{(1/3)}$$
(A material safety factor of $\gamma_m = 1.25$ is included)
$$f_{cu} \leq 40 \text{ N/mm}^2, \; 100 \, A_s/(b_v d) \leq 3, \; 400/d \geq 1, \text{ i.e. } d \leq 400$$

Table 18.6 shows a comparison between EC2 and BS 8110 values for v_c for $f_{cu} = 30 \text{ N/mm}^2$

Table 18.6 v_c values for EC2 and BS 8110, $f_{cu} = 30 \text{ N/mm}^2$

ρ%	Code	v_c N/mm^2						
		d, mm						
		150	200	250	300	400	500	600
0.15	EC 2	0.53	0.51	0.49	0.47	0.44	0.40	0.36
	BS8110	0.46	0.42	0.40	0.38	0.36	0.36	0.36
0.25	EC 2	0.55	0.53	0.51	0.49	0.45	0.41	0.38
	BS8110	0.54	0.50	0.48	0.45	0.42	0.42	0.42
0.5	EC2	0.59	0.57	0.55	0.53	0.49	0.45	0.41
	BS8110	0.68	0.63	0.60	0.57	0.53	0.53	0.53
0.75	EC 2	0.63	0.61	0.59	0.56	0.52	0.48	0.43
	BS8110	0.78	0.73	0.69	0.66	0.61	0.61	0.61
1.00	EC 2	0.67	0.65	0.63	0.60	0.56	0.51	0.46
	BS8110	0.86	0.80	0.76	0.72	0.67	0.67	0.67
1.50	EC 2	0.76	0.73	0.70	0.68	0.63	0.57	0.52
	BS8110	0.98	0.91	0.86	0.83	0.77	0.77	0.77
2.00	EC 2	0.84	0.81	0.78	0.75	0.70	0.64	0.58
	BS8110	1.08	1.01	0.95	0.91	0.85	0.85	0.85
2.50	EC 2	0.84	0.81	0.78	0.75	0.70	0.64	0.58
	BS8110	1.16	1.08	1.02	0.98	0.91	9.91	0.91
3.00	EC 2	0.84	0.81	0.78	0.75	0.70	0.64	0.58
	BS8110	1.24	1.15	1.09	1.04	0.97	0.97	0.97

18.6.3 Total Shear Capacity

Symbols inside brackets are BS 8110 symbols.

$$V_{Rd3} \, (V) = \text{Applied shear force}$$
$$V_{cd} \text{ or } V_{Rd1} \, (V_c) = \text{Shear force resisted by reinforced concrete} = v_c \, b_v \, d$$
$$V_{wd} \, (V_s) = \text{Shear resistance provided shear reinforcement}$$
$$V_{Rd3} \, (V) = V_{cd} \text{ or } V_{Rd1} \, (V_c) + V_{wd} \, (V_s)$$

18.6.4 Shear Reinforcement in the Form of Links

EC2: $A_{sw} = 1.28(v - v_c)\dfrac{b_v s}{f_{yk}}$

where $f_{yk} = f_{yv}$ and $A_{sw} = A_{sv}$ in BS 8110 notation.
This includes the factor of safety of 0.87 on f_{yk}.
The shear force carried by the links is V_{wd} (V_s).

$$V_{wd}(V_s) = (v - v_c)b_v d = 0.781 A_{sw} f_{yk} \frac{d}{s}$$

BS 8110:

$$A_{sv} = (v - v_c)\frac{b_v s}{0.95 f_{yv}}$$

$$V_{wd}(V_s) = (v - v_c)b_v d = 0.95 A_{sv} f_{yv} \frac{d}{s}$$

18.6.5 Maximum Permitted Spacing of Links

In EC 2, the rules for maximum permitted link spacing are a function of the ratio of design shear force to maximum permitted shear force. BS 8110 adopts a much simpler rule.

EC 2:
Link Spacing should not exceed the smaller of:

a. 0.8d or 300 mm if V_{sd} (V) $\leq$ 0.2 V_{Rd2} (V_{max})
b. 0.6d or 300 mm if 0.2 V_{Rd2} (V_{max}) < V_{sd} (V) $\leq$ 0.67 V_{Rd2} (V_{max})
c. 0.3d or 200 mm if V_{sd} (V) > 0.67 V_{Rd2} (V_{max})

BS 8110
Link Spacing should not exceed 0.75 d.

18.6.6 Minimum Area of Links

EC2:

$$\frac{A_{sw}}{s\, b_v} = \rho_w$$

In order to compare with the equations in BS 8110, using the values of ρ_w in Table 18.7, the equation for minimum shear steel can be expressed as follows.

$$\frac{A_{sw}}{s\, b_v} = \frac{C}{0.87 f_{yv}}$$

Table 18.7 shows the value of C.
BS 8110:

$$\frac{A_{sw}}{s\,b_v} = \frac{0.4}{0.95 f_{yv}}, \therefore \rho_w = \frac{0.4}{0.95 f_{yv}}$$

Table 18.7 Values of ρ_w for minimum stirrups

Concrete strength class in f_{cu}	$(f_{yk})f_{yv} = 250$ N/mm^2		$(f_{yk})f_{yv} = 460$ N/mm^2	
	ρ_w %	C	ρ_w %	C
C15 and C 25	0.15(0.17)	0.33	0.08(0.09)	0.32
C30 to C45	0.22(0.17)	0.48	0.12(0.09)	0.48
C50 to C60	0.28(0.17)	0.61	0.14(0.09)	0.56

Note: Numbers in brackets for ρ_w refer to BS 8110 values.
The value of C for BS 8110 is 0.4 in all cases.

EC2 requires a greater amount of minimum links than BS 8110

18.6.7 Example of Shear Design

A simply supported T-beam carries an ultimate load of 38 kN/m over a span of
6.5m. The supports are 200 mm wide. The T-beam has:
web width, b_w (b_v) = 250mm, overall depth, h = 450 mm
It is reinforced by 5T20 bars in tension at mid-span and curtailed to 3T20 towards
the support. Design 8 mm high yield steel shear links required at a distance of d
from the face of the support. The materials are grade 30 concrete and grade 460
reinforcement.
Note: EC 2 uses b_w for web width and BS 8110 uses b_v.
Assume 30 mm cover. Effective depth
$$d = 450 - 30 - 8 - 20/2 = 402 \text{ mm}$$
$$\text{Reaction} = 38 \times 6.5/2 = 123.5 \text{ kN}$$
Shear force at d from the face of support:
$$V = 123.5 - 38 \times (200/2 + 402) \times 10^{-3} = 104.42 \text{ kN}$$
Design shear force V_{Rd3} (V) = 104.42 kN
$$v = V_{Rd3} \text{ (V)}/ (b_w\ d)$$
$$v = 104.42 \times 10^3/ (250 \times 402) = 1.04 \text{ N/mm}^2$$
$$A_s = 3T20 = 943 \text{ mm}^2$$
$$100\ A_s/ (b_v d) = 100 \times 943/ (250 \times 402) = 0.94$$

EC2:

$$v_{max} = 0.24 \times (0.7 - 30/250) \times 30 = 4.18 \text{ N/mm}^2$$
$$V_{Rd2} \text{ (}V_{max}\text{)} = b_w\ d\ v_{max} = 250 \times 402 \times 4.18 \times 10^{-3} = 420.1 \text{ kN}$$
$$v < v_{max}$$

$$100 \ A_s/ \ (b_v d) = 0.94 < 2.0$$
$$d = 402 < 600 \ mm$$
$$v_c = 0.03 \ f_{cu}^{(2/3)} \ (1.6 - d/1000) \ \{1.2 + 0.4 \ (100 \ A_s/b_v d)\}$$
$$v_c = 0.03 \times (30)^{(2/3)} \times (1.6 - 402/1000) \times \{1.2 + 0.4 \times 0.94\}$$
$$v_c = 0.55 \ N/mm^2$$
$$v_c < v < v_{max}$$

Design Shear links: 2 leg 8 mm links, $A_{sv} = 100 \ mm^2$

$$A_{sw} = 1.28(v - v_c)\frac{b_v s}{f_{yk}}$$

$$100 = 1.28(1.04 - 0.55)\frac{250 \times s}{460} \ , \ s = 293 \ mm$$

Is $[V_{sd} \ (V) = 104.42] \leq [0.2 \ V_{Rd2} \ (V_{max}) = 0.2 \times 420.1 = 84.0]$, No.

Is $[0.2 \ V_{Rd2} \ (V_{max}) = 84.0] < [V_{sd} \ (V) = 104.42] \leq [0.67 \ V_{Rd2} \ (V_{max}) = 280.1]$, Yes

Link spacing is smaller of ($0.6d = 241$ mm) or 300 mm
Therefore maximum spacing is limited to 240 mm.

Minimum links: $f_{yv} = 460 \ N/mm^2$ and $f_{cu} = 30 \ N/mm^2$, $A_{sv} = 100 \ mm^2$, $b_v = 250$ mm

$$\frac{A_{sw}}{s \ b_v} = 0.0012$$

$s = 333$ mm. Say 300 mm
Shear resisted by minimum links:

$$V_{wd} (V_s) = (v - v_c)b_v d = 0.781 \ A_{sw} \ f_{yk} \ \frac{d}{s} = 0.781 \times 100 \times 460 \times \frac{402}{300} \times 10^{-3} = 48.1 kN$$

Assuming $v_c = 0.55 \ N/mm^2$, shear resisted by concrete:

$$V_{cd} \ or \ V_{Rd1} \ (V_c) = v_c \ b_v \ d = 0.55 \times 250 \times 402 \times 10^{-3} = 55.3 \ kN$$

Total shear force V resisted in the region of minimum shear links is

$$V_{Rd3} \ (V) = V_{Rd1} \ (V_c) + V_{wd} \ (V_s) = 55.3 + 48.1 = 103. \ 4 \ kN$$

Shear force at d from the face of support = 104.42 kN
Provide four links at 240 mm from the centre of support at either end and the rest at 300 mm c/c.

BS 8110:
$$v_{max} = 0.8\sqrt{30} = 4.38 \ N/mm^2$$
$$v < v_{max}$$
$$100 \ A_s/ \ (b_v d) = 0.94 < 3.0$$
$$d = 402 > 400 \ mm, \ 400/d = 1$$
$$f_{cu} = 30 < 40 \ N/mm^2$$
$$v_c = (0.79/1.25) \ (100 \ A_s/b_v d)^{(1/3)} \ (400/d)^{(1/4)} \ (f_{cu}/25)^{(1/3)}$$

$$v_c = (0.79/1.25) \times (30/25)^{(1/3)} \times (1.0)^{(1/4)} \times (0.94)^{(1/3)} = 0.66 \text{ N/mm}^2$$

$$v_c < v < v_{max}$$

Design Shear links: 2 leg 8 mm links, $A_{sv} = 100 \text{ mm}^2$

$$A_{sw} = 100 = (v - v_c) \frac{b_v s}{0.95 f_{yk}} = (1.04 - 0.66) \frac{250 \times s}{0.95 \times 460} , \quad s = 460 \, mm$$

Maximum spacing is limited to 0.75 d = 300 mm.

Minimum links: $f_{yv} = 460 \text{ N/mm}^2$, $A_{sv} = 100 \text{ mm}^2$, $b_v = 250 \text{ mm}$

$$A_{sv} = 0.4 \frac{b_v s}{0.95 f_{yv}}$$

$$s = 437 \text{ mm} > (0.75d = 301 \text{ mm})$$

Shear resisted by the minimum links:

$$V_{wd}(V_s) = (v - v_c) b_v d = 0.95 A_{sv} f_{yv} \frac{d}{s}$$

$$V_{wd}(V_s) = 0.95 \, A_{sw} f_{yk} \frac{d}{s} = 0.95 \times 100 \times 460 \times \frac{402}{300} \times 10^{-3} = 58.6 \, kN$$

Assuming $v_c = 0.55 \text{ N/mm}^2$, shear resisted by concrete:

$$V_c = v_c b_v d = 0.55 \times 250 \times 402 \times 10^{-3} = 55.3 \text{ kN}$$

Total shear force V resisted in the region of minimum shear links is

$$V = V_c + V_s = 58.6 + 48.1 = 106.7 \text{ kN}$$

Shear force at d from the face of support = 104.42 KN
Provide minimum links at 300 mm c/c throughout the beam.

The main difference between the two designs is in the region approximately at d from the face of the support, where closer spacing of the links is needed according to EC2.

18.7 PUNCHING SHEAR

18.7.1 Location of Critical Perimeter

In EC2, critical perimeters are at a constant distance from the loaded area while in BS 8110 critical perimeters have the same shape as the loaded area.

In EC 2 the location of the first critical perimeter is at a *constant* distance of 1.5 d from the loaded area. BS 8110 considers rectangular perimeters at a distance of 1.5 d from the loaded area. For a rectangular loaded area a x b, the critical perimeters are (Fig. 18.1):

EC2: perimeter, u = 2(a + b) + 3π d

BS 8110: perimeter, $u = 2(a + b) + 12d$

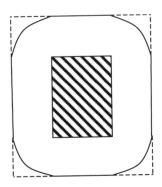

Fig. 18.1 Critical shear perimeters in EC2 and BS 8110.

18.7.2 Maximum Permissible Shear Stress, V_{max}

$v_{max} = V / (u_0 d)$, where u_0 = perimeter of loaded area.
EC2: $v_{max} \leq 0.9\sqrt{f_{ck}} \approx 0.81\sqrt{f_{cu}}$
BS 8110: $v_{max} \leq 0.8\sqrt{f_{cu}}$ or 5 N/mm^2

18.7.3 Permissible Shear Stress, v_c

The values are as discussed in section 18.5.2. Generally enhancement of shear strength close to supports is not permitted.

EC2: The code recommends that d should be the average of the effective depth in two directions and the reinforcement ratio $\rho = A_s/ (b_v d)$ used in the expression for v_c is given by

$$\rho = \text{lesser of } [\sqrt{\rho_x \rho_y}] \text{ and } 0.015$$

In order to allow for combined effect of bending and shear, the design effective shear

$$V_{eff} = \alpha V$$

where

$\alpha = 1.50$ for corner column,
$\alpha = 1.40$ for edge column,
$\alpha = 1.15$ for internal column.

BS 8110: BS 8110 assumes that the reinforcement ratio used in calculation of v_c is the average of values in two directions. The multiplication factor α for calculating design effective shear V_{eff} is given by

$\alpha = 1.25$ for corner and edge columns for bending parallel to free edge.
$\alpha = 1.40$ for corner columns for bending perpendicular to free edge.

$$\alpha = 1.15 \text{ for internal column.}$$

18.7.4 Shear Reinforcement

EC 2: The maximum shear capacity is limited to $v = 1.6 \, v_c$. Shear reinforcement is required if

$$v_c < v \leq 1.6 \, v_c$$

Shear reinforcement provides shear capacity of V_s where

$$V_s = 0.87 \, f_y \sum A_{sv}$$

where $\sum A_{sv}$ is the sum of cross-sectional areas of all stirrups with in a distance of 1.5 d or 800 mm which ever is smaller from the edge of the loaded area. Further critical perimeters at a distance of 1.5 d from the outside of the shear reinforced area must be checked for need for shear reinforcement.

BS 8110: See Chapter 5, section 5.1.8.

18.7.5 Example

A flat slab 250 mm thick overall is reinforced by 16 mm diameter bars at 175 mm both ways. Assuming $f_{cu} = 30$ N/mm^2 and $f_y = 460$ N/mm^2, design shear reinforcement to prevent punching failure at an edge column. The axial force in the 300 mm square column is 350 kN. Take cover = 30 mm.

Effective depths:

$$d_x = 250 - 30 - 16/2 = 212 \text{ mm}$$
$$d_y = 250 - 30 - 16 - 16/2 = 196 \text{ mm}$$
$$d_{average} = 204 \text{mm}$$
$$A_{sx} = A_{sy} = 1150 \text{ mm}^2/\text{m}$$
$$\rho_x = A_{sx}/(1000 \times d_x) = 0.00543$$
$$\rho_y = A_{sx}/(1000 \times d_y) = 0.00587$$

EC2:

$$\rho = \sqrt{\rho_x \, \rho_y} = 0.00564 < 0.015$$
$$100 \, A_s/(b_v d) = 0.56 < 2.0$$
$$d = 204 < 600 \text{ mm}$$
$$v_c = 0.03 \, f_{cu}^{(2/3)} (1.6 - d/1000) \{1.2 + 0.4 (100 \, A_s/b_v d)\}$$
$$v_c = 0.03 \times (30)^{(2/3)} \times (1.6 - 204/1000) \times \{1.2 + 0.4 \times 0.56\} = 0.58 \text{ N/mm}^2$$
$$V = 350 \text{ kN}$$

Edge column: $\alpha = 1.4$

$$V_{eff} = 1.4 \times 350 = 490 \text{ kN}$$

Check shear around the column perimeter:

$$u_0 = 4 \times 300 = 1200 \text{ mm}$$
$$v_{max} = 490 \times 10^3/(1200 \times 204) = 2.00 \text{ N/mm}^2 < (0.81 \sqrt{f_{cu}} = 4.44 \text{ N/mm}^2)$$

Depth of slab is adequate.

Calculate shear stress at a perimeter 1.5 d from loaded area:

$$u = 2(300+300) + 2\pi(1.5d) = 3123 \text{ mm}$$
$$v = 490 \times 10^3/(3123 \times 204) = 0.77 \text{ N/mm}^2$$

$$(v_c = 0.58) < (v = 0.77) < (1.6v_c = 0.93)$$

Slab needs shear reinforcement.
Design of shear reinforcement:
$$V_c = ud \ v_c = 3123 \times 204 \times 0.58 \times 10^{-3} = 369.5 \text{ kN}$$
$$V_s = 0.87 \ f_y \ \Sigma A_{sv} = V - V_c = 490 - 369.5 = 120.49 \text{ kN}$$
$$\Sigma A_{sv} = 120.49 \times 10^3/(0.87 \times 460) = 301 \text{ mm}^2$$
Minimum steel area required is 60% of that required for beams. From Table 18.7,
$$\rho_w = 0.6 \times 0.0012 = 0.00072$$

$$\rho_w = \frac{\Sigma A_{sv}}{[A_{crit} - A_{Load}]}$$

A_{crit} = Area inside the critical perimeter, A_{load} = Loaded area
$$A_{crit} - A_{load} = 2 \times 300 \times 1.5d + 2 \times 300 \times 1.5d + 2\pi (1.5d)^2$$
$$= 4 \times 300 \times 306 + 2 \pi (306)^2 = 0.956 \times 10^6 \text{ mm}^2$$
Minimum area of links required:
$$\Sigma A_{sv} = 0.00072 \times 0.956 \times 10^6 = 688 \text{ mm}^2 > (301 \text{ mm}^2 \text{ required})$$

Provide 8 mm diameter, 2-leg links in two perpendicular directions. Area of one link = 100 mm². Provide 8 links in all.

BS 8110:
$$\text{Average } 100 \ A_s/ (b_v d) = 0.565 < 3.0,$$
$$400/d = 1.96 > 1.0$$
$$v_c = (0.79/1.25) (100 \ A_s/b_v d)^{(1/3)} (400/d)^{(1/4)} (f_{cu}/25)^{(1/3)}$$
$$v_c = (0.79/1.25) \times (0.565)^{(1/3)} \times (1.96)^{(1/4)} \times (30/25)^{(1/3)} = 0.66 \text{ N/mm}^2$$
$$V = 350 \text{ kN}$$

Edge column: $\alpha = 1.25$
$$V_{eff} = 1.25 \times 350 = 437.5 \text{ kN}$$
Check shear around the column perimeter:
$$u_0 = 4 \times 300 = 1200 \text{ mm}$$
$$v_{max} = 437.5 \times 10^3/ (1200 \times 204) = 1.79 \text{ N/mm}^2 < (0.8\sqrt{f_{cu}} = 4.38 \text{ N/mm}^2)$$
Depth of slab is adequate.

Calculate shear stress at a perimeter 1.5 d from loaded area:
$$u = 2(300+300) + 12d = 3648 \text{ mm}$$
$$v = 437.5 \times 10^3/ (3648 \times 204) = 0.59 \text{ N/mm}^2$$
$$(v_c = 0.66) > (v = 0.59)$$

No shear reinforcement is necessary.

18.8 COLUMNS

This section deals with the design of short columns only.

18.8.1 Short or Slender Column?

EC 2:

The slenderness of a column is defined by λ the ratio of effective length ℓ_0 to the radius of gyration r.

$$\lambda = \ell_0/r, \ \ell_0 = \beta \, \ell_{col}, \ r = \sqrt{(I/A)},$$

I = second moment of area of the section about the axis being considered,
A = cross sectional area.
ℓ_{col} = height of the column between the centres of restraint
β = coefficient which is a function of the ratios of column to beam stiffnesses k_a and k_b at the top and bottom respectively of the column.

$$k_a \ or \ k_b = \frac{\Sigma \dfrac{I_{col}}{\ell_{col}}}{\Sigma \dfrac{\alpha I_{beam}}{\ell_{beam}}}$$

The summation sign Σ indicates that all columns and all beams framing in to the joint at a or b should be included. The coefficient α depends on the fixity at the end of the beam remote from the joint. $\alpha = 1.0$ for continuous end and 0.5 for simply supported end and 0 for a cantilever.

If $\lambda \leq \lambda_{min}$ then the column is considered short and slenderness effects can be ignored.

λ_{min} is greater of 25 or $15/\sqrt{v_u}$
where

$$v_u = N_{sd}/ (A_c \ f_{cd}),$$
$$N_{sd} = \text{Design axial load on column}$$
$$A_c = \text{Area of cross section}$$
$$f_{cd} = f_{ck}/1.5 \approx 0.54 \ f_{cu}$$

BS 8110:

The code gives in Tables 3.19 and 3.20, β values for different end conditions. Three types of conditions are defined for braced columns and four types for unbraced columns. The effective length ℓ_e of the column is equal to $\beta \, \ell_{col}$. A Column is short if ℓ_e /h and ℓ_e/ b are both less than 15 for braced column and 10 for unbraced column.

In Part 2, the following equations are given:
Braced Columns:

$$\beta = \text{lesser of } [\{0.7 + 0.05(k_a + k_b)\} \text{ or } \{0.85 + \text{minimum of } (k_a \text{ and } k_b)\}]$$

18.8.2 Example

Design the column shown in Fig. 18.2.
Design axial load N = 1800 kN, f_{cu} = 30 N/mm²
Column = 400 × 400 mm, height between floor, ℓ_{col} = 3.5 m

EC2:

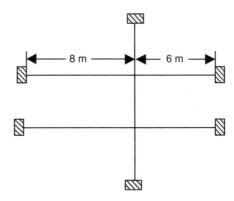

Fig. 18.2 Example for checking slenderness ratio.

(i) Calculate slenderness ratio, λ

$$I_{col} = 400^4/12 = 2133 \times 10^6 \text{ mm}^4$$
$$A = 400^2 = 16 \times 10^4 \text{ mm}^2 ,$$
$$r = 400/\sqrt{12} = 115.5 \text{ mm}$$

Beams : T-beams but only the rib 300 x 500 is considered.
$$I_{beam} = 300 \times 500^3/12 = 3125 \times 10^6 \text{ mm}^4$$
$$k_a = k_b = (2133/3.5 + 2133/3.5)/(3125/8 + 3125/6) = 1.34 > 0.4$$

From the Nomogram, β = 0.8
$$\ell_o = 0.8 \text{ x } 3.5 = 2.8 \text{ m}$$
$$\lambda = 2.8 \times 10^3/115.5 = 24.4$$

(ii) Calculate λ_{min}

$$N_{sd} = 1800 \text{ kN}$$
$$A_c = 400^2 = 16 \times 10^4 \text{ mm}^2$$
$$v_u = 1800 \times 10^3/ (16 \times 10^4 \times 0.54 \times 30) = 0.69$$
$$15/\sqrt{v_u} = 18.06$$
$$\lambda_{min} = \text{greater of } (25 \text{ or } 18.06) = 25.0$$
$$(\lambda = 24.4) < (\lambda_{min} = 25)$$
$$\text{Column is short.}$$

BS 8110:
(i) Simplified approach
Braced column with end conditions at both top and bottom is 1.

$$\beta = 0.75$$
$$\ell_e = 0.75 \times 3.5 = 2.625 \text{ m}$$
$$\ell_e/h = 2.625 \times 10^3/400 = 6.57 < 15$$
Column is short.

(ii) Using the equations of Part 2

$$\beta = \text{lesser of } [\{0.7 + 0.05(1.34+1.34)\} \text{ or } \{0.85 + 0.05 \times 1.34)\}]$$
$$\beta = 0.834 < 1.0$$
$$\ell_e = 0.834 \times 3.5 = 2.92 \text{ m}$$
$$\ell_e/h = 2.92 \times 10^3/400 = 7.30 < 15$$

Column is short.

18.9 DETAILING

This section gives a short summary of some significant aspects of detailing in EC2.

18.9.1 Bond

Quality of bond depends, apart from obvious things, on the dimensions of the member and the position and inclination of reinforcement during concreting.
a. If the bar is inclined to the horizontal between 45° and 90° to the horizontal, then the bond on the bar is taken as good.
b. If the bar is inclined to the horizontal between 0° and 45° to the horizontal and the direction of concreting is normal to the bar, then for a slab of depth h, good bond can be assumed in the following cases:
• if $h \leq 250$ mm
• for bars in the lower half of the slab depth if $250 < h < 600$ mm
• for bars in the depth below 300 mm from the top if $h \geq 600$ mm
In cases of 'poor' bond, the bond stress is taken as 0.7 the values for 'good' bond. A table of bond stress, f_{bd} vs. cylinder strength is provided. The bond strength varies from 3.0 N/mm² for $f_{cu} \approx 24.3$ N/mm² to 4.0 N/mm² for $f_{cu} \approx 40.5$ N/mm².

18.9.2 Anchorage Lengths

Basic anchorage length ℓ_b is given in terms of the bar diameter ϕ by

$$\ell_b = \frac{0.87 f_y}{f_{bd}} \frac{\phi}{4}$$

The required anchorage length $\ell_{b, net}$ is given by

$$\ell_{b,net} = \alpha_a \ell_b \frac{A_{s,required}}{A_{s,provided}} \geq \ell_{b,minimum}$$

$\alpha_a = 1.0$ for straight bars
$\alpha_a = 0.7$ for curved bars in tension if the cover perpendicular to the plane of curvature is at least 3ϕ.
$\ell_{b,minimum}$ = greatest of $[0.3\ell_b, 10\ \phi, 100\ \text{mm})$ for bars in tension
$\ell_{b,minimum}$ = greatest of $[0.6\ell_b, 10\ \phi, 100\ \text{mm})$ for bars in compression

18.9.3 Longitudinal Reinforcement in Beams

(i) Minimum longitudinal reinforcement
Minimum required steel percentage is given by

$$100\frac{A_s}{bd} \geq 0.15 \text{ for } f_y = 460 \text{ N/mm}^2.$$

$$100\frac{A_s}{bd} \geq 0.24 \text{ for } f_y = 250 \text{ N/mm}^2.$$

b = average width in the tension zone.

(ii) Maximum tension or compression reinforcement
Except at laps, the maximum tension or compression reinforcement is given by

$$100\frac{A_s}{A_c} \leq 4.0$$

(iii) In monolithic construction
In monolithic construction, even where simple support has been assumed, at least 25% of the maximum span moment steel should be provided.

(iv) Minimum steel at support
At least 25% of the bottom steel in the span should continue up to the supports.

18.10 REFERENCES

British Cement Association, 1993, *Concise Eurocode for the design of concrete buildings*.
British Cement Association, 1994, *Worked examples for the design of concrete buildings*.
Beeby, A.W. and Narayanan, R.S. 1995, *Designers' handbook to Eurocode 2, Part 1.1: Design of concrete structures*, (Thomas Telford).
Narayanan, R.S (Ed.), 1994, *Concrete Structures: Eurocode 2 and BS 8110 compared*. (Longman Scientific and Technical).
Mosley, W.H. Hulse, R. and Bungey, J.H. 1996, *Reinforced concrete design to Eurocode 2*. (Macmillan Press).

CHAPTER 19

DEFLECTION AND CRACKING

In Chapter 6, 'Deemed to satisfy' rules such as
a. using minimum ratios of span to depth to ensure deflection criteria at serviceability limit state
b. restricting maximum spacing of tension reinforcement to satisfy crack width criteria at serviceability limit state
were given. In normal design structures are designed for ultimate limit state and by satisfying the 'Deemed to satisfy' clauses, their satisfaction at serviceability limit states are ensured. Only in rare cases is detailed calculation of deflection and crack widths required, the exception being design of liquid retaining structures which are governed by crack width considerations (See Chapter 17). The object of this chapter is to discuss these detailed calculations.

19.1 DEFLECTION CALCULATION

19.1.1 Loads on the Structure

The design loads for the serviceability limit state are set out in BS 8110: Part 2, clause 3.3. The code distinguishes between calculations
1. to produce a best estimate of likely behavior
2. to comply with serviceability limit state requirements which may entail taking special restrictions into account
   In choosing the loads to be used the code again distinguishes between characteristic and expected values. For best estimate calculations, expected values are to be used. The code states that
1. for dead loads characteristic and expected values are the same
2. for imposed loads the expected values are to be used in best estimate calculations and the characteristic loads in serviceability limit state requirements (in apartments and office buildings 25% of the imposed load is taken as permanently applied)
   Characteristic loads are used in deflection calculations.

19.1.2 Analysis of the Structure

An elastic analysis based on the gross concrete section may be used to obtain moments for calculating deflections. The loads are as set out in 19.1.1.

19.1.3 Method for Calculating Deflection

The method for calculating deflection is set out in BS 8110: Part 2, section 3.7. The code states that a number of factors which are difficult to assess can seriously affect results. Factors mentioned are
1. inaccurate assumptions regarding support restraints
2. that the actual loading and the amount that is of long-term duration which causes creep cannot be precisely estimated
3. whether the member has or has not cracked
4. the difficulty in assessing the effects of finishes and partitions
   The method given is to assess curvatures of sections due to moment and to use these values to calculate deflections.

19.1.4 Calculation of Curvatures

The curvature at a section can be calculated using assumptions set out for a cracked or uncracked section. The larger value is used in the deflection calculations. Elastic theory is used for the section analysis.

19.1.5 Cracked Section Analysis

The assumptions used in the analysis of cracked section are as follows:
1. Strains are calculated on the basis that plane sections remain plane;
2. The reinforcement is elastic with a modulus of elasticity of 200 kN/mm^2;
3. The concrete in compression is elastic;
4. The modulus of elasticity of the concrete to be used is the mean value given in BS 8110: Part 2, Table 7.2;
5. The effect of creep due to long-term loads is taken into account by using an effective modulus of elasticity with a value of $1/(1 + \phi)$ times the short-term modulus from Table 7.2 of the code, where ϕ is the creep coefficient;
6. The stiffening effect of the concrete in the tension zone is taken into account by assuming that the concrete develops some stress in tension. The value of this stress is taken as varying linearly from zero at the neutral axis to 1 N/mm^2 at the *centroid of the tension steel* for short term loads and reducing to 0.55 N/mm^2 for long term loads.
   To show the method for calculating curvature, consider the doubly reinforced rectangular beam section shown in Fig. 19.1(a). The strain diagram and stresses and internal forces in the section are shown in 19.1(b) and 19.1(c) respectively.
The terms used in the figure are defined as follows:

f_c = stress in the concrete in compression
f_{sc} = stress in the compression steel
f_{st} = stress in the tension steel
f_{ct} = stress in the concrete in tension at the level of the tension steel
1 N/mm^2 for short-term loads; (0.55 N/mm^2 for long-term loads)
A_s = area of steel in tension
A_s' = area of steel in compression

$$x = \text{depth to the neutral axis}$$
$$h = \text{depth of the beam}$$
$$d = \text{effective depth}$$
$$d' = \text{inset of the compression steel}$$
$$C_c = \text{force in the concrete in compression}$$
$$C_s = \text{force in the steel in compression}$$
$$T_c = \text{force in the concrete in tension}$$
$$T_s = \text{force in the steel in tension}$$

The following further definitions are required:

$$E_c = \text{modulus of elasticity of the concrete}$$
$$E_s = \text{modulus of elasticity of the steel}$$
$$\alpha_e = \text{modular ratio}, E_s / E_c$$

Note that for long-term loads the effective value of E_c is used.

E_{eff} = effective modulus of elasticity of the concrete for long term

$$E_{eff} = E_c / (1 + \phi)$$
$$\phi = \text{creep coefficient}$$

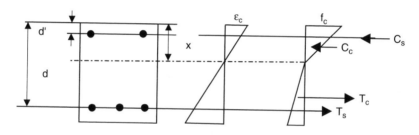

Fig.19.1

If the maximum stress in concrete is f_c, the corresponding strain ε_c in concrete is

$$\varepsilon_c = f_c / E_c$$

Assuming full bond, the strains in compression and tension steels are

$$\varepsilon_{sc} = \varepsilon_c (x - d')/x$$
$$\varepsilon_s = \varepsilon_c (d - x)/x$$

The stresses in compression and tension steels are

$$f_{sc} = E_s \varepsilon_{sc}$$
$$f_s = E_s \varepsilon_s$$

Substituting for strains in steel in terms of concrete strain,

$$f_{sc} = E_s \varepsilon_{sc} = E_s [\varepsilon_c (x - d')/x]$$
$$= E_s (f_c/E_c) (x - d')/x$$
$$f_{sc} = \alpha_e f_c (x - d')/x$$

Similarly

$$f_s = E_s \varepsilon_s = E_s [\varepsilon_c (d - x)/x]$$
$$= E_s (f_c/E_c) (d - x)/x$$
$$f_s = \alpha_e f_c (d - x)/x$$

The concrete stress f_t in tension at the *bottom face* is

$$f_t = f_{ct} (h - x)/(d - x)$$

The internal forces due to compression in concrete, compression steel, tension steel and tension in concrete are given by

$$C_c = 0.5 f_c \, b \, x$$
$$C_s = \alpha_e \, f_c \, A_s' \, (x - d')/x$$
$$T_s = \alpha_e \, f_c \, A_s \, (d - x)/x$$
$$T_c = 0.5 f_{ct} \, b \, (h - x)^2/(d - x)$$

For equilibrium, the sum of the internal forces is zero:

$$C_c + C_s = T_s + T_c$$

Substituting for the forces in terms of stresses

$$0.5 f_c \, b \, x + \alpha_e \, f_c \, A_s' \, (x - d')/x = \alpha_e \, f_c \, A_s \, (d - x)/x + 0.5 f_{ct} \, b \, (h - x)^2 / (d - x)$$

Multiplying through out by x

$$0.5 f_c \, b \, x^2 + \alpha_e \, f_c \, A_s' \, (x - d') = \alpha_e \, f_c \, A_s \, (d - x) + 0.5 f_{ct} \, b \, x \, (h - x)^2/(d - x)$$

$$(19.1)$$

The sum of the moments of the internal forces about the neutral axis is equal to the applied moment M

$$M = 0.67 \, C_c \, x + C_s \, (x - d') + T_s \, (d - x) + 0.67 \, T_c \, (h - x)$$
$$M = 0.33 \, f_c \, b \, x^2 + \alpha_e \, f_c \, A_s' \, (x - d')^2 /x + \alpha_e \, f_c \, A_s \, (d - x)^2/x$$
$$+ \, 0.33 \, f_{ct} \, b \, (h - x)^3 / (d - x)$$

$$(19.2)$$

For a given value of M, equations 19.1 and 19.2 can be solved simultaneously by successive trials to obtain the values of f_c and x. Note that the area of concrete occupied by the reinforcement has not been deducted in the expressions given above.

Let M_c be the moment resisted by tension in concrete. M_c is given by

$$M_c = 0.33 \, f_{ct} \, b \, (h - x)^3 / (d - x) \qquad\qquad (19.3)$$
$$M - M_c = 0.33 \, f_c \, b \, x^2 + \alpha_e \, f_c \, A_s'(x - d')^2 /x + \alpha_e \, f_c \, A_s(d - x)^2/x$$
$$M - M_c = f_c \, [0.33 \, b \, x^3 + \alpha_e \, A_s'(x - d')^2 + \alpha_e \, A_s(d - x)^2]/x$$
$$M - M_c = f_c \, I/x$$

where

$$I = 0.33 \, b \, x^3 + \alpha_e \, A_s'(x - d')^2 + \alpha_e \, A_s(d - x)^2 \qquad\qquad (19.4)$$

I is called second moment of area of cracked transformed section.

$$f_c = [(M - M_c) \, I] /x \qquad\qquad (19.5)$$

The compressive strain ε_c in concrete is

$$\varepsilon_c = f_c/E_c$$

The curvature 1/r is

$$1/r = \varepsilon_c \, /x = (M - M_c)/ \, (E_c \, I)$$

The value of E_c to be used depends whether the loads are short term or long term. Solutions are required for both short- and long-term loads.

19.1.5.1 Simplified Approach

Solution of equations 19.1 and 19.2 simultaneously by successive trials to obtain the values of f_c and x is an onerous task. The solution of problem can be considerably simplified without too great a loss of accuracy, if in considering equilibrium in the axial direction the contribution of f_{ct} is ignored. With this assumption equation 19.1 simplifies to

$$0.5f_c\, b\, x^2 + \alpha_e\, f_c\, A_s'(x - d') = \alpha_e\, f_c\, A_s\, (d - x)$$

Dividing through out by f_c

$$0.5\, b\, x^2 + \alpha_e\, A_s'(x - d') = \alpha_e\, A_s\, (d - x) \qquad (19.6)$$

This is a quadratic equation in x from which the value of x can be determined without the need for complicated analysis.

$$f_c = [(M - M_c)\, I]\, /x$$

19.1.6 Uncracked Section

For an uncracked section concrete and steel are both considered to act elastically in tension and compression. The analysis is similar to the cracked section analysis except that the area of concrete below the neutral axis is uncracked. The stress at the bottom face of the beam is $f_c\, (h - x)/x$ instead of $f_{ct}\, (h - x)/ (d - x)$ in a cracked section. The internal forces are given by

$$C_c = 0.5f_c\, b\, x$$
$$C_s = \alpha_e\, f_c\, A_s'\, (x - d')/x$$
$$T_s = \alpha_e\, f_c\, A_s\, (d - x)/x$$
$$T_c = 0.5f_c\, b\, (h - x)^2/x$$

For equilibrium, the sum of the internal forces is zero:

$$C_c + C_s = T_s + T_c$$
$$0.5f_c\, b\, x + \alpha_e\, f_c\, A_s'\, (x - d')/x = \alpha_e\, f_c\, A_s\, (d - x)/x + 0.5f_c\, b\, (h - x)^2/x$$

Multiplying through out by x/f_c

$$0.5\, b\, x^2 + \alpha_e\, A_s'\, (x - d') = \alpha_e\, A_s\, (d - x) + 0.5\, b\, (h - x)^2$$

Simplifying

$$\alpha_e\, A_s'(x - d') = \alpha_e\, A_s\, (d - x) + 0.5\, b\, h^2 - b\, h\, x$$

Solving for x

$$x = \{0.5\, b\, h^2 + \alpha_e\, (A_s\, d + A_s'\, d')\}/ [b\, h + \alpha_e\, (A_s + A_s')] \qquad (19.7)$$

The sum of the moments of the internal forces about the neutral axis is equal to the external moment M

$$M = 0.67\, C_c\, x + C_s\, (x - d')\ + T_s\, (d - x) + 0.67\, T_c\, (h - x)$$
$$M = 0.33\, f_c\, b\, x^2 + \alpha_e\, f_c\, A_s'\, (x - d')^2\, /x + \alpha_e\, f_c\, A_s\, (d - x)^2/x$$
$$+\, 0.33\, f_c\, b\, (h - x)^3/x$$
$$M = [0.33\, b\, x^3 + \alpha_e\, A_s'\, (x - d')^2 + \alpha_e\, A_s\, (d - x)^2$$
$$+\, 0.33\, b\, (h - x)^3]\, (f_c/x)$$

Simplifying

$$M = [0.33\, b\, h^3 + \alpha_e\, A_s'\, (x - d')^2 + \alpha_e\, A_s\, (d - x)^2$$
$$-\, b\, h\, x\, (h - x)]\, (f_c/x)$$
$$M = I\, (f_c/x)$$

$$I = [0.33 \, b \, h^3 + \alpha_e \, A_s' \, (x - d')^2 + \alpha_e \, A_s \, (d - x)^2 - b \, h \, x \, (h - x)] \qquad (19.8)$$

19.1.7 Long-Term Loads: Creep

The effect of creep must be considered for long-term loads. Load on concrete causes an immediate elastic strain and a long-term time–dependent strain known as creep strain. The strain due to creep may be much larger than that due to elastic deformation. On removal of the load, most of the strain due to creep is not recovered.

Creep is discussed in BS 8110: Part 2, section 7.3. The creep coefficient ϕ is used to evaluate the effect of creep. Values of ϕ depend on the age of loading, effective section thickness and ambient relative humidity. The code recommends suitable values for indoor and outdoor exposure in the UK and defines the effective section thickness for uniform sections as *twice* the cross-sectional area divided by the exposed perimeter.

In deflection calculations, for calculating the curvature due to the long-term loads creep is taken into account by using an effective value for the modulus of elasticity of the concrete equal to

$$E_{eff} = E_c/ (1 + \phi)$$

E_c is the short-term modulus for the concrete, values of which at 28 days are given in BS 8110: Part 2, Table 7.2.

19.1.8 Shrinkage Curvature

Concrete shrinks as it dries and hardens. This is termed drying shrinkage and is discussed in of BS 8110: Part 2, section 7.4. The code states that shrinkage is mainly dependent on the ambient relative humidity, the surface area from which moisture can be lost relative to the volume of concrete, and the mix proportions. It is noted that certain aggregates produce concrete with a higher initial drying shrinkage than normal.

Values of drying shrinkage strain ε_{cs} for plain concrete which depend on the effective thickness and ambient relative humidity may be taken from BS 8110: Part 2, Fig. 7.2. A plain concrete member shrinks uniformly and does not deflect laterally. Reinforcement prevents some of the shrinkage through bond with the concrete and if it is asymmetrical as in a singly reinforced beam this causes the member to curve and deflect. More shrinkage occurs at the top of the doubly reinforced beam because the steel area is less at the top than at the bottom.
BS 8110: Part 2, Clause 3.7, gives the following equation for calculating the shrinkage curvature $1/r_{cs}$.

$$1/r_{cs} = \varepsilon_{cs} \, \alpha_e \, S_s/I$$

where

$\qquad \alpha_e = E_s/E_{eff}$ is the modular ratio,
$\qquad \varepsilon_{cs}$ is the free shrinkage strain,

$E_{eff} = E_c/(1 + \phi)$ is the effective modulus of elasticity

I corresponds to cracked section (Equation 19.4) or uncracked section (Equation 19.7) depending on which value is used to calculate the curvature due to the applied loads. S_S is the first moment of area of the reinforcement about the centroid of the cracked or gross section.

19.1.9 Total Long-Term Curvature

BS 8110: Part 2, section 3.6, gives the following four step procedure for assessing the total long-term curvatures of a section:

1. Calculate the instantaneous curvatures under the total load and under the permanent load;
2. Calculate the long-term curvature under the permanent load;
3. Add to the long-term curvature under the permanent load the difference between the instantaneous curvatures under the total and the permanent load;
4. Add the shrinkage curvature.

19.1.10 Deflection Calculation

Curvature is equal to the second derivative of deflection with respect to distance along the span. Deflection can be calculated directly by integrating the curvature using a numerical integration technique. The code gives the following simplified method as an alternative.

The deflection a is calculated from
$$a = KL^2 (1/r_b)$$
where

L is the effective span of the member,

$1/r_b$ is the curvature at the point of maximum moment which is at mid-span in the case of beams and at support in the case of cantilevers. K is a constant which depends on the shape of the bending moment diagram

19.1.10.1 Evaluation of constant K

The method of calculating K is illustrated by a few examples. Expressions for deflection of an elastic beam are given in books on Structural Analysis.

Example 1: Simply supported beam carrying uniformly distributed total load W
The deflection Δ at mid-span is given by
$$\Delta = 5WL^3/ (384\ EI)$$
Bending moment M at mid-span is
$$M = W\ L/8$$
Replacing the load W by moment M
$$\Delta = 5ML^2/ (48\ EI) = K\ L^2/ (EI)$$
$$K = 5/48$$

Example 2. Cantilever carrying uniformly distributed total load W
The deflection Δ at the tip is given by
$$\Delta = WL^3/ (8\ EI)$$
Bending moment M at support is
$$M = W\ L/2$$
Replacing the load W by moment M
$$\Delta = ML^2/ (4\ EI) = K\ L^2/ (EI)$$
$$K = 1/4$$

Example 3: An intermediate span beam carrying uniformly distributed total load W with support moments of M_A and M_B.
The deflection Δ at mid span is given by
$$\Delta = 5WL^3/ (384\ EI) - (M_A + M_B)L^2/(16\ EI)$$
Bending moment M at mid-span is
$$M = W\ L/8 - (M_A + M_B)/2$$
Replacing the load W by moment M
$$\Delta = 5\{M + (M_A + M_B)/2\}\ L^2/ (48\ EI) - (M_A + M_B)L^2/(16\ EI)$$
$$\Delta = \{5\ M/48 - (M_A + M_B)/96\}\ L^2/ (EI)$$
$$\Delta = K\ L^2/ (EI)$$
$$K = 5/48 - (M_A + M_B)/ (96\ M)$$
$$K = 5/48\ (1 - \beta/10)$$
$$\beta = (M_A + M_B)/ (10\ M)$$
BS 8110, Part 2 in Table gives values of K for different bending moment distributions.

19.2 EXAMPLE OF DEFLECTION CALCULATION FOR T-BEAM

A simply supported T-beam of 6 m span carries a dead load including self-weight of 14.8kN/m and an imposed load of 10 kN/m. The T-beam section has the tension reinforcement designed for the ultimate limit state and the bars in the top to support the links. The dimensions of the beam in Fig. 19.2 are: web width $b_w = 250$ mm, flange width $b = 1450$ mm, total depth $h = 350$ mm, flange thickness $h_f = 100$ mm, effective depth $d = 300$ mm, inset of compression steel $d' = 45$ mm, compression steel 2T16, $A_s' = 402$ mm^2, tension steel 3T25, $A_s = 1472$ mm^2. The materials are grade C30 concrete and grade 460 reinforcement. Calculate the deflection of the beam at mid-span.

(a) Moments
The deflection calculation will be made for characteristic dead and imposed loads to comply with serviceability limit state requirements. The permanent load is taken as the dead load plus 25% of the imposed load as recommended in BS 8110: Part 2, clause 3.3.
$$\text{total load} = 14.8 + 10 = 24.8\ \text{kN/m}$$
$$\text{permanent load} = 14.8 + 0.25 \times 10 = 17.3\ \text{kN/m}$$

The moments at mid-span are

Total load: $M_T = 24.8 \times 6^2/8 = 111.6$ kNm

Permanent load: $M_P = 17.3 \times 6^2/8 = 77.85$ kNm

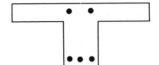

Fig.19.2 T-beam

(b) Instantaneous curvatures for the cracked section: accurate analysis

The instantaneous curvatures for the total and permanent loads are calculated first. The static modulus of elasticity from BS 8110: Part 2, Table 7.2, is $E_c = 26$ kN/mm² for grade 30 concrete. For steel $E_s = 200$ kN/mm² from BS 8110: Part 1, Fig. 2.2. The modular ratio α_e

$$\alpha_e = 200/26 = 7.69.$$

The stress in the concrete in tension at the level of the tension steel is 1.0 N/mm². The neutral axis is assumed to be in the flange as shown in Fig. 19.3. This is checked on completion of the analysis.

The stresses in the steel in terms of the concrete stress f_c in compression are

$$f_{sc} = 7.69 \, f_c \, (x - 45)/x$$
$$f_{st} = 7.69 \, f_c \, (300 - x)/x$$
$$A_s' = 402 \text{ mm}^2$$
$$A_s = 1472 \text{ mm}^2$$

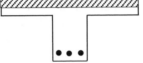

Fig. 19.3

$$C_c = 0.5 \, f_c \times 1450 \, x$$
$$C_s = A_s' \times f_{sc} = 402 \times 7.69 \, f_c \, (x - 45)/x$$
$$T_s = A_s \times f_{st} = 1472 \times 7.69 \, f_c \, (300 - x)/x$$
$$T_{c \, Rib} = 0.5 \times [(350 - x) / (300 - x)] \times 1.0 \times (350 - x) \times 250$$
$$T_{c \, Flange} = 0.5 \times [(100 - x)/ (300 - x)] \times 1.0 \times (1450 - 250) \times (100 - x)$$

For equilibrium in the axial direction:

$$C_c + C_s - T_s - T_c = 0$$

For moment equilibrium:

$$M = C_c \times 0.67 \, x + C_s \times (x - 45) + T_s \times (300 - x) + T_{c \, Rib} \times 0.67 \times (350 - x)$$
$$+ T_{c \, Flange} \times 0.67 \times (100 - x)$$

The solution for x and f_c can be obtained as follows.

- Assume a value of x

- From equilibrium in the axial direction, calculate f_c
- From moment equilibrium, calculate f_c.
- If the two f_c values are different change the value of x and go to step 2 and repeat till the two values are same.

Table 19.1 shows the summary results of calculation.
The solution is

$$x = 63.9 \text{ mm}$$
$$f_c = 8.7 \text{ N/mm}^2$$

The neutral axis lies in the flange as assumed.
The compressive strain ε_c in the concrete is

$$\varepsilon_c = f_c/E_c = 8.7/(26 \times 10^3) = 3.346 \times 10^{-4}$$

The curvature for the total loads is

$$1/r = \varepsilon_c /x = 3.346 \times 10^{-4}/63.9 = 5.24 \times 10^{-6}$$

Table 19.1 Determination of x and f_c for M = 11.6 kNm

x	f_c from Axial	f_c from Bending	Difference
62.0	20.13	8.44	11.69
62.5	14.88	8.51	6.36
63	11.80	8.58	3.22
63.5	9.79	8.65	1.14
63.8634	8.71	8.70	0.01
64.5	7.30	8.78	− 1.48

(c) Instantaneous curvature for permanent loads
For the instantaneous curvature for permanent loads the applied moment is 77.85 kNm. Table 19.2 shows the summary of calculations for determining the values of x and f_c.

Table 19.2 Determination of x and f_c for M = 77.85 kNm

x	f_c from Axial	f_c from Bending	Difference
64	8.36	5.87	2.49
64.2	7.90	5.88	2.02
64.4	7.49	5.90	1.59
64.6	7.12	5.92	1.20
64.8	6.78	5.94	0.85
65.0	6.48	5.96	0.52
65.2	6.20	5.97	0.23
65.365	5.99	5.99	0.00
65.6	5.71	6.01	−0.30
65.8	5.49	6.03	−0.53

The solution is
$$x = 65.4 \text{ mm}$$
$$f_c = 6.0 \text{ N/mm}^2$$
The neutral axis lies in the flange as assumed.
The compressive ε_c strain in the concrete is
$$\varepsilon_c = f_c/E_c = 6.0/(26 \times 10^3) = 2.3077 \times 10^{-4}$$
The curvature for the total loads is
$$1/r = \varepsilon_c /x = 3.346 \times 10^{-4}/65.4 = 3.53 \times 10^{-6}$$

(d) Long-term curvature under permanent loads
The creep coefficient ϕ is estimated using data given in BS 8110: Part 2, section 7.3.
effective area $A_c = 2[(250 \times 350) + (1200 \times 100)] = 4.15 \times 10^5 \text{ mm}^2$
exposed perimeter $= 1450 + (1450 - 250) + (2 \times 350) + 250 = 3600$
effective section thickness $= 4.15 \times 10^5/3600 = 115 \text{ mm}$
The relative humidity for indoor exposure is 45%. If the age of loading is 14 days, say, when the soffit form and props are removed, the creep coefficient ϕ from Fig. 7.1 of the code is
$$\phi = 3.5$$
The effective modulus of elasticity is
$$E_{eff} = E_c/ (1 + \phi) = 26/ (1 + 3.5) = 5.78 \text{ kN/mm}^2$$
The modular ratio:
$$\alpha_e = 200/5.78 = 34.6.$$
Assuming that the neutral axis is in the web as shown in Fig. 19.4,

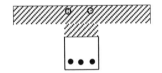

Fig.19.4

$$C_{c1} = 0.5 \ f_c \times 1450 \ x$$
$$C_{c2} = - (1450 - 250) \times (x - 100) \times 0.5 \times f_c \times (x - 100)/x$$
$$C_s = A_s' \times f_{sc} = 402 \times 34.6 \ f_c \ (x - 45)/x$$
$$T_s = A_s \times f_{st} = 1472 \times 34.6 \times f_c \ (300 - x)/x$$
$$T_{c \ Rib} = 0.5 \times [(350 - x) / (300 - x)] \times 0.55 \times (350 - x) \times 250$$
Note: C_{c1} is the compressive force in a rectangle of area $b \ x$, $x > h_f$.
C_{c2} is the negative compression in a rectangle of area $(b - b_w) \ (x - h_f)$
For equilibrium in the axial direction:
$$C_c + C_s - T_s - T_c = 0$$
For moment equilibrium:
$$M = C_{c1} \times 0.67 \ x + C_{c2} \times 0.67 \ (x - 100) + C_s \times (x - 45) + T_s \times (300 - x)$$
$$+ T_{c \ Rib} \times 0.67 \times (350 - x)$$
Table 19.3 shows the summary of calculations.
The solution is

$$x = 113.59 \text{ mm}$$
$$f_c = 3.33 \text{ N/mm}^2$$

The neutral axis lies in the web as assumed.

The compressive ε_c strain in the concrete is

$$\varepsilon_c = f_c/E_c = 3.33/(5.78 \times 10^3) = 5.761 \times 10^{-4}$$

The curvature for the permanent loads is

$$1/r = \varepsilon_c /x = 5.761 \times 10^{-4}/113.59 = 5.07 \times 10^{-6}$$

The moment M_c due to tensile stress in concrete is

$$M_c = 0.5 \times f_{ct} \times (h - x)^3/(d - x) \times b_w(2/3)$$
$$M_c = 0.5 \times 0.55 \times (350 - 113.59)^3/ (300 - 113.59) \times 250 \times (2/3)$$
$$M_c = 3.248 \times 10^6 \text{Nmm}$$
$$I = (M - M_c) x/f_c$$
$$I = (77.\,86 \times 10^6 - 3.248 \times 10^6) \times 113.59/3.333 = 2.543 \times 10^9 \text{ mm}^4$$

Table 19.3 Determination of x and f_c for $M = 77.86$ kNm

x	f_c from Axial	f_c from Bending	Difference
113.2	3.77	3.32	0.44
113.25	3.70	3.32	0.38
113.30	3.64	3.32	0.32
113.35	3.58	3.33	0.26
113.40	3.53	3.33	0.20
113.45	3.47	3.33	0.14
113.50	3.42	3.33	0.09
113.55	3.37	3.33	0.04
113.59	3.33	3.33	0.00
113.60	3.32	3.33	−0.01
113.65	3.27	3.33	−0.06

(e) Curvature due to shrinkage

The value of drying shrinkage for plain concrete is evaluated from of BS 8110: Part 2, Fig. 7.2, for an effective thickness of 115 mm (calculated in (d) above) and a relative humidity of 45%. The 30 year shrinkage value is

$$\varepsilon_{cs} = 420 \times 10^{-6}$$

The effective modulus for an age of loading of 14 days is

$$E_{eff} = 5.78 \text{ kN/mm}^2$$
$$\alpha_e = 34.6$$

The moment of inertia of the cracked section is calculated using the depth to the neutral axis determined in (d) above.

$$I = 2.543 \times 10^9 \text{ mm}^4$$

The first moment of area of the reinforcement about the centroid of the cracked sections is

$$S_s = A_s' (x - d') + A_s (d - x)$$

$$= 402 \times (113.59 - 45) + 1472 \times (300 - 113.59) = 3.019 \times 10^5 \text{mm}^3$$

The shrinkage curvature is

$$1/r_{cs} = \varepsilon_{cs} \times \alpha_e \times S_s/I$$
$$1/r_{cs} = 420 \times 10^{-6} \times 34.6 \times 3.019 \times 10^5/2.543 \times 10^9$$
$$1/r_{cs} = 1.7252 \times 10^{-6}$$

(f) Final curvature

The final curvature $1/r_b$ is the instantaneous curvature under the total load (from (b) above) minus the instantaneous curvature under the permanent load (from (c) above) plus the long-term curvature under the permanent load (from (d) above) plus shrinkage curvature (from (e) above):

$$1/r_b = (5.24 - 3.53 + 5.07 + 1.725) \times 10^{-6}$$
$$1/r_b = 8.51 \times 10^{-6}$$

(g) Beam deflection

For a simply supported beam carrying a uniform load, $K = 5/48$. (from 19.1.10.1)

$$\text{Deflection } a = K L^2 (1/r_b)$$
$$(5/48) \times 6000^2 \times 8.51 \times 10^{-6} = 31.9 \text{ mm}$$
$$\text{permissible deflection} = 6000/250 = 24 \text{ mm or } 20 \text{ mm}$$

The beam does not meet deflection requirements for the creep and shrinkage conditions selected. The beam is satisfactory when checked by span/d ratio rules. This example shows that controlling deflection through span/d ratio might not always be satisfactory.

(h) Instantaneous curvatures for the cracked section: Approximate method

In the simplified approach the main assumption made is to neglect the contribution by the tensile strength of concrete in considering equilibrium in the axial direction. Using the equations from (b) above

$$C_c = 0.5 \, f_c \times 1450 \, x$$
$$C_s = A_s' \times f_{sc} = 402 \times 7.69 \, f_c \, (x - 45)/x$$
$$T_s = A_s \times f_{st} = 1472 \times 7.69 \, f_c \, (300 - x)/x$$
$$T_{c \, Rib} = 0.5 \times [(350 - x) / (300 - x)] \times 1.0 \, x \, (350 - x) \times 250$$
$$T_{c \, Flange} = 0.5 \times [(100 - x)/ (300 - x)] \times 1.0 \times (1450 - 250) \times (100 - x)$$

For equilibrium in the axial direction:

$$C_c + C_s - T_s = 0$$

Multiplying through out by x/f_c and simplifying

$$0.5 \times 1450 \, x^2 + 402 \times 7.69 \, (x - 45) - 1472 \times 7.69 \, (300 - x) = 0$$

Simplifying

$$x^2 + 19.877 \, x - 4875.88 = 0$$
$$x = 60.6 \text{ mm ('Exact' value} = 63.9 \text{ mm)}$$
$$T_{c \, Rib} = 0.5 \times [(350 - x) / (300 - x)] \times 1.0 \, x \, (350 - x) \times 250 = 43.73 \text{ kN}$$
$$T_{c \, Flange} = 0.5 \times [(100 - x)/ (300 - x)] \times 1.0 \, x \, (1450 - 250) \times (100 - x) = 3.89 \text{ kN}$$

For moment equilibrium:

$$M = C_c \times 0.67 \, x + C_s \times (x - 45) + T_s \times (300 - x) + T_{c \, Rib} \times 0.67 \times (350 - x)$$
$$+ T_{c \, Flange} \times 0.67 \, x \, (100 - x)$$

$$M_c = T_{c\,Rib} \times 0.67 \times (350 - x) + T_{c\,Flange} \times 0.67 \times (100 - x)$$
$$M_c = (43.73 \times 0.67 \times 289.4 + 3.89 \times 0.67 \times 39.4) \times 10^{-3} = 8.54 \text{ kNm}$$
$$M = 111.6 \text{ kNm}$$
$$M - M_c = 111.6 - 8.54 = 103.1 \text{ kNm}$$
$$M - M_c = C_c \times 0.67 \, x + C_s \times (x - 45) + T_s \times (300 - x)$$
$$= (f_c/x) \, [0.33 \times 1450 \times x^3 + 402 \times 7.69 \times (x - 45)^2 + 1472 \times 7.69 \times (300 - x)^2]$$
$$I = [0.33 \times 1450 \times x^3 + 402 \times 7.69 \times (x - 45)^2 + 1472 \times 7.69 \times (300 - x)^2]$$
$$I = 757.07 \times 10^6 \text{ mm}^4$$
$$f_c = (M - M_c)x/I = 103.1 \times 10^6 \times 60.6 / (757.07 \times 10^6)$$
$$f_c = 8.3 \text{ N/mm}^2 \text{ ('Exact' value} = 8.7 \text{ N/mm}^2)$$

The compressive strain ε_c in the concrete is
$$\varepsilon_c = f_c/E_c = 8.3/(26 \times 10^3) = 3.192 \times 10^{-4}$$

The curvature for the total loads is
$$1/r = \varepsilon_c / x = 3.192 \times 10^{-4}/60.6 = 5.27 \times 10^{-6} \text{ (Exact value} = 5.24 \times 10^{-6})$$

Clearly the 'approximate' values are very close to the exact values. The simplified approach is recommended.

(i) Curvature for the uncracked section

In the uncracked section calculation, f_{ct} is replaced by $f_c(d - x)/x$. The rest of the calculations remain as before. The neutral axis can be assumed to be in the web. From (d) above but using $\alpha_e = 7.69$

$$C_{c1} = 0.5 \, f_c \times 1450 \, x$$
$$C_{c2} = - (1450 - 250) \times (x - 100) \times 0.5 \times f_c \times (x - 100)/x$$
$$C_s = A_s' \times f_{sc} = 402 \times 7.69 \, f_c \, (x - 45)/x$$
$$T_s = A_s \times f_{st} = 1472 \times 7.69 \, f_c \, (300 - x)/x$$
$$T_{c\,Rib} = 0.5 \times [(350 - x) / (300 - x)] \times \{f_c \times (300 - x)/x\} \times (350 - x) \times 250$$

Simplifying
$$T_{c\,Rib} = 0.5 \times (350 - x)^2 \, \{f_c / x\} \times 250$$

For equilibrium in the axial direction:
$$C_{c1} + C_{c2} + C_s - T_s - T_c = 0$$

Multiplying through out by x/f_c and simplifying
$$0.5 \times 1450 \, x^2 - (1450 - 250) \times (x - 100)^2 \times 0.5 + 402 \times 7.69 \, (x - 45)$$
$$- 1472 \times 7.69 \, (300 - x) - 0.5 \times (350 - x)^2 \times 250 = 0$$

Simplifying
$$(2.219 \, x - 248.48) \times 10^5 = 0$$
$$x = 112 \text{ mm}$$

For moment equilibrium:
$$M = C_{c1} \times 0.67 \, x + C_{c2} \times 0.67 \, (x - 100) + C_s \times (x - 45) + T_s \times (300 - x)$$
$$+ T_{c\,Rib} \times 0.67 \times (350 - x)$$
$$M = [0.33 \times 1450 \times x^3 - 0.33 \times 1200 \times (x - 100)^3 + 402 \times 7.69 \, (x - 45)^2$$
$$+ 1472 \times 7.69 \times (300 - x)^2 + 0.33 \times 250 \times (350 - x)^2](f_c/x)$$
$$I = [0.33 \times 1450 \times x^3 - 0.33 \times 1200 \times (x - 100)^3 + 402 \times 7.69 \, (x - 45)^2$$
$$+ 1472 \times 7.69 \times (300 - x)^2 + 0.33 \times 250 \times (350 - x)^2]$$
$$I = 2.22 \times 10^9 \text{ mm}^4$$

$$x = 113.59 \text{ mm}$$
$$f_c = (M/I)x = [111.6 \times 10^6/(2.22 \times 10^9)]113.6 = 5.71 \text{ N/mm}^2$$

The neutral axis lies in the web as assumed.
The compressive ε_c strain in the concrete is

$$\varepsilon_c = f_c/E_c = 5.63/(26.0 \times 10^3) = 2.165 \times 10^{-4}$$

The curvature for the permanent loads is

$$1/r = \varepsilon_c /x = 2.165 \times 10^{-4}/113.6 = 1.906 \times 10^{-6}$$

As is to be expected, this is much less than 5.24×10^{-6} calculated in (b) above for cracked section. Deflection calculation based on cracked section.

19.3 CALCULATION OF CRACK WIDTHS

Calculation of crack widths in connection with the design of structures retaining aqueous liquids was discussed in Chapter 17. In this section determination of crack widths in connection with beams will be discussed.

19.3.1 Cracking in Reinforced Concrete Beams

A reinforced concrete beam is subject to flexural cracks on the tension face when the tensile strength of the concrete is exceeded. Primary cracks form first and with increase in moment secondary cracks form. Cracking has been extensively studied both experimentally and theoretically. The crack width at a point on the surface of a reinforced concrete beam has been found to be affected by two factors:

1. the surface strain found by analyzing the sections and assuming that plane sections remain plane and
2. the distance of the point from a point of zero crack width at the neutral axis and the surface of longitudinal reinforcing bars. The larger this distance is, the larger the crack width will be.

Referring to Fig. 19.5 the critical locations for cracking on the beam surface are
1. at C equidistant between the neutral axis and the bar surface
2. at B equidistant between the bars
3. at A on the corner of the beam

19.3.2 Crack Width Equation

The calculation of crack widths is covered in BS 8110: Part 2, section 3.8. The code notes that the width of a flexural crack depends on the factors listed above. It also states that cracking is a semi-random phenomenon and that it is not possible to predict an absolute maximum crack width.

The following expression is given in BS 8110: Part 2, clause 3.8.3, to determine the design surface crack width.

$$\text{design crack width} = \frac{3\,a_{cr}\,\varepsilon_m}{1+2\dfrac{(a_{cr}-c_{min})}{(h-x)}}$$

The code states that this formula can be used provided that the strain in the tension reinforcement does not exceed $0.8f_y/E_s$. The terms in the expression are defined as follows:

a_{cr} distance of the point considered to the surface of the nearest longitudinal bar

ε_m = average strain at the level where the cracking is being considered (this is discussed below)

c_{min} = minimum cover to the tension steel

h = overall depth of the member

x = depth of the neutral axis

The average strain ε_m can be calculated using the method set out for determining the curvature in BS8110: Part 2, section 3.6, and section 19.1. The code gives an alternative approximation in which

1. the strain ε_1, at the level considered is calculated ignoring the stiffening effect of the concrete in the tension zone (the transformed area method is used in this calculation)

2. the strain ε_1, is reduced by an amount equal to the tensile force due to the stiffening effect of the concrete in the tension zone acting over the tension zone divided by the steel area

3. ε_m for a rectangular tension zone is given by

$$\varepsilon_m = \varepsilon_1 - \frac{b_t\,(h-x)(a'-x)}{3\,E_s\,A_s\,(d-x)}$$

where

b_t = the width of the section at the centroid of the tension steel

a' = the distance from the compression face to the point at which the crack width is required

The code adds the following comments and requirements regarding use of the crack width formula:

1. A negative value of ε_m indicates that the section is not cracked;

2. The modulus of elasticity of the concrete is to be taken as *one-half* the instantaneous value to calculate strains;

3. If the drying shrinkage is very high, i.e. greater than 0.0006, ε_m should be increased by adding 50% of the shrinkage strain. In normal cases shrinkage may be neglected.

19.4 EXAMPLE OF CRACK WIDTH CALCULATION FOR T-BEAM

The beam chosen is the same as the one used in section 19.2. The total moment at the section due to service loads is 111.6kNm. The materials are grade 30 concrete and grade 460 reinforcement. Determine the crack widths at the corner A, at the centre of the tension face B and at C on the side face midway between the neutral axis and the surface of the tension reinforcement as shown in Fig. 19.5.

The approximate method set out in section 19.2 (h) is used in the calculation. The properties of the cracked section are computed first. The values for the moduli of elasticity are as follows:

Reinforcement = E_s = 200kN/mm^2

$$\text{Concrete} = E_c = 0.5 \times 26 = 13 \text{ kN/mm}^2$$
$$\text{Modular ratio} = \alpha_e = 200/13 = 15.4$$
$$C_c = 0.5 \text{ } f_c \times 1450 \text{ } x$$
$$C_s = A_s' \times f_{sc} = 402 \times 15.4 \text{ } f_c \text{ } (x - 45)/x$$
$$T_s = A_s \times f_{st} = 1472 \times 15.4 \text{ } f_c \text{ } (300 - x)/x$$
$$T_{c\,Rib} = 0.5 \times [(350 - x) / (300 - x)] \times 1.0 \times (350 - x) \times 250$$
$$T_{c\,Flange} = 0.5 \times [(100 - x)/ (300 - x)] \times 1.0 \times (1450 - 250) \times (100 - x)$$

For equilibrium in the axial direction:

$$C_c + C_s - T_s = 0$$

Multiplying through out by x/f_c and simplifying

$$0.5 \times 1450 \text{ } x^2 + 402 \times 15.4 \text{ } (x - 45) - 1472 \times 15.4 \text{ } (300 - x) = 0$$

Simplifying

$$x^2 + 39.806 \text{ } x - 9764.45 = 0$$
$$x = 80.9 \text{ mm}$$

For moment equilibrium ignoring contribution from the tensile stress in concrete:

$$M = C_c \times 0.67 \text{ } x + C_s \times (x - 45) + T_s \times (300 - x)$$
$$M = 111.6 \text{ kNm}$$
$$M = C_c \times 0.67 \text{ } x + C_s \times (x - 45) + T_s \times (300 - x)$$
$$= (f_c/x) \text{ } [0.33 \times 1450 \times x^3 + 402 \times 15.4 \times (x - 45)^2 + 1472 \times 15.4 \times (300 - x)^2]$$
$$f_c = 6.7 \text{ N/mm}^2$$

The compressive strain ε_c in the concrete is

$$\varepsilon_c = f_c/E_c = 6.7/(13 \times 10^3) = 5.137 \times 10^{-4}$$

Strain ε_s at steel level

$$\varepsilon_s = (\varepsilon_c /x) \text{ } (d - x)$$
$$\varepsilon_s = 5.137 \times 10^{-4} \times (300 - 80.9)/80.9 = 1.391 \times 10^{-3}$$

Strain ε_1 at the bottom face is

$$\varepsilon_1 = \varepsilon_s \text{ } (h - x)/ (d - x)$$
$$\varepsilon_1 = 1.391 \times 10^{-3} \times \frac{(350 - 80.9)}{(300 - 80.9)} = 1.708 \times 10^{-3}$$

The strain reduction due to the stiffening effect of the concrete in the tension zone, where a' = h = 350mm, is

$$= \frac{b_t \text{ } (h - x)(a' - x)}{3 \text{ } E_s \text{ } A_s \text{ } (d - x)}$$
$$= \frac{250 \times (350 - 80.9) \times (350 - 80.9)}{3 \times 200 \times 10^3 \times 1472 \times (300 - 80.9)} = 0.094 \times 10^{-3}$$

The average strain at the bottom face is therefore
$$\varepsilon_m = (1.708 - 0.094)10^{-3} = 1.614 \times 10^{-3}$$
$$\text{Cover } c_{min} = 37.5\text{mm}$$

Crack width at A:

Vertical or horizontal distance from the surface of beam to centre of bar is
$$= \text{Cover} + \text{bar diameter/2}$$
$$= 37.5 + 25/2 = 50 \text{ mm}$$
Distance from the surface of the bar to point A is
$$a_{cr} = \sqrt{(50^2 + 50^2)} - 25/2 = 58.2 \text{ mm}$$

$$\text{Crack width} = \frac{3\,a_{cr}\,\varepsilon_m}{1 + 2\dfrac{(a_{cr} - c_{min})}{(h - x)}}$$

$$= \frac{3 \times 58.2 \times 1.614 \times 10^{-3}}{1 + 2\dfrac{(58.2 - 38.5)}{(350 - 80.9)}} = 0.21\,mm$$

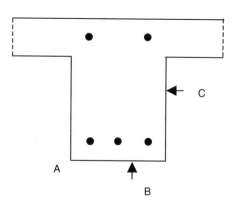

Fig.19.5

(b) Crack width at B

The bars are spaced at
$$(250 - 50 - 50)/2 = 75 \text{ mm centres.}$$
$$a_{cr} = \sqrt{\{(75/2)^2 + 50^2\}} - 25/2 = 50\text{mm}$$
$$\varepsilon_m = 1.614 \times 10^{-3}$$

$$\text{Crack width} = = \frac{3 \times 50 \times 1.614 \times 10^{-3}}{1 + 2\dfrac{(50 - 37.5)}{(350 - 80.9)}} = 0.22 \text{ mm}$$

(c) Crack width at C

C is midway between the neutral axis and the surface of the reinforcement. Vertical distance between the neutral axis and the upper surface of the bar is

= 350 − 80.9 (neutral axis depth) − 37.5 (cover) − 25 (bar diameter) = 206.6 mm
Vertical distance a' to the point from compression face
$$a' = 80.9 + 206.6/2 = 184.2 \text{ mm}$$
Strain ε_1 at C
$$\varepsilon_1 = (\varepsilon_c /x)(a' - x)$$
$$\varepsilon_1 = 5.137 \times 10^{-4} \times (184.2 - 80.9)/80.9$$
$$\varepsilon_1 = 0.656 \times 10^{-3}$$
The strain reduction due to the stiffening effect of the concrete in the tension zone, where a' = h = 350mm, is
$$= \frac{b_t (h-x)(a'-x)}{3 E_s A_s (d-x)}$$
$$= \frac{250 \times (350-80.9) \times (184.2-80.9)}{3 \times 200 \times 10^3 \times 1472 \times (300-80.9)} = 0.036 \times 10^{-3}$$
The average strain at C is therefore
$$\varepsilon_m = (0.656 - 0.036)10^{-3} = 0.62 \times 10^{-3}$$
$$\text{Cover } c_{min} = 37.5 \text{mm}$$
Horizontal distance from the vertical surface of beam to centre of bar is
$$= \text{Cover} + \text{bar diameter}/2$$
$$= 37.5 + 25/2 = 50 \text{ mm}$$
Vertical distance from C to centre of bar is
$$206.6/2 + 25/2 = 115.8 \text{ mm}$$
Distance from the surface of the bar to point C is
$$a_{cr} = \sqrt{(115.8^2 + 50^2)} - 25/2 = 113.6 \text{ mm}$$
$$\text{Crack width} = \frac{3 a_{cr} \varepsilon_m}{1 + 2\dfrac{(a_{cr} - c_{min})}{(h-x)}}$$
$$= \frac{3 \times 113.6 \times 0.62 \times 10^{-3}}{1 + 2\dfrac{(113.6-37.5)}{(350-80.9)}} = 0.135 \, mm$$
All crack widths are less than 0.3 mm and are thus satisfactory.

19.5 REFERENCE

Ghali, A, Favre, R. and Elbadry, M., 2002, *Concrete Structures: Stresses and Deformations*, (E & FN Spon).

ADDITIONAL REFERENCES

In addition to references given in the body of the text, the following additional references will be found useful.

Prestressed Concrete
Collins, M.P. and Mitchell, D. 1991, *Prestressed concrete structures* (Prentice-Hall).

Gilbert, R.I. and Mickleborough, N.C. 1990, *Design of Prestressed Concrete* (Unwin-Hyman).

Menn, C. 1990, *Prestressed Concrete Bridges* (Birkhauser).

Reinforced Concrete
Bennett, D.F.H. and MacDonald, L.A.M. 1992, *Economic assembly of reinforcement* (Reinforced Concrete Council/British Cement Association).

Booth, Edmund (Ed.). 1994, *Concrete structures in earthquake regions* (Longman).

Kotsovos, M.D. and Pavlovic, M.N. 1995, *Finite element analysis for limit state design* (Thomas Telford).

Kotsovos, Michael D. and Pavlovic, M.N. 1999, *Ultimate limit state design of concrete structures: A New approach* (Thomas Telford)

MacGregor, J.G. 1992, *Reinforced Concrete: Mechanics and Design* (Prentice-Hall).

Mosley, W.H., Bungey, J.H. and Hulse, R. 1999, Reinforced concrete design (Palgrave).

O'Brien, E.J. and Dixon, A.S. 1995, *Reinforced and Prestressed concrete design* (Longman).

O'Brien, E.J. and Keogh, D.L. 1999, *Bridge Deck Analysis* (E & FN Spon).

Rowe, R. *et al.*, 1987, *Handbook to British Standard BS 8110: 1985* (A view point Publication).

Tubman, J. 1995, *Steel reinforcement* (CIRIA)

Matrix Analysis
Bhatt, P. 1999, *Structures* (Longman).

INDEX

Also available from Taylor & Francis

Steel Structures, 3rd edition

Hassan Al-Nageim, T.J. MacGinley

Taylor & Francis	Hb: 0–415–30156–4
	Pb: 0–415–30157–2

Fundamentals of Durable Reinforced Concrete

Mark G. Richardson

Spon Press	Hb: 0–419–237–801

Prestressed Concrete Design, 2nd edition

M.K. Hurst

Spon Press	Hb: 0–419–21800–9

Non-linear Mechanics of Reinforced Concrete

K. Maekawa, H. Okamura, A. Pimanmas

Spon Press	Hb: 0–415–27126–6

Reinforced Concrete Designer's Handbook, 10th edition

Charles E. Reynolds and James C. Steedman

Spon Press	Hb: 0–419–14530–3
	Pb: 0–419–14540–0

Information and ordering details

For price availability and ordering visit our website **www.sponpress.com**
Alternatively our books are available from all good bookshops.